MODERN CONSTRUCTION AND GROUND ENGINEERING
EQUIPMENT AND METHODS
SECOND EDITION

**By Frank Harris:**

*Modern Construction Management* Third edition (Blackwell), coauthor R. McCaffer

*Management of Construction Equipment* Second edition (Macmillan), coauthor R. McCaffer

# MODERN CONSTRUCTION AND GROUND ENGINEERING EQUIPMENT AND METHODS
## SECOND EDITION

**FRANK HARRIS**
*Professor of Construction Studies*
*University of Wolverhampton*

Prentice
Hall

*An imprint of* **Pearson Education**

Harlow, England · London · New York · Reading, Massachusetts · San Francisco
Toronto · Don Mills, Ontario · Sydney · Tokyo · Singapore · Hong Kong · Seoul
Taipei · Cape Town · Madrid · Mexico City · Amsterdam · Munich · Paris · Milan

**Pearson Education Limited,**
Edinburgh Gate,
Harlow, Essex CM20 2JE,
England

and Associated Companies throughout the world.

*Visit us on the World Wide Web at:*
http://www.pearsoneduc.com

First published 1989
Second edition 1994

**British Library Cataloguing in Publication Data**
A catalogue entry for this title is available from the British Library.

ISBN 0–582–23657–6

**Library of Congress Cataloging-in-Publication Data**
A catalog entry for this title is available from the Library of Congress.

Harris, Frank, 1944–
    Modern construction and ground engineering equipment and methods /
Frank Harris.—2nd ed
        p.   cm.
    Includes bibliographical references and index.
    1. Earthmoving machinery.   2. Excavating machinery.
3. Construction equipment.   I. Title.
TA725.H37   1994
624—dc20                                                          94-9606 CIP

10   9   8   7   6   5   4
04   03   02   01   00

Typeset by 16 in 10/12pt Monotype Times
Produced by Pearson Education Asia Pte Ltd.
Printed in Singapore (CNC)

# Preface

During the past ten to fifteen years national economies have been subjected to rapidly increased trade flows, resulting in keener competition in the design, production and distribution of goods. The construction industry has not been immune from these effects, as demonstrated by the need for both designers and contractors to seek work in the international market place.

This growing demand for higher productivity in the economy is being reflected by an increasing mechanisation of construction work. Consequently it is essential that students in particular are aware of the applications and management of construction equipment, and also that experienced professionals have a readily accessible source of information on modern construction technology.

This book has therefore been written for students and practitioners of civil engineering, construction and building. Specifically it will find a market amongst working contractors, engineers, builders, quantity surveyors, architects, specification writers, equipment and materials manufacturers, project managers, and insurance and legal advisors.

A special effort has been made to present the principles in deciding on methods of construction, devising temporary works and selecting appropriate equipment. The text is fully augmented by diagrams, tables and performance data, which together with worked examples, should help the reader to prepare cost plans and estimates of work.

To this end the second edition contains an extensive new section on ground engineering methods to augment the coverage in the original volume of the major categories of construction equipment and methods common to general contracting, dealing in order with:

1. Ground engineering equipment and methods.
2. Excavating and materials handling plant.
3. Sand and aggregate production, concrete batching and prestressed concrete.

4. Road pavement construction with the emphasis directed towards pavement construction equipment.
5. Bridgework, concentrating on temporary works structures, falsework and formwork for deck construction.

# Acknowledgements

The author is especially grateful to friends and colleagues for their helpful comments on some of the specialist topics. The book has also benefited from project studies undertaken by former students. I am especially indebted to Professor R. Seeling at Aachen Technische Hochschule, West Germany who provided the inspiration and much information for the writing of this book during my study leave.

Special thanks are extended to Janet Redman who performed the arduous task of redrafting the original diagrams and sketches.

# Conversion factors

| | |
|---|---|
| 1 mile | $= 1.609$ km |
| 1 in$^2$ | $= 645.2$ mm$^2$ |
| 1 ft$^2$ | $= 0.0929$ m$^2$ |
| 1 yd$^2$ | $= 0.8361$ m$^2$ |
| 1 yd$^3 = 0.765$ m$^3$ | $= 765$ litres |
| 1 UK gallon | $= 4.546$ litres |
| 1 lbf | $= 4.448$ newtons (N) |
| 1 lb | $= 0.4536$ kg |
| 1 kgf | $= 9.807$ N |
| 1 lbf/ft$^2$ | $= 47.88$ N/m$^2$ |
| 1 hp | $= 745.7$ watts (W) |
| 1 mph | $= 1.609$ km/h |
| 1 atm (bar) | $= 14.7$ lbf/in$^2 = 0.1$ N/mm$^2 = 32$ ft water |

The units in the text reflect the current parallel use of both customary and metric systems. Many manufacturers also still make machines in imperial sizes, e.g. excavator buckets are often denominated in cubic yards and engine power quoted as horse power units. These have been maintained with the metric equivalents provided only as an aid.

# Notes

Inevitably when using mechanical equipment safety must be considered. It is, however, beyond the scope of this book to cover all the safety aspects and associated regulations with respect to the operation and maintenance of machinery.

Nevertheless, some general literature is provided in the Bibliography at the end of this book. More detailed guidance is usually available in the safety legislation of the country of operation.

In the case of the UK, the new legislation has brought Health and Safety requirements into line with other European Community member countries, particularly with respect to the need to provide and take specific measures during both design and construction. Work equipment, manual handling, health and safety at work, the control of hazardous substances, and other aspects are comprehensively set out in the new documents listed in the bibliography.

Interestingly, however, the new proposals 'Construction (Design and Management) Regulations and Approved Code of Practice' issued by the HSC attempts to place much of the responsibility for health and safety firmly with the client most notably during the pre-contract stage. A safety planning supervisor is to be appointed by the client to consider safety issues and form a safety plan covering progress of the project from inception to completion and even maintenance thereafter. In this manner, most notably, the designer, will be advised on safety matters arising from the design likely to affect the work on site and subsequent life of the building. This third party may be further retained by the client during the construction phase, for example to evaluate design changes. Also the principle contractor will be required to produce a safety plan for the construction phase which will need to complement that prepared by the Safety Planning supervisor for the whole project.

# Foreword

## THE NEED FOR CONSTRUCTION EQUIPMENT

It is estimated that 35–45% of the cost of building work is spent on materials, and in civil engineering the corresponding value sometimes approaches 35%. Traditionally, particularly in house building, much of the material has been manhandled, but as labour costs rise in relation to the costs of using mechanical equipment, it is inevitable that increased productivity will be sought on the construction site by improved applications of machinery. Witness the increasing use of concrete pumps, conveyors and not least tower cranes on all forms of construction work. However, while there is likely to be a continuing search for improved efficiency, the construction site cannot of course be operated like a factory, because of the unique character of each project. The site itself is only a temporary feature, equipment must be installed, moved around and finally dismantled in a relatively short period of time. Therefore, unless there is a continuity of work for an item on another project, there is little advantage to be gained from investing in large numbers of machines. But as applications are extended then the possibility of transfer of units from site to site will improve and facilitate more economic utilisations. It is this aspect which demands the attention of site management when balancing labour and equipment resources, and every opportunity must be taken to improve the methods of working. If this objective is firmly upheld then mechanical equipment can be deployed: (a) on increasing production and reducing costs, (b) for replacing manual operations, particularly the heavy tasks, (c) for maintaining production where labour is scarce or too expensive to employ, and (d) for reducing materials wastage.

Whenever the strategy of mechanisation has been followed, in general there has been a reduction in the amount of labour employed without a decline in the volume of production, and a corresponding improvement in materials wastage figures.

## AUTOMATION

Generally the advances made by the manufacturing and engineering industries with respect to computer aided machinery such as robots, CNC machines, etc., have not been so easily achieved with construction equipment, largely because of the rigours of the construction site and the awkward nature of 'one off' designs. Nevertheless some progress has been made, particularly on earthmoving equipment. For example, audio sensor devices have been used to recognise unique engine vibration patterns, enabling individual trucks in a fleet to be recorded and counted when passing the sensor. Laser devices are now also commonplace in the guidance of tunnel boring machines (TBMs), micro tunnelling cutting heads, etc. Cameras are routinely used to explore the condition of sewer pipes and to guide repair equipment. Furthermore automatic welding equipment is now commonplace for the joining of long stretches of steel piping, in gas or oil work.

Laser equipment is also being applied in trenching work to control the depth of excavation through sensors on the digging arm of the backhoe. The tipping height on front end loader shovels can be similarly limited. Indeed sensors and laser beams have long been attached to road paving machines for controlling the depth of the different road base materials.

Some developments have also begun in automating cranes. For example, experiments with sensor devices linked into computer software have been programmed to control boom traversing velocity and angles, trolley and hook speeds etc. Ultimately material stockpiles and delivery systems could be co-ordinated with real production team performance and so provide a fully automated environment. The first fully proven systems are likely to be associated with tower crane situations.

A few other exotic examples of automation and robots have been demonstrated for bricklaying, laying of large concrete bays, spray painting and plastering, multi head drilling of tunnel faces, deep well drilling, etc.

In summary, much construction equipment could be designed to work remotely. However, the high cost, and more importantly, the variability of the construction site, together with the nature of the tasks undertaken will preclude all but the most straightforward of operations, as clearly most construction work is still too variable to enable the economic fitting of automatic guidance devices at present.

# CONTENTS

# SECTION 1

# GROUND ENGINEERING METHODS

### Compressed-air supply

Compressed air is used to drive many items of construction plant, including drills, tunnelling equipment, pile hammers, etc. Principles and practices of production and distribution of compressed-air supplies are explained and described.

### Rock drilling

Percussive, rotary-percussive, 'down-the-hole', rotary and diamond drilling methods and equipment are described and compared. Criteria in selecting drilling rigs, bits and flushing mediums are set out and expected rates of penetration tabulated.

### Soft ground drilling

Continuous flight augering, intermittent augering, and grabbing methods are described. Drilling bits for application in soil to soft rock are compared and the need for and installation of protective borehole linings is explained. Performance characteristics for each are enumerated.

### Grouting

A theory of grout penetration is developed and methods of designing grouting patterns are illustrated. Appropriate types of grout for use in soils and rocks are recommended. Selecting grouting equipment and methods, estimation of grout quantities and *in situ* testing for permeability are explained.

### Ground de-watering

Principles and theory of pumping from wells and sumps are explained. Numerical examples illustrate methods of calculating the size of wells and well layouts. Installing procedures for deep wells, well points, vacuum wells, recharge wells and sumps are described.

### Explosives and rock blasting

The theory of explosions is analysed and formulae for calculating quantities of explosive for use in quarrying, tunnelling and demolition work are derived. Types, selection and use of explosives are described including fuse and electrically detonated systems.

### Tunnelling methods

Classical, shield and rotary machine methods are examined. Propping systems and the functions and installation of linings are explained. Consideration is given to pipe jacking and thrust boring techniques used for small diameter tunnels and sewers. Rates of advance, dealing with ground water and bad ground, are highlighted. Reference is also made to submerged tubes and compressed-air working.

### Pile-driving and sheet piling

Theoretical formulae are derived to aid selection of pile hammer sizes. Drop, single-acting, double-acting, diesel, hydraulic and vibratory pile hammers are described and rated. Piling equipment and techniques are compared, including pile caps and helmets, dollies, leaders and frames. Methods of installing and extracting sheet piles are given particular attention, including work over water.

### Shoring systems

Open excavations, trench timbering, sheeting, linings, slide rails and drag boxes are described and compared, with special emphasis given to output rates and labour and plant requirements. The construction and application of single- and double-walled sheet piled coffer-dams, piled walls, diaphragm walls, injected membranes, and ground anchors are described and illustrated. The design, construction and sinking procedures for caissons, including box, open and pneumatic types are discussed and explained.

# 1

# INTRODUCTION

## 1.1 SOIL CLASSIFICATION AND IDENTIFICATION

Many of the topics in this section require an understanding of soil and rock mechanics and properties.

Therefore, before continuing further it is perhaps prudent to define some of the terms in the text as they may differ from descriptions used in codes of practice. Indeed it seems unlikely that a precise standard terminology will ever be universally recognised.

The civil engineer commonly refers to soil and rock. Soil is a naturally occurring material composed of mineral grains of various types and sizes. A major feature of a soil is the ability to separate the grains by agitation in water. Rock, however, is a much stronger material which can generally only be broken up by the application of strong crushing forces. Rocks are described as igneous, sedimentary or metamorphic according to how they were formed.

In the United Kingdom a widely accepted classification system for soils and rocks can be found in BS 5930:1981—*Code of Practice for Site Investigations*. In the USA there are two systems. These are the Unified Soil Classification System generally used by engineers and government agencies and the AASHO (American Association of State Highway Officials) System commonly adopted for road design and construction. Elements of these systems are summarised in Figs 1.1a and 1.1b with some slight changes in terminology from the official text.

Many readers will have undertaken quite comprehensive courses as students, and have a good understanding of fundamental soil mechanics; for others, further reading may be necessary. In particular *Soil Mechanics in Engineering Practice* by Terzaghi and Peck and *Foundation Design and Construction* by Tomlinson are recommended.

Fig. 1.1(a)  Soil types, characteristics and terminology

| Soil description | | Particle size (mm) | Method of assessing compaction/strength | |
|---|---|---|---|---|
| Boulders | | 200–250 | difficult to determine in the field | |
| Cobbles | | 60–200 | | |
| gravel | coarse | 20–60 | Loose – can be excavated with a spade or 50 mm wooden peg can be easily driven | non-cohesive |
| | medium | 6–20 | Compact – requires a pick for excavation and 50 mm wooden peg is hard to drive | |
| | fine | 2–6 | | |
| sand | coarse | 0.6–2.0 | | |
| | medium | 0.2–0.6 | | |
| | fine | 0.06–0.2 | | |
| silt | coarse | 0.02–0.06 | Soft – easily remoulded in the fingers | |
| | medium | 0.006–0.02 | Firm – can be remoulded by strong pressure in the fingers | |
| | fine | 0.002–0.006 | | |
| clay | very fine only | less than 0.002 | Very soft – extrudes between fingers when squeezed in the hand | cohesive |
| | | | Soft – moulded by light finger pressure | |
| | | | Firm – moulded by strong finger pressure | |
| | | | Stiff – very difficult to mould | |
| | | | Hard – can be indented by thumb nail and is brittle | |
| peats and organic clay, silts and sands | | varies | Firm – fibres compressed together | |
| | | | Spongy – very compressible and open structure | |
| | | | Plastic – can be moulded in hand and smears the fingers | |

**Fig. 1.1(b)** Main rock types and characteristics

Descriptions applied when secondary constituents are present

| Description | Content of secondary soil | Term (secondary-primary) |
| --- | --- | --- |
| slightly | up to 5% | e.g. slightly silty sand |
| — | 5% to 15% | e.g. silty sand |
| very | 15% to 35% | e.g. very silty sand |
| and | 35% to 50% | e.g. clay and silt |

| Rock description | | Compressive strength (N/mm²) | Rock type |
| --- | --- | --- | --- |
| very weak | soft | under 1.25 | |
| weak | | 1.25–5 | |
| moderately weak | | 5–12.5 | Shales, mudstones, poorly compacted sedimentary rocks, highly weathered igneous rocks, boulder clay |
| moderately strong | medium | 12.5–50 | |
| strong | | 50–100 | Sandstone, limestone, most metamorphic weathered igneous rocks |
| very strong | hard | 100–200 | Quartzite, granites, most extrusive igneous rocks |
| extremely strong | very hard | 200+ | Fine-grained granites, most intrusive igneous rocks and horn felsites |

*Note:* These are generalisations; sometimes a soft rock may be tougher and more difficult to penetrate than a hard rock.

# 2

# COMPRESSED AIR

## 2.1 INTRODUCTION

Compressed air is required on construction sites to power rock and
concrete breaking equipment, winches, drilling equipment, pumps, small
tools, concreting tools, and to provide ventilation and pressurised
working areas.

Depending upon the pressure and quantity of compressed air
required, equipment can be obtained ranging from small portable
compressors to large semi-permanent plant.

## 2.2 PRINCIPLES OF COMPRESSING AIR

According to Boyle's Law, the pressure ($p$) of a gas varies inversely
with its volume ($v$) provided the temperature is kept constant, i.e.
$p \times v =$ constant. Thus for a given mass of gas of absolute pressure $p_1$
and volume $v_1$, if changed to $p_2$ and $v_2$ the following relationship holds:

$$p_1 \times v_1 = p_2 \times v_2$$

This process of expansion or compression of a gas is called 'isothermal'
and is assumed to take place without any change of temperature.
However, if no heat is allowed to enter or leave the gas during
expansion or compression, thereby allowing a temperature change to
occur, the process is called 'adiabatic' ('isentropic'). Since higher
compression is obtained under isothermal conditions (Fig. 2.1) every
effort should be made to remove the heat produced by the process. In
practice this cannot be fully achieved, and the production of
compressed air will fall between isothermal and isentropic. This is
called a polytropic process, which follows the law

$$pv^n = \text{constant (k)}$$

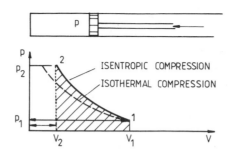

**Fig. 2.1** Isothermal and isentropic pressure-volume diagrams

in which the effect of temperature is provided for by $n$, where

$$n = 1 \quad \text{for isothermal compression}$$

and $\quad n = 1.4$ for isentropic compression.

## 2.3 ACTION IN A COMPRESSOR

The usual type of compressor used in construction work operates on the displacement principle, but where large quantities of air and constant flow are required, a dynamic compressor is often favoured. In the displacement type, the pressure rise is obtained by enclosing a volume of gas in a confined space, with subsequent reduction of its volume by mechanical action. A reciprocating piston, rotary screw or rotary vane provides this form of compression. In a dynamic compressor, compression of the gas is obtained by imparting kinetic energy. A centrifugal pump falls into this category.

## 2.4 DISPLACEMENT PRINCIPLE OF COMPRESSING AIR

The principles of the displacement type of air compressor may be followed by reference to Fig. 2.2. A piston (d) is driven up and down in a cylinder (a). On the downward stroke, valve (1) is open and air enters the cylinder through the inlet (c). During the upward stroke valve (1) is closed and the air compressed. The discharge valve (2) opens when the pressure inside the cylinder reaches that of the receiver (b), the receiver being a large vessel for storing the compressed air.

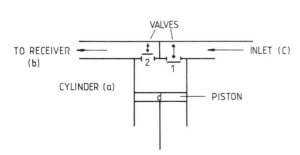

**Fig. 2.2** Principles of the displacement-type compressor

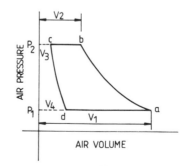

**Fig. 2.3** Single stage pressure–volume diagram

The process is maintained under as near isothermal conditions as possible, by removing the heat produced with a surrounding jacket of circulating water or similar.

The pressure–volume diagram of the process is shown in Fig. 2.3. At the bottom of the stroke, the volume of air in the cylinder is $v_1$, at atmospheric pressure $p_1$. As the piston moves upwards the air is compressed until the pressure $p_2$ of the air in the receiver is obtained, when the volume is $v_2$. The air is then delivered at constant pressure $p_2$ into the receiver and the piston continues until it reaches the top of the stroke. A small quantity ($v_3$) of air at pressure $p_2$ represents the clearance volume between piston and cover, which expands to $v_4$ on the downward stroke, finally reaching atmospheric pressure when the suction valve opens to admit a fresh supply of air. The cycle is then repeated, until the desired pressure in the receiver is obtained. The total work done on the air during compression, delivery and recharging is represented by area abcda.

The work done ($w$) can be represented by

$$dw = -v\delta p$$

but $$pv^n = k$$

and for isothermal conditions $n = 1$. Neglecting the work done by the small quantity of compressed air in the clearance volume (Fig. 2.4),

$$dw = -\int_{p_1}^{p_2} \frac{k}{p} \delta p$$

work done

$$w = -k \log_e p + \text{constant (c)}$$

when $$p = p_1$$

$$w = 0$$

thus $$c = k \log_e p_1$$

when $$p = p_2$$

$$w = -k \log_e p_2 + k \log_e p_1$$

therefore

$$w = -k \log_e \frac{p_2}{p_1} \qquad (2.1)$$

but $$k = pv = p_1 v_1 \qquad (2.2)$$

If $p_1 = 1.01$ bar ($0.101$ N/mm²) i.e. atmospheric pressure, and $v_1 =$ m³ per min., and disregarding the minus sign, then the power required to

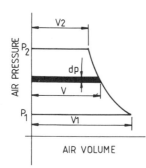

**Fig. 2.4** Single stage pressure–volume diagram without a clearance volume

compress $v_1$ from atmospheric pressure to $p_2$ is

$$\text{Power} = 10^6 \times 0.101 \times v_1 \log_e \frac{p_2}{0.101} \text{ Nm per min.}$$

Thus    $$\text{Power} = 10^3 \times 0.101 \times \frac{v_1}{60} \log_e \frac{p_2}{0.101} \text{ kW} \qquad (2.3)$$

(1 kilowatt = 1000 Nm per s).

### Example

To compress 2.8 m³ (100 ft³) of free air per minute from atmospheric (0.101 N/mm²) to (0.7 N/mm² (100 p.s.i.) indicated on the gauge (i.e. 8.01 bar absolute) requires a compressor with a theoretical power value:

$$\text{Theoretical power} = 10^3 \times 0.101 \times \frac{2.8}{60} \log_e \frac{0.801}{0.101}$$

9.7 kW or 13.1 HP

*Note*: Air at atmospheric pressure is usually referred to as 'free air'. In practice the conditions will not be entirely isothermal and, for example, the power requirement would theoretically increase to about 13 kW if isentropic conditions were assumed ($n \simeq 1.4$).

Taking into account mechanical losses, the required power of a compressor delivering air at 7 bar above atmospheric (gauge) is approximately:

$$\text{Power (kW)} = 6 \times v_1$$

where $v_1$ is the volume of free air in m³/min, drawn in for compressing.

### Multi-staging

To operate as near isothermally as possible, the compression of air to 7 bar (gauge) (typical for powering construction equipment) is

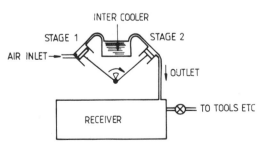

**Fig. 2.5**  Two-stage displacement-type compressing

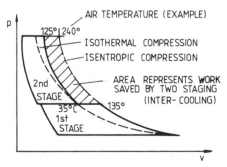

**Fig. 2.6**  Two-stage pressure–volume diagram

performed in two stages (Fig. 2.5). After stage 1 the air is immediately cooled before entering the stage 2 compressor. The pressure volume diagram is shown in Fig. 2.6 where the shaded area represents the work saved by two-staging. For even higher pressure rises, additional compressing stages are required.

### Influence of altitude

The density of air decreases with increasing altitude and thus for a compressor operated above sea level, $k$ in eqn. (2.2) should be reduced. For example, to compress $2.8 \text{ m}^3$ ($100 \text{ ft}^3$) of air per minute from $0.05 \text{ N/mm}^2$ (atmospheric pressure) to $0.801 \text{ N/mm}^2$ (absolute pressure) requires a compressor with a power value of

$$10^3 \times 0.05 \times \frac{2.8}{60} \log_e \frac{0.801}{0.05} = \underline{6.47 \text{ kW}}$$

where $k = p_1 v_1 = 0.05 \times 2.8$

$p_1 = 0.05$

$p_2 = 0.801$

$v_1 = 2.8$.

However, the efficiency of the diesel motor to produce the power

**Table 2.1** Effect of altitude on compressed-air production

| Altitude (m) | Reduction in power needed (%) | Loss of power in motor | |
|---|---|---|---|
| | | Diesel (%) | Electric (%) |
| 1000 | 5 | 6 | 0 |
| 2000 | 10 | 15 | 8 |

unfortunately decreases with height and therefore counteracts the advantages. Table 2.1 illustrates typical values.

## 2.5 DISPLACEMENT COMPRESSOR TYPES

### (i) Reciprocating compressor

A typical arrangement is illustrated in Fig. 2.2 as described earlier. Double staging can be obtained by passing air from stage 1 to stage 2 compressed through a cooler as shown in Fig. 2.5.

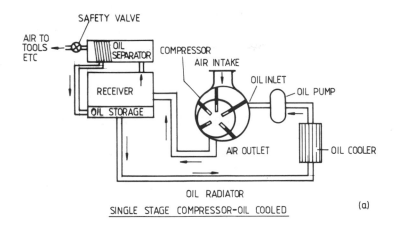

SINGLE STAGE COMPRESSOR—OIL COOLED (a)

**Fig. 2.7** Air cooled and oil cooled compressor

TWO-STAGE COMPRESSOR — AIR COOLED

(b)

**Fig. 2.8**  Screw feed compressor

**Fig. 2.9**  Vane compressor

### (ii)  Screw feed compressor (Fig. 2.8)

This is a new development comprising two counter-rotating screw rotor elements. The rotors do not touch and therefore lubrication is not required within the compression cylinder. Unbalanced mechanical forces and inlet and outlet valves are avoided and therefore high rates of rotation can be achieved.

### (iii)  Vane compressor (Fig. 2.9)

Comprises a rotor with radially adjusting blades mounted eccentrically in the cylinder. Air on entering is trapped in the space between the vanes and the cylinder wall, and as the rotation takes place (at 1500 to 1800 rpm) the air volume decreases until the discharge port to the receiver is reached.

Cooling is usually obtained by injecting small amounts of oil into the compression space, which is subsequently separated from the air and passed through a cooler, before being reintroduced (Fig. 2.7). Up to 8 bar pressure is commonly obtained in a single stage.

## 2.6   REGULATING A COMPRESSOR

The demand on the compressor depends upon the size and number of tools and equipment drawing from the receiver. The rate of production of compressed air can be regulated by varying the speed of the motor driving the compressor, whereby the throttle is automatically regulated by the air pressure in the receiver. This method is often used with electric- or turbine-driven compressors, but can also be applied to diesel engines. For most diesel-powered compressors, however, the production of compressed air is regulated by the simple on–off principle. When the pressure in the receiver reaches the maximum set, the compressor is stopped, and is only restarted when the pressure falls below a certain minimum value – the difference usually being about 10% of maximum pressure. This principle is even more basic when using the reciprocating type of compressor, where the suction valve is simply left open when maximum pressure in the receiver is obtained. The air sweeps in and out within the cylinder and thus little power is consumed.

## 2.7   THE RECEIVER

A receiver is required between the outlet from the compressor chamber and the connection to the tools and equipment, and serves to:

(i) Store the compressed air and so equalise air pressure variations discharged from the compressor caused by changes in demand.
(ii) Increase the cooling effect.
(iii) Assist in removal of water vapour and oil separation, where oil is the cooling fluid.

The size of the vessel depends upon the output of the compressor and the regulating method. A large vessel is required when regulation involves stopping and starting the drive motor and compressor, otherwise, when the demand varies slightly, too much starting and stopping results, causing high wear and tear on the moving parts. The approximate size of the vessel required can be determined from the following formula:

Continuous running (e.g. open valve or speed regulator)      $Q \simeq 0.1 \, V$
Stop-start regulation                                        $Q \simeq 0.16 \, V$

where $Q$ is in $m^3$, and $V$ is in $m^3/min$ and represents the free air capacity of the compressor.

*Note*: A safety valve must be fitted to the receiver, set to blow off if pressure builds up by accident or malfunction.

## 2.8  COMPRESSED-AIR LINES

Pressure losses are incurred in transferring the compressed air to the tools, and they depend upon:

  (i) Length of pipe – increasing pipe length increases the friction losses.
 (ii) Air pressure at point of entry.
(iii) Pipe diameter – the larger the diameter, the smaller the friction loss.
(iv) Rate of flow of air – greater flow increases the friction loss.
 (v) Bends, fittings, valves, etc.

If the sum of these frictional resistances were to exceed the initial pressure, then there would be no air flow at all. Construction

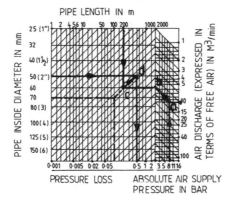

**Fig. 2.10**  Diagram to calculate pipe diameters

equipment is usually designed to operate at about 7 bar (100 psi) above atmospheric pressure. The efficiency of such equipment falls very rapidly as the pressure decreases, and therefore the air delivery pipe is usually designed to limit the pressure drop to within 10% of the supply pressure. The formula for estimating the drop in pressure for a particular length of pipeline is given by

$$dp = \frac{f \cdot v^{1.85}}{d^5 p} \tag{2.4}$$

where   $dp$ = the pressure drop in bar
        $f$ = coefficient ($82 \times 10^3$ for steel pipes)
        $l$ = pipe length (m)
       $d$ = pipe diameter (mm)

$p$ = initial absolute pressure in bar
$v$ = volume rate of low (m$^3$/min) of free air.

For estimating purposes a chart (Fig. 2.10) can be used to determine the pressure loss for combinations of (i), (ii), (iii) and (iv) above, based on eqn. (2.2).

### Example

Calculate the pressure loss for a 200 m length of 50 mm diameter pipe, resulting from delivering 10 m$^3$/min (free air) compressed to 7 bar, i.e. 8 bar absolute.

### SOLUTION

The path through the graph (Fig. 2.10) starts by entering at 10 m$^3$/min and proceeding horizontally to meet the vertical line drawn up from 8 bar at (a), then proceeds parallel to the sloping line at (b). This line is then projected horizontally to (c), where it meets the line drawn vertically down from 200 m. Proceed along the parallel line, to meet the line drawn horizontally from 50 mm diameter at (d). The vertical projection from (d) gives the pressure loss of 0.5 bar. An example illustrating the diagram for a 70 mm diameter pipe is shown dotted.

### Other losses

Stop valve is equivalent to about 16 m of pipe friction loss.
One 90° bend is equivalent to about 5 m of pipe friction loss.
One 30° bend is equivalent to about 3 m of pipe friction loss.

These loss values should be added to the pipe length for use with the graphical method described above.

## 2.9 PIPE LAYOUT

It can be seen from the chart (Fig. 2.10) that pressure losses can be relatively high for small diameter pipes and therefore the pipe length should be kept to a minimum. In practice, pipes for construction application are generally within 25–150 mm diameter, but for flexibility, textile reinforced rubber hose about 25 mm diameter is mostly used in short lengths to snap connect to the tools and equipment. Where the compressor cannot be located near the work place, a long pipe of sufficient diameter should be laid down and flexible connections made to it. Such pipes when laid outdoors should preferably be buried to avoid freezing of the condensate in winter (otherwise ice may form and cause blockage), and also for protection against damage caused by site traffic.

## 2.10   COMPRESSOR RATING

Compressors are rated in terms of the volume of free air per minute taken in for compressing. They are available as portable, towed or permanent units with capacities ranging from 1 m$^3$/min to more than 100 m$^3$/min. It can be seen from eqn. (2.3) that an increase in the output pressure requires additional power to compress a given quantity of free air. Thus the free air rating is usually stated at a particular delivery pressure, e.g. 7 bar (gauge). Two stage compression through an intercooler is commonly used to produce air at 7 bar gauge pressure to minimise the adiabatic effects.

## 2.11   TYPICAL AIR CONSUMPTIONS BY CONSTRUCTION EQUIPMENT

|  | *Free air consumed* (m$^3$/min) *delivered at 7 bar gauge pressure* |
|---|---|
| Heavy concrete breaker | 2.5 |
| Medium concrete breaker | 1.5 |
| Light concrete breaker | 1.0 |
| Clay digger (spade type) | 0.9–1.25 |
| Picks | 1–1.2 |
| Hand held sinker drills | 2–4 |
| Feedleg drills | 4–6 |
| Rig mounted rotary and rotary-percussive drills | 5–10 |
| Drill hole flushing | Up to 5 |
| Vibrators | 1–2 |
| Small tools | 0.5–1 |
| Hoists and winches | Up to 10 |
| Pumps | 2–5 |
| Pile hammers | 1–50 |

*Note*: Higher working pressures may be necessary when operated in compressed air chambers, e.g. tunnelling.

# 3

# ROCK DRILLING

Rock boring is required on a variety of civil engineering projects
including:

(i) core sampling for geological investigations
(ii) confining blasting charges in quarrying and tunnelling
(iii) rock bolting and anchoring
(iv) grouting.

There are basically three methods of producing holes in rock, each
suited to a particular application. These are:

(i) rotary drilling
(ii) rotary–percussive drilling
(iii) percussive drilling.

A fourth method, involving intense heat concentrated on a confined
part of the rock, is occasionally used. Other developments are
progressing with laser beams and high pressure water jets. Such
methods are at present only in the pilot stage and it may be some
years before they are sufficiently reliable to be adopted by the
construction industry.

## 3.1  CHOICE OF METHOD

For many construction applications, such as in quarrying, rock bolting,
grouting and tunnelling, where relatively shallow holes less than 50
metres deep and up to 100–150 mm diameter are required,
rotary-percussive equipment provides a light, manoeuvrable and
efficient method in medium to very hard rock. For larger diameters or
when boring to greater depths and accuracy, or where soil or soft rock
is to be encountered, rotary drilling machinery may be preferred.
Percussive drilling only is very slow, but because it is a simple and
basic method, it is sometimes more convenient to use this technique in

**Table 3.1**  Comparison of drilling methods

| Method | Rock type | | | | Normal application | |
| --- | --- | --- | --- | --- | --- | --- |
| | Soft | Medium | Hard | Very hard | Max. borehole dia. (mm) | Max. borehole depth (m) |
| Percussive | **** | **** | **** | **** | 400 | No limit |
| Rotary–percussive | | | | | | |
|     Drifter drilling | * | ** | *** | **** | 150 | 40 |
|     Down hole drilling | *** | **** | **** | **** | 200‡ | 250 |
| Rotary | | | | | | |
|     Cutting | **** | **** | * | — | 600 | 5000 |
|     Crushing | *** | **** | **** | *** | 600§ | 5000 |
|     Abrasive | | | | | | |
|     (diamond drilling) | — | ** | **** | **** | 150 | 2000† |
| Thermal | — | * | **** | **** | 150 | 5 |

\* = poor    \*\*\*\* = good
† For core sampling, otherwise as for crushing method
‡ Bits are available up to 800 mm diameter for special applications
§ Roller heads up to 6 m diameter are available for special tasks

soils investigation for breaking through a minor rock obstruction. The merits of the various methods are shown in Table 3.1 and the classification of the rocks appropriate for each is given in Table 1.2. (Although the methods are designed for rock drilling, soils can be accommodated provided adequate flushing is available.)

## 3.2  ROTARY DRILLING

Rotary drilling (Fig. 3.1) relies on a high feed thrust applied down the drill stem, to force the edges of the bit into the rock surface. High torque and rotation of the drill shaft then cause cracking and chipping, and rock fragments are broken away. The rotary drill is supported on a mast above the hole, and additional drill rod extensions are inserted into the rotor to deepen the borehole. Either a compressed air or more usually a hydraulic or an electric motor is used to power the unit.

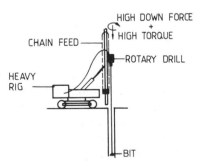

**Fig. 3.1**  Principles of rotary drilling

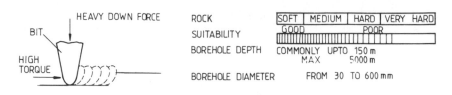

| ROCK | SOFT | MEDIUM | HARD | VERY HARD |
|---|---|---|---|---|
| SUITABILITY | GOOD | | POOR | |

BOREHOLE DEPTH    COMMONLY  UPTO  150 m
                  MAX         5000 m

BOREHOLE DIAMETER    FROM  30  TO  600 mm

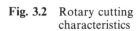

**Fig. 3.2**  Rotary cutting characteristics

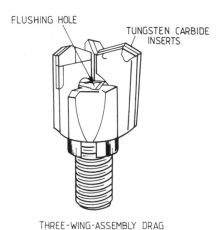

**Fig. 3.3**  Winged bit for rotary cutting

THREE-WING-ASSEMBLY DRAG BIT

Flushing of the drill hole may be carried out with water or compressed air. For versatility in construction work, rigs are generally truck-mounted.

### 3.3  ROTARY DRILLING METHODS (also see directional drilling in Chapter 10)

#### Rotary cutting (Fig. 3.2)

Winged bits (Fig. 3.3) with tungsten carbide inserts, to aid penetration and durability, are suitable for holes up to about 150–200 mm diameter. For wider borings it is common practice to follow the primary bit with staged reamers (Fig. 3.8).

Current bit design limits the method to applications in soft rock and requires a feed force up to 0.5 kN/mm bit diameter to facilitate adequate penetration of the inserts. Thus a 200mm diameter bit in medium hard rock would require a rig capable of delivering 10 tonnes (100 kN) of feed force. The torque and feed force required to sheer the rock plane depends upon the borehole diameter and rock strength. The bit is rotated within the range 50–150 rpm: the harder the rock, the slower the rotation speed.

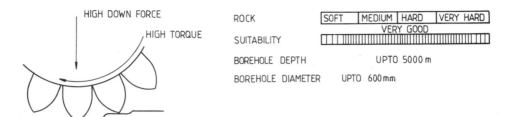

Fig. 3.4    Rotary crushing characteristics

### Rotary crushing (Fig. 3.4)

#### TOOTHED AND BUTTON ROLLER BITS

During the last 20 years development of roller bits by the oil industry has been updated for more general application(s) by construction users and today they are commonly chosen for medium to hard rock. Like the rotary cutting method, a high feed force must be applied to the bit, but instead of a cutting edge, steel cones with tungsten carbide tipped teeth (Fig. 3.5) or buttons (Fig. 3.6) rotate on their own axis and the rock is crushed and chipped away. Because the drill stem is also rotated, a fresh part of the hole base is always worked on. The button-type bit is favoured for harder rocks.

The roller bit method relies on overcoming the compressive strength of the rock, and high feed pressures up to 1.5 kN/mm of bit diameter over the borehole cross-section may be required in hard rock. Thus for a 200 mm diameter borehole a feed force of 30 tonnes (300 kN) may be necessary and 75 tonnes for a hole of 500 mm diameter bore. Bits, however, are available in smaller sizes down to 75 mm, requiring less feed force. Only light truck-mounted rigs are then needed.

Fig. 3.5    Rotary crushing bit

Fig. 3.6    Roller cone bit

### Effect of feed force

The penetration rate is directly dependent upon the feed force applied to the drill bit. However, as the force is increased there is an optimum at which the resistance to turning of the bit exceeds the advantages gained from the extra down-force, as shown in Fig. 3.7.

### Factors limiting the depth of drilling

The limit to the depth of hole that can be bored by rotary methods is largely governed by the quality of drill bit, the bit life (rods must be withdrawn to change the bit), strength of the drill rod, the capability of the feed equipment (as relatively high torque and down feed forces are required) and the flushing efficiency. Also, a long drill stem is heavy and so provision of sufficient hoisting capability is another important limiting factor.

Fig. 3.7    Effect of feed force on rate of penetration

### Curved boreholes

Recent developments in drilling technology now make it possible to
deflect the borehole from the vertical (Fig. 3.8). Steering is obtained by
inserting a specially designed wedge-shaped piece directly behind the
bit. Techniques such as mechanical plumbs, electronic dip meters,
acid-etch, photography, etc. are then used to measure the deflection so
that adjustments can be made to the directional path.

The latest developments now have the rotary motor located at the
bottom of drill string, powered by the flushing medium. In this manner
twisting of the drill stem can largely be eliminated.

## 3.4  ROTARY DRILLING RIGS

The drilling rig (Fig. 3.9) comprises:

  (i) Rotary drill (usually hydraulically driven) with variable revolution
     rate control.
 (ii) Hydraulic or chain system to provide down-thrust to the drill
     stem and bit.
(iii) Suitable support mast to raise and support the drill stem.
(iv) Swivel barrel loader to position the drill stem insertions.
 (v) Winch to raise the drill stem and bit.
(vi) Centralising chuck to hold the stem and casing on line when
     drilling.
(vii) A vehicle to support the engine, drilling equipment, winches,
     compressor and flushing pump.

Typical data for drilling rigs are given in Table 3.2.

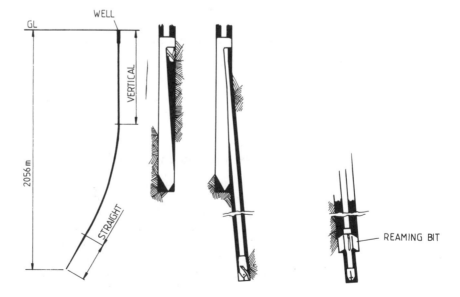

**Fig. 3.8**  Deflecting a drill hole from vertical

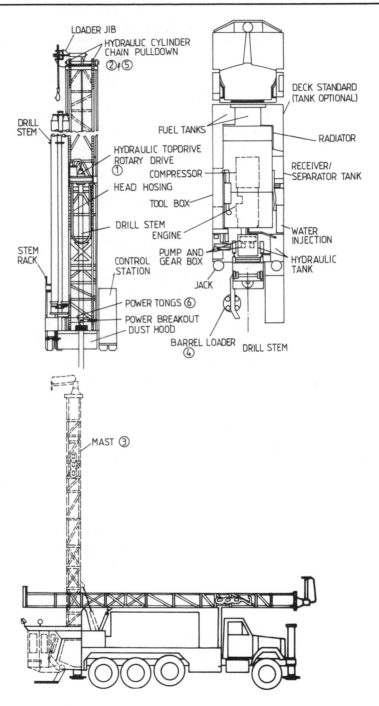

**Fig. 3.9** Truck-mounted rotary drilling rig

Currently, rigs can be equipped to produce boreholes 5000–6000 m deep. However, for most construction applications, medium to light rigs, truck-, wheel- or crawler-mounted, are usual, with the larger vehicles being suitable for boreholes up to 200 mm diameter and for drilling to depths of up to 100–200 m. The truck-mounted version is

**Table 3.2**  Rotary drilling rigs

| | Rig for boreholes | |
| | up to 125 mm diameter | up to 250 mm diameter |
|---|---|---|
| Torque | 6 kN.m | 10–12 kN.m |
| Feed force | 50 kN | 180 kN |
| Hoist capacity | 50 kN | 100 kN |
| Rotary speed | 25–150 rpm | 25–150 rpm |
| Rotary motor power | 30 HP (22 kW) | 100 HP (75 kW) |
| Hoist speed | 30 m/min fast, 6 m/min slow | 5–30 m/min |
| Feed speed | 0–5 m/min | 2–16 m/min |
| Mast | Up to 7 m stems | Up to 9 m stems |
| Angle drilling | 0–45° | 0–30° |
| Total weight | | Approx. 28 tonnes |

*Notes*
(1) Air flushing              – up to 20 m³/min
    Mud or water flushing     – up to 600 l/min
(2) Rotary drill motors and rigs are often suitable for use with 'down the hole' drills as a rotary–percussive method

**Table 3.3**  Production data for rotary drilling performance – 200 mm diameter borehole

| Rock classification | Bit loading (kN/mm bit diameter) | Penetration rate (m/h) | Bit life (m of hole) |
|---|---|---|---|
| Soft | 0.3–0.6 | 20–45 | 1500–6000 |
| Medium | 0.5–0.8 | 10–20 | 600–1500 |
| Hard | 0.7–1.0 | 5–10 | 150–600 |
| Extremely hard | 0.9–1.3 | 2–10 | 50–150 |

particularly useful, because of the advantage of its mobility. Most of the modern rotary rigs can be quickly converted to deal with soils and different rock hardness, simply by changing the bit or even switching to the rotary–percussive technique.

## 3.5  DIAMOND DRILLING (ROTARY ABRASION) (Fig. 3.10)

Diamond drilling is mostly used in mineral exploration where core samples of the rock are required for analysis. The method is suitable for borings of up to about 150 mm diameter.

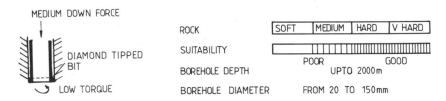

**Fig. 3.10**  Diamond drilling characteristics

Drilling takes place on a similar principle to other rotary methods, and requires very high feed force, but relatively lower torque than rotary drilling. Borehole depths of over 2000 m are not unusual with rotary speeds of up to 1000 rpm. The method relies on abrasion, using a diamond-impregnated or surface-set bit.

## 3.6    DIAMOND BITS (Fig. 3.11)

### Method of action of diamond bit

The general opinion is that the combined pressure and rotation applied to a single diamond causes plastic deformation of the rock. In hard, brittle rock the compressive stresses break the material along the zone of maximum sheer stress and the small fragments must then be

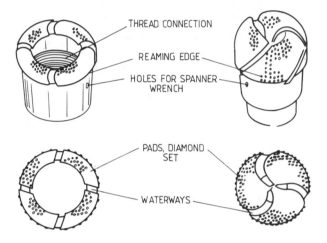

**Fig. 3.11**   Diamond bits

continuously and efficiently removed by a flushing fluid to prevent obstruction. Thus the rate of penetration of the bit is influenced by the rotation speed, feed pressure and pumping rate of the flushing medium.

### ROTATION

For diamond surface-set bits, rotary speeds of 50–150 m/min are used; slightly higher speeds of 100–200 m/min are possible with impregnated bits. Thus for a 50 mm diameter bit, 150 m/min represents approximately 1000 rpm. It has been shown by Marx that for a constant feed pressure, the theoretical rate of penetration is proportional to the rotation speed of the bit. Thus, as the diameter of the bit is reduced, the revs per minute must be increased to maintain an adequate rotation speed at the radius of the diamond surface. However, the penetration rates also depend upon the rock hardness, for example a rotation speed of 120 to 180 m/min is suitable in soft formations, whereas 60 to 120 m/min is preferred for a hard formation. The rates of penetration may not, of course, be the same for both.

## FEED FORCE

Each diamond in the bit must transmit a stress which exceeds the strength of the rock, and clearly, therefore, there is a close relationship between bit load and penetration rate.

Thus     $G \leqslant \dfrac{P}{d \times a} \leqslant D$

where   $D$ = diamond strength
   $G$ = rock strength
   $P$ = bit load
   $d$ = number of diamonds
   $a$ = contact surface area of diamond

The following loads per carat are recommended for different rock types:

| Rock type | Load per carat ($N$) |
|---|---|
| Granite | 200–1000 |
| Quartz | 400–2000 |
| Basalt | 400–1800 |
| Basalt lava | 100–600 |
| Sandstone | 200–1000 |
| Shale | 200–400 |
| Limestone | 20–1200 |

Thus for a given size of diamond, i.e. stones per carat and density of stones in the cutting surface, the required feed load on the bit can be determined.

In soft to medium hard rock, diamonds of 20 to 90 stones per carat are selected, while for hard to very hard rock 200 to 250 stones per carat are necessary, i.e. the harder the rock, the smaller the diamond and greater the number of diamond stones.

### Diamond loss and wear

The edges of active diamonds get rounded as they are subject to harsh erosion forces. This is compounded by high rates of vibration, which cause some shattering and sheering of the diamonds. Insufficient cooling by the flushing fluid may also cause burring of the diamond. The carat loss is related to the length of borehole drilled by the bit, as shown in the approximate data given in Table 3.4.

Table 3.4   Diamond loss and bit life

| Rock type | Caret loss per m of penetration | Bit life (m) |
|---|---|---|
| Basalt | 0.2 | 10–60 |
| Dolomite | 0.03 | 25–60 |
| Granite | 0.2 | 10–30 |
| Limestone | 0.05 | 25–150 |
| Sandstone | 0.05 | 25–150 |
| Quartz | 0.5 | 5–30 |

### 3.7  DIAMOND DRILLING EQUIPMENT

The drilling equipment is similar to that required for normal rotary boring, except that the feed force is usually delivered through hydraulic rams rather than chain feed. The drill may be mounted on wheels, tracks or skids. The rig shown in Fig. 3.12 comprises a diesel engine or electric motor, skids or tracks and base frame (1), main hoist and cat head to handle the drill rods (2), hydraulically-powered rotary drill and hydraulic chuck to supply the downfeed force (3), and drill rods. Typical data to handle 900 m of drill rods of 40 mm dimeter are as follows:

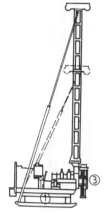

**Fig. 3.12**  Diamond drilling rig

| | |
|---|---|
| Torque | 1.5 kN.m |
| Feed force | 45 kN |
| Hoist capacity | 60 kN |
| Rotary speed | 100–1500 rpm |
| Engine power | 40 HP (30 kW) |
| Drill rod diameter | up to 45 mm |
| Hoist speed | 30 to 150 m/min. 3 m or 6 m long drill rods |
| Mast | up to 8 m |
| Feed length | 0.5 m |
| Angle drilling | 0° to 360° |
| Flushing | Water or air supply are additional requirements. |

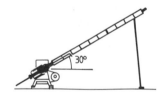

**Fig. 3.13**  Angle drilling

It can be seen in Fig. 3.13 that the chuck and mast can be turned to accommodate angle drilling. Drill rod extensions are raised into position by the mast winch, and the chuck is simply unclamped and repositioned when its full 0.5 m range has been reached during the drilling operation. The new extension is thereby passed down and through the rotary unit as drilling proceeds.

### 3.8  CORE SAMPLING

Core sampling requires special tubular edged bits. In hard, compact formations the single tube core barrel is often used (Fig. 3.14). However, for less firm material the double tube core barrel is preferred (Fig. 3.15), the advantage being that the flushing water is shielded from the core and so minimises the washing action. The inner tube is normally of the 'swivel type', thus reducing the risk of breaking the core.

Unfortunately, with these methods the whole of the drill stem must be withdrawn from the borehole to remove the core sample (300–400 mm long), thus resulting in unproductive time. To overcome this problem the wire line core barrel was introduced, making it possible to take the core sample from the barrel without pulling the string of drill rods (Fig. 3.16).

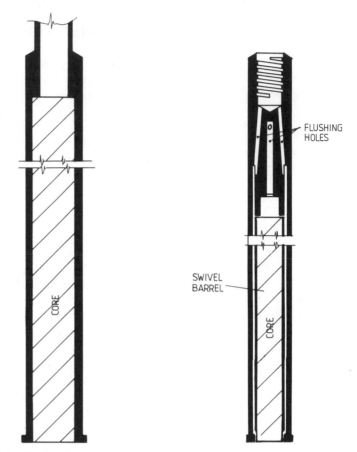

**Fig. 3.14**  Single tube core barrel.    **Fig. 3.15**  Double tube core barrel.

Core samples up to 150 mm diameter can be taken with diamond bits. In soft, abrasive rock or in mixed formations where large diamonds of 8–15 stones per carat would be necessary, it is sometimes possible to use tungsten carbide particles mixed with special alloy-tipped bits instead of diamonds, as these are less expensive.

Core drilling bits have a tendency to form a slightly tapered hole, but with the insertion of a reaming shell (Fig. 3.17) the correct gauge is maintained and so a new bit can be inserted into the borehole without damage.

## 3.9  CASING SHELLS FOR ROTARY AND DIAMOND DRILLING

Where the borehole must pass through unstable rock or soil, or where the flushing medium would be absorbed into the surrounding formation, casing tubes may be required. The method basically involves following the drill bit and stem with a casing bit and lining tube of a slightly larger diameter. Casing shoes (Fig. 3.18) for drilling have

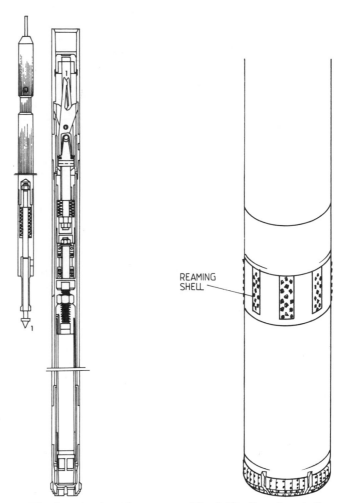

**Fig. 3.16**  Wire line core barrel.       **Fig. 3.17**  Reaming shell.

diamond-tipped cutting segments (Fig. 3.19) on the exterior lip, while the interior is smooth. The inner diameter must be sufficient to permit free passage of the drill bit, stem and core sample.

## 3.10   ROTARY–PERCUSSIVE DRILLING (Fig. 3.20)

Where medium to hard rock is to be encountered, the rotary-percussive method is often favoured because the rig is light and good rates of penetration can be obtained. The method is used for blast holes, rock anchors, grouting holes, wells, etc. In rotary-percussive drilling the drill bit is supplied with both a percussive and a rotary action rather than with a high feed force as in rotary boring. The force of the blow causes penetration of the bit inserts into the rock surface to form a crater, which is subsequently disturbed by the turning motion. The percussive

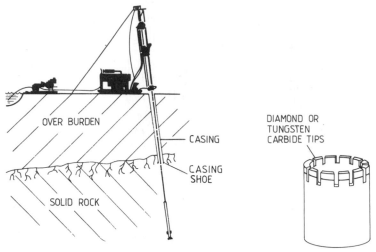

**Fig. 3.18**   Hole lining or casing.    **Fig. 3.19**   Casing shoes.

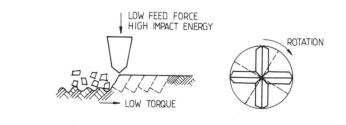

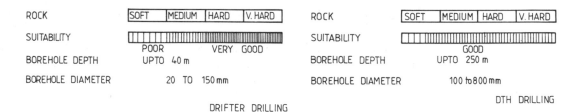

| ROCK | SOFT | MEDIUM | HARD | V. HARD |
|---|---|---|---|---|

SUITABILITY  POOR  VERY GOOD

BOREHOLE DEPTH  UPTO 40 m

BOREHOLE DIAMETER  20 TO 150 mm

DRIFTER DRILLING

| ROCK | SOFT | MEDIUM | HARD | V. HARD |
|---|---|---|---|---|

SUITABILITY  GOOD

BOREHOLE DEPTH  UPTO 250 m

BOREHOLE DIAMETER  100 to 800 mm

DTH DRILLING

**Fig. 3.20**   Rotary–percussive drilling characteristics

action provides a considerably greater force than would be achieved with the same load applied without impact. The broken rock fragments are removed with either air or water flushing introduced under pressure down the drill stem and out through the base of the bit.

Currently the two basic methods are drifter drilling (Fig. 3.21), where the drill is located at ground level, and 'down the hole' drilling (Fig. 3.22), where the percussive part of the drill actually follows the bit down the borehole.

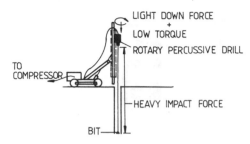

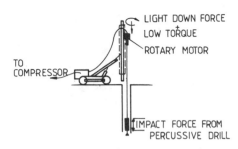

**Fig. 3.21**  Principles of rotary–percussive drifter drilling

**Fig. 3.22**  Principles of rotary–percussive 'down the hole' drilling

### 3.11   ROTARY–PERCUSSIVE DRILLING EQUIPMENT

The drilling rig (Fig. 3.23(a)) consists of:

  (i) A compressed-air or hydraulic driven rotary–percussive rock drill.
 (ii) Chain feed to maintain the feed force on the bit.
(iii) Mast or leaders to support and guide the drill.
(iv) Centralising chuck to hold the drill rod.
 (v) Telescopic boom.
(vi) Tracks or similar.
(vii) Compressed-air supply.

Tracks are usually selected for surface drilling and the boom can be tilted and repositioned to accommodate vertical, horizontal and angle drilling as shown in Fig. 3.23(b).

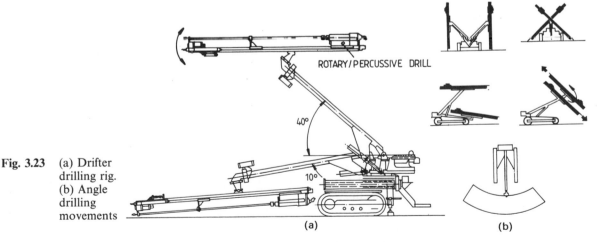

**Fig. 3.23**  (a) Drifter drilling rig. (b) Angle drilling movements

### Compressed air drifter drill (Fig. 3.24)

Operation of the drill is generally by means of compressed-air, whereby kinetic energy provides the percussive blow, as follows:

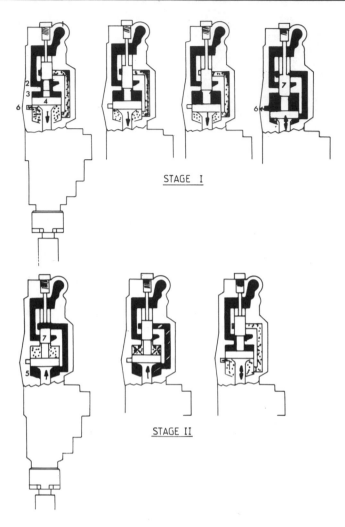

STAGE I

STAGE II

**Fig. 3.24** Drifter drill
working
principle

*Stage 1*. Compressed air enters 1, through the support port 2 and into the cylinder 3, pushing the piston 4 forward. The exhaust port 6 meanwhile allows the air in the lower part of the cylinder at atmospheric pressure to be expelled until the exhaust port 6 is blocked.

*Stage 2*. The piston continues to move downwards under its own momentum to uncover the exhaust port 6. Meanwhile the compressed air is shut off by the piston control head 7. During this phase the control head 7 allows compressed air to enter the lower part of the cylinder 5, causing the piston to move upwards until the cycle is completed, ready to start again. Thus a single blow has been delivered to the striking bar.

Rotation of the drill bit is achieved by one of two basic methods, the difference being in the mechanism used for turning the drill rod. In the earlier models the piston travels up and down in the cylinder along a helical spline. The spline is part of a rifle bar which operates a ratchet which allows rotation in one direction only.

Unfortunately with this type of mechanism, when the resistance to turning of the drill is greater than the torque output, the piston stroke is forced to shorten, with a consequent loss of power and efficiency. To redress this disadvantage, modern drifter drills have a separate air motor to provide the turning effect. As a result, higher blow rates can be achieved.

For both types, approximately 1500 blows per minute at 100–150 rpm are required to obtain acceptable penetration rates.

### 3.12 HYDRAULIC DRIFTER DRILL

Hydraulically-powered 'down the hole' (DTH) and drifter drills are now proving popular, which manufacturers claim to be more efficient than the equivalent compressed-air drills i.e. up to twice the penetration rate.

The hydraulic rotary motor can be varied from 0–300 rpm. Because the hydraulic percussive mechanism allows the use of a slim piston, greater energy flow through to the stem is obtained, thereby raising drilling performance. Either compressed-air or water may be used for flushing.

The latest versions are designed in dual purpose mode capable of carrying out purely rotary drilling duties as well as rotary-percussive.

### 3.13 'DOWN THE HOLE' (DTH) DRILL (Fig. 3.25)

The conventional drifter drill described above becomes less efficient as the length of borehole increases. This is due to loss of impact energy into the drill stem itself, and, more importantly, impact energy is dissipated as heat at each drill rod connection. Practical tests have shown that the energy loss across a rod joint is approximately 10% of the energy in the rod before the joint. Thus the rate of penetration will decrease as the depth of borehole increases.

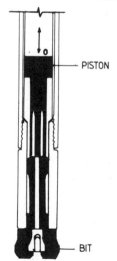

**Fig. 3.25** 'Down the hole' drill working principle

To overcome these difficulties a 'down the hole' (DTH) drill was developed (Fig. 3.22). The hydraulic or compressed-air rotary motor remains above ground level while the bit is followed down the hole by its compressed-air driven pneumatic hammer, to produce a virtually constant drilling rate. Unfortunately, because of restrictions on the practical size of piston, etc., the minimum size of borehole is about 100 mm diameter, thus restricting the available power to the drill. Thus for shallow holes of this approximate diameter, the selection of a more powerful drifter drill would allow much faster drilling rates to be obtained.

However, because of the loss of efficiency with depth of drilling when using the drifter (Fig. 3.26), a break-even point with the DTH drill would eventually be reached since there is virtually no deterioration in drilling rate with the DTH machine, and so for deep holes a lower-powered DTH drill would be more efficient than the larger drifter drill.

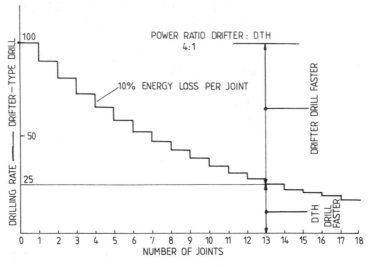

**Fig. 3.26**   Comparison of drifter and DTH drilling rates

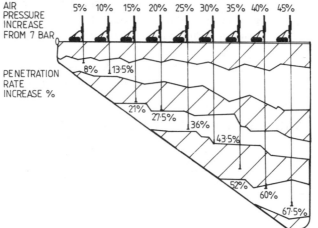

**Fig. 3.27**   Effect of air pressure on penetration rates

## 3.14   EFFECT OF AIR PRESSURE

Most drifter drills are operated with air pressure at 100–150 psi (7–10 bar), but greater energy per blow with an improved rate of penetration can be achieved (Fig. 3.27) by increasing the air pressure and 250 psi (18 bar) is not uncommon, especially on DTH drills. However, with the drifter method, earlier failure results from the increase in impact velocity producing considerable extra stress in the drill stem. Similarly with the modern high powered hydraulic drifter drills.

## 3.15   BITS

The penetration rate for rotary–percussive drilling is largely dependent upon rate and weight of blow, speed of rotation, length of the bit's

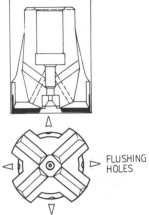

cutting edge, and flushing efficiency. Thus there is an optimum spacing of the blow interval and rotation speed to give the most economic rate of penetration for a particular type and hardness of rock. Modern developments in tungsten carbide inserts have improved the durability of the bit, and 3 or 4 winged or cross bits (Fig. 3.28) can be used for drifter drills in most types of rock, but are particularly suited to fractured material. Other bit types (Fig. 3.29) include:

Single chisel  – homogeneous rock
Double chisel – medium-hard rock
Crown bit     – very hard rock
Button bit    – medium to very hard rock.

**Fig. 3.28**  Tungsten carbide tipped rotary–percussive winged bit

The button bit (Fig. 3.30) is often selected for the DTH drill.

Button bits allow intervals between regrinding which are 4–5 times longer than for insert bits and they also tend to give better penetration rates. But insert bits are more resistant to gauge wear since they have a cemented carbide surface on the periphery. They also tend to produce straighter holes than button bits.

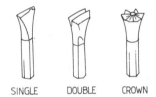

**Fig. 3.29**  Types of bit used in rotary–percussive drilling

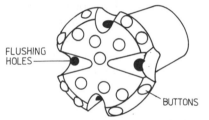

**Fig. 3.30**  Button bit for 'down the hole' drilling

### 3.16  DRILL RODS FOR ROTARY AND ROTARY-PERCUSSIVE DRILLING METHODS

The drill rod is the connecting component between the drive motor on the rig and the drilling bit. Consequently, torque and feed force are transferred through the drill rod to the cutting component.

The drill rod is hollow to allow the flushing medium to be pumped down through the drill string to the bit. It is therefore essential that the threaded joints for coupling the rods together do not allow any leakage to occur.

Rods are made in lengths to suit the particular drilling operation. For most construction operations, 3 m long rods are sufficient, but for large-scale work such as well holes etc., rods up to 7 m long are not unusual, when loading is assisted with winches.

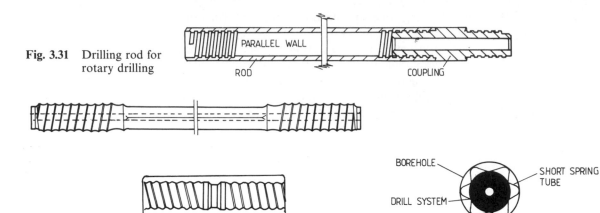

**Fig. 3.31** Drilling rod for rotary drilling

PARALLEL WALL

ROD

COUPLING

**Fig. 3.32** Drilling rod for rotary–percussive drilling

**Fig. 3.33** Spacer device for drilling rods

BOREHOLE

SHORT SPRING TUBE

DRILL SYSTEM

For rotary and 'down the hole' drilling the drill stems are usually round with a large flushing hole (Fig. 3.31), but in drifter drilling, because the blow energy is transferred down the rods, a strong and rigid stem with high bending stiffnesses is required and the flushing hole is usually not larger than 10–15 mm for a 35 mm diameter rod (Fig. 3.32). Such rods may be round or hexagonal in cross-section; a hexagonal rod helps to churn the chippings in the flushing medium and is therefore less prone to jamming.

In order to try to maintain accurate direction in deep drilling, the gap between the hole wall and the drill stem can be stabilised with guide tubes (Fig. 3.33). An accuracy of about 100 mm deviation per 33 m of drilling can normally be expected. Drifter drilling is least accurate.

### 3.17 FLUSHING MEDIUM FOR ROTARY, DIAMOND AND ROTARY–PERCUSSIVE DRILLING METHODS

A flushing medium is used for two functions:

(i) To remove the cuttings from the bottom of the hole to keep the bit area free.
(ii) To cool the bit.

For percussive, rotary-percussive, or rotary methods, either water, mud or air may be chosen. Generally, water or mud is used in unstable ground, deep drilling or confined spaces and air is used where dust does not pose an environmental hazard, such as in quarrying operations. However, much of the dust and particles can be directed into a collector (Fig. 3.34) if necessary. With compressed-air drifter drills using air flushing, the flushing pressure should not exceed that required to power the drill if the medium is fed axially through the drill, as this could cause 'back up' in the machine. Thus pressures

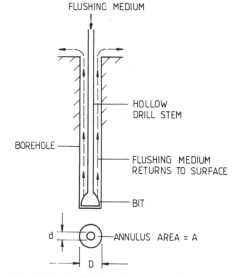

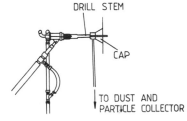

**Fig. 3.34**  Dust and particle collector          **Fig. 3.35**  Drill-hole flushing action

would normally have to be less than 100 psi (7 bar) gauge. Alternatively, where greater pressures are required, for example in deep holes, a separate swivel can be attached to the drill string and pressures of up to 200 lb/in² gauge or more (14 bar) can then be achieved.

If cleaning is to be effective, then for water or mud the flushing speed should be between 0.4 and 1.0 m/s (0.3–0.6 m/s for diamond drilling) depending upon the size and density of the chippings. For air the ideal rate should be 15–30 m/s in order to lift and carry the particles. By reference to Fig. 3.35, the quantity of flushing medium passing through the annulus area is

$$Q = \frac{\pi}{4}(D^2 - d^2) \times U \times 60$$

where   $Q = \text{m}^3/\text{min}$
         $U = \text{m/s}$
         $d = $ drill pipe outside diameter in m
         $D = $ borehole diameter in m

Thus for a 100 mm diameter hole, produced by a drifter drill using air flushing with 35 mm diameter rod (couplers), when $U = 15$ m/s

$$Q = \frac{\pi}{4}(0.1^2 - 0.035^2) \times 15 \times 60$$

$$= 6 \text{ m}^3/\text{min (214 ft}^3/\text{min)}$$

If air is adopted as the flushing medium, the gauge air pressure is usually 7 bar. For a 13 mm diameter hole in the drilling rods, only about 2.25 m$^3$/min of air at this pressure will reach the bit when passed down a 30 m stem. Therefore the corresponding annulus velocity is

$$U = \frac{2.25}{\frac{\pi}{4}(0.1^2 - 0.035^2) \times 60}$$

$$= 5.4 \text{ m/s}$$

As a consequence, the chippings must be broken down into very small pieces before they can be carried away, thereby slowing down the drilling rate.

Flushing is less of a problem with the DTH drill, because the drill rods are much larger in diameter, thus reducing the size of the annulus. Furthermore, the hole in the rods which serves to supply air to drive the piston can be greater and subsequently used for flushing. Higher rates of penetration per kW of power supply are therefore possible with DTH than with drifter drilling. For similar reasons flushing is also generally not a significant problem with the rotary drilling method.

## 3.18  DRILLING MUDS AND FOAMS

The consistency of drilling mud depends upon the hole stability, the density and size of cuttings and the rate of penetration, but it is usually of a creamy to custard consistency. Natural mud is generally a bentonite clay which exhibits thixotropic characteristics, but it can also be made from biodegradable organic polymers. To reduce the quantity of flushing medium required, the mud is passed through tanks (Fig. 3.36) – a suction pit and a settling pit. Mud is pumped from the suction pit down the hole and is returned to the settling pit, where the cuttings are deposited. Mud then flows back to the suction pit.

The pits are commonly dug in the ground and the volume of the pits should be at least three times the anticipated drill hole volume. The mud density and viscosity should be continuously tested during use.

Foam may be used as an alternative to mud since it usually has greater carrying capacity than water-based muds and produces a lower hydrostatic head.

Foam is made by forming an emulsion between water and a foaming agent, which can be natural liquid soap or polymer-based. The consistency used will depend on the amount of support required and density of cuttings to be lifted, but it is commonly of 'shaving foam' consistency. The velocity of the foam is between 0.23 and 0.75 m/s. Normal foam is usually a 1% volume mix.

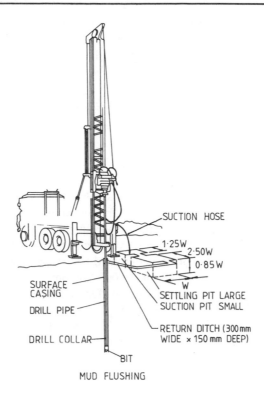

SUCTION HOSE

1·25W
2·50W
0·85W
W

SURFACE
CASING

DRILL PIPE

DRILL COLLAR

BIT

SETTLING PIT LARGE
SUCTION PIT SMALL

RETURN DITCH (300mm
WIDE × 150 mm DEEP)

MUD FLUSHING

**Fig. 3.36** Flushing medium settling tanks

## 3.19 OTHER FORMS OF ROTARY–PERCUSSIVE DRILLING EQUIPMENT

### Drills and drilling rigs for tunnelling applications

Drilling holes in the face or sides of a tunnel requires a more versatile and manoeuvrable rig than that used primarily for vertical drilling. Furthermore, to achieve high levels of production it is often necessary to produce several boreholes at the same time. Multi-head rigs have been developed to suit this need, supported either by rails (Fig. 3.37) or on rubber-tyred wheels (Fig. 3.38). This method, using a rotary-percussive action, is usually referred to as drifter drilling, as opposed to bench drilling in quarrying.

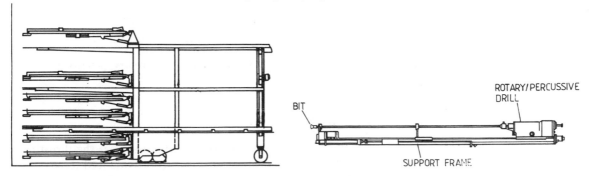

BIT

ROTARY/PERCUSSIVE
DRILL

SUPPORT FRAME

**Fig. 3.37** Rail-mounted multi-boom drifer drilling rigs

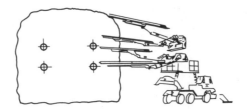

**Fig. 3.38**   Wheel-mounted multi-boom drifter drilling rigs

### Hand drills (Fig. 3.39)

The hand-held drill is very similar in appearance to the concrete breaker, except for the rotary action; it weighs 10–20 kg, and is suitable for drilling shallow and small diameter holes, up to about 30 mm diameter, to depths not exceeding approx. 8 m.

It operates on the combined rotary-percussive principle at approximately 145 rpm/2000 blow pm and requires about 1–2 m³/m free air supply (including air for flushing). However, water may also be used as a flushing medium.

All machines are fitted with a detachable muffler, but the operator should also wear ear defenders.

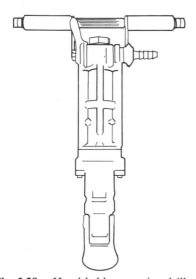

**Fig. 3.39**   Hand held percussive drill

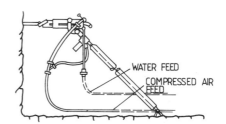

**Fig. 3.40**   Jack and feedleg rotary–percussive drill

### Jack and feedleg drills

These are similar to hand drills, except that the drill uses the thrusting action of a feed leg, hydraulically controlled to provide feed force. Reaction to the thrust is either against an opposite wall or from an operator's board (Fig. 3.40). Higher penetration rates can be achieved because of the extra control offered by the feed leg, which can also be reversed to pull out the drill rod in case of jamming. The equipment is fairly light (90 kg) and is fitted with handles which make it easily transportable.

### Jammed bits

When using rotary-percussive methods, there is a tendency that the bit will get stuck, especially if it is of the older type, without tungsten carbide inserts. Such bits can only penetrate a metre or so, depending upon the rock hardness, before requiring replacement. To overcome the jamming problem it is advisable to taper such holes in steps by changing the bit size, as shown in Fig. 3.41. Practice has demonstrated that the diameter of the hole should be reduced by about 3 mm for each metre of borehole length.

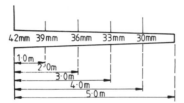

**Fig. 3.41**  Taper borehole

## 3.20  ROTARY–PERCUSSIVE DRILLS DATA

(i) Drifter drills

| Hole diameter (mm) | Drill weight (kg) | Air consumption (m³/min at 7 bar) | Feed type |
|---|---|---|---|
| up to 30 mm | 10 | 1.5 | Hand |
| up to 40 mm | 20 | 3.0 | Hand |
| up to 40 mm | 40 | 4.5 | Feed leg |
| up to 60 mm | 90 | 7.5 | Chain feed |
| up to 90 mm | 150 | 12.0 | Chain feed |
| up to 130 mm | 250 | 18.0 | Chain feed |

(ii) DTH drills

| | | | |
|---|---|---|---|
| up to 100 mm | 40 | 8* | Chain feed |
| up to 150 mm | 80 | 13 | Chain feed |

* Excluding air flushing requirement.

### 3.21 PRODUCTION DATA FOR ROTARY–PERCUSSIVE DRILLING

**Table 3.5**  Drilling performance – 100 mm diameter drifter drill

| Rock type | Bit loading (N/mm bit dia) | Penetration rate (m/h) | Bit life (m of hole) |
|---|---|---|---|
| Soft | 100–200 | 30–50 | 300–500 |
| Hard | 100–200 | 10–20 | 100–350 |
| Very hard | 100–200 | 5–10 | 50–150 |

The 100 mm diameter DTH drill, with about $\frac{1}{4}$ the power of the drifter drill, achieves approximately $\frac{1}{3}$ the above rates of penetration, when drilling holes down to 30 m or so.

### 3.22 (iii) PERCUSSIVE DRILLING

#### Rope-operated rig (Fig. 3.42)

When rope-operated rigs are beng used in soil investigation, pile-driving, etc., it is sometimes more expedient to use a chisel suspended from a rope and operated by a winch to break through a thin seam or intrusion of soft rock.

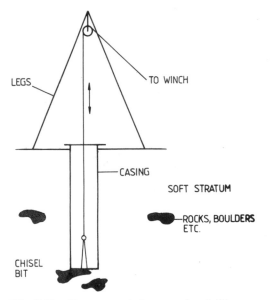

**Fig. 3.42**  Rope-operated percussive drilling

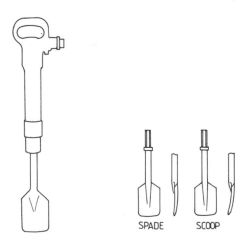

**Fig. 3.43**  Compressed air driven hand tools

#### Hand tools (Fig. 3.43)

Small powered tools are required for concrete breaking, roadworks, picks, spades, etc. These are usually light, 30–40 kg, and use up to 1–2 m³/min air supply. However, these small tools are increasingly using hydraulic powering. The method relies purely on percussion; rotary action is not included.

# 4 ROCK BLASTING

## 4.1 INTRODUCTION

The use of explosives to break up rock and hard soils is required in tunnelling, quarrying and for excavation works where mechanical plant cannot perform the task economically. Explosives are also used in the demolition of many types of structure, such as bridges, foundations, high rise buildings, chimneys, cooling towers, etc.

## 4.2 EXPLOSION IN SOIL AND ROCK

The effect of an explosion is to convert a chemical substance into a gas which then produces enormous pressure. The process takes place rapidly at 2000–6000 m/s shattering the rock adjacent to the explosive and exposing the surrounding area to stress.

The nature of the wave formations presents the following phenomena:

On ignition of a spherical charge placed in a solid medium such as rock, a shock wave is propagated, causing crushing and possibly liquefaction of the adjacent rock. Radial cracks are also formed which fade out with increasing distance from the wave front, as the energy is dispersed to a greater volume of the medium. In most applications of explosives in construction work, a free face is usually present and a more complex pattern of wave formation is produced, as shown in Fig. 4.1. On meeting the free surface (Fig. 4.1(a)) the initial shock wave travelling out from the centre of the explosion is reversed in direction and moves away from the epicentre $0'$ (Fig. 4.1(b)). On reaching the explosion cavity, the wave front is again reflected, which, together with the force of the expanding gases, raises the medium into a cupola (Fig. 4.1(c)). If the charge is sufficiently powerful, then material is forced outwards to produce a pattern of shattering and cracking of the rock as illustrated in Fig. 4.2.

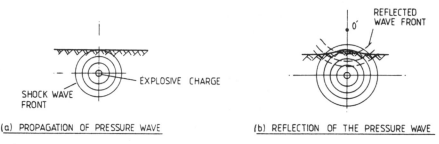

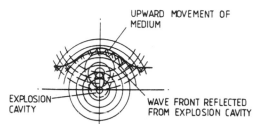

**Fig. 4.1**   Shock waves from an underground explosive charge

(c) SWELLING OF THE SOIL ABOVE THE CHARGE

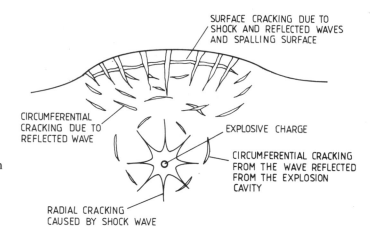

**Fig. 4.2**   Cracking and shattering pattern from an underground explosive charge

## 4.3   BLASTING THEORY

An explosive charge placed at a depth $B$ (called the burden) below a free surface is assumed to throw out materials to form a conically shaped crater (Fig. 4.3) with 45° side slopes. The energy required to impart the initial velocities will be proportional to the mass of the body of material in the cone, i.e. $B^3$. If gravitational forces are present, the energy required to lift the mass to surface level is proportional to the product of the mass and the distance moved ($B$), i.e. $B^4$. The energy required to overcome the frictional forces between the surface of the crater and surrounding medium is proportional to the crater surface area, i.e. $B^2$.

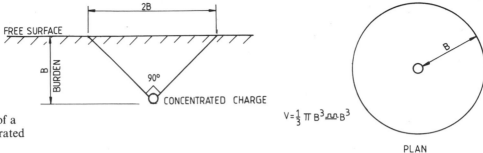

**Fig. 4.3**  Effects of a
concentrated
charge

Thus, the total weight of charge required is given by

$$Q = c_2 B^2 + c_3 B^3 + c_4 B^4 \qquad (4.1)$$

(*Note*: The use of $c_0$ and $c_1$ has been avoided in order to maintain the same nomenclature as other writers.)

### Benching

Equation (4.1) has been shown by Langefors to be appropriate for a spherical charge (Fig. 4.3) and for a section of cylindrical charge of length $B$, where $B$ is much smaller than the full length of charge ($h$) (Fig. 4.4). Thus to tear the burden in the case of benching, the charge in the drill hole may be considered to consist of two partial charges:

(i) a concentrated spherical charge at the bottom
(ii) a cylindrical charge for the rest of the column (neglecting end effects).

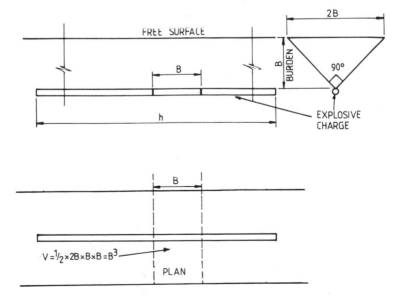

**Fig. 4.4**  Effects of a
cylindrical charge

### (i) BOTTOM CHARGE (Fig. 4.5)

The concentrated bottom charge $Q_1$ given by eqn. (4.1) is

$$Q_1 = c_2 B^2 + c_3 B^3 + c_4 B^4 \tag{4.2}$$

where   $Q_1$ is in kg of a given explosive
$B$ is the burden (depth of charge) in metres
$c_2, c_3, c_4$ are empirical coefficients which depend on the bench height ($H$), burden ($B$) and length of explosive charge ($h$)

whereby

$$c_i = k_i \left( \frac{H}{B}, \frac{h}{B} \right)$$

If the parameters are chosen such that $H = B$ and the charge is concentrated so that $h/B \simeq 0$, then

$$c_i = c_i(1, 0) \text{ and so } c_2(1, 0), \; c_3(1, 0), \; c_4(1, 0)$$

These may be represented by new constants

$$a_2, a_3 \text{ and } a_4$$

Thus the quantity of charge $Q_1$ needed to tear the burden is given by

$$Q_1 = a_2 B^2 + a_3 B^3 + a_4 B^4 \tag{4.3}$$

Values for $a_2$ and $a_3$ can be determined from test blasts with $B$ at say 0.5 m and 1 m respectively and the simultaneous equations solved. This task is made simpler by ignoring $a_4$ which produces less than 1% of $Q_1$ for a bench arrangement.

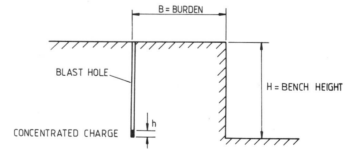

**Fig. 4.5** Fixed base blasting with a concentrated charge

### (ii) COLUMN CHARGE (Fig. 4.6)

For a high bench, a section of length $H = B$ of the column can be considered to be cylindrically charged (see Fig. 4.6). Thus the weight of charge ($Q_2$) required to tear the burden is given by

$$Q_2 = b_2 B^2 + b_3 B^3 + b_4 B^4 \tag{4.4}$$

where $b_i$ is the special case of $k_i$ (for a cylindrical charge).

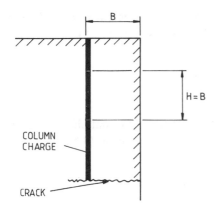

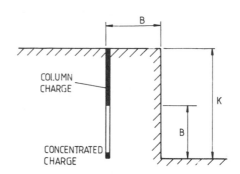

**Fig. 4.6**  Free base blasting with a cylindrical charge

**Fig. 4.7**  Combined concentrated and cylindrical charge for bench blasting

The quantity of charge required per metre ($q$) for a section of bench of height equal to the burden ($B$) is therefore

$$\frac{Q_2}{B} = q = b_2 B + b_3 B^2 + b_4 B^3 \tag{4.5}$$

The total charge for a bench of height $k$ (Fig. 4.7) is therefore a summation of the concentrated charge at the base $Q_1$, and the column charge ($q$) which is assumed to be proportional to ($k - B$) (neglecting end effects). Thus

$$Q_T = Q_1 + q(k - B) \tag{4.6}$$

$$= a_2 B^2 + a_3 B^3 + a_4 B^4 + (b_2 B + b_3 B^2 + b_4 B^3)(k - B) \tag{4.7}$$

Langefors indicates that tests have demonstrated that for benching, $b_2 = 0.4a_2$ and $b_3 = 0.4a_3$ irrespective of rock type, thus rearranging (4.7),

$$Q_T = 0.4a_2 \left( \frac{k}{B} + 1.5 \right) B^2 + 0.4a_3 \left( \frac{k}{B} + 1.5 \right) B^3$$

$$+ \left( a_4 + b_4 \left( \frac{k}{B} - 1 \right) \right) B^4 \tag{4.8}$$

where $a_2$, $a_3$ and $a_4$ are determined for a bottom charge in case (i).

In practice the bottom charge cannot be concentrated and has to be distributed in the drill hole uniformly from the bottom upwards and therefore the full effect of a concentrated bottom charge cannot be realised. Tests by Langefors indicate that in this type of loading the charge effect at the base is only about 60% of the theoretical value with a concentrated charge. However, it was shown that the breaking power could be increased to almost 90% of the theoretical value by drilling the hole to a depth $0.3B$ below the base and loading the bottom charge to a height of $1.3B$. The bottom charge distributed in this manner was equivalent to $Q_T$ in equation (4.8) for a bench height of $k = 2B$. Thus the required uniformly distributed bottom charge ($Q_b$) can be calculated from (4.8)

for        $k = 2B$, as

$$Q_b = 1.4a_2B^2 + 1.4a_3B^3 + a_4B^4 + b_4B^4 \tag{4.9}$$

When the bench height ($k$) exceeds $2B$, the additional charge required can be calculated as a column charge from equation (4.5). Thus the column charge $Q_c$ is

$$Q_c = q(k - 2B) = (b_2B + b_3B^2 + b_4B^3)(k - 2B)$$

and since

$$b_2 = 0.4a_2, \; b_3 = 0.4a_3$$

$$Q_c = 0.4\left(\frac{k}{B} - 2\right)(a_2B^2 + a_3B^3) + b_4(k - 2B)B^3 \tag{4.10}$$

adding equations (4.9) and (4.10),

$$Q_T = Q_b + Q_c = 0.4a_2\left(\frac{k}{B} + 1.5\right)B^2 + 0.4a_3\left(\frac{k}{B} + 1.5\right)B^3$$

$$+ \left(a_4 + b_4\left(\frac{k}{B} - 1\right)\right)B^4 \tag{4.11}$$

where    $Q_T$ = Total charge

which also corresponds with equation (4.8).

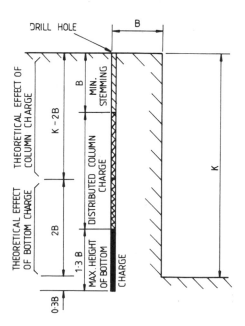

**Fig. 4.8**  Distribution and effect of explosive for bench blasting

*Note*: $a_2$ and $a_3$ are determined experimentally as described in case (i) and $b_4$ is very small. (For usual cases with a standard dynamite these are approximately 0.07 kg/m², 0.4 kg/m³, 0.004 kg/m⁴ respectively.) The distribution of the complete charging is shown in Fig. 4.8.

### Effect of charge strength

Clearly, for the case when $k \leqslant 2B$, the bottom charge would cause excessive throw and the burden must therefore be reduced (and the charge recalculated). Alternatively the use of an explosive allowing a more concentrated charge at the bottom would be more favourable, as described under sections (i) and (ii). For these reasons, the maximum height of the bench is generally limited to about 30 m in hard rock as a safety precaution.

The use of a stronger explosive can be incorporated into equations (4.9), (4.10) and (4.11) by the introduction of a factor $1/w$ where

$$\frac{1}{w} = \frac{\text{specific energy (unit weight strength) of the standard explosive}}{\text{specific energy (unit weight strength) of the explosive used}}$$

therefore

$$Q'_b = \frac{1}{w} Q_b \qquad\qquad (4.12)$$

$$Q'_c = \frac{1}{w} Q_c \qquad\qquad (4.13)$$

$$Q'_T = \frac{1}{w} Q_T \qquad\qquad (4.14)$$

**Fig. 4.9**  Single row short delay blasting

### Effect of hole spacing

Where a row of drill holes is involved, the volume of material available for removal by each blast is less than for a single drill hole. For example in Fig. 4.9, with $B = S$ there is a degree of overlapping and tests have demonstrated that the charge in each drill hole may be reduced by about 20%.

Thus   $Q''_T = \dfrac{0.8}{w} Q_T$ when $B = S$

For other spacings $Q_T$ has been found to be proportional to the ratio $S/B$.

Thus
$$Q_T'' = \frac{0.8}{w} \cdot \frac{S}{B} \times Q_T \qquad (4.15)$$

Generally the ratio is held within the bounds $0.9 \leqslant S/B \leqslant 1.3$. At the lower limit fragmentation is poor, whilst at the upper limit uneven tearing may result.

### Effect of slope and fixity

The charge required for a vertical bench with a free base (Fig. 4.6) is about 75% of that for a fixed base (Fig. 4.5) and is incorporated into the charge formula by a degree of fixity ($f$) factor. Also, less energy is required to tear the burden of a sloping face, because of the larger angle at the base.

Langefors suggests the following values for $f$

|  | *Vertical bench* | *Sloping bench* 3:1 | 2:1 |
|---|---|---|---|
| Free base | 0.75 | 0.75 | 0.75 |
| Fixed base | 1.00 | 0.9 | 0.85 |

Thus
$$Q_T''' = f \times \frac{0.8}{w} \times \frac{S}{B} \times Q_T \qquad (4.16)$$

### Effect of hole diameter

The hole diameter has little effect on the degree of loosening as it is the total quantity of charge, particularly the bottom charge, which determines the breakage.

### EXAMPLE

A drill hole of diameter $d$ (mm), with the bottom charge extending $1.3B$ from the base, contains explosive of unit weight $P$ kg/m$^3$, thus

$$\frac{P}{10^6} \times \frac{\pi d^2}{4} \times 1.3B = Q_b''' = f \times \frac{0.8}{w} \times \frac{S}{B}(1.4a_2B^2 + 1.4a_3B^3)$$
$$+ (a_4 + b_4)B^4 \qquad (4.17)$$

When $f = 1$, $S/B = 1$, $w = 1$, $P = 1000$, $a_2 = 0.07$, $a_3 = 0.4$ and terms with $B^4$ are ignored

$$\frac{d^2}{0.846 \times 36^2} = B^2\left(\frac{0.07}{B} + 0.4\right)$$

or $\qquad B^2 \simeq \dfrac{1}{0.86} \times \left(\dfrac{d}{36}\right)^2$, since $\dfrac{0.07}{B}$ is very small,

$$B \simeq \frac{1}{\sqrt{0.86}} \times \frac{d}{36} \simeq 0.03d$$

where $B$ is in m and $d$ in mm e.g. when $d = 50$ mm $\underline{B = 1.5 \text{ m}}$.

### Throw and scattering (Fig. 4.10)

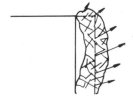

An increase in the charge causes increased cracking and shattering of the rock. In addition material will be thrown forward, including fragmentation and scattering. Calculations can be made to determine the extent of the effects and the reader is referred to Langefors and Kihlstroem for a detailed discussion.

**Fig. 4.10** Throw of debris

### Example to determine total quantity of charge (Fig. 4.11)

Assume $P = 1000 \text{ kg/m}^3$, $S/B = 1$, $w = 1$, $a_2 = 0.07 \text{ kg/m}^2$, $a_3 = 0.4 \text{ kg/m}^3$ and terms with $B^4$ can be ignored. $B = 1.5$ m and $k = 4.5$ m.

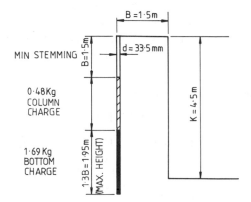

**Fig. 4.11** Example of distribution of explosive charges

(i) Bottom charge
From equation (4.9)

$$Q_b''' = f \times \frac{0.8}{w} \times \frac{S}{B}(1.4a_2 B^2 + 1.4a_3 B^3)$$

$$= 0.8(0.225 + 1.89) = \underline{1.69 \text{ kg}}.$$

(ii) Diameter of drill hole

$$\frac{P}{10^6} \times \frac{\pi d^2}{4} \times l = 1.69 \text{ if } l = 1.3B = 1.3 \times 1.5 = 1.95 \text{ m}$$

$$d^2 = \frac{1.69}{1.95} \times \frac{4}{\pi} \times 1000 = 1104$$

$$d = \sqrt{1104} = \underline{33.5 \text{ mm drill hole}}$$

(iii) Column charge

$$Q_c''' = f \cdot \frac{0.8}{w} \cdot \frac{S}{B} \left[ 0.4 \left( \frac{k}{B} - 2 \right)(a_2 B^2 + a_3 B^3) + b_4(k - 2B)B^3 \right]$$

$$= 0.8 \times 0.4 \times (0.16 + 1.35) = \underline{0.48 \text{ kg}}$$

## 4.4 BLASTING PATTERNS

When the charge requirements have been calculated, the blast hole pattern can be designed. The method of firing is performed in one of two ways:

(i) by simultaneous shots
(ii) by short delay blasting

### Simultaneous method

In this method a number of holes are fired simultaneously and it is therefore ideal for breaking out a long bench of rock. The technique requires less accuracy than delay shots, and a slight variation in hole spacing and direction of the line of the drill hole can be accepted. Excessive ground vibration, scatter and overbreak may occur, but when quarrying for rough stone these problems may not be important.

### Short delay method

### (i) SINGLE-ROW SHORT DELAY BLASTING

Practice has demonstrated that by delaying the shots in sequence, the result is improved fragmentation, reduced throw and a slightly lower consumption of explosive. Langefors and Kihlstroem suggest that the delay interval $t$ can be related to the burden ($B$) by the formula

$$t = cB$$

where    $t$ is in milliseconds
         $B$ is in m
$c$ is a constant $\simeq 3$ milli seconds/metre

Several different orders of sequence have been developed depending upon the conditions, but for a homogeneous material, with accurately timed and spaced drill holes, the sequence in Fig. 4.9 is often used.

### (ii) MULTIPLE-ROW SHORT DELAY BLASTING

The principle is similar to instantaneous single-row blasting, but with several rows shot in a delay sequence (as shown in Fig. 4.12). The effect is to produce a more intense fragmentation, behind the first row, and significantly reduce flying rock. This method is often used for foundation work and also in quarrying, where large quantities of rock are required to keep the work gangs and all equipment fully utilised.

In blasting large excavations, a preliminary area must be loosened first to produce a free face for subsequent benches. This first, or sinking, cut should be heavily charged as only a single free face is available and the shots should be delayed in sequence as shown in Fig. 4.13. The area may be subsequently expanded, section by section, using either single row delays, i.e. removing the material progressively (Fig. 4.9) or with the multiple-row technique (Fig. 4.14). Material from the sinking cut may either be excavated or left in place. In the multiple-row technique, when the subsequent firings take place the ground will have been loosened sufficiently to allow tearing of the burden.

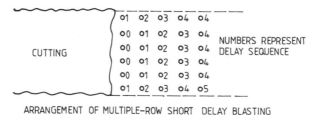

**Fig. 4.12** Arrangement of multiple-row short delay blasting

ARRANGEMENT OF MULTIPLE-ROW SHORT DELAY BLASTING

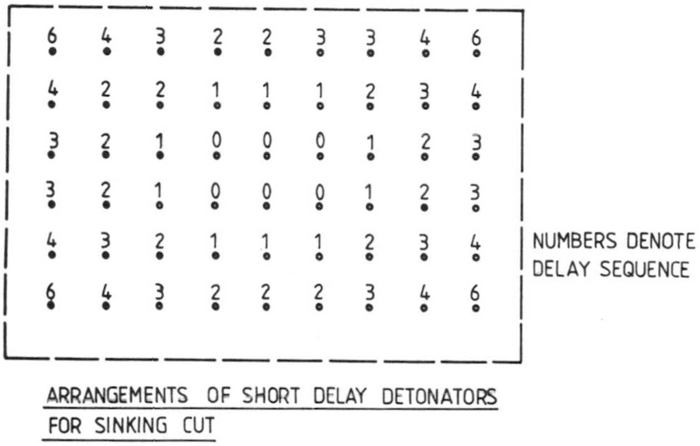

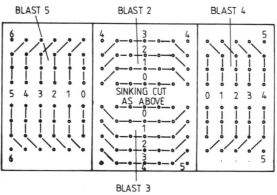

**Fig. 4.13** Short delay blasting for a sinking cut

ARRANGEMENTS OF SHORT DELAY DETONATORS FOR SINKING CUT

**Fig. 4.14** Arrangement of delays for a large excavation

TYPICAL ARRANGEMENT OF SHOTHOLES AND DELAYS FOR A LARGE EXCAVATION

### Prevention of overbreak

The above blasting methods unfortunately produce overbreak and irregularities. A smoother wall for cuttings and tunnels can be obtained by pre-split blasting. This technique consists of lightly charged and well stemmed shots placed around the perimeter of the excavation at close centres (500 mm). The drill hole is usually of small diameter (50 mm max.) and must be accurately drilled, to give even effect.

The pre-splitting charge is exploded, followed either instantaneously or with a slight delay by the main charges.

## 4.5   TUNNEL BLASTING

The principles of calculating the required charges in tunnel blasting are similar to those described for benching. The charges are arranged in a predetermined pattern and released in a planned sequence as in short delay blasting.

The first holes in the sequence produce the loosening, thus providing the free face for subsequent opening up of the heading. Two principal methods have been developed, namely wedge cuts and parallel hole cuts. The latter is easier to carry out, but is not well suited to soft rock.

### Wedge cuts (Fig. 4.15)

The holes are arranged regularly to facilitate drilling and are set with short delays usually progressing outward to form a fan, wedge or conical section at the centre of the face, which is thereby released first to provide further free faces for the subsequent rows of charges.

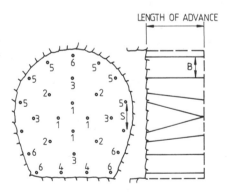

Fig. 4.15   Arrangement of wedge/fan cut for a tunnel heading

NUMBERS REPRESENT DELAY SEQUENCE

### Parallel cuts (Fig. 4.16)

In recent years successful results have been obtained by blasting towards several closely spaced, uncharged, empty holes, or alternatively towards a single, large (100–150 mm diameter) uncharged hole. The

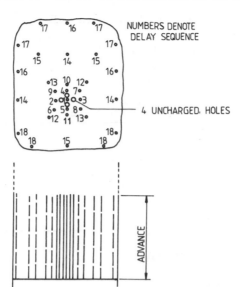

NUMBERS DENOTE
DELAY SEQUENCE

4 UNCHARGED HOLES

ADVANCE

**Fig. 4.16** Arrangement of
a parallel cut for
a tunnel heading

effect is to produce a progressive enlargement along the full advance.
This method has the advantage that the holes can be drilled parallel to
each other with a considerable saving in time and expertise. However,
in soft rock or where the advance becomes excessive, the rock tends to
jam and the opening-up effect is hindered.

### Length of advance

The length of drive to be selected depends upon the span of rock which
can remain unsupported during the time time taken to install the
supports. However, by using a modern method of temporary support
such as guniting, rock bolting, ribbing, (described in Chapter 10) full
face advance within the range 2–7 m is commonly obtained. As a rule
of thumb, however, the length of advance is usually not greater than
the face width.

   In very poor strength rock, where the propping could not be fixed
over the full face in the time available before a collapse occurs, it may
be necessary to divide the face up into sections, whereby the roof would
be propped and the sides removed in stages.

   Generally full face tunnelling produces the lowest costs, but where
the face cross-sectional area exceeds about 50 m$^2$, hole drilling with the
modern multi-headed jumbo drills and subsequent blasting are best

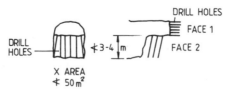

DRILL HOLES

FACE 1

DRILL
HOLES

3–4 m

FACE 2

X AREA
≮ 50 m$^2$

**Fig. 4.17** Use of benching
in tunnelling

achieved from a bench as shown in Fig. 4.17. The height of the bench should not be less than 3–4 m, and further benches should be introduced where the bench height exceeds about 10 m. A typical operating cycle is shown in Table 4.1.

**Table 4.1** Time cycle in hours per blasting sequence

|  | Hours |
|---|---|
| Boring, charging, blasting | 3.0 |
| Ventilation pause | 0.25 |
| Loading–transport | 3.25 |
|  | 6.50 |
| Rock bolting | 1–4 |

## 4.6 BLASTING DATA

### Benching

| Operation | Drilling required (m) per m³ of rock blasted | | |
|---|---|---|---|
|  | Soft | Medium | Hard |
| Quarrying | 0.3 | 0.4 | 0.5 |
| Large excavations | 0.7 | 1.1 | 1.5 |
| Narrow excavations | 1.0 | 1.5 | 2.0 |

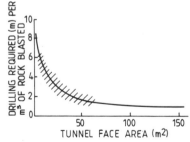

**Fig. 4.18** Length of drilling plotted against tunnel heading face area

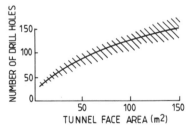

**Fig. 4.19** Number of drill holes plotted against tunnel heading face area

### Tunnelling

Figures 4.18 and 4.19 illustrate comparable output data for tunnelling in hard rock, where it can be seen that for small diameter tunnels considerably more drilling is required than for large diameter sections or in normal benching. This is mostly due to the degree of fixity of the burden, and the need to open up from a single free face cut section for each and every advance.

## 4.7 CONSUMPTION OF EXPLOSIVE (approximate)

### Benching

| Operation | High explosive (kg) per m³ of rock blasted | | |
|---|---|---|---|
| | Soft | Medium | Hard |
| Quarrying | 0.1 | 0.2 | 0.3 |
| Large excavations | 0.3 | 0.45 | 0.6 |
| Narrow excavations | 1.0 | 1.5 | 2.5 |

### Tunnelling

Figure 4.20 illustrates comparable explosives data for tunnelling in hard rock.

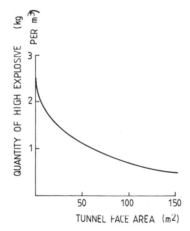

**Fig. 4.20** Quantity of explosive plotted against tunnel heading face area

## 4.8 UNDERWATER EXPLOSIVES

When water-resistant explosives are necessary and because of the difficult working conditions, e.g. oversize boring, poor packing of explosive, etc., energy loss occurs, resulting in a requirement of 150 to 400% more explosive than for blasting on land.

## 4.9 TYPES OF EXPLOSIVE

An explosive charge on detonation reacts to form a large volume of gas at high temperature. The release of the gas is almost instantaneous and produces high pressure and shock waves (see earlier description on page 42) which continues until the reaction is complete.

Many types of explosive of varying strength, detonation times, etc., have been developed and they may be classified into:

  (i) slow explosives
 (ii) high or quick explosives
(iii) initiating explosives.

### (i) Slow explosives

The most common type is known as black powder, which consists of a mixture of saltpetre (potassium nitrate), charcoal and sulphur.

Black powder can be ignited by exposure to an open flame and must be closely confined to maximise the effect of the relatively low gas pressure produced. The velocity of detonation is slow, about 600 m/s and thus there is little crushing of the rock near the centre of the explosion. The main effect is a bursting action causing displacement and fragmentation as the gas is slowly released from the reaction. Its uses are mainly limited to quarrying and in safety fuse cord.

### (ii) High explosives

These are not very sensitive to detonation by spark or flame and require initiating explosives. When detonated in this way, the reaction takes place at up to 9000 m/s with accompanying high gas pressure. The surrounding rock is then crushed, cracked and shattered. This type of explosive is therefore ideally suited to quarrying and construction applications. The most common types of high explosive are:

(a) nitroglycerine
(b) ammonium nitrate plus fuel oil (AN/FO)
(c) slurries

They have the following properties:

 (i) density – approx. 1.5 g/cm$^3$ for gelatins, 1.1 for powders.
(ii) specific energy 1000–2000 kcal/kg.

### (a) NITROGLYCERINE

Is produced by the action of nitric and sulphuric acids on glycerine. The resulting mixture is highly sensitive to impact and temperature and can be detonated by the action of a gentle blow. Inert substances such as charcoal, sodium nitrate or nitrocellulose may be added to improve stability but these generally reduce the weight strength of the explosive. Nitroglycerine explosive compounds are often referred to as Dynamite when composed of glycerine-ethylene, glycol nitrates, ammonium nitrate, wood chippings and inert substances. The compound has a dry, granular consistency. Dynamites for use in water are best gelatinised;

blasting gelatin containing a mixture of nitroglycerine and nitrocellulose was formerly popular although more powerful gelatins are now available.

### (b) AMMONIUM NITRATE

Nitroglycerine-based explosives are fairly expensive and a mixture of ammonium nitrate and oil (AN/FO) provides an alternative substance. The compound is mostly used in powder form and can be poured directly into the drill hole. However a powder loader using compressed air greatly speeds up the operation. AN/FO is not water-resistant, and therefore is not suitable for use in wet drill holes, but this may be overcome by using a gelatinised form, possibly containing a nitroglycerine additive.

The gelatins are usually manufactured in paper or cardboard cartridges for ease of handling and loading.

### (c) SLURRIES

These consist of aluminium particles, or TNT (trinitrotoluene), plus an inert stabilising compound, all suspended in water. This type of explosive has a slightly greater weight strength than AN/FO and can be used in wet conditions, but it is slightly more expensive.

### (iii) Initiating explosives

Because high explosives are not very sensitive to detonation by spark or flame, an initiating explosive is required to set off the charge. These are extremely sensitive and easily exploded, producing sufficient shock and temperature rise to induce a reaction in the high explosive. Such explosives are used in small quantities to form detonators.

## 4.10    DETONATION OF EXPLOSIVES

The initiating explosive in a detonator may be ignited either with a safety fuse cord (Fig. 4.21) or electrically (Fig. 4.22).

### Safety fuse cord

This has a black powder core sheathed in plastic. The cord is designed to burn at a rate of approximately 100 m/s and a short length is led to the detonating cap which is embedded in the high explosive. When black powder is itself the explosive compound, then the fuse cord is sufficient to initiate the charge and a detonator is not required. Although safety fuse cord can be ignited by flame, it is more usual when several shots are to be fired to link the free end of each safety

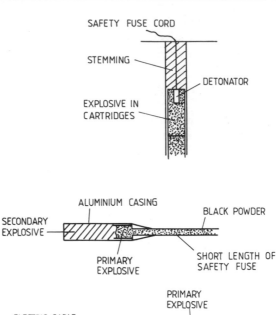

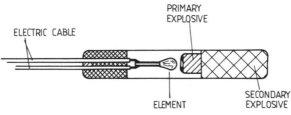

**Fig. 4.21** Safety-cord fired detonator (black powder fuse)

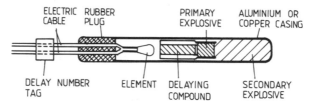

**Fig. 4.22** Electrically fired detonator

**Fig. 4.23** Electrically fired delay detonator

fuse cord to igniter cord, which is then led back to a safe location. Igniter cord can be obtained with slow (3.5 m/s), moderate (33 m/s) or fast (50 m/s) burning rates.

### Detonating caps (blasting caps)

(i) *Non-electric* (*Fig. 4.21*) – comprises an aluminium tube containing a powerful initiating explosive, such as pentaerythritol, and a priming charge of lead oxide. Caps are manufactured in different strengths, the common sizes being No. 6 and No. 8, which give efficient detonation with most types of blasting explosives.

With this type of detonator, time delays are introduced by varying the length of the safety fuse cord.

(ii) *Electric detonators* (*Figs 4.22 and 4.23*) – are also made in varying strengths, with the initiating explosives contained in an aluminium or copper casing (Fig. 4.22). The element in the cap is simply connected by wires to an electric source, which subsequently sets off the primary explosive. Delays ranging from about 8 to 8000 milliseconds can be incorporated (Fig. 4.23).

## 4.11  CIRCUIT DESIGN

In simultaneous blasting the caps can be arranged in series as shown in Fig. 4.24. But for delayed charges, obviously the circuit must be maintained after the first shot, and therefore a parallel circuit is required. Figure 4.25 illustrates an arrangement for single-row delays and Fig. 4.26 for multiple-row delays. The $R_i$ values indicate the resistance in ohms of each cap and with a knowledge of the current ($I$) needed to fire each cap (about 1.5 amps) the total volts can be

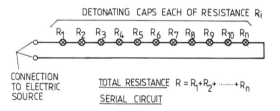

**Fig. 4.24**  Serial row detonator arrangement

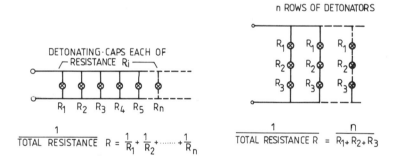

**Fig. 4.25**  Single row parallel circuit detonator arrangement

**Fig. 4.26**  Parallel circuit multiple row detonator arrangement

calculated from the formula $V = IR_{\text{total}}$. An exploder (source power) of the required size is connected to other ends of the electric wires.

### Detonating fuse

As an alternative to the use of safety fuse cord and its associated detonating cap, detonating fuse can be used to fire the main explosive.

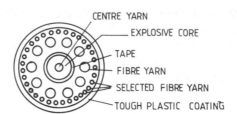

**Fig. 4.27** Detonating fuse (high explosive fuse)

It consists of a central core of pentaerythritol-tetranitrate explosive wrapped in tape. The outer casing is made up from strands of yarn and sheathed in plastic (Fig. 4.27). This type of fuse cord produces a high velocity of detonation (6000 m/s) and must itself be initiated by a detonating cap. It can be used for firing simultaneous charges (Fig. 4.28) or with delays by the insertion of detonating relays (Fig. 4.29). Because the detonating cord can be run alongside or through the high explosive cartridges or powder in the drill hole (Fig. 4.30) there is less risk of unscheduled ignition than with a single blasting cap.

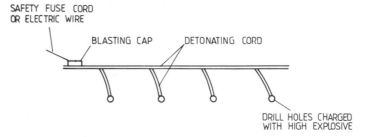

**Fig. 4.28** Simultaneous blasting arrangement with detonating fuse

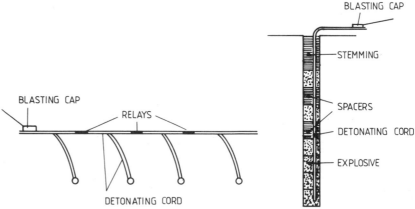

**Fig. 4.29** Delay blasting arrangement with detonating fuse

**Fig. 4.30** Distribution of explosive and detonating fuse

### 4.12  LOADING THE DRILL HOLE

The hole should first be cleaned out with compressed air. When using powder explosive, pneumatic equipment can speed up the loading operation, but unless special equipment is available, explosive supplied in cartridges must be laboriously loaded into the drill hole and pressed firmly down with a long pole, the objective being to produce as high a density of packing as possible (P).

The top part of the drill hole should be plugged with damp sand or clay, which acts to confine the energy of the explosion. Ideally the length of stemming should not be less than the width of the burden.

### 4.13  VIBRATIONS FROM BLASTING

The US Bureau of Mines (USBM) recommends that vibration levels resulting from blasting operations be limited below a peak particle velocity of 51 mm/s if structural damage is to be avoided. Several researchers have published results of the effects of vibrations and these are summarised in Table 4.2

A useful formula for estimating the likely particle velocity resulting from an explosive charge is given by USBM as

$$V = k \frac{D}{\sqrt{E}} - B$$

where   $E$ = explosive charge (kg)
$V$ = particle velocity (mm/s)
$D$ = distance from shot to recording stations (m)
$k$ and $B$ are empirical co-efficients which can be determined from field trials with instrumentation investigations

**Table 4.2**  Structural damage as related to peak particle velocity

| Peak particle velocity (mm/s) | Langefors (Sweden) | Edwards (Canada) | Bureau of Mines (USA) |
|---|---|---|---|
| 250 | Serious cracks | Damage | Major damage (Fall of plaster, serious cracking) |
| 200 | Cracks | | Minor damage (Fine plaster cracks, opening of old cracks) |
| 150 | Small cracks and fall of loose plaster | | |
| 100 | Caution | Caution | Caution |
| 75 | | | |
| 50 | No noticeable cracks | | |
| 25 | | | |
| 00 | | Safe limit | Safe limit |

### Air blast

If part of the explosion escapes to the air, the expanding gases push back the air causing a local increase in air pressure.

Windows can be broken when this over-pressure lies in the range of 0.05 to 0.14 kg/cm$^2$.

# 5

# SOFT GROUND DRILLING

Boreholes are necessary in soft ground for:

 (i) soils investigation
 (ii) ground anchorages
(iii) foundations, piles, both in situ and precast
(iv) water wells
 (v) horizontal conduits
(vi) piled walls

The two principal methods for drilling holes are rotary, or auger, boring and conventional grabbing with a bucket. The choice of type depends upon the ground conditions, the diameter and depth of borehole required and the cost and availability of the equipment. A summary of the capabilities of the various methods is shown in Table 5.1.

**Table 5.1**  Methods of producing boreholes in soft ground

| Method | Max. depth (m) | Max. dia. (mm) | Soil type |
|---|---|---|---|
| Soil investigation boring | When bedrock is reached | 150–200 | All |
| Continuous flight augering | 30 | Up to 600 | Firm uniform soils |
| Intermittent flight augering | 75 | Up to 2500 | Firm uniform soils |
| Rotary boring with buckets | 50 | Up to 1500 | Free-flowing soils |
| Rotary boring with belling buckets | 30–40 | Up to 6000 | Cohesive soils |
| Grabbing | 100 | 500–2000 | Difficult soils and those containing small boulders |
| Circulation drilling | 200 | 300–2000 | Soft rock |

## 5.1   SOILS INVESTIGATION TECHNIQUES

In civil engineering and building, soils investigation is often undertaken to determine the suitability of the soil to support foundations; the size of, and depth to which to sink or drive piles; groundwater conditions, etc.

Boreholes for sampling and soil testing must be large enough to accommodate the sampling equipment, and 150–200 mm diameter is usually sufficient to do this. The borehole is generally terminated on reaching bedrock. Up to this stage the methods used include:

  (i) shell and auger boring
 (ii) percussive drilling
(iii) wash boring

Simple equipment is necessary, which can be quickly set up, transported, and operated by at most 2 men.

### (i) Shell and auger boring

The machine comprises a three- or four-legged derrick, sometimes fitted with detachable wheels to facilitate transportation, and winch and clutch capable of lifting about 4000 kg. The height from ground to centre line of sheave is approximately 5 m with an overall weight of 1000 kg (Fig. 5.1).

### (a) THE SHELL CUTTER

The shell cutter shown in Fig. 5.2(a) consists of a steel tube fitted with a cutting edge, which is hung from the rig and allowed to fall. A core is cut which is retained in the cylinder. The cutter is then raised and the material removed. As the process is repeated, the borehole is gradually formed. In sandy soils a small clack valve is incorporated into the cylinder to prevent the material flowing out.

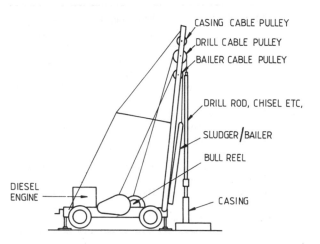

**Fig. 5.1** Soils investigation rig

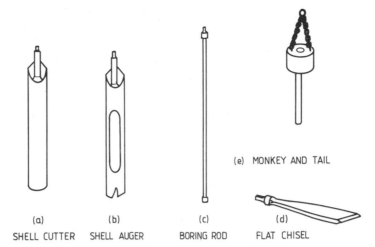

**Fig. 5.2** Soils investigation borehole equipment

| (a) | (b) | (c) | (d) |
| --- | --- | --- | --- |
| SHELL CUTTER | SHELL AUGER | BORING ROD | FLAT CHISEL |

(e) MONKEY AND TAIL

### (c) THE AUGER

For shallow work (4–5 m max. depth) an auger is often preferred (Fig. 5.2(b)). This consists of a hollow tube, connected to boring rods, which is rotated manually at ground level. The soil is removed from openings in the side of the cylinder wall.

The auger is used for coring out clay and other fairly soft strata.

The advantages of these systems are that the equipment is simple to erect and use and that undisturbed samples can be obtained, thus relieving the operator of having to identify the strata in the field.

### (ii) Percussive drilling

During the investigation a thin rock layer or small boulder may be encountered. The shell or auger is then replaced by a chisel of approximately 680–700 kg weight (Fig. 5.2(d)). The continuous pounding by this chisel breaks up the rock. The cuttings and slurry are removed by baler or shell cutter. The process is applied until the operator is satisfied that the particular stratum is in fact bedrock or otherwise.

### (iii) Wash boring (Fig. 5.3)

The wash boring method uses a light tripod, a small winch and a small pump, and sometimes a small power-driven capstan is provided.

Casing for the borehole is first driven into the grund. The soil is then removed from inside the casing by chiselling with the hollow drill pipe supported by the winch. The soil is loosened by a combination of the chiselling action and the stream of water or drilling mud issuing from the lower end of the wash pipe.

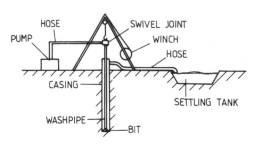

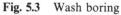

**Fig. 5.3**   Wash boring

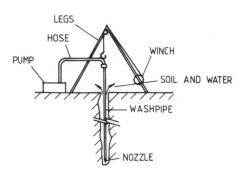

**Fig. 5.4**   Wash probing

The loosened debris is carried up the annular space between the casing and wash pipe into a settling tank.

Unfortunately the samples are disturbed samples and are therefore suitable only for identification of the various soil strata encountered i.e. unsuitable for laboratory testing.

The main advantage of this method is that the equipment is easy to assemble and operate. However, it can only be used to good effect in softer type soils, such as sands and clays.

Although the equipment is simple, it requires an experienced operator to interpret the debris from the borings. The method is also very slow.

An even cruder alternative is wash probing, (Fig. 5.4), which gives a rough indication of the position of the change between a soft or loose soil and that of a more compact soil. The casing is not necessary, the change being determined by the operator from the feel of the wash pipe.

### Lining tubes

In most non-cohesive soils the borehole will probably need a lining. This consists of sections of tubular casings driven from the surface by means of a 'monkey and tail' operated from a winch (Fig. 5.2(e)).

## 5.2   CONTINUOUS FLIGHT AUGERING

The rotary boring machine designed for drilling rock, described in Chapter 3, can also be fitted with a continuous flight auger (Fig. 5.5) and bits suitable for boring in soft ground. Flushing is not required.

The spiral flight on the auger has a pitch designed to bring the spoil to the surface. By inserting spiral sections, boreholes of up to 600 mm diameter can generally be sunk to depths of 30 m. However, depending upon the hoist capacity, depths of up to 100 m with smaller diameters have been obtained.

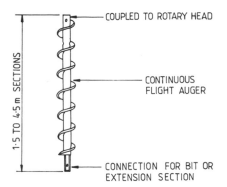

**Fig. 5.5** Continuous flight augering

COUPLED TO ROTARY HEAD

CONTINUOUS FLIGHT AUGER

CONNECTION FOR BIT OR EXTENSION SECTION

1·5 TO 4·5 m SECTIONS

This method is best suited to boulder-free foundations, such as sand, gravel, clay, slate, chalk, limestone and coal. For harder foundations, the spiral auger is removed and the rig is converted to operate with either roller or drag bits (see Chapter 3) as a conventional rotary drill.

The method is unsuitable where free-flowing materials are encountered which cannot be retained on the flighting, such as water-bearing gravels, sands and silts. For these formations a grabbing or rotary bucket technique is usually more appropriate. A lining tube is used where the sides of the borehole are likely to collapse.

Typical uses of this method are for sub-surface exploration, well drilling and boreholes for ground anchors.

It should be noted that the diamond drilling rig is unsuitable, because of its high rotation speeds and relatively low torque availability.

### Augers (Fig. 5.5)

The augers for continuous augering are generally made in 1.5, 3, 4.5 and 6 m sections and are added as the drilling progresses.

### Drilling heads

For common earth, sand and some gravels, a finger bit (Fig. 5.6(a)) of up to 300 mm diameter is suitable, but in clay soils the

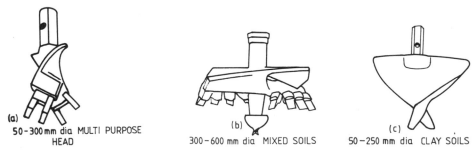

(a)
50-300 mm dia MULTI PURPOSE HEAD

(b)
300-600 mm dia MIXED SOILS

(c)
50-250 mm dia CLAY SOILS

**Fig. 5.6** Auger bits for soft to medium soils

specially-designed clay bit is preferred (Fig. 5.6(c)). For larger diameters the earth bit is more appropriate (Fig. 5.6(b)). The method can also be used in frozen soils and soft to medium hard rocks, with rock augers (Fig. 5.7) but an increased feed force may be necessary to produce adequate penetration rates.

(a) 75–250mm dia. – MEDIUM HARD FORMATIONS

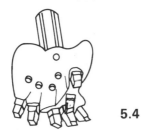

(b) 100–250 mm dia. – HARD FORMATIONS

**Fig. 5.7**   Rock auger bits

## 5.3   INTERMITTENT AUGERING

While the continuous flight auger has adequately serviced the need for relatively shallow and small diameter boreholes, the demand for large diameter borings, especially for foundation piles, has required a modified augering procedure. The intermittent method uses a short flight auger or drilling bucket and holes of up to approximately 2.5 m diameter can be drilled, to depths of 50 m.

## 5.4   EQUIPMENT

### Drilling rig

The design and construction of the rig is similar to the rotary drilling machine used for rock boring, but no flushing medium is needed. The basic machine consists of either a truck (Fig. 5.8) or crawler crane (Fig. 5.9) supporting a mast, telescoping kelly bar and drill head; a drilling table and separate diesel engine to power the drilling units.

The details of the arrangement are shown in Fig. 5.10. The mast guides the telescoping kelly bar, the latter being attached to the crane or truck base through two sets of hydraulic rams, and a base unit

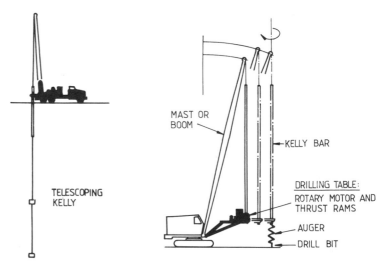

**Fig. 5.8**   Truck-mounted intermittent auger rig

**Fig. 5.9**   Crane-mounted intermittent auger rig

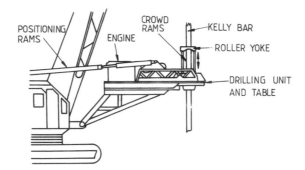

POSITIONING RAMS
CROWD RAMS
KELLY BAR
ENGINE
ROLLER YOKE
DRILLING UNIT AND TABLE

**Fig. 5.10** Drilling unit for intermittent augering

connected to the crane slewing ring. Two ropes are used, one to operate the kelly bar, the other to handle any casing tubes. Power is supplied from a separate diesel engine to drive a hydraulic system which provides the turning movement and crowding force at the turntable. High rotation speeds of up to 200 rpm are possible, depending upon the torque and gear selected, and depths down to 100 m have been achieved, but more usually the telescoping action of the kelly is available only to 50 m depth, with additional kellies having to be added manually thereafter.

### Auger and drill bit

This method is suitable for non conglomerate soils, and drill bits similar to the continuous flight augering method are used. However, unlike the continuous auger, the intermittent method requires only a short section of flighting above the drill bit (Fig. 5.11). The kelly bar must therefore be raised to ground level when the auger is fully laden. The material is discharged by simply spinning the soil off the flighting.

The flight auger can bore holes of up to about 2.5 m diameter but unfortunately is unsuitable for free-flowing materials and mud, etc., which cannot be retained on the flighting. In such circumstances a drilling bucket is necessary (Fig. 5.12). This has a conical base with two openings to allow the spoil to enter. Flaps cover the openings to retain the spoil when the bucket is raised from the hole. The bucket is fitted with two sets of digging teeth mounted across the base, and holes of up to approximately 2 m diameter can be bored.

### Belling buckets

In stable soils, the diameter of a borehole produced by a rotary method can be extended to about three times that of the hole, with special purpose belling tools.

A hole is drilled in the normal way with an auger or bucket, which is then replaced by the belling bucket (Fig. 5.13). With this bucket standing on the bottom of the hole, the belling arms are gradually opened out as the bucket is rotated. However, there is a tendency for part of the belled roof to fall away before concrete can be placed (e.g.

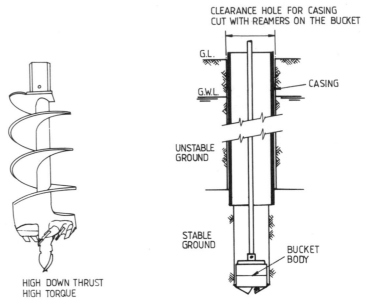

Fig. 5.11  Intermittent flight auger and bit

Fig. 5.12  Intermittent bucket auger

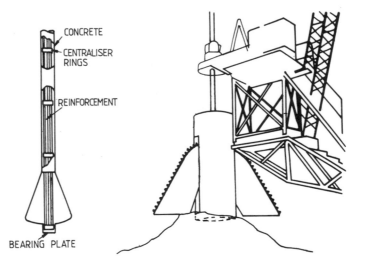

Fig. 5.13  Under-reamed borehole and belling equipment

in situ piling), and therefore good practice requires an inspector to be lowered down the borehole in a safety skip (Fig. 5.14) and the work examined, cleaned up, etc. before concrete placing begins.

## Borehole lining

In many situations the ground will be stable and self-supporting while drilling takes place. However, where the ground would collapse or

**Fig. 5.14** Inspecting an under-reamed borehole

where inflowing water would cause problems of stability, the sides of the borehole require supporting. This may be achieved by several techniques, such as casing or fluid pressure methods. In general, casing is preferred, as shown in Fig. 5.12. In non cohesive soil the casing may be installed and extracted with a vibrating method described for sheet piling (page 134). Alternatively the casing may be screwed in by:

(i) using the kelly
(ii) oscillating with a special hydraulic oscillator, or
(iii) driving with a drop hammer suspended from the mast of the crane or rig.

Because of the limited headroom under the drilling table, it is usual to install the casing in sections which are progressively screwed together as the borehole depth is extended. Often a clearance hole can be drilled before casing is required, thus allowing the lead casing to be of greater length than the extension sections.

**Table 5.2** Rotary boring rig data (telescoping type)

| Auger or bucket diameter (m) | Max.* drilling depth (m) | Power unit (kW) | Max. rpm | Crowd or feed force (kN) | Torque (kN.m) | Equivalent crane size (tonnes) |
|---|---|---|---|---|---|---|
| 0.7 | 30 | 70 | 200 | 120 | 17 | 15 |
| 0.9 | 30 | 80 | 160 | 120 | 25 | 20 |
| 1.5 | 35 | 90 | 130 | 130 | 35 | 30 |
| 2.0 | 40 | 110 | 105 | 150 | 55 | 40 |
| 2.5 | 45 | 160 | 70 | 150 | 120 | 50 |

Output 10–20 m per hour. Accuracy 1 in 75 vertically.
* Can be extended to approximately 100 m with additional drill stems

### Characteristics of the auger boring method

TORQUE

The main characteristic of this type of rotary boring method compared to small diameter rotary rock boring is the need to generate very high torque because of the large diameter boreholes involved. Indeed, as an auger cuts through the earth, the torque required to keep it rotating varies continuously, as illustrated by the example shown in Fig. 5.15. Clearly it is therefore highly desirable to have sufficient power in reserve to respond to a rapidly changing torque need.

## CROWD AND HOIST

The weight of the kelly bar provides some downward thrust but for work in medium to hard soils, additional down-force must be provided by the hydraulic feed.

## 5.5  GRABBING METHODS

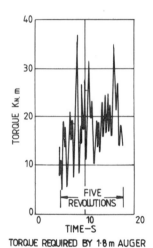

TORQUE REQUIRED BY 1·8 m AUGER

**Fig. 5.15**  Torque required for auger boring

When poor ground, or obstructions are frequently encountered in a soil such as a conglomerate, then the most economical solution may be to dispense with the rotary system and concentrate on a grabbing or chiselling technique, such as the Benoto method. The Benoto machine consists of a basic frame supporting a mast, which houses a grabbing bucket, engine and winches (Fig. 5.16). The borehole casing is both oscillated and forced downwards, using hydraulic rams (Fig. 5.17).

In soft soils the lining is pressed into the soil ahead of the grab, while in harder materials the casing must first be under-reamed. Like the casing in the augering method, sections of pipe, 2, 4 or 6 m long, are screwed together. Boreholes from 0.5 to 2 m diameter can be obtained and rakes of up to 12° from the vertical are possible. Depths of 100 m are achievable with an accuracy of up to 1 in 200 vertically.

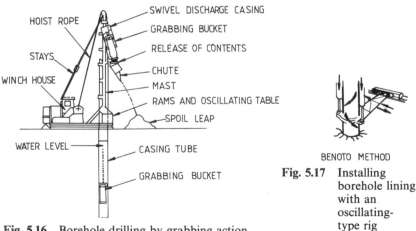

**Fig. 5.17**  Installing borehole lining with an oscillating-type rig

**Fig. 5.16**  Borehole drilling by grabbing action

### The grabbing bucket (Fig. 5.18)

The grab is controlled by a single line from the winch (or alternatively by a hydraulic kelly bar). A simple device on the line opens the bucket for discharge of the contents. Typical bucket sizes are shown in Table 5.3. The rate of excavation is slow and depends upon the material and size of grab. The method is suitable for most types of material and can

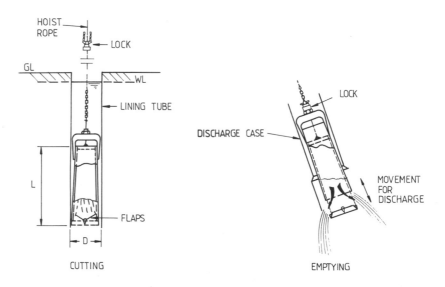

**Fig. 5.18**  Grabbing bucket

**Table 5.3**  Rope grab data (*see Fig. 5.18*)

| D (mm) | L (mm) | Weight (kg) |
|--------|--------|-------------|
| 350 | 1500 | 450 |
| 450 | 1500 | 450 |
| 550 | 1750 | 450 |
| 650 | 1750 | 450 |
| 800 | 2500 | 1000 |
| 950 | 3200 | 1700 |
| 1200 | 3800 | 2500 |
| 1500 | 5300 | 7250 |

even be used as a rock chisel. In soft to medium soil, a 1 m$^3$ capacity grabbing bucket could excavate 2–5 m$^3$ per hour.

### The Benoto rig

| | |
|---|---|
| Total weight | 30 tonnes |
| Engine | 100 kW |
| Angle boring | Up to 12° |

## 5.6 CIRCULATION DRILLING (Fig. 5.19)

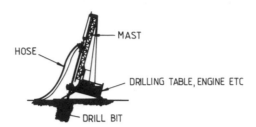

**Fig. 5.19** Circulation drilling

The installation of lining tubes in very deep borings may be technically difficult and very expensive. This problem may be overcome by introducing fluid pressure into the borehole to stem the flow of incoming water and to support the soil.

There are three basic circulation drilling methods:

  (i) direct circulation
 (ii) reverse circulation
(iii) reverse circulation with air lift

Reverse circulation is best suited to large diameter boreholes.

### (i) Direct circulation (Fig. 5.20)

Drilling fluid is pumped down a hollow drill pipe, around the drill bit and back to the surface in the annular space around the drill pipe. Cuttings are carried to the surface by the flow.

The volume of drilling fluid necessary to give a satisfactory raising velocity within the annular space depends on the volume of that space. This, in turn, determines the capacity of the pump needed (see Flushing, page 34).

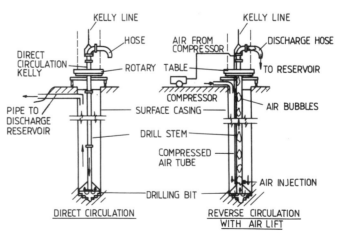

**Fig. 5.20** Direct circulation drilling method

**Fig. 5.21** Reverse circulation drilling with air lift

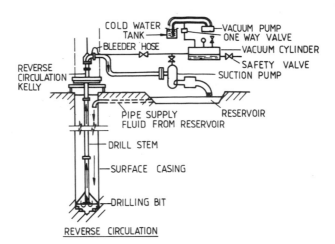

**Fig. 5.22** Reverse circulation with vacuum pump

### Reverse circulation (Fig. 5.22)

In this counter-flow system, drilling fluid is introduced into the annular space around the drill pipe where it moves down and around or through the drill bit. The fluid – along with excavated material – is then drawn back up to the surface through the inside of the drill pipe by a high capacity, low head, centrifugal pump. This system must be primed with a vacuum pump and the hole kept full of fluid at all times.

### (ii) Reverse circulation with air lift (Fig. 5.21)

Drilling fluid is introduced into the bore hole as with the pumped reverse circulation method. The driving force to return the flow, and cuttings, to the surface is created by injecting air into the drill string. The composite mixture of fluid and air bubbles, which is lighter in weight than the fluid outside the drill pipe, forms a pressure differential which creates the flow.

An adequate fluid level must be maintained in the hole to provide the necessary hydrostatic head.

This method is simple and robust and requires no pumps. Much deeper boreholes can be obtained by this method than by direct or reverse circulation. Approximate consumption of compressed-air is

| *Pipe diameter* (mm) | *Free air* (m³/min) |
|---|---|
| 150 | 4.5 to 6 |
| 200 | 6 to 10 |
| 300 | 15 to 20 |

### (iii) Circulation fluid

In sandy soils the interstices may be partially sealed and the grains stabilised by adding bentonite to the pump water, to produce a thixotropic solution weighing approximately 1200–1400 kg/m³. Usually a 4% solution by volume is sufficient but up to 10% may be required in very difficult conditions.

Concrete can be deposited in the bentonite through a tremie tube and the displaced fluid drawn off. Sometimes it may be necessary to install a short section of lining casing, especially when drilling over water, to give a firm support, near the top of the borehole.

**Rig data**

| | |
|---|---|
| Engine | Up to 100 kW |
| Rotation speeds | Up to 60 rpm |
| Circulation rates | Up to 8000 l/m |
| Torque | Up to 200 kN.m. |

This method is principally designed for deep boreholes of 300–2000 mm diameter, in uniform soils and consolidated materials. The bit used is similar to that required for a drilling bucket, with two or three rows of teeth mounted across a circular ring (Fig. 5.23).

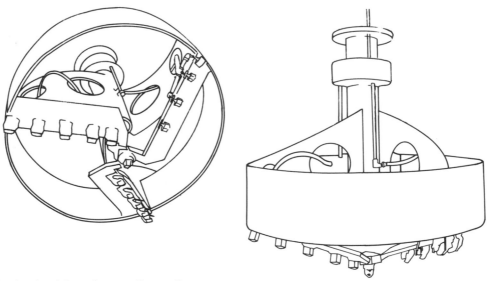

**Fig. 5.23** Cutting head for soft to medium soils

**Outputs (i.e. rate of penetration)**

| | |
|---|---|
| Common earth | 3–4 m/h |
| Sands and clays | 2–3 m/h |
| Gravel | 1.5 m/h |

2 man operation.

### 5.7  DRILLING ROCK

**Fig. 5.24**  Roller bits for rock drilling

Although not a common feature in civil engineering work, roller bits (Fig. 5.24) have been developed to sink mine shafts through rock strata. The bits are similar in construction to those described under rock drilling and tunnelling in Chapters 3 and 10 and diameters exceeding 6 m are available.

Feed force of up to 0.5 kN/mm bit diameter may be required. A formula developed by the Hughes company gives torque required as:

$$T = 5252 \times k \times D^{2.5} \times W^{1.5}$$

where  $D$ = bit diameter in inches
  $T$ = torque in lb. ft.
  $W$ = load per inch on the bit in 1000 lb.
  $k$ = formation constant, e.g. $4 \times 10^{-5}$ for granite.

### Output

|  |  |
|---|---|
| Medium hard rock | 0.1–0.5 m/h |
| Soft rock | 0.5–1 m/h |

Typical data for soft/medium rock drilling are given in Table 5.4.

**Table 5.4**  Various drilling requirements for effective penetration when circulation drilling with a full hole bit in medium soft rock formations

| Hole diameter (mm) | Flushing rate (l/m) | Weight of drill stem (tonnes) | Rotary speed (rpm) |
|---|---|---|---|
| 750 | 6000 | 20 | 30–50 |
| 1000 | 8000 | 25 | 25–35 |
| 1250 | 10000 | 38 | 20–30 |
| 1500 | 12000 | 40 | 15–25 |
| 1750 | 16000 | 47 | 10–20 |
| 2000 | 20000 | 55 | 10–20 |
| 2250 | 24000 | 61 | 10–15 |
| 2500 | 28000 | 68 | 10–15 |
| 3000 | 32000 | 80 | 7–12 |
| 4000 | 36000 | 100 | 5–10 |

# 6

# DE-WATERING OF GROUND

## 6.1 METHODS OF DEALING WITH GROUND WATER

In order to carry out construction work below surface levels it is normally necessary for the working area to be reasonably free from standing water. Therefore the ground water flow must either be blocked or carried away. The type of soil, the height of the water table, the depth of the excavation and its shape all influence the choice of method of dealing with water flow in and around the excavation works and a variety of methods is available. For example, the water may be removed by pumping, or it may be isolated from the works by providing a barrier of injected material to fill the soil interstices, or the water in soil voids may be frozen, or compressed-air may be used to provide a pressurised chamber to balance the water head in the ground.

It can be seen in Fig. 6.1 that for the majority of granular soils, pumping methods are most appropriate. For medium to coarse gravels however, permeability of the material is usually too high for pumping, and cement grouting may be necessary. At the other extreme, clay is an impermeable soil and therefore water seepage would not be a major problem, any surface run-off could be adequately removed from a sump. Silty soils are particularly troublesome and often cannot be effectively de-watered by pumping. Either electro-osmosis, or more usually, ground freezing or grouting should be considered. Freezing and grouting are more adequately covered in Chapters 7 and 9 and compressed-air in Chapters 9 and 10.

This chapter deals with pumping methods only.

## 6.2 THEORY OF DE-WATERING SOILS

The objective of de-watering is to lower the water table in the vicinity of an excavation to provide a relatively dry and stable working area. Pumping from wells positioned outside the excavation boundary is

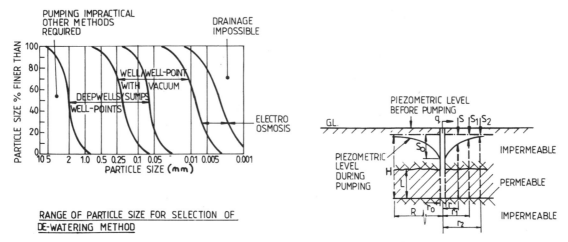

**Fig. 6.1**  Range of particle size for selection of de-watering method

**Fig. 6.2**  Flow to a fully penetrating well in an artesian aquifer

usually the preferred technique. The type of soil and the position of impermeable strata have a marked effect on the rate of pumping which is possible and theoretical models have been developed to estimate the discharge from a well in different soil configurations. However, to simplify the formula for practical applications, the flow into a well is usually assumed to be either confined or unconfined.

### Steady confined flow to a single well

Darcy's Law states that the quantity of water flowing under laminar conditions is proportional to the hydraulic gradient, i.e.

$$q \propto -\frac{ds}{dr} \tag{6.1}$$

Thus for the well shown in Fig. 6.2,

$$q = -k2\pi r L \frac{ds}{dr} \tag{6.2}$$

where   $r$ = radius from centre of well to a given point
   $L$ = thickness of permeable soil
   $q$ = theoretical quantity of flow per unit time through a cross-sectional area at radius $r$
   $k$ = constant = coefficient of permeability of the soil, defined as quantity of water that will flow through a unit cross-section of porous material in unit time under a hydraulic gradient of unity.
   $-\dfrac{ds}{dr}$ = hydraulic gradient at radius $r$.

After integrating,

$$s = -\frac{q}{2\pi kL} \log_e r + C$$

Thus the draw-down at $S_1 = -\frac{q}{2\pi kL} \log_e r_1 + C$

and draw-down at $S_2 = -\frac{q}{2\pi kL} \log_e r_2 + C$

Therefore $\qquad S_1 - S_2 = \frac{q}{2\pi kL} \log_e \frac{r_2}{r_1}$ (6.3)

The equation is more generally written as:

$$S = \frac{q}{2\pi kL} \log \frac{R}{r}$$

This is Dupuit's formula, where $R$ is the radius at the tangent point between the pumped piezometric level (draw-down curve) and the original water table.

### Steady unconfined flow to a single well (Fig. 6.3)

Using Darcy's formula

$$q = 2\pi rh \times k \times \frac{dh}{dr}$$ (6.4)

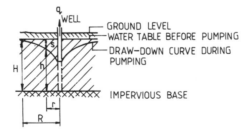

**Fig. 6.3** Flow to a fully penetrating well in an unconfined aquifer

rearranging

$$hdh = \frac{q}{2\pi k} \times \frac{dr}{r}$$

and integrating

$$h^2 = \frac{q}{\pi k} \log_e r + C$$

This formula is generally written as

$$H^2 - h^2 = \frac{q}{\pi k} \log_e \frac{R}{r} \qquad (6.5)$$

which is the Dupuit formula.

The draw-down at a radius $r$ is $S = H - h$ and therefore, substituting for $h$ into (6.5)

$$S = \frac{q}{\pi k (2H - S)} \log_e \frac{R}{r}$$

For small draw-downs in deep aquifers, $S$ is small compared to $H$ and can be neglected and so

$$S = \frac{q}{2\pi k H} \log_e \frac{R}{r} \qquad (6.6)$$

Thus the difference in draw-down between two such points $S_1$ and $S_2$ can be shown by

$$S_1 - S_2 = \frac{q}{2\pi k H} \log_e \frac{r_2}{r_1} \qquad (6.7)$$

This is Thiem's formula.

### Partially penetrating wells in deep aquifers (Fig. 6.4)

Excavations are often fairly shallow, perhaps 2–15 m in depth. Therefore where the impermeable stratum is relatively deep the Dupuit-Forcheimer formula is modified to:

$$q = \frac{\pi k (T^2 - (T - H + h_0)^2)}{\log_e R - \log_e r_0} \times \sqrt{\frac{H}{T}} \times \sqrt[k]{\frac{2T - 1}{T}} \qquad (6.8)$$

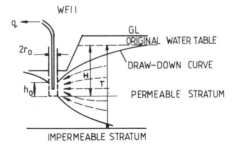

**Fig. 6.4** Flow to a well founded above an impermeable stratum (Jureka)

### Other confined soil configurations

The Dupuit formulae described for a well founded in a completely confined or unconfined stratum can be further developed to embrace layered impervious and porous strata. This is beyond the scope of this book and the reader is recommended to follow *Groundwater Recovery* by Huisman for further information.

## 6.3 DETERMINING WELL DISCHARGE

In order to assess accurately the discharge from a well and its associated draw-down curve, for example in water supply calculations or where complex non-uniform soil strata are involved, it is necessary to carry out field measurement on a trial borehole and take measurements from stand pipes (Fig. 6.5). The method requires pumping until steady state conditions are obtained, which may take many weeks.

However, for estimating purposes, the above Dupuit formulae can be used to give the approximate pumping rate necessary to produce an assumed draw-down level. Thus, for given values of $k$, $H$, $h_0$, $R$ and $r_0$, the pumping rate from a well may be calculated.

Approximate coefficients of permeability based on field experience are given in Table 6.1. Typical values for $R$ are shown in Fig. 6.6.

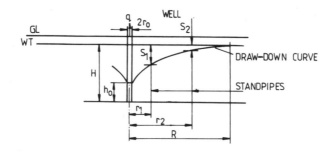

**Fig. 6.5** Determining well discharge from pumping trials

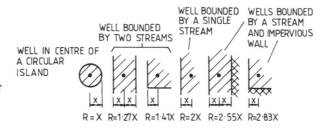

**Fig. 6.6** Approximate values of maximum radii of draw-down curves (Huisman)

**Table 6.1** Soils and their approximate values of permeability

| Soil | Grain size (mm) | Coefficient of permeability (mm/s) |
|---|---|---|
| Medium gravel | 4–7 | $>30$ |
| Fine gravel | 2–4 | 30–60 |
| Uniform sand | 0.11 | 2–0.05 |
| Graded sand | 0.1–0.3 | $1–10^{-2}$ |
| Silty sand | Varies | $10^{-1}–10^{-2}$ |
| Clayey sand | Varies | $10^{-2}–10^{-3}$ |
| Silt | 0.05–0.002 | $10^{-3}–10^{-5}$ |
| Clay | 0.002 | $10^{-5}–10^{-6}$ |

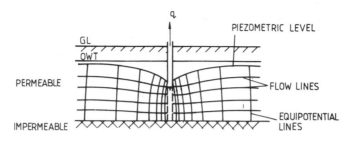

**Fig. 6.7** Flow net of gravity well

### Well losses

The Dupuit formulae assume that the water flows in horizontal planes. At large radii from the well, Fig. 6.7 shows this assumption to be reliable. Near the well, however, the flow is curvi-linear and the free surface and Dupuit's assumed curve do not coincide. An estimate of difference in height ($m_0$) between these two curves at the well periphery (Fig. 6.8(a)) is given by Ehrenberger as:

$$m_0 \simeq \frac{H}{2}\left(1 - \frac{h_0}{H}\right)^2 \tag{6.9}$$

Also, in practice, additional losses are caused by entrance friction at the screen ($m_2$) and through the filter ($m_1$) (Fig. 6.8(b)). Therefore the actual rate of discharge from the well is likely to be considerably less than the theoretical.

Sichardt has suggested that the actual hydraulic gradient ($i$) at the entrance to the well is related to the coefficient of permeability ($k$) by

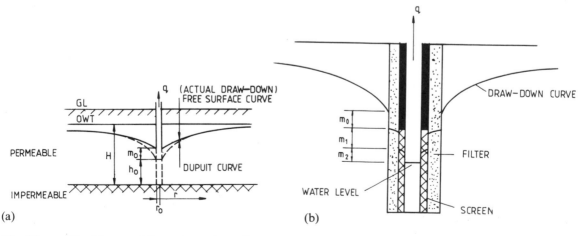

(a)     (b)

**Fig. 6.8**     Head losses at the entrance to well

$$i = \frac{1}{15\sqrt{k}} \text{ for the temporary wells}$$

and $\quad i = \dfrac{1}{30\sqrt{k}}$ for permanent, long standing wells.

From Darcy's law, the entrance velocity $v = ki$. Therefore for a temporary well

$$v = \frac{\sqrt{k}}{15}$$

Thus for a well with radius $r_0$ and screen depth $h_w$ the actual discharge obtained would be

$$q_w = 2\pi r_0 h_w \frac{\sqrt{k}}{15} \tag{6.10}$$

Consequently, to obtain the desired draw-down level, either a greater well radius than the theoretical, or several wells, are required.

## 6.4 DISCHARGE FROM AN UNCONFINED WELL SYSTEM

In de-watering of construction works it is often more economical to install a group of small wells rather than a single large well. Interference between two or more wells occurs when the cones of draw-down overlap. The draw-down at any point on the composite cone of depression (i.e. the combined draw-down curve of several wells) is then equal to the sum of the draw-downs at that point for each well, assuming it to be pumped separately as shown in Fig. 6.9.

Thus at any radius $r$ from a well in a group of wells, the draw-down at $r$ is

$$S = \sum_{i=1}^{i=n} S_i$$

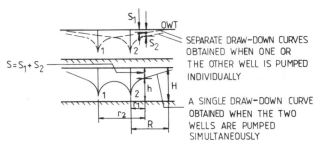

**Fig. 6.9** Interference between adjacent draw-down curves

From equation (6.5),

$$H^2 - h^2 = \sum_{i=1}^{i=n} (H^2 - h_i^2) = \sum_{i=1}^{i=n} \frac{q_i}{\pi k} \log_e \frac{R}{r_i}$$

If $q_1 = q_2 \ldots = q_n = q$ i.e. discharge is the same for all wells where $q$ is the discharge from a single well, then

$$H^2 - h^2 = \frac{q}{\pi k} (n \log_e R - (\log_e r_1 + \log_e r_2 \ldots \log_e r_n))$$

Thus

$$H^2 - h^2 = \frac{q}{\pi k} (n \log_e R - \log_e r_1 \times r_2 \times \ldots r_n) \qquad (6.11)$$

Wells arranged in a circle

$$r_1 = r_2 = \ldots r_n$$

Therefore

$$H^2 - h^2 = \frac{nq}{\pi k} (\log_e R - \log_e r) \qquad (6.12)$$

If the wells are arranged around the excavation (see Fig. 6.10) the required radius of the circle $x \simeq \sqrt{\dfrac{b \times l}{\pi}}$

where   $b$ = breadth of excavation
$l$ = length of excavation.


### Example (Fig. 6.10)

Wells are founded in a circular pattern around an excavation 30 m long and 25 m wide. The height of the water table above an impermeable stratum is 9 m. The depth of the excavation is 5 m and the water table lies 1 m below ground level. The coefficient of permeability of the soil $k = 0.002$ m/s. Determine the required number of wells.

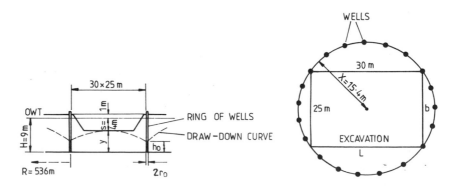

**Fig. 6.10**   Design example

## SOLUTION

From equation (6.11) and when $r = x$

$$H^2 - h^2 = \frac{nq}{\pi k} \log_e R - \log_e x)$$ (A)

(i) $x = \sqrt{\dfrac{b \times l}{\pi}} = \sqrt{\dfrac{25 \times 36}{\pi}} = 15.4$ $\therefore \log_e x = 2.73.$

(ii) Values of $R$ can be obtained from Fig. 6.6, but in open country, Sichardt has shown that for an excavation

$$R \simeq 3000 \times S \times \sqrt{k}$$

where $k$ is in m/s.

Thus $R = 3000 \times 4 \times \sqrt{0.002} = 536$ m

as shown in Fig. 6.10

$$\log_e R = 6.28$$

$$h = H - S = 9 - 4 = 5 \text{ m}.$$

(iii) Therefore substituting in (A)

$$nq = \frac{\pi k (H^2 - h^2)}{\log_e R - \log_e x} = \frac{\pi \times 0.002 (9^2 - 5^2)}{6.28 - 2.73} = 0.1 \text{ m}^3/\text{s}.$$

(iv) If the effective length of screening on a well of diameter $r_0$ is $h_0$

then the capacity of the well $q = 2\pi r_0 h_0 \times \dfrac{\sqrt{k}}{15}$

Assuming the effective length of screening is approximately $0.4H$ and $r_0$ is 100 mm then, substituting values into the equation

$$q = 2 \times \pi \times 0.1 \times 0.4 \times 9 \times \frac{\sqrt{0.002}}{15} = 0.0067 \text{ m}^3/\text{s or } 6.7 \text{ l/s}$$

Therefore the number of wells required is

$$n = \frac{0.1}{0.0067} = \underline{15 \text{ wells}} \text{ plus approx. 5 wells for reserve}$$

$$= 20 \text{ wells}$$

(v) Well diameter: Sichardt recommends that the minimum spacing of wells should be about $10 r_0 \pi \simeq 32 r_0$.
Thus with $x = 15.4$ m (Fig. 6.10),

$$\text{spacing of wells} = \frac{2\pi \times 15.4}{20} = 4.8 \text{ m} > 32 \times 0.1 > 3.2.$$

Thus the selected well diameter of 200 mm is suitable. A slightly smaller diameter would also be acceptable, although the costs of producing additional boreholes and the extra pumps must be taken into account.

## 6.5 PUMPING FROM OPEN EXCAVATIONS

### Open sumps

The system of pumping from an open sump is popular because the costs of installation and maintenance of the equipment are relatively low compared to those for wells, and because the system is applicable to most soils. Unfortunately, the method draws water into the excavation, because the sump is usually located inside the excavation itself. As a consequence, the tendency is to wash out the banks when working in fine soils. Nevertheless with due care, coupled with support for the sides of the excavation, pumping may be adequate in providing a reasonably water-free working area. Dense and well-graded granular soils, hard fissured rocks and surface run-off from clays are favourable to open pumping, but problems with slope stability are likely to occur in loose granular soils, soft granular silts and soft rock.

### Open excavation in an unconfined soil founded on an impermeable stratum (Fig. 6.11) Steady state condition

The theory of pumping from sumps can be developed along similar lines to wells on the assumption of steady flow from a confined or unconfined aquifer.

**Fig. 6.11** Open excavation in an unconfined soil founded on an impermeable stratum

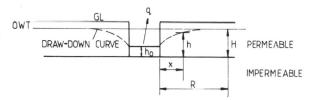

According to Darcy, the rate of flow of water through a unit length of sump wall is

$$q = kh\frac{dh}{dx} \quad \text{or} \quad hdh = \frac{q}{k}dx$$

and integrating

$$\frac{h^2}{2} = \frac{qx}{k} + C$$

when $x = 0$, $h = h_0$

$$\therefore \quad C = \frac{h_0^2}{2}$$

when $x = R$, $h = H$ and so

$$\frac{H^2}{2} = \frac{qR}{k} + \frac{h_0^2}{2}$$

Therefore

$$q = \frac{k}{2R}(H^2 - h_0^2) \tag{6.13}$$

Jureka suggests that for steady state conditions length of draw-down curve from the sump face to the tangent to the water table be represented by the equation

$$R = 2H\sqrt{kH} \quad \text{where } k \text{ is in m per day.}$$

**Open excavation in an unconfined soil above an impermeable stratum (Fig. 6.12) (steady state condition)**

The simple situation above is rarely found in practice and it is more usual for an excavation to be located above an impermeable stratum. Jureka recommends that eqn. (6.13) be adjusted to take into account seepage through the base as follows:

$$q = \frac{kH}{2}\left[\frac{H}{R} + \frac{\pi}{\log_e\dfrac{d}{\pi b} + \dfrac{\pi R}{2d}}\right] \tag{6.14}$$

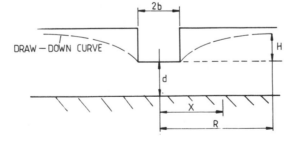

**Fig. 6.12** Open excavation in an unconfined soil above an impermeable stratum

**Open excavation in a confined soil above an impermeable stratum (Fig. 6.13)**

$$q = k\left[\frac{(2s - m)}{R}m + \frac{\pi s}{\log_e\dfrac{d}{\pi b} + \dfrac{\pi R}{2d}}\right] \tag{6.15}$$

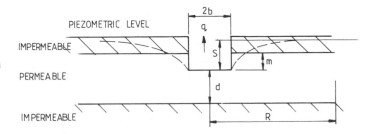

**Fig. 6.13**  Open excavation in a confined soil above an impermeable stratum

As with wells, the actual rate of flow ($q_s$) into the sump will be significantly lower than the theoretical value $q$.

*Note*: During the early stages of pumping, flows may be considerably higher until steady conditions are obtained.

### Closed excavations founded above an impermeable stratum

The flow of water entering an excavation can be reduced by enclosing the sides, for example with sheet piles. Ideally the piles should be driven down to an impermeable stratum to seal the excavation completely, as shown in Fig. 6.14. Where this is impracticable, seepage through the base must be expected (Fig. 6.15) and the rate of flow per unit length is given by:

$$q = bki \tag{6.16}$$

where $i$ is the hydraulic gradient as determined from flow net analysis.

However this formula is only accurate when $d$ is much less than $D$. If this relationship is altered by driving the piles deeper, then the rate of flow in the formula should be reduced by deducting the following values from $q$.

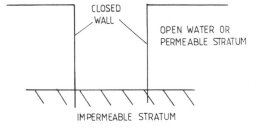

**Fig. 6.14**  Closed excavation founded on an impermeable stratum

| $\dfrac{d}{D}$ | 0.1 | 0.2 | 0.3 | 0.4 | 0.5 | 0.6 | 0.7 | 0.8 | 0.9 | 1.0 |
|---|---|---|---|---|---|---|---|---|---|---|
| % deducted from $q$ | 5 | 10 | 15 | 20 | 25 | 30 | 40 | 50 | 65 | 100 |

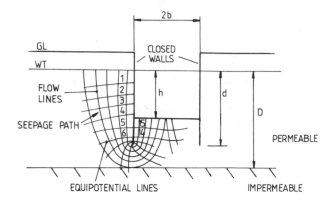

**Fig. 6.15** Closed excavation founded above an impermeable stratum

## 6.6 'PIPING' OR 'BOILING' OF THE SOIL INSIDE THE EXCAVATION

Piping will occur when the upward force of the water issuing is greater than the weight of the particles.

The upward force of the water $= \gamma_w z$ per unit area

The weight of the particles in water $= \gamma_w \left( \dfrac{P_p - 1}{1 + e} \right)$ per unit area

where   $\gamma_w$ = unit weight of water
  $P_p$ = specific gravity of the particles
  $e$ = voids ratio.
  $z$ = seepage head at a cell
  $a$ = length of column of particles

Therefore piping occurs when $z > \dfrac{P_p - 1}{1 + e} \cdot a$.

### Example (see Fig. 6.15)

For sand, $P_p \fallingdotseq 2.8$ and $e = 0.8$.

$$\frac{P_p - 1}{1 + e} = \frac{2.8 - 1}{1.8} = 1.$$

If $h = 21$ m, $n = 15$, the minimum number of cells in a path, and the side length ($a$) of cell 15 is 1 m, then:

$$z_{15} = \frac{h}{n} = \frac{21}{15} = 1.4 > 1 \times 1$$

and piping will occur.

In practice a factor of safety significantly in excess of 1 is desirable, because the exit velocity in highly permeable soils may also cause movement of the soil particles.

Piping usually does not occur in clays and silty clays, because seepage either does not take place or is quite small. It is also unlikely to occur in gravels because the high permeability tends to allow extensive drawdown and thereby produces a naturally long seepage path. Piping occurs most often in loose fine sands which have relatively high permeabilities with concentrated draw-down curves, and whose small grains are also susceptible to forces caused by the flow of water. In these conditions excavation work can be most hazardous and, rather than the use of sumps, external de-watering of the soil may be required, together with cut-off walls. Unless these methods are effective, soil particles will be taken up by the flow into the excavation and the surrounding banks may suffer subsidence.

## 6.7  SUMP DESIGN

A sump is positioned in the deepest part of the excavation and preferably away from the main works. A small ditch cut around the base of the excavation falling towards the sump will help to keep the area reasonably clear of standing water. If the work is to continue for a period then it often pays to use porous pipes in a gravel fill in the ditch.

The sump (Fig. 6.16) base should be approximately 1 m below the bottom of the excavation, with the walls protected with timber, loose trench sheets, a perforated oil drum, etc. The bottom of the sump can be stabilised with a 0.5 m layer of graded granular fill, e.g. sand to coarse gravel. If this aspect is overlooked, fines in the soil may be washed through and damage the pump.

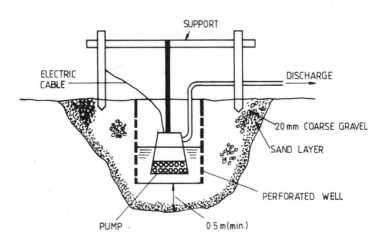

**Fig. 6.16** Sump arrangement

## 6.8   WELL POINTING

In permeable strata where soil stability might be endangered by using sumps, the single stage well pointing system is a preferred method for shallow excavations not exceeding 6–7 m deep. For greater depths, multi-stage well pointing or a single, large well should be considered. The objective of the well point system is to produce a cone of depression in the water table so that excavation can take place in relatively dry conditions. The system consists of a number of individual well points, each comprising a jetting/riser pipe, 40–50 mm diameter, drilled with a ring of inlet ports at the bottom (Fig. 6.17). A strainer about 1 m long is placed over the tube to cover the ports. The riser is connected at surface level to a header pipe, about 150 mm diameter, which in turn is connected to a suction pump.

### Installation

The well point is fitted with a rubber ball valve placed inside the jetting shoe (see Fig. 6.17) and during installation the top end of the riser/jetting pipe is connected to a jetting hose and water under pressure is forced through the well point. Usually maximum pressure of $1 \text{ N/mm}^2$ (10 bar or 150 psi) is sufficient in most types of soil, with a pump capable of delivery of 100–120 1/min.

   The procedure simply requires that an operator place the point in the desired position (Fig. 6.18). The water pressure is turned on and the washing or jetting action causes the pipe to penetrate the soil. The water is turned off when the correct depth is located. The operation is then repeated until all well points are in position.

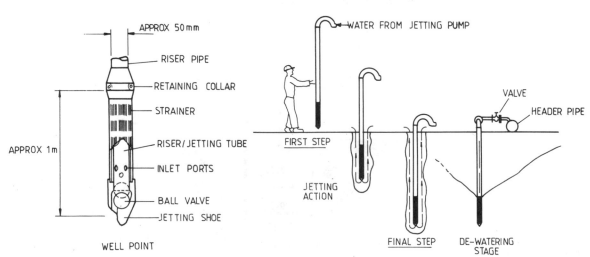

**Fig. 6.17**   Well point

**Fig. 6.18**   Installing well points and riser pipes

### 6.9   WELL POINTING ARRANGEMENTS

**Progressive systems (Fig. 6.19)**

For trench work a header pipe is placed along the line of the proposed trench outside the track of the excavator. Well points are progressively installed ahead of the excavation. By inserting isolating valves in the header pipe and repositioning the suction pump, the excavation may be progressively backfilled. The trenches can be timbered, sheet piled or battered in the usual manner to improve stability as demanded.

The number of well points needed can be theoretically calculated as described on page 85, but in practice 750 mm centres are typical in loose gravel or coarse sand, while the spacing may be increased in fine running sand.

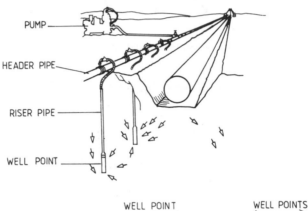

Fig. 6.19   Progressive well point system

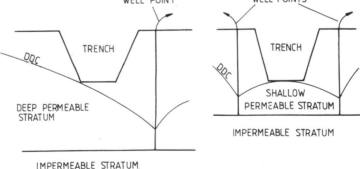

Fig. 6.20   Effects of depth of foundation and impermeable stratum on choice of de-watering arrangement

In deep permeable soil a single row of well points may be adequate, but if the trench is founded close above an impermeable stratum then well points both sides of the trench may be necessary to obtain draw-down below the bottom of the excavation (Fig. 6.20). In these latter conditions it may be impossible to secure a completely dry base, as some water will inevitably pass between points.

An installation gang of approximately 4 men is usually sufficient to keep ahead of a trenching gang.

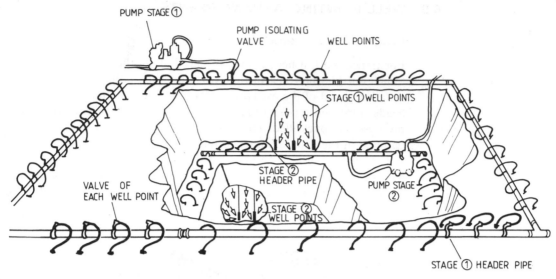

**Fig. 6.21**   Ring well pointing system (also shows staged rings)

### Ring system (Fig. 6.21)

In this system the well points remain in position over the full construction period. The header pipe is placed at surface level around the perimeter of the excavation to form a closed ring. Well points are then sunk into position. Where more than a single pump is required to cope with the rate of flow, each pump should be located to house an equal number of points either side.

### Staged ring system

In theory, atmospheric pressure limits suction to about 10.35 m but because of leakages in pipe joints etc. the maximum depth for a single stage system is about 6–7 m. Well pointing of deeper excavations, however, can be achieved with additional rings, as shown in Fig. 6.21. Each stage lifts within the limiting depth, and water is pumped to surface level with a separate pump located at the stage level. The method requires a wide area to accommodate the stages and when the costs of the additional excavation work are taken into account, the alternative method of using deep wells might be a cheaper alternative.

### General points in well point operation

(i)   Install one or more 'dummy' wells (i.e. open-ended pipes) to monitor the draw-down level.

(ii)  It is undesirable to commence excavation work before the water table has reached the design depth and remained steady for a period.

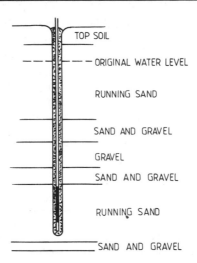

**Fig. 6.22** Sanding in well points

TOP SOIL

ORIGINAL WATER LEVEL

RUNNING SAND

SAND AND GRAVEL

GRAVEL

SAND AND GRAVEL

RUNNING SAND

SAND AND GRAVEL

(iii) Pumping should be continued until permanent structures are completely above standing water level. Hence standby pumps are required.

(iv) If output from the pumps is small, but the vacuum gauge has a high reading, then the well points are probably blocked. This problem can sometimes be overcome by opening the header pipe to atmospheric pressure and allowing the water to flow back to ground.

(v) If vacuum is lost, with an accompanying rise in the water table, then air leaks into the circuit are present. The source of leakage can sometimes be found by isolating the individual valves, one at a time on each of the riser pipes.

(vi) Initially the discharge may be discoloured for a few minutes and then clear. If the discoloration persists, then fines are being drawn into the flow, which may cause damage to nearby structures.

(vii) Where permeable soil is interspersed with clay layers, the water table will be lowered, but also water may flow along the impermeable strata and into the excavation. In these conditions the well points should be 'sanded in' as shown in Fig. 6.22. By enlarging the well diameter, a surround of filter material will conduct the flow more efficiently. 'Sanding in' should also be employed in fine silts, whereby the filter medium may assist in preventing the fines passing through the strainer of the well points.

**Output data**

The maximum capacity of a single well point is approximately 1.0–1.51 l/s, and to try to achieve this output the spacing is altered to suit the permeability of the soil, for example:

Clay and other impermeable soil – cannot be de-watered with well
                                                  points

Silty sands of low permeabilty  – 1.5 m centres
Fine to coarse sands  – 0.85–1.0 m centres
Sandy gravel  – 0.25–0.85 m centres
Coarse gravel  – 0.3 m centres
Very porous ground  – well pointing is unsuitable

Jureka in Germany gives typical field results with a well pointing
system, as shown in Table 6.2.

**Table 6.2**  Well pointing field results in various soils

| Surface area (m²) | Number of points | Draw-down depth (m) | Soil permeability (mm/s) | Total discharge (l/s) | | Duration of project (days) |
|---|---|---|---|---|---|---|
| | | | | Start | Finish | |
| 1500 | 50 | 2 | $1.0 \times 10^{-1}$ | 15 | 10 | 100 |
| 1500 | 100 | 4 | $3.0 \times 10^{-2}$ | 20 | 12 | 100 |
| 2000 | 75 | 3 | $2.5 \times 10^{-1}$ | 13 | 10 | 200 |
| 2000 | 100 | 15 | $5.0 \times 10^{-1}$ | 23 | 18 | 250 |
| 2000 | 300 | 13 | $1.0 \times 10^{-3}$ | 7 | 4 | 150 |
| 2500 | 175 | 1 | $1.0 \times 10^{-2}$ | 4 | 3 | 350 |
| 3000 | 100 | 3 | $2.5 \times 10^{-1}$ | 20 | 10 | 275 |

## 6.10  DEEP WELLS (150–300 mm DIAMETER)

In situations where staging of well point systems is required (see Fig.
6.21) or where the soil permeability is too high for well pointing
practicability, deep wells should be considered (Fig. 6.23), the main
difference being that a submersible pump is located at the bottom of
the well, thereby avoiding the limitations imposed by suction. An
appropriate size of pump can be chosen to deal with the rate of flow
and the well diameter formed accordingly. Such wells are used for
depths greater than about 7.5 m and have been successfully operated at
depths greater than 100 m.

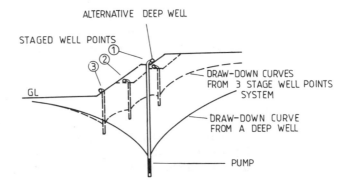

**Fig. 6.23**  Comparison of
well points and
deep wells

### Well construction (Fig. 6.24)

The well is usually bored with one of the rotary boring methods described in Chapter 5 and a temporary outer casing is driven to give stability where demanded by the conditions. When the borehole has reached the required depth a perforated well liner is placed into position and plugged at the well bottom to produce a reasonably good seal and stable base. Layers of filter material are placed around the casing to keep out fines.

For the best results the filter material should be cylindrically layered in thicknesses of about 100 mm. The layer near the screen should have

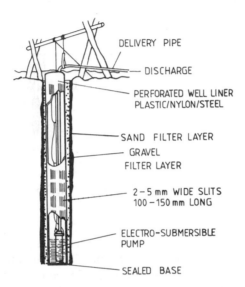

**Fig. 6.24** Installation of deep wells

sizes of 5–20 mm diameter, surrounded by a layer with 0.5–2 mm grains. The screen itself should be made in plastic, wood, or mild steel sheeting with the perforations placed horizontally or vertically.

To minimise clogging the slits should be as wide as possible; 2–5 mm is usual, depending upon the size of the surrounding filter material or soil. (A filter medium is not required in a coarse gravel stratum.) Finally, a submersible pump is installed and the discharge pipe led into a nearby stream.

### Output data for shallow tube wells

Typical field results from shallow tube wells are given in Table 6.3.

## 6.11  ELECTRO-OSMOSIS AND VACUUM WELLS

Soils such as silts and silty-clays are virtually impossible to drain with normal pumping methods, because the capillary forces acting on the

**Table 6.3** Shallow tube wells with results in various soils (Jureka)

| Surface area (m²) | Draw-down depth (m) | Number of wells | Soil permeability (mm/s) | Total discharge (l/s) | | Duration of project (days) |
|---|---|---|---|---|---|---|
| | | | | Start | Finish | |
| 250 | 3.5 | 10 | $2.0 \times 10^{-1}$ | 10 | 5 | 75 |
| 300 | 3.5 | 8 | $2.5 \times 10^{-1}$ | 20 | 10 | 180 |
| 400 | 3.5 | 15 | 1.5 | 40 | 30 | 200 |
| 600 | 4.0 | 15 | 2.5 | 40 | 25 | 180 |
| 1000 | 6.0 | 50 | 2.5 | 120 | 60 | 175 |
| 2000 | 2.0 | 50 | $5 \times 10^{-1}$ | 30 | 20 | 350 |

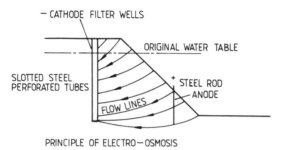

**Fig. 6.25** Principles of electro-osmosis de-watering

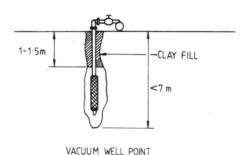

**Fig. 6.26** Vacuum well pointing

pore water prevent free flow under gravity. The osmosis technique may therefore be considered as a possible alternative.

This method, shown in Fig. 6.25, consists of steel rod anodes and filter wells as cathodes. The positively charged water thus flows towards the filter wells. Unfortunately the system has not been widely used in the UK and comprehensive cost data from various soils is not available.

The method is an alternative which may be considered along with vacuum wells or well points (Fig. 6.26), grouting and ground freezing. So-called 'vacuum wells' involve the top part of the riser tube being surrounded with clay to form a seal. The 'vacuum' accelerates the pore water movement.

## 6.12 RECHARGE WATER

A lowering of the water table reduces buoyancy of the soil, which may cause settlement of weak, compressible strata. Where settlement must be kept to a minimum, recharge water should be provided. A simple recharge trench (Fig. 6.27) is commonly used in construction works, but reversed well points or recharge wells are alternative methods. However, care is required when choosing the appropriate method. For

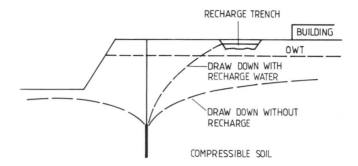

**Fig. 6.27**  Recharge water from a trench

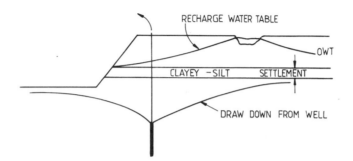

**Fig. 6.28**  Example of inappropriate recharging

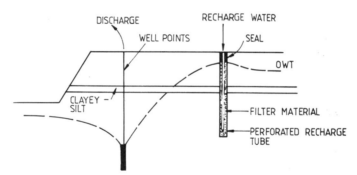

**Fig. 6.29**  Recharge water from wells

example, in Fig. 6.28 the recharge water would not effectively penetrate the compressible layer of clay-silt and might actually increase the load on this layer. Figure 6.29 illustrates the application of a recharge well to try to avoid this effect. The required capacity of a recharge well or trench can be calculated on similar principles to those described earlier for wells and sumps, but with water flowing away from rather than to the well.

## 6.13  PUMPING EQUIPMENT

### Lift and force pump (Fig. 6.30)

The lift pump depends upon atmospheric pressure for its working principle and comprises a cylinder (F) fitted with a piston (A). A valve

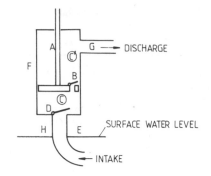

**Fig. 6.30** Principles of the lift and force pump

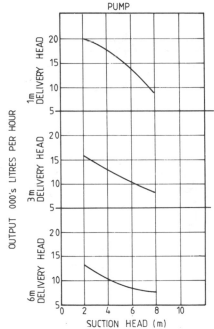

**Fig. 6.31** Output characteristics of a 75 mm diameter connection (2 kW) diaphragm pump

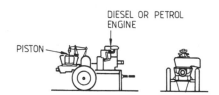

**Fig. 6.32** Diaphragm pump

is located at the entrance of the cylinder and a valve (B) on the piston. During the upward stroke valve (B) is closed and (D) is opened, thus from Boyle's Law where $p \times v =$ constant, as $v$ is increased the pressure of the air in (C) falls. The pressure of the atmosphere acting on the water surface (E) causes water to flow up the pipe (H) and into (C). On the downward stroke (D) closes and (B) opens and the water is released into (C′). Repetition of the process discharges the water through (G) on the upward stroke to produce a force pump.

In practice, lift and force pumps are operated well within the maximum theoretical atmospheric pressure (taken to be 10.35 m of water) because of leakages in joints etc., and 6–7 m is the more normal limiting suction lift. Such pumps are also designed to deliver against a similar head, to give a combined suction and delivery head of about 14 m.

A pump of given power rating produces a rate of discharge which varies with the head as shown in Fig. 6.31.

Such pumps are commonly used in shallow excavations and manufactured with a diesel engine as shown in Fig. 6.32. They are extremely robust and most suited to the variable conditions typical of construction work. 75 mm or 100 mm diameter pipes are usual, fitted with a strainer on the suction end. Such pumps are usually designed (i) to pump on 'snore' i.e. air and water mixed during low flows and (ii) to pick up suction without priming.

### Centrifugal pump

Water may be raised by the centrifugal rotation of a vane wheel, operating with a reversed turbine action (Fig. 6.33). The pump consists of an impeller surrounded by a spiral casing. If the height of the pump above the water surface is located within the lift permitted by atmospheric pressure, water will be drawn inside the casing as the impeller is rotated, imparting additional kinetic energy to it. However, because the casing is spiral in shape, the velocity of the water decreases with increasing area of flow, and on reaching the delivery pipe the velocity will be relatively low, whereas the pressure will be high. In this manner, water can be pumped against extremely high delivery heads.

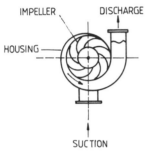

**Fig. 6.33** Principles of the centrifugal pump

### Self-priming pump (Fig. 6.34)

Centrifugal pumps are commonly used with well pointing, and in sumps they should be of the self-priming type. A 150 mm diameter pump is usually suitable for average well pointing requirements with characteristics typical of those shown in Fig. 6.35. Where a long suction line and many well points are involved, leaks in the connections and air in the water and ground might require a combined centrifugal water pump and vacuum pump to handle the large quantities of air (see Fig. 5.23, and compressed air, Chapter 2) and a float chamber to separate the air from the water. Such an arrangement is virtually essential in vacuum well pointing.

### Submersible pumps (Fig. 6.36)

The electric submersible centrifugal pump has gained wide acceptance because of its portability and its ability to deal with water in both sumps and deep wells. Priming is not required and models are now manufactured to suit wells of as little as 250 mm diameter. Larger units are available up to 100 kW power. The characteristics of a typical pump suitable for a deep well are shown in Fig. 6.37. The main disadvantage with this type of pump is the wear caused to the impeller

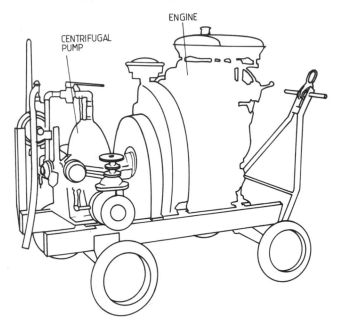

Fig. 6.34 Self-priming centrifugal pump

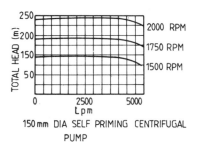

150 mm DIA SELF PRIMING CENTRIFUGAL PUMP

Fig. 6.35 Output characteristics of 150 mm diameter connection self-priming centrifugal pump

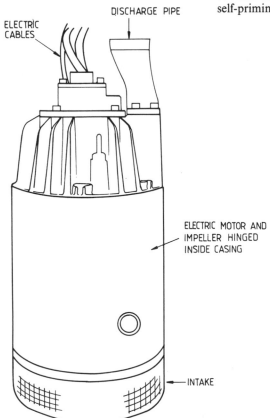

Fig. 6.36 Submersible centrifugal pump

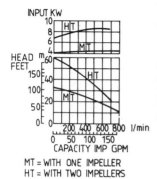

MT = WITH ONE IMPELLER
HT = WITH TWO IMPELLERS

**Fig. 6.37** Output characteristics of 75 mm diameter connection (3 kW) submersible pump

blades by granular particles carried in the water, which results in loss of capacity.

## Stand-by pumps

It is emphasised that a pump breakdown can be expensive if the works become flooded and damage occurs. Therefore 100% reserve should always be immediately available.

# 7 GROUTING

Grouting is the term used to describe the method of injecting a fluid
substance into rock fissures or into a soil either to provide improved
stability or to reduce permeability. The grout substance can be selected
from a range of materials such as cement, clay suspensions in water,
chemical solutions, or even emulsions of bitumen and water (Fig. 7.1).
The grout is selected to meet the requirements of a particular situation
and different techniques have been developed for the various grout
types. Grouting is an expensive process and therefore every effort
should be made to investigate the nature of the ground before the work
commences.

## 7.1 GROUTING APPLICATIONS

Grouting is used both for temporary and permanent works and has
applications in:

(i) Sealing pockets and lenses of permeable or unstable soil or rock
prior to excavation of a tunnel heading (Fig. 7.2) or alternatively
grouting a stratum from ground level (Fig. 7.3).

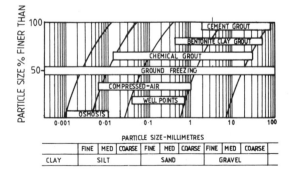

**Fig. 7.1** Soil particle size
and the choice of
grouting or
de-watering
method

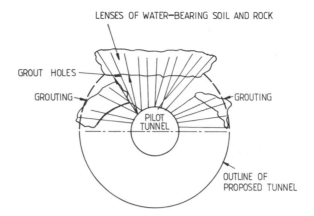

**Fig. 7.2** Grouting from a pilot tunnel

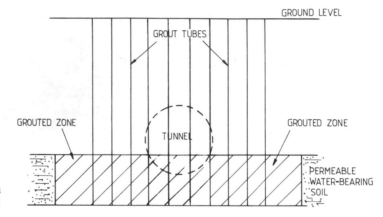

**Fig. 7.3** Grouting a tunnel line from the surface level

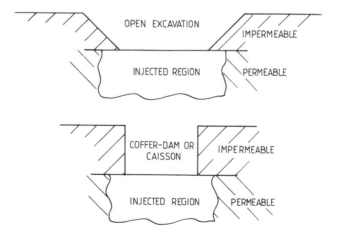

**Fig. 7.4** Grouting a permeable stratum at the base of an excavation

    (ii) Sealing the base of an excavation, coffer-dam or caisson founded on a permeable stratum (Fig. 7.4).

   (iii) Grouting in ground anchors (Fig. 7.5) for sheet pile walls, concrete pile walls, retaining walls, stabilising rock cuttings, tunnels, etc.

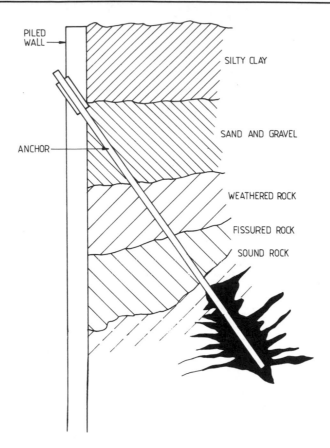

PILED WALL

SILTY CLAY

SAND AND GRAVEL

ANCHOR

WEATHERED ROCK

FISSURED ROCK

SOUND ROCK

**Fig. 7.5**   Grouting in ground anchors

(iv) Repairing: (a) the ground underneath a foundation may be strengthened with the injection of a suitable grout;
(b) a new damp proof course can be formed in a layer of brickwork by using a chemical grout;
(c) cracks and structural defects on building masonry can be filled;
(d) pavements and sunken slabs can be raised, etc.

(v) Filling the void between the lining and rock face in tunnel works (Fig. 7.6).

(vi) Forming a 'grout curtain' in layers of permeable strata below a dam and so effectively sealing off any flow from the storage side. This method may be applied either before construction takes place or alternatively to effect repairs to an existing structure (Fig. 7.7).

(vii) Grouting up the tendons in prestressed post tensioned concrete.

(viii) Sealing the gap with a non-shrinkable grout between the surface of a concrete foundation and the base plate of a stanchion, etc.

(ix) Producing mass concrete structures and piles by introducing cement after the aggregate is in position (Fig. 7.8).

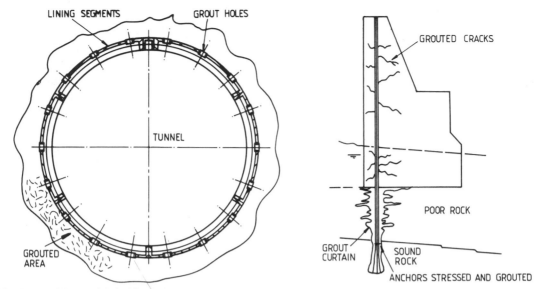

**Fig. 7.6** Filling voids behind tunnel lining

**Fig. 7.7** Forming a grout curtain below a dam

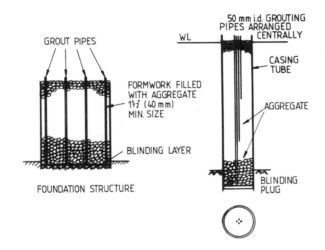

**Fig. 7.8** Producing foundations and piles (Colcrete)

## 7.2 PRINCIPLES OF GROUTING

The principle of grouting is to introduce a substance to fill the voids in a soil or the fissures in a rock by pumping fluid down a small diameter tube placed in a drill hole. The depth of the boring controls the thickness of the layer to be grouted and as each stage is completed the borehole is lengthened and a further layer is grouted, and so on until the design depth is reached. (However, modern developments allow the borehole to be made to the full depth in one operation, with subsequent use of packers to obtain stage grouting; this technique and others are discussed later.)

### Penetration of grout

If the medium into which grout is to be injected can be assumed to have equal permeability in all directions, then for a Newtonian fluid (see Fig. 7.16) the radius of penetration of the grout into the soil is assumed to be spherical spreading out from the point of injection (Fig. 7.9). The mathematical analysis according to Maag is as follows:

$$q = A \times v$$

where $A$ is the surface area of a sphere of grout of radius $r$, and
$v$ is the velocity of flow of the grout.

From Darcy's Law

$$v = -k_g \frac{dh}{dr}$$

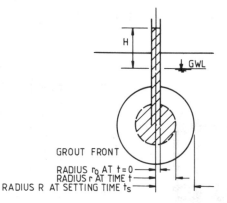

**Fig. 7.9**  Principal of grout penetration

H

GWL

GROUT FRONT

RADIUS $r_0$ AT t = 0

RADIUS $r$ AT TIME t

RADIUS R  AT SETTING TIME $t_s$

where $k_g$ is the coefficient of permeability of the soil with a grout flow in length/unit time.

$-\dfrac{dh}{dr}$ is the rate of change of fluid head.

Thus from Fig. 7.9,

$$q = -4\pi r^2 k_g \frac{dh}{dr}$$

or

$$-\int_H^0 dh = \int_{r_0}^R \frac{q}{4\pi r^2 k_g} \times dr$$

By integration

$$-h = \frac{q}{4\pi r k_g} + C$$

when $r = r_0$ the radius of the grout tube then $h = H$ the effective head of grout. Therefore

$$C = -\frac{q}{4\pi r_0 k_g} - H$$

when $r = R$, the limiting radius of grout penetration $h = 0$. Therefore

$$C = \frac{-q}{4\pi R k_g}$$

Substituting for $C$, then

$$\frac{q}{4\pi R k_g} = H + \frac{q}{4\pi r_0 k_g}$$

$$q = \frac{4\pi k_g H}{\left(\dfrac{1}{R} - \dfrac{1}{r_0}\right)} \tag{7.1}$$

But if the fluid increases the radius of the sphere by $dr$ in time $dt$, and $n$ is the porosity of the soil (i.e. ratio of voids to total volume), then

$$q = 4\pi r^2 \frac{dr}{dt} \times n$$

Integrating,

$$\int_{r_0}^{R} nr^2 \, dr = \frac{q}{4\pi} \int_{0}^{T} dt$$

$$\frac{nr^3}{3} = \frac{qt}{4\pi} + C$$

when $r = 0$, $t = 0$ and so

$$C = \frac{nr_0^3}{3}$$

when $r = R$, $t = T$ and so

$$C = \frac{nR^3}{3} - \frac{qT}{4}$$

Substitution for $C$ then

$$\frac{qT}{4\pi} = \frac{n}{3}(R^3 - r_o^3) \tag{7.2}$$

Substituting for $q$ and solving (7.1) and (7.2)

$$\frac{4\pi k_g H}{\left(\dfrac{1}{R} + \dfrac{1}{r_0}\right)} = \frac{n}{3}(R^3 - r_0^3)$$

but $1/R$ is negligible compared to $1/r_0$

and
$$k_g = k_w \frac{\eta_w}{\eta_g} \times \frac{\gamma_g}{\gamma_w} \tag{7.3}$$

where   $k_w$ = coefficient of permeability of soil with water flow,
$\eta_w, \eta_s$ are the respective dynamic viscosities of water and grout,
e.g. in Newton-seconds per $m^2$,
$\gamma_w, \gamma_g$ are the respective unit weights of water and grout e.g. in
$N/m^3$.

Therefore

$$R = \left( \frac{3 \times r_0}{n} \times k_w \times \frac{\eta_w}{\eta_g} \times \frac{\gamma_g}{\gamma_w} \times H \times T + r_o^3 \right)^{1/3} \tag{7.4}$$

or
$$R = \left( \frac{3 \times r_0}{n} \times k_w \times \frac{v_w}{v_g} \times H \times T + r_o^3 \right)^{1/3} \tag{7.5}$$

where $v$ represents kinematic viscosity.

In practice the viscosity of the grout changes with time and eventually gels. Thus it can be seen from equation (7.4) that if all the other factors remain constant, the radius of penetration $R$ is reduced as the viscosity of the grout increases.

However, it is also apparent that the pumping pressure (or hydraulic head $H$) may be increased to counteract this effect. Other influential factors are the grout density, soil porosity, soil permeability, and the diameter of the grout pipe. The soil permeability in particular is affected by the particle size, arrangement and distribution, the continuity of pores and formation stratification. Furthermore, the permeability in the horizontal direction generally exceeds that in the vertical and the resulting injected volume tends to be cylindrical (Fig. 7.10) rather than the theoretically assumed sphere.

The presence of ground water, non uniform soils, fissures in rock, etc. often cause distortions to the theory and therefore the injection pattern, pumping pressures, etc. must be tempered by experience based on information obtained from a full site investigation.

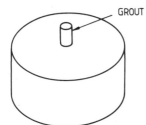

GROUT HOLE

**Fig. 7.10**   Effects of soil permeability on grout penetration

## 7.3   PRE-GROUTING SITE INVESTIGATION

To obtain satisfactory grouting of the soil or rock strata it is essential to carry out a full site investigation before commencing operations. This may be expensive, but without it the extent of required grouting work could only be guessed. The site investigation may be undertaken separately from the grouting contract and in many cases this is preferable in order that full and proper testing can take place, unhindered by influences such as the contractors' need to perform the grouting task in the most economic time programme. The investigation should include a geological survey, and investigation drilling.

### Geological survey

This consists of examining the general geology of the area, using mapping methods, supplemented by exploratory drilling to establish fissure systems, faults, folds, etc. The final geological map should provide information on the extent of soil and rock formations, zones of weakness, the dip and strike, etc. of the strata.

### Site investigation

The site investigation involves a second and more detailed exploration of the precise location where grouting is to be concentrateed, and is used to determine the criteria for estimating the type, quantity and extent of the grouting necessary. Core samples are taken during borehole drilling for analysis in the laboratory to determine rock or soil strength, grain size, permeability, porosity, etc. An important aspect of the investigation is in situ permeability tests on the strata, which yield more reliable data than tests made on individual samples in the laboratory, and therefore facilitate more precise estimation of the required pumping pressure, grouting pattern and type of grout.

### In situ permeability tests

The approximate permeability of a stratum may be determined by several methods, using either a constant or variable head of water. The method involves sinking a borehole through the stratum. The drill stem of the boring equipment is then used to maintain a head of water above the ambient pore pressure in the ground.

The rate of flow at different time intervals is measured, to obtain a projection of the steady state rate of flow. In the case of a constant head, the coefficient of permeability ($k$) is subsequently calculated from a formula of the form

$$k = \frac{q}{H} \times \text{factor } (F)$$

where   $q$ is the steady state rate of flow, and
        $H$ is the hydraulic head.

The factor ($F$) is dependent upon the shape of the exit and values can be obtained either by reference to Hvorslev or Gibson.

It is usual to log the soil at 1 m intervals over the entire length of borehole as shown in Fig. 7.11. This is achieved with the aid of two packers (i.e. plugs) as illustrated in Fig. 7.12.

In rock strata, because of the presence of fissures, it is common practice to determine an equivalent permeability by the Lugeon test. Units of Lugeon are the flow in litres per minute absorbed by 1 m of drill hole at an injection pressure of 1 N/mm$^2$ above ambient pore pressure.

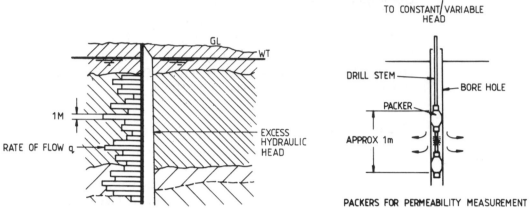

**Fig. 7.11**　In situ permeability test logging　　　**Fig. 7.12**　In situ permeability testing method

One unit of Lugeon is equivalent to a conventional permeability ($k_w$) of approximately $1 \times 10^{-4}$ mm/s.

With in situ values of permeability ($k_w$) determined for water, the equivalent value for a grout ($k_g$) may be calculated from eqn. (7.3). By assimilating other data from the site investigation, the theoretical extent of grout penetration may be ascertained from eqn. (7.4).

## 7.4　SPECIFICATION FOR GROUTING

The results of a borehole permeability log indicate the zones to be grouted, the type of grout to choose, and the likely extent of grout penetration under a given pressure. Practical experience, however, has demonstrated that spacings of grout holes in different soil types should be similar to those given in Table 7.1, but this will very much depend upon the viscosity of the grout being used.

Once the length, breadth and depth of the zone to be grouted and the grout hole spacing are fixed, the quantity of grout required can be calculated. The total volume required is dependent upon the voids ratio

**Table 7.1**　Grout hole spacing

| Coefficient of permeability (soil–water) (mm/s) | Grid spacing (m) | Soil types |
|---|---|---|
| >1 | 6 | Fissured rocks |
| 1 to $1 \times 10^{-1}$ | 3 | Medium/coarse sands and gravels |
| $<10^{-1}$ | 0.5–1 | Fine sands |

*Note*
Generally after adequate grouting, permeability can be reduced below $1 \times 10^{-3}$ mm/s

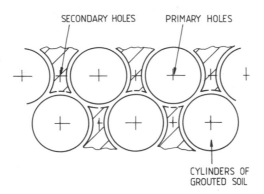

**Fig. 7.13** Grouting pattern

of the mass to be injected. However, exact estimates are difficult to obtain because of the necessity of altering viscosity, pumping rates and pressures to accommodate the unknown influencing factors such as ground water conditions, fissures, etc. during the execution of the work. However, for a simple example of a grouted sphere of 1 m radius of voids ratio 0.6, the volume of grout required

$$= \frac{4}{3} \times \pi \times 1^3 \times \frac{0.6}{1.6} = 1.57 \text{ m}^3 \text{ or } 1570 \text{ litres.}$$

It can be seen in Fig. 7.13 that flow from the primary holes travels out on a radial front while flow from secondary and tertiary holes fills the spaces between adjacent primary injections. Estimates of grout volumes should be made on this basis.

## 7.5   GROUT DESIGN

Grout is injected into soil, rock or other material to:

(i) Reduce the flow of ground water
(ii) Increase the strength of the material

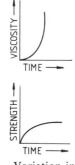

**Fig. 7.14** Variation in grout strength and viscosity with time

   To obtain these effects the grout must set and harden, but not so quickly that pumping becomes impracticable. Additives are therefore frequently required to control the viscosity (gelling time) and subsequent rate of hardening. The viscosity and strength of grouts generally increase with respect to time as shown in Fig. 7.14. For example, cement- or clay-based grouts may take 28 days or more to reach almost full strength. Most important, however, during injection is the fluid viscosity, which gradually changes as chemical reactions take place to transform the grout from a liquid into a gel (i.e. initial set). Some grouts react in minutes, while for others maximum viscosity is only obtained after about 24 hours.

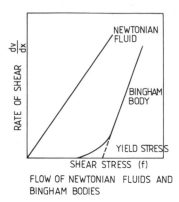

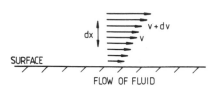

**Fig. 7.15**   Principles of viscosity

**Fig. 7.16**   Newtonian and Bingham fluid viscosity characteristics

## Viscosity

When a fluid is disturbed its various parts move at different velocities, which come to rest when the disturbance is removed. For example, for a liquid in motion on the plane solid surface shown in Fig. 7.15, the layer in contact with the surface is at rest, while the layers above move at increasing velocities to produce a velocity gradient $dv/dx$. According to Newton there is a shear stress between any two layers, and the relationship between this stress ($f$) and the velocity gradient is called the coefficient of dynamic viscosity i.e.

$$\frac{f}{\dfrac{dv}{dx}} \text{ (Expressed as units of poise or } \frac{1}{100} \text{ poise, i.e. cP)}$$

1 cP = 10 N s/m$^2$.

The viscosity of water at 15°C is approximately 1 cP.

If change in viscosity caused by hardening or temperature is ignored, then a plot of rate of shear $dv/dx$ against shear stress produces a straight line, as shown in Fig. 7.16, for a purely viscous liquid such as water or a thin chemical solution. Such liquids are known as Newtonian liquids.

For suspensions in water such as cement or clay, and some chemical precipitates, the curve no longer passes through the origin and a yield value is exhibited. Such suspensions are called Bingham fluids and extra force must be applied to overcome the inherent shear strength. Some Bingham fluids, e.g. bentonite suspensions, are thixotropic, and behave like a fluid when in movement, but form a 'jelly' when stationary. Usually the transformation is quite reversible and may be repeated.

Grouts which exhibit the above characteristics fall into the following categories:

**Table 7.2**  Grout types and applications

| PFA | Mass filling in very coarse soils and rock fissures |
|---|---|
| Cement | Mass filling in very coarse soils and rock fissures plus ground strengthening |
| Clays | Mass filling in medium coarse soils and impermeability improvement |
| Clay/cement | Similar to clays, plus added strength |
| Emulsions | Impermeability improvement |
| Solutions, single shot | Permeability and/or strength improvement in medium coarse soils |
| Solutions, double shot | As for single shot, with additional control of gel time. Also suitable in fine soils |

(with Suspensions bracketing PFA, Cement, Clays, Clay/cement)

(a) suspensions of solid particles in water, such as clays, cements, bentonite, plaster, PFA (pulverised fuel ash), lime, etc.;
(b) emulsions, such as bitumen in water;
(c) solutions, which react after injection to form insoluble precipitates.

The principles to follow in choosing the grout are:

(i) the grout must be able to penetrate the voids of the mass to be injected (e.g. a cement grout or flocculating chemical grout may get filtered in a fine sand).
(ii) the grout should be resistant to chemical attack when in place.
(iii) the grout should be able to develop sufficient shear strength to withstand the hydraulic gradient imposed during injection and on flowing ground water. Most grouts are suitable to meet this latter consideration.

A wide selection of grouts is available and Table 7.2 sets out recommendations for applications in various situations (see Fig. 7.1).

## 7.6  GROUT TYPES

### PFA grout (Table 7.3)

Pulverised fuel ash reacts with lime and water to produce a stable cementitious material. The particle size is similar to cement and therefore is mainly used in highly permeable materials (i.e. greater than 1 mm/s) such as fissured rock, gravels and coarse sands.

**Table 7.3**  Physical properties of PFA grouts

| Type of grout | Water/ solids | PFA/ cement | 28 day crushing strength $(N/mm^2)$ | Density $(kg/m^3)$ | Setting time (h) | Viscosity (cP) |
|---|---|---|---|---|---|---|
| PFA | 0.5 | — | 0.2 | 1100–1500 | 24 | 400 |
| PFA/Cement | 0.5 | 20 | 0.8 | 1100–1500 | 24 | 500 |
| PFA/Cement | 0.5 | 10 | 2 | 1100–1500 | 24 | 750 |
| PFA/Cement | 0.5 | 5 | 4 | 1100–1500 | 24 | 1000 |

A water:solids ratio of about 0.50 produces a free-flowing grout, but at 0.35 a thick slurry is formed which is a little too viscous for optimum pumping.

When set, the hardened grout has good resistance to sulphate attack and shrinkage is negligible.

Crushing strength can be improved by adding cement, and PFA-cement ratios by dry weight of 5:1 to 20:1 are common.

### Cement grout

The average specific surface area (i.e. fineness) of ordinary Portland cement (OPC) particles is approximately $30 \, mm^2$ per gram, the minimum being $20 \, mm^2/g$. Therefore like PFA, OPC grouts are suitable only in fissured rocks, gravels and coarse sands.

The water:cement ratio may be varied from about 0.6:1 to 3:1 depending upon the ground conditions and required strength.

Rapid hardening cement is finer than OPC and produces a quicker setting time and high early strength, and therefore may be preferred to OPC in ground with high flowing water.

High alumina cement also has rapid strength gain and offers good resistance to attack by sulphates and dilute acids. Supersulphated cement has a fineness of about $60 \, mm^2/g$ and is therefore suitable for penetrating finely fissured rocks.

In very coarse materials and fragmented rocks, sand is often added as a filler as dictated by the pumping conditions.

Cement grouts are mainly selected for ground strengthening and after 28 days, crushing strengths of up up $60 \, N/mm^2$ can be obtained depending upon the water:cement ratio.

### Clay grouts (Table 7.4)

**Table 7.4**  Physical properties of clay grouts

| Type of grout | Water/ solids | Clay/ cement | Density $(kg/m^3)$ | Crushing strength $(N/mm^2)$ | Setting time (h) | Comments |
|---|---|---|---|---|---|---|
| Clay– chemical | 20 | — | 1100 | — | Varies | Fluid |
| Clay–cement | 7 | 0.5 | 1100 | 0.2 | 24 | Thin slurry |
| Cement– bentonite clay | 0.3 | 0.05 | 1600 | 5 | 24 | Thick slurry |

Clay is a complex compound made up of minute mineral particles less than 0.002 mm in diameter, and is thus suitable for injection into medium coarse sands and other soils with a permeability of 1 to $10^{-1}$ mm/s. Clay grout behaves like a Bingham fluid and gels when undisturbed. However, poor strength characteristics are exhibited and so it is mainly used to reduce permeability.

Kaolinite or Illite based clays produce low viscosities and are therefore preferred as filler grouts. Viscosity can be reduced by the addition of a dispersing chemical such as sodium phosphate. A rigidifying agent such as sodium silicate can then also be included to improve the shear strength when set. However, the final shear strength is likely to be poor and the material can be displaced by a hydraulic gradient of 3–4 units.

Clay grouts tend to flocculate when in contact with acid water and may get filtered if the soil is too fine.

Cement may be added to clay to increase its shear strength but because of the relatively large particle size of cement it is limited to use in coarse grained soils or fissured rocks. This type of mixture is called a clay–cement grout, because of the predominance of the clay. Where the clay is a smaller proportion of the total, the grout is referred to as a cement–clay grout. In this latter case bentonite clay is sometimes added in small quantities to reduce the tendency of the cement particles to settle out before full penetration is reached.

### Bitumen

Hot bitumen at about 150°C is an effective grout in highly permeable soils containing flowing water which might wash away clay or cement grouts during injection. As hot bitumen makes contact with the colder ground water, its outside surface skins over, while the inside stays hot and fluid and remains capable of being injected over extensive distances, especially in fractured or fissured rock strata. When the bitumen solidifies, the void is completely filled to provide an effective water seal.

Alternatively, bitumen emulsions containing 50–60% bitumen, 1% emulsifier, and water are suitable for application in fine to medium sands. A coagulant is often included to precipitate the bitumen, the resulting particles gradually building up in the soil pores under the action of pumping. Bituminous grouts are less common today, as developments in chemical grouts prove more economical.

### Chemicals

Like cements, chemical grouts rely on a chemical reaction to produce a material of continuous structure when set. But, unlike cement particles suspended in water, a chemical grout is formed from two separate solutions which react together to produce the gel, which subsequently hardens. The effect can either be obtained by injecting the two fluids one after the other, known as 'two shot' grouting, or alternatively by a single injection of a chemical containing an accelerator to induce gelling.

## TWO SHOT GROUTING

This method was developed by Joosten and consists of an initial injection of sodium silicate followed by one of calcium chloride. The

**Table 7.5** Physical properties of examples of single shot chemical grouts

| Type | Viscosity (cP) | Gel time (min) | Specific gravity | Application (soil permeability) (mm/s) | Crushing strength (N/mm²) |
|---|---|---|---|---|---|
| Sodium silicate–sodium bicarbonate | 1.5 | 0.1–300 | 1.02 | $10^{-1}$ | 0.5 |
| Chrome–lignin | 3.0 | 5–1000 | 1.10 | $10^{-1}$ | 0.5 |
| Sodium silicate–amide | 5–50 | 5–300 | 1.10 | $10^{-1}$ | 5 |
| Resinous polymers | 1.0–1.3 | 1–300 | 1.01 | $10^{-2}$ | 0.5 |

gelled compound behaves like a Bingham fluid, and because of its relatively high viscosity (100 cP) compared to thin suspension grouts, the grout hole spacings must be close together, 700 mm being typical. As a consequence, high pumping pressures are necessary. The method is limited to use in medium coarse sands (permeability 10–1 mm/s or higher). The hardened gel has a strength comparable with cement grout, i.e. approximately 8 N/mm² after 28 days.

### SINGLE SHOT (Table 7.5)

To overcome the viscosity disadvantages of the two shot method, Guttman modified the Joosten process by diluting the sodium silicate solution with sodium bicarbonate solution (resulting viscosity 20 cP and lower strength), but more recent developments have concentrated on a 'one shot' method whereby the chemical additives slowly react with the sodium silicate to form a gel. Many varieties are now available, possessing a range of properties from low to high strength, viscosity, etc., which allows grouting of fine sands with a permeability of about $10^{-1}$ mm/s.

Other types of chemical grouts have been developed in recent years, for example, resin-based compounds can be used in very fine soils of permeability $10^{-2}$–$10^{-3}$ mm/s.

Common examples of one shot chemical grouts are given in Table 7.5. However, it is emphasised that chemical grouting is more expensive than cement and clay grouting.

## 7.7 GROUTING PROCEDURES

### Hole pattern

Ideally, the spacing of the grout holes should be set out on a grid pattern such that the radius of penetration is sufficient to cause slight overlapping between adjacent holes. A second and subsequent half-size grid is then injected to fill the spaces between adjacent columns (see

Fig. 7.13). A third, quarter-size grid is sometimes required to achieve an acceptable reduction in permeability.

### Grout consistency

The grouting speed and rate of pumping are governed by the relationship

Available power from pump $\propto$ pumping pressure $\times$ quantity of grout delivered per unit time.

Before grouting starts it is good practice to flush clean the drill hole with water and indeed to pump water into the soil or rock to clean out cavities or to lubricate the soil particles to aid grout flow.

At the commencement of pumping operations, grout is delivered at the maximum achievable rate. However, with suspension grouts pumping is normally started with a thin mix and the concentration gradually increased until the pressure begins to rise, with a corresponding fall in delivery rate. The viscosity is subsequently maintained at this consistency and the pressure increased up to the limit imposed by the overburden, to avoid heave and leakage to the surface (this should be calculated by reference to soil unit weight). The process is continued until the calculated volume of grout has been injected.

## 7.8 METHODS OF WORKING

The stratum is divided into zones depending upon the permeability tests obtained during the site investigation and each zone is separately grouted. This may be achieved by several methods as follows.

### Grouting from the bottom upwards (Fig. 7.17)

A grout hole 50–75 mm diameter is drilled to full depth using one of the methods described in Chapters 3 and 5. In rock strata and rigid soils a self expanding packer (Fig. 7.18) is placed directly above the lowest zone and grout is pumped in. The packer is then raised to the next zone and the procedure repeated, thereby grouting the drill hole successively upwards. In soft or unstable soils the drill hole must often be lined with a casing to give support and to provide a good seal between the packer and borehole walls. The casing is progressively raised with the packer as shown in Fig. 7.17b. In fissured rock, a particularly permeable stratum can be isolated using a double packer (Fig. 7.19).

### Grouting from the top downwards

Grouting commences in the top zone, the drill hole is then deepened and the next zone grouted. The procedure is repeated until the full

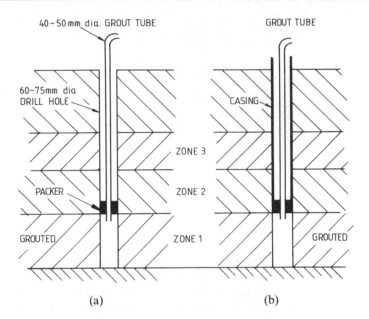

**Fig. 7.17** Grouting from the bottom upwards method

depth has been treated. This method is appropriate for shallow working, as the grouted zones then provide a seal and prevent leakages to the surface.

### Circulation grouting (Fig. 7.20)

This method operates on the principle of grouting from the top downwards. A drill hole is bored to the depth of the first zone and grout is pumped down the grout pipe and returned up the drill hole. In this way, clogging is reduced. The hole is then deepened and the procedure repeated.

### Tube-à-Manchette grouting (Fig. 7.21)

While the double packer method is practicable in rocks, for zonal grouting in alluvial soils it has been found difficult to effect a seal on the lower packer (the casing tube provides a seal for the upper packer). This problem has been overcome by the method developed by E. Ischy in Switzerland.

The equipment comprises a grout pipe located inside a sleeve pipe. The grouting zone is isolated between two packers and grout is pumped through holes in the sleeve pipe, set at about 0.3 m intervals, covered with a rubber band. The void between the borehole walls and sleeve pipe is filled with a thin cement layer to form a seal. During grouting, the grout is forced locally through the seal and enters the soil at the required depth.

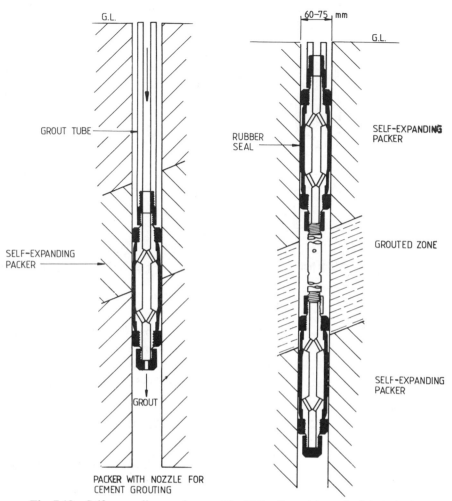

G.L.

GROUT TUBE

SELF-EXPANDING
PACKER

GROUT

PACKER WITH NOZZLE FOR
CEMENT GROUTING

60–75 mm

G.L.

RUBBER
SEAL

SELF-EXPANDING
PACKER

GROUTED ZONE

SELF-EXPANDING
PACKER

**Fig. 7.18**  Self-expanding packer   **Fig. 7.19**  Grouting an isolated stratum
using self-expanding packers

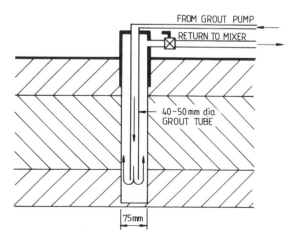

FROM GROUT PUMP

RETURN TO MIXER

40–50 mm dia.
GROUT TUBE

75 mm

**Fig. 7.20**  Circulation
grouting method

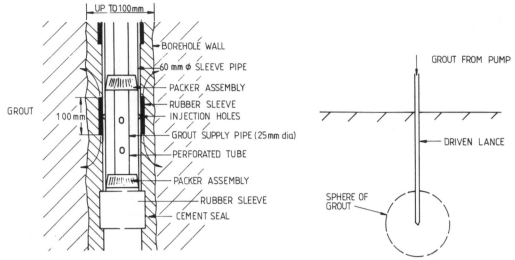

**Fig. 7.21**   Tube-à-Manchette grouting method          **Fig. 7.22**   Point grouting method

### Point grouting

In shallow work, perhaps 10–12 m deep, grouting is injected at predetermined levels from a driven lance (Fig. 7.22). Further grouting may take place as the tube is withdrawn. The technique is commonly used with chemical grouts, where close spacings are required.

### Jet grouting (Fig. 7.23)

Unlike conventional injection methods, jet grouting is used as a replacement technique. Soils ranging from silt to clay and weak rocks can be treated. The method requires a borehole approximately 150 mm

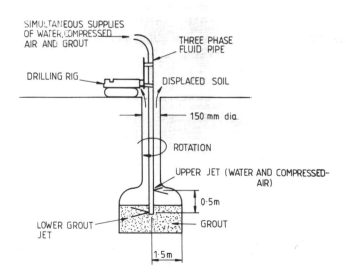

**Fig. 7.23**   Jet grouting method

diameter, into which a drill pipe is lowered. The pipe is specially designed to convey simultaneously pumped water, compressed air and grout fluid. Near the bottom end of the pipe, two nozzles are located 500 mm apart. The upper nozzle (1.8 mm diameter) delivers water at about 400 bar, surrounded by a collar of compressed air at 7 bar to produce a cutting jet. The lower nozzle (7 mm dimeter) delivers the grout at approximately 40 bar. The grouting action requires the stem to be slowly raised, whereby the excavated material produced from the jetting action is replaced by the grout and forced to the surface. The jet of water has an effective reach of about 1.5 m and thus by rotating the stem, a column of replaced earth may be formed.

### JETTING DATA

| | |
|---|---|
| Columns | 0.5–2 m diameter |
| Depth | In excess of 15 m |
| Grouting rates | 180–200 l/min |
| Water | 70 l/min at 400 bar |
| Withdrawal rate | 50 mm/min |

**Table 7.6**  Jet grouting with a water cement slurry

| Soil | W/C ratio | Compressive strength (N/mm$^2$) | Final co-efficient of permeability (mm/s) |
|---|---|---|---|
| Granular | 1 | 5–10 | $10^{-5}$–$10^{-8}$ |
| Cohesive | 1 | 1–5 | $10^{-5}$–$10^{-8}$ |

## 7.9  GROUTING EQUIPMENT (Fig. 7.24)

The basic items required are:

(i) Measuring tank – to control the volume of grout injected
(ii) Mixer – to mix the grout ingredients
(iii) Agitator – to keep the solid particles in suspension until pumped (not required for chemical grouts)
(iv) Pump (usually of the piston or diaphragm type) – e.g.

| | Pressure (N/mm$^2$) | Delivery (l/min) |
|---|---|---|
| Small pump | 0.8 | 120 |
| | 1.5 | 45 |
| Large pump | 3.5 | 450 |
| | 10 | 130 |

(v) Grout pipe, packers, casing tubes, flow meters, pressure gauges, etc.

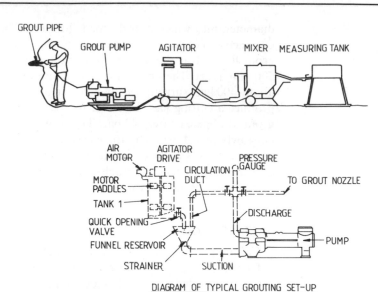

GROUT PIPE
GROUT PUMP    AGITATOR    MIXER   MEASURING TANK

AIR MOTOR    AGITATOR DRIVE
CIRCULATION DUCT    PRESSURE GAUGE
MOTOR PADDLES    TO GROUT NOZZLE
TANK 1
QUICK OPENING VALVE    DISCHARGE
FUNNEL RESERVOIR    PUMP
STRAINER    SUCTION

**Fig. 7.24**  Grouting
equipment

DIAGRAM OF TYPICAL GROUTING SET-UP

## 7.10   FINAL PERMEABILITY TESTS

After completion of grouting, further boreholes may be drilled and core samples taken and examined for grout penetration. Permeability of the treated ground may be determined from water tests conducted in a similar manner to the tests carried out during the site investigation.

# 8

# PILING METHODS

## 8.1 INTRODUCTION

Many building projects and virtually all civil engineering construction works require the use of piling equipment during the ground engineering phase. Piles must be installed to support foundations, sheet pile walls are necessary to secure excavations, coffer-dams are needed to provide a dry site for river and coastal works, etc.

Usually the piles are driven into position with a pile hammer, except for those types of pile which are formed in situ. During pile-driving both pile and hammer must be temporarily held in place either from a crane jib, pile frame or from leaders. Also, piles for temporary works must often be removed after completion of the work, and a pile extractor is therefore required. All these aspects will be considered in turn in this chapter.

## 8.2 PILE HAMMERS

Types of hammer
- (i) Drop
- (ii) Single-acting steam, compressed air, or diesel
- (iii) Double-acting steam, compressed air, or diesel } impact hammers
- (iv) Hydraulic
- (v) Vibratory
- (vi) Hydraulic sheet pile type

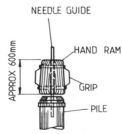

**Fig. 8.1** Hand-driven hammer

NEEDLE GUIDE

HAND RAM

APPROX 600mm

GRIP

PILE

### Drop Hammer

Simple hand-driven hammers (Fig. 8.1) have been used to drive light timber piles to shallow depths in soft earth, but today this method would seldom be selected. The more common choice is the power-assisted drop hammer (Fig. 8.2). The hammer weight is hung

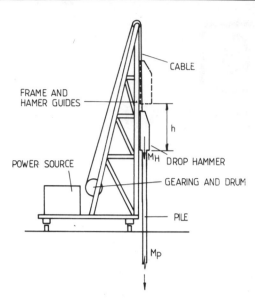

**Fig. 8.2** Drop hammer

from a rope or cable running over a pulley to a rope drum. The whole arrangement is supported on a strong frame or leader. The hammer is released by a manual trip and drops under free fall. It is raised by winching powered by a diesel or electric motor and suitable reduction gearing. About 10–20% of the potential energy is lost in the drag of the cable. This method is fairly slow, but is simple, requires little maintenance or specialist operators and is commonly used for all types of pile-driving.

## 8.3 THEORETICAL PRINCIPLES OF THE FREE FALL HAMMER (INELASTIC BODIES)

### Blow efficiency (Fig. 8.2)

The impact energy delivered by the blow is equal to the potential energy of the hammerhead, i.e.

$$E_i = M_H \times h \times g \tag{8.1}$$

where   $E_i$ = impact energy per blow
$M_H$ = mass of the hammerhead or ram
$h$ = height of drop
$g$ = force of gravity

Also, according to the principles of impact momentum the sum of momenta of hammerhead and pile before impact is equal to the combined momentum of pile and hammerhead after impact.

Thus   $V(M_H + M_p) = M_H V_H + M_p V_p \tag{8.2}$

where $M_H$ = mass of the hammerhead
$M_p$ = mass of the pile
$V_H$ = velocity of hammerhead before impact
$V_p$ = velocity of pile before impact
$V$ = common velocity of hammerhead and pile immediately after impact (true only for the inelastic condition)

But $V_H = \sqrt{2gh}$ and $V_p = 0$

Thus
$$V = \frac{M_H \times \sqrt{2gh}}{(M_H + M_p)} \tag{8.3}$$

The kinetic energy after impact is $E_k$ and

$$E_k = \tfrac{1}{2}(M_H + M_p) \times V^2 \tag{8.4}$$

by substituting $V$ from (8.3) into (8.4)

$$E_k = \frac{M_H^2 \times g \times h}{M_H + M_p} \tag{8.5}$$

The efficiency of the blow $\eta$ is

$$\eta = \frac{\text{transmitted energy } E_k}{\text{input energy } E_i} \tag{8.5a}$$

Thus
$$\eta = \frac{M_H^2 \times g \times h}{(M_H + M_p)} \times \frac{1}{M_H \times h \times g}$$

$$= \frac{M_H}{M_H + M_p} \text{ or } \frac{W_H}{W_H + W_p} \tag{8.6}$$

*Note*: $W = M \times g$.

Thus it can be seen that when the weight of the pile and hammerhead are equal, the blow efficiency is only 50%. Good practice therefore requires that for effective pile-driving the weight of the hammerhead should be at least equal to the weight of the pile. When the ratio of hammerhead weight to pile weight falls below $\tfrac{1}{3}$, the driving effectiveness is seriously impeded.

It can also be seen in equation (8.1) that the input energy is proportional to the product of hammerhead weight and height of fall and may be increased by changing one or both of these factors. The energy transmitted is affected as follows: The energy wasted is $E_i - E_k$

$$E_i - E_k = 0.5 M_H V_H^2 - 0.5(M_H + M_p)V^2 \tag{8.7}$$

Substituting for $V$

$$E_i - E_k = 0.5 M_H V_H^2 - 0.5(M_H + M_p)\frac{M_H^2 V_H^2}{(M_H + M_p)^2}$$

$$= \frac{M_H M_p}{2(M_H + M_p)} \times V_H^2 \tag{8.8}$$

$$= \frac{M_H M_p}{(M_H + M_p)} \times gh \tag{8.9}$$

For a partially elastic pile with a coefficient of resistance ($e$), the corresponding equation is

$$E_i - E_k = \frac{(1 - e^2) M_H \cdot M_p}{2(M_H + M_p)} \times V_H^2 \tag{8.10}$$

*Note*:
  Totally inelastic  $e = 0$
  Totally elastic   $e = 1$

Thus a heavy hammerhead with a low velocity produces a higher blow efficiency than a light hammer with high velocity.

Also, when driving piles wich are easily shattered, e.g. concrete, it is sensible to choose a heavy hammer with a low drop, thus reducing the impact velocity for a given level of input energy.

Because the drop hammer can supply a very heavy blow, it is suited to piles with high end bearing or which must overcome high soil resistance.

## DROP HAMMER DATA

  Weight      250–3000 kg
  Height of drop   Up to 3 m
  Rate of blows   5–15 blows per minute

### Single-acting steam or compressed-air hammer

Single-acting steam or compressed-air hammers drive piles in a similar manner to the drop hammer, but the hammer is raised by steam or compressed air rather than by winching.

An example of a typical steam hammer is shown in Fig. 8.3. It consists of a part hollow piston rod and sliding cylinder. Steam or air is admitted into the piston rod through a valve, by means of a lever which is either manually operated to control the hammer speed, or set automatically to give a preselected rate. The cylinder is thus raised up the piston shaft as the air enters the chamber. The lower part of the piston rod, which is solid, passes through the base of the cylinder to rest on the pile head, thus providing a steadying effect to the pile and exerting a slight downward force. To induce the hammer blow the lever is released, thus shutting off the air inlet valve and opening the exhaust valve to cause the cylinder to fall onto the anvil at the base of the piston rod.

A safety valve is usually incorporated in the cylinder wall to prevent the hammer over-running on the upward stroke. Also, difficulties may occur in cold weather when the air or steam condenses on the inner wall of the cylinder and feed pipes, thus reducing the efficiency of the

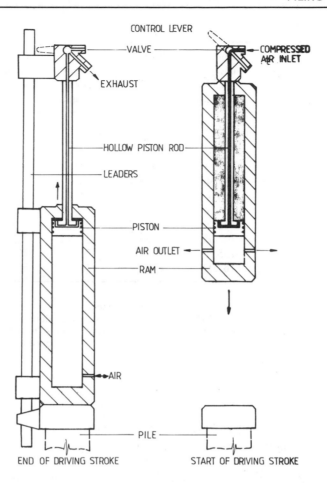

**Fig. 8.3**  Single-acting air or steam hammer

unit. The top of the piston rod is usually attached by rope to a winching mechanism, for raising and lowering the hammer when changing piles. The whole unit must be attached to leaders or a pile frame on a sliding guide, to provide directional control during the driving phase.

A higher striking rate is achieved compared with the drop hammer, but like the latter, a heavy blow (i.e. high impact energy) is delivered and it is thus not suited to driving light, thin-walled or sheet piles. Typical data are given in Table 8.1.

It can be seen from the data in Table 8.1 that an appropriate ram weight may be selected to match a given weight of pile as described for the drop hammer. Indeed, for high soil resistances, hammers are currently available exceeding 12 tonnes weight.

### Single-acting diesel hammer

In recent years the diesel hammer has proved to be a successful and in many cases a better alternative to the drop, single-acting steam or

**Table 8.1** Single-acting hammer data

| Overall height (m) | Ram weight (kg) | Stroke or drop height (m) | Impact energy per blow (kN.m) | Total operating weight (kg) | Blows per minute | Compressed-air consumption (m³/min of free air) | Steam consumption per hour (kg) |
|---|---|---|---|---|---|---|---|
| 4.2 | 1000 | 1.25 | 12.50 | 1400 | 35–45 | 15 | 400 |
| 4.2 | 1500 | 1.25 | 18.75 | 1900 | 35–45 | 18 | 500 |
| 4.5 | 2000 | 1.25 | 25.00 | 2500 | 35–45 | 20 | 600 |
| 4.5 | 2500 | 1.25 | 31.25 | 3000 | 35 | 22.5 | 700 |
| 4.6 | 3000 | 1.25 | 37.50 | 3700 | 35 | 25.5 | 770 |
| 4.6 | 4000 | 1.25 | 50.00 | 4800 | 25–30 | 28.5 | 910 |
| 5.0 | 5000 | 1.35 | 67.50 | 5900 | 25–30 | 37.0 | 1090 |
| 5.1 | 6000 | 1.35 | 81.00 | 7000 | 20–25 | 40.0 | 1360 |
| 5.1 | 8000 | 1.35 | 108.00 | 9000 | 20–25 | 42.0 | 1360 |
| 5.1 | 10000 | 1.35 | 135.00 | 11000 | 20 | 42.0 | 1360 |
| 5.1 | 12000 | 1.35 | 162.00 | 13800 | 20 | 51.0 | 1360 |

*Notes*

(i) The blow rate can be increased to about 50 per minute on some models.

(ii) Single-acting hammers are available up to 140 0000 kg ram weight for special duties.

(iii) A comparable range of differential-acting hammers are still manufactured. These hammers use the steam or air pressure acting on the piston in addition to gravity. Also, a higher rate of blows is obtained compared to the single-acting version.

compressed-air hammer, because the available input energy per blow is roughly doubled for a comparable weight of ram. The action is similar to that of the air or steam hammer, but is started by first raising the ram, which is automatically tripped to fall at the top of the stroke (Fig. 8.4(a)). As the ram falls, fuel is injected onto the impact block, and after passing the exhaust ports the air trapped in the cylinder is compressed. The impact of the ram on the impact block delivers energy to the pile but also causes combustion of the highly compressed air/fuel mixture, thus imparting further energy to the pile (Fig. 8.4b). The explosion causes the piston to move upwards and the gases are expelled through the exhaust ports (Fig. 8.4c). The cycle will continue until the fuel supply is cut off. Typical data are given in Table 8.2.

A diesel hammer ram weighs less than the equivalent single-acting steam or compressed-air hammer and does not require a steam boiler or compressed-air supply. However, starting may be difficult in soft ground, when the impact of the falling ram may not be sufficient to atomise the fuel. There is also a tendency for the stroke to increase to the maximum with increasing penetration resistance. While this can be an advantage, care should be taken not to damage the pile head.

Like other single-acting hammers the diesel hammer is most suitable where a very heavy blow (i.e. high impact energy) is required. The weight of the ram should be at least equal to the weight of the pile for efficient pile-driving.

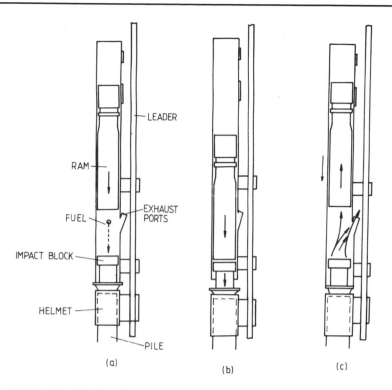

**Fig. 8.4**  Diesel hammer

**Table 8.2**  Single-acting diesel hammer data

| Ram weight (kg) | Stroke (m) | Impact energy per blow (kN.m) | Blows per minute | Total operating weight (kg) | Fuel consumption (l/h) | Overall height (m) |
|---|---|---|---|---|---|---|
| 200 | 0.6–1.3 | 1.5–2.5 | 60–70 | 350 | 0.5 | 2.0 |
| 400 | 0.8–1.6 | 2.2–5.0 | 50–60 | 650 | 0.8 | 2.4 |
| 500 | 1.2–2.0 | 12.5 | 40–60 | 1250 | 3.5 | 3.8 |
| 1000 | 1.2–2.0 | 30.0 | 40–60 | 2500 | 6.5 | 4.2 |
| 1500 | 1.2–2.0 | 37.5 | 40–60 | 3000 | 6.5 | 4.5 |
| 2250 | 1.2–2.0 | 33.5–67.0 | 35–60 | 5000 | 7.0 | 5.0 |
| 3000 | 1.2–2.0 | 45.5–91.0 | 35–50 | 6000 | 8.0 | 5.2 |
| 3500 | 1.2–2.0 | 55.0–110.0 | 35–50 | 8000 | 11.5 | 5.2 |
| 4500 | 1.2–2.0 | 70.0–140.0 | 35–50 | 9000 | 16.5 | 5.2 |
| 5500 | 1.2–2.0 | 90.0–180.0 | 35–45 | 12000 | 20.0 | 5.9 |

## Double-acting steam or compressed-air hammer (Fig. 8.5a)

The double-acting hammer usually employs air or steam, which is admitted to the upper and lower cylinders alternately by means of a valve actuated by the piston. In this way both free fall impact and additional energy from the release of compressed air into the upper cylinder are obtained on the downward stroke. By switching the air supply to the lower cylinder, the piston is raised and the air in the upper cylinder expelled, ready to repeat the cycle. The stroke of double-acting hammers is quite small, usually less than 0.3 m. They are

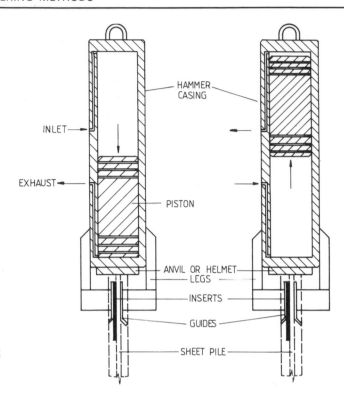

**Fig. 8.5**  (a) Double-acting
air or steam
hammer

designed with relatively light rams for pile-driving in loose soil and for
light sheet piles and thin-walled piles, which offer relatively little
resistance to penetration. Only sufficient impact energy is provided to
avoid undue damage to the top of the pile. It can be seen from Tables
8.1 and 8.3 that the largest double-acting air hammer is approximately
equivalent in size and energy to medium range single-acting hammers.
For heavier piles, double-acting diesel hammers are available with
similar energy delivery characteristics to the single-acting versions.

Double-acting hammers run at considerably greater speeds than
single-acting diesel or hydraulic hammers, thus ensuring that the pile is
maintained in almost continuous motion, which is particularly
advantageous in overcoming static friction in granular soils.
Double-acting hammers, however, are less effective in cohesive soils
such as clay, where a heavier free fall hammer may be more suitable.
Also, because of the rapid blow delivery rate, concrete piles may suffer
head damage by shattering. Finally, double-acting hammers can be
operated successfully under water, when fitted with an exhaust
extension.

### Hydraulic impact hammer

Hydraulic impact hammers (Fig. 8.5b) are gaining in popularity with
the increasing availability of hydraulic powering on modern
construction equipment. Similar advantages to double-acting air

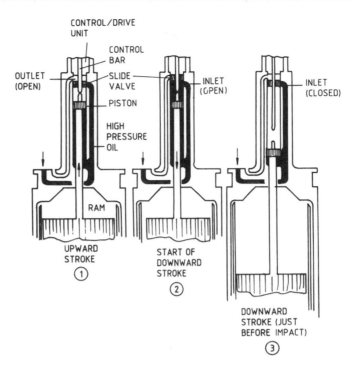

**Fig. 8.5** (b) Hydraulic impact hammer

**Table 8.3** Double-acting hammer data

| Weight of ram (kg) | Input energy per blow (kN.m) | Total operating* weight (kg) | Overall† height (m) | Air consumption (m³/min) | Steam consumption (kg/h) | Blows per min |
|---|---|---|---|---|---|---|
| 10 | 0.15 | 65 | 0.7 | 2 | 140 | 500 |
| 20 | 0.22 | 150 | 0.8 | 2 | 140 | 500 |
| 30 | 0.50 | 300 | 1.5 | 3 | 200 | 400 |
| 90 | 1.40 | 700 | 1.5 | 7 | 270 | 300 |
| 180 | 3.50 | 1300 | 1.6 | 11 | 360 | 275 |
| 260 | 5.70 | 2300 | 1.8 | 13 | 480 | 225 |
| 720 | 12.00 | 3200 | 2.5 | 17 | 610 | 145 |
| 1400 | 18.00 | 5000 | 2.8 | 21 | 680 | 105 |
| 2300 | 26.00 | 6400 | 3.4 | 25 | 950 | 95 |

\* Including legs for driving sheet piling
† Excluding legs
Air pressure 6.5 bar (90 psi)
Steam pressure 10 bar (140 psi)

hammers can be enjoyed as the hydraulic fluid is used to both raise the ram and provide additional driving force to gravity. Hammers for construction duties vary from about 15–200 kN.m of impact energy with corresponding ram weights 3000 to 10 000 kg. Larger hammers for offshore work are available up to 3000 kN.m or so. The blow rate may be increased up to about 100 blows per minute with similar stroke lengths to single acting hammers.

### 8.4   THEORETICAL ENERGY DELIVERY OF DOUBLE-ACTING HAMMER

$$E_i = [W_H + (A_p \times p)] \times h \qquad (8.8)$$

where
$E_i$ = impact energy per blow
$W_H$ = ram weight
$A_p$ = surface area of piston
$p$ = average pressure in cylinder
$h$ = stroke

Energy delivered per minute = $E_i \times n$, where $n$ is the number of blows per minute.

The input energy ($A_p \times p \times h$) indicated in eqn. 8.8, simply increases $E_i$ compared to the free fall condition.

Thus from equation (8.5a) it can be seen that the blow efficiency ($\gamma$) may be increased by using a heavy hammer ($M_H$) relative to the pile weight ($M_p$).

### Vibratory pile-driver

In recent years there has been a demand for a quiet method of driving piles, for example near hospitals and in residential areas. For sheet piles and other types of steel pile a vibratory technique has proved successful in granular type soils.

The vibratory pile-driver (Fig. 8.6) consists of an electric or hydraulic motor coupled to two eccentrically and separately mounted cams which rotate in synchronised opposition to each other. The whole unit is housed in a steel casing and for operation is suspended from the lifting rope of a crane. A suspension bracket, spring-mounted to the pile casing, eliminates upward vibration into the rope.

The vibrator may be attached to the pile by means of remotely controlled, hydraulically-operated grips.

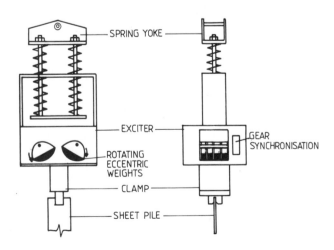

**Fig. 8.6**   Vibratory pile-driver

**Table 8.4** Vibratory pile data

| Max. dynamic force (kN) | Suspended weight (kg) | Power (kW) | Max frequency (Hz) | Max. crane pull for use as an extractor (tonnes) | Duties |
|---|---|---|---|---|---|
| 200 | 1200 | 50 | 35 | 7 | Light duty sheet piles etc. |
| 400 | 2000 | 100 | 25 | 20 | Light duty sheet piles, open-ended tubes up to 500 mm diameter |
| 650 | 4000 | 175 | 25 | 30 | Medium duty sheet piles, columns, open-ended tubes up to 6 tonnes wt |
| 1400 | 10000 | 350 | 25 | 40 | Heavy duty sheet piles, beams, open-ended tubes up to 20 tonnes wt |
| 2000 | 16000 | 550 | 20 | 80 | Heavy columns, large diameter tubes, generally heavy sections |
| 4000 | 30000 | 800 | 20 | 80 | Very heavy sections |

*Note*
Vibratory pile-drivers may be powered electrically from the main or a generating set (400–450 V), or hydraulically from a diesel engine

The action on the pile is produced by the rotation in unison of the two eccentric cams. During each cycle the centrifugal forces act downwards at 0° and 360°, and upwards at 180° from the vertical, while at all other positions the opposing horizontal components of their respective forces are cancelled. The pile is thus momentarily shaken up and down, thereby reducing the static friction between the soil and pile. Thus when the end bearing area of the pile is small, the downward force produced by the self-weight of the driver and pile causes the pile to sink into the soil.

The effectiveness of the vibrator can be improved by regulating the frequency (i.e. altering the rotation speed of the cams) to suit the particular soil conditions and to avoid resonance in nearby structures. The amplitude of vibration may also be adjusted by altering the mass of the eccentric weights.

The vibratory pile-driver is most efficient in water-bearing sands, and is effective in most loose to medium sands and gravels. Some difficulty may be encountered in very dense granular material, where the downward dynamic force may not be sufficient to produce adequate penetration of the pile. The method is less effective in cohesive soil. However, the Bodine Sonic pile-driver, and more recent hammers, which operate at frequencies up to 150 hertz (9000 cpm) and thereby

put the pile at its resonant frequency, have been used in fairly stiff clay. The result is a minute and momentary alternate increase and decrease in the cross-section thickness of the pile thus breaking the cohesion between pile and soil. The downward driving force is simply supplied by the pile-driver and
self-weight of the pile.

The main advantage of the vibratory pile-driver is the relative lack of noise and absence of exhaust fumes. In addition, the pile head suffers little damage compared to hammer methods. The method can also be used to extract piles, by simply providing an upwards pull of perhaps 15–25 tonnes from the crane rope.

### Hydraulic sheet pile-driver

While the vibratory pile-driver is suitable for 'silent' driving most types of pile in sands, a quiet pile-driver is often required for cohesive soils. In the early 1960s Taylor Woodrow Construction Ltd. set about solving the problem for sheet piling, and developed the Taywood Pilemaster (Fig. 8.7). The Pilemaster consists of an electric motor, hydraulic pumps, fuel tank, etc. mounted on a crosshead fabricated from steel plate. Eight hydraulic jacks are attached to the base flange of the crosshead. Each jack works on the principle as shown in Fig. 8.8, where oil at a pressure of approximately 62 N/mm$^2$ (620 bar or 9000 psi) is introduced into the cylinder through a valve, thereby raising or forcing the piston downwards for pile-driving. Each jack is connected to a friction plate by a fork-shaped high tensile forged steel connector (Fig. 8.9). The friction plate is then bolted to the pile. The connector head may be rotated to accommodate attachment to most

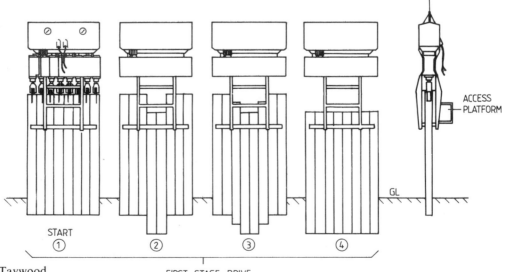

**Fig. 8.7** Taywood Pilemaster

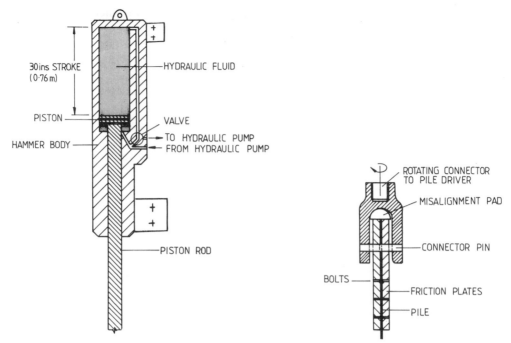

**Fig. 8.8** Hydraulic pile-driver

**Fig. 8.9** Pile coupling to the Taywood Pilemaster

shapes of pile and the hemispherical seating permits up to 3°
misalignment between the pile-driver and piles.

### PILE-DRIVING METHOD

The Pilemaster drives in panels of seven or eight piles, and bends or
junctions must usually be overcome with specially shaped piles. Piles
are normally pitched in a temporary frame to give initial support (see
Fig. 8.39). The pile-driver itself weighs about 10–12 tonnes and is hung
from a crane jib. Guide legs maintain the machine in alignment with
the pile, and driving usually starts with the centre piles, working out to
the end piles in the panel (see Fig. 8.7). When all the piles have been
driven the full stroke of the rams, the Pilemaster is lowered to the new
level and the procedure repeated.

The initial reaction for driving the first pair of piles is obtained from
the weight of the Pilemaster and the weight of the panel of piles
attached to it. The friction and restraint provided by the first pair are
then used to give additional restraint to drive the next pair. A
progressive accumulation of frictional restraint is therefore generated as
the piles penetrate deeper into the soil, until the restraint is almost
entirely frictional.

Support from the crane may be dispensed with when the panel
becomes self standing, thereby freeing lifting capacity for pitching the
adjacent panel. If the driving is to continue to ground level, piles

should be initially pitched in a shallow trench to facilitate removal of the friction plate.

The Pilemaster has proved successful in loose to medium dense fine granular sands and in cohesive soils such as clay. Coarse granular material offers too high a toe resistance for efficient pile-driving.

The load supplied to each jack is about 250 tonnes, which enables driving depths of up to 20 m in favourable soils.

The driving speed varies according to the driving resistance encountered, e.g. 0.5 m/min at high pressure, to 1 m/min at low pressure operation. The Pilemaster can be operated with most types of sheet pile and with suitable support rakes approaching 1 in 5 may be obtained. In addition the machine may be used to extract sheet piles from all types of soil.

## PILEMASTER DATA

| | |
|---|---|
| Pile weight | 10–12 tonnes |
| Power | 30 kW |
| Voltage | 400–450 volts |
| Driving force per ram | 200–250 tonnes |
| Length of pile | 15–20 m |

### Pile extractors

The most common type of extractor operates on the principle of the inverted pile hammer using steam or compressed air. In the example shown in Fig. 8.10, air is admitted to the upper cylinder via a valve causing upward movement of the casing until the base of the cylinder meets the underside of the ram, thereby transferring an upward blow to the pile. The whole unit is hung from a crane or frame via a spring mechanism to prevent vibrations being transferred to the lifting rope.

**Table 8.5**  Pile extractor data

| | Input energy per blow (kN/m) | Weight of ram (kg) | Air* consumption (m³/min) | Steam† consumption (kg/h) | Blows per min |
|---|---|---|---|---|---|
| Light duty | 0.35 | 10 | 3.5 | — | 1000 |
| | 0.75 | 20 | 5.5 | — | 800 |
| | 1.50 | 40 | 10.0 | — | 700 |
| | 3.00 | 80 | 10.0 | — | 400 |
| Heavy duty | 8.00 | 500 | 10.0 | 350 | 150 |
| | 11.00 | 750 | 10.0 | 500 | 140 |
| | 16.00 | 1750 | 13.0 | 600 | 120 |
| | 30.00 | 3500 | 16.0 | 800 | 90 |

* 6.5 bar (90 psi)
† 10 bar (140 psi)

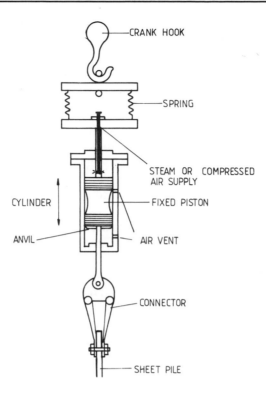

**Fig. 8.10** Pile extractor

Pile extractors are available for removing light trench sheeting up to to timber and heavy sheet and H piles.

Other methods of pile extraction include vibration and hydraulic jacks.

### Pile-driving assisted by water jet

Work done in Japan has demonstrated the application of a water jet in assisting the driving of sheet piles through mudstone (Fig. 8.11) with a vibratory hammer. Water is pumped through a 25 mm diameter pipe attached to the side of the pile with a water jet, giving a 30° angle of

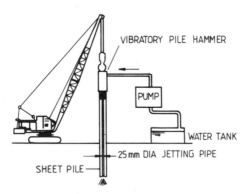

**Fig. 8.11** Jetting assistance for sheet pile-driving

spread. The method can be used to break up the hard material ahead of the tip of the sheet pile, and it is claimed that the rate of driving may be doubled compared with using the vibro hammer by itself. Water flow of $5-15 \, \text{m}^3/\text{h}$ is necessary.

### Noise

Noise is a particularly unpleasant feature of pile hammers and manufacturers have given some attention to noise reduction, but as yet with only moderate success. The principal method of reducing noise is totally to enclose the hammer, leaders and pile in a large soundproof box, the leaders being used to form the box framework. Clearly, this method is cumbersome, and cannot be easily adapted for sheet piling. More satisfactory methods use vibration or hydraulic methods, but these are generally more expensive.

## 8.5  PILE HELMETS AND DOLLIES

### Helmets

A pile helmet is required to distribute the blow from the hammer evenly to the head of the pile, to cushion the blow and to protect the pile head itself.

Helmets with a relatively thin base can be used with solid piles (Fig. 8.12) because the load is transferred across the whole pile head. However, pipes, sheet piling and other thin-walled piles require a much thicker base (Fig. 8.13) to ensure that the impact energy is directly transferred into the pile wall. The helmet should have about 10 mm clearance around the pile (Fig. 8.14) to avoid damage to the pile head.

### Dollies

The dolly is placed in a recess in the helmet and acts to cushion the blow (Fig. 8.14). The effect is to extend the impact time by storing the impact energy in the dolly, as indicated in Fig. 8.15. By selecting the appropriate material, a better transfer of energy from hammer to pile can be obtained in different ground conditions.

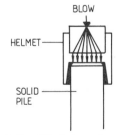

**Fig. 8.12**  Pile helmet for solid piles

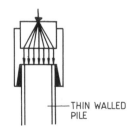

**Fig. 8.13**  Pile helmet for thin-walled piles

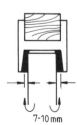

**Fig. 8.14**  Dolly used to cushion the blow

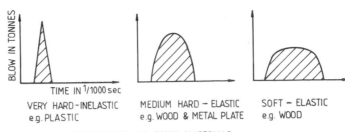

**Fig. 8.15** Impact times for dolly materials (Delmag)

IMPACT TIMES FOR DOLLY MATERIALS

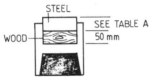

**Fig. 8.16** Composite dolly comprising steel plate and resin bonded fabric

| TABLE A | STEEL PLATE DIMENSIONS FOR DOLLIES | |
|---|---|---|
| BLOW ENERGY kNm | DOLLY SURFACE AREA (m²) | STEEL THICKNESS (mm) |
| 35 | 0·16 | 100 |
| 100 | 0·30 | 150 |
| 150 | 0·40 | 200 |
| 200 | 0·70 | 300 |

**Fig. 8.17** Composite dolly comprising steel plate and wood

**Fig. 8.18** Wooden dolly

For medium to heavy penetration resistances, a plastic or resin-bonded fabric dolly with a steel plate is preferred (Fig. 8.16). Alternatively, a wooden dolly made from oak or similar material capped with a steel plate can be used; this combination tends to give more recoil (Fig. 8.17). For light to medium driving, a simple wooden dolly of beech or elm is adequate (Fig. 8.18).

Plastic dollies may last for several hundred piles in moderate driving conditions, but unfortunately they produce a much harsher effect on the pile head, which may cause damage. Wooden dollies, however, tend to disintegrate quickly (i.e. 2–5 piles) if used in heavy driving conditions, and if not replaced when smoking and burning is observed, much of the impact energy will be absorbed, thus reducing the effectiveness of the blow.

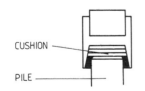

**Fig. 8.19** Packing used to protect pile head

### Packing (Fig. 8.19)

Concrete piles and other materials are likely to suffer damage from the force of impact and require a cushion placed between the pile head and underside of the helmet. In this way, peak impact forces are avoided and the forces are transmitted uniformly into the pile.

Packing plates (Fig. 8.20) of 25–30 mm minimum thickness made of soft plywood are suitable in soft ground conditions; alternatives are paper bags filled with sawdust. For hard driving conditions asbestos fibre has proved adequate.

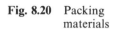

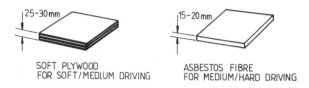

**Fig. 8.20** Packing materials

SOFT PLYWOOD
FOR SOFT/MEDIUM DRIVING

ASBESTOS FIBRE
FOR MEDIUM/HARD DRIVING

## 8.6 LEADERS AND PILE FRAMES

In order to drive piles accurately, it is necessary to pitch and hold the pile in position, support the hammer and guide both the pile and hammer along the required drive angle. The choice of method varies according to weight of hammer and pile, ground conditions, length of pile, and manoeuvrability demanded. Depending upon these factors pile frames, hanging leaders or rope-suspended leaders may be selected.

### Hanging leaders (Fig. 8.21)

Hanging leaders consist of a sturdy lattice-framed mast attached to the jib of a crane. The rake of the mast may be adjusted to suit forward, backward and lateral piling angles, thereby allowing piling to take place on uneven or sloping ground. The mast is moved into the desired position by means of hydraulically-controlled rams attached between the base of the jib and mast. The machine requires separate winches to lift and pitch the pile and to hold the hammer. The hammer is usually attached to guides on the leader as shown in Fig. 8.22. Guides are also necessary to control the direction of the pile during driving (Fig. 8.23) especially when using slender piles, when flexing may occur. Wherever possible, it is desirable to attach the helmet to the leader guides, both

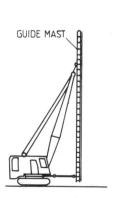

**Fig. 8.21** Hanging leaders

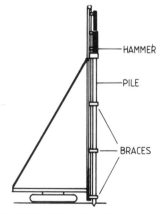

**Fig. 8.22** Use of guides to secure pile

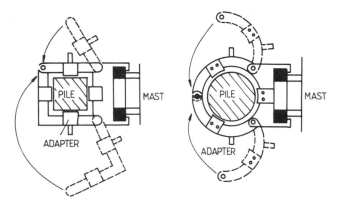

**Fig. 8.23** Guide clamp details (Delmag)

to provide directional restraint and to avoid losing the helmet during recoil in hard driving conditions.

The crawler-mounted hanging leader is particularly useful for driving on uneven or poor ground and in awkward positions. The weight of the hammer, pile and leaders must be within the lifting capacity of the crane, thereby limiting choice of the method to medium size duties.

## EXAMPLES OF HANGING LEADERS DATA

|  | *Small rig* | *Large rig* |
|---|---|---|
| Height (H) | 12 m | 30 m |
| Hammer | 800 kg | 13 000 kg |
| Pile | 700 kg | 15 000 kg |
| Leader | 650 kg | 15 000 kg |
| Forward and rear rake | 1:3 | 1:3 |
| Lateral rake | 1:3 | 1:3 |

### Pile frame (Fig. 8.24)

When piles are either too long or too heavy for hanging leaders, or where it is convenient to mount the pile rig on rails, the pile frame is often the most suitable method to produce accurate directional control. The method is similar to that for hanging leaders and small economical rigs or large frames are available.

Modern frames consist of a leader supported by two adjustable struts to allow forward and backward raking. The mast can be extended at the bottom to facilitate pile-driving over water.

The base frame rests on four swivel steel castors so that the rigs may be moved around on hard ground or pads. However, for installing rows of piles, the manoeuvrability of the rig can be improved by placing the frame on a turntable to permit 360° turning.

The rig is equipped with winches and the whole unit can be erected in about $\frac{1}{2}$ day by four workers.

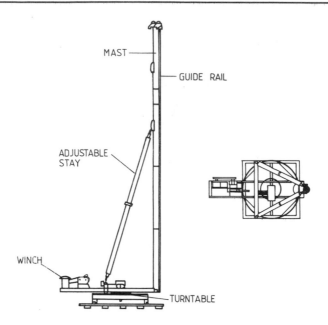

**Fig. 8.24**  Pile frame

## EXAMPLES OF PILE FRAME DATA

|  | *Small frame* | *Large frame* |
|---|---|---|
| Height | 12 m | 50 m |
| Weight of hammer, plus pile | 6000 kg | 40 000 kg |

### Rope-suspended leaders (Fig. 8.25)

Piles supported in temporary framework (Fig. 8.26) are usually driven from rope-suspended leaders. Indeed, such leaders may be used to pitch and drive piles without the temporary support (Fig. 8.27) but there is then more danger that the pile will veer off the required direction if obstructions are encountered. Rope-suspended leaders are suitable for driving on a forward or lateral rake (Fig. 8.28).

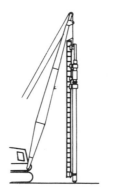

**Fig. 8.25**  Rope suspended leaders

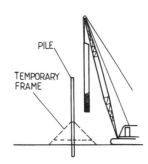

**Fig. 8.26**  Temporary frame to hold a pile

**Fig. 8.27** Pile supported by rope suspended leader

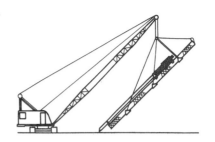

**Fig. 8.28** Pile driven on the rake with rope suspended leaders

## EXAMPLE OF ROPE-SUSPENDED LEADERS DATA

|  | *Small* | *Large* |
|---|---|---|
| Height | 12 m | 30 m |
| Weight of hammer | 1000 kg | 10 000 kg |
| Weight of leader | 600 kg | 7000 kg |
| Weight of pile | Varies | Varies |

Like the hanging leader, it size is governed by the lifting capacity of the crane.

## 8.7   HANDLING PILES

### Dragging

A pile should never be dragged along the ground from the top of the leader as this may cause overturning of the rig. Always use the bottom pulling technique (Fig. 8.29).

### Pitching

The pile should be raised into position carefully and lifted from about the third point (Fig. 8.30). These requirements are especially important for concrete piles which can easily crack.

### Transporting

To avoid fracture or cracking, the pile should be lifted and carried at the points shown in Fig. 8.31.

If flexing is excessive, extra pick-up points can be provided from a temporary beam (Fig. 8.32).

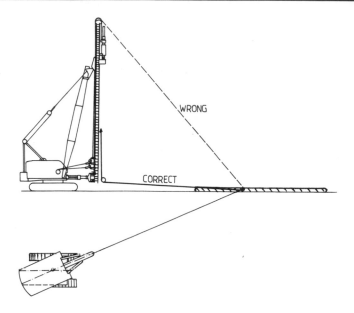

**Fig. 8.29** Dragging a pile ready for positioning in the frame

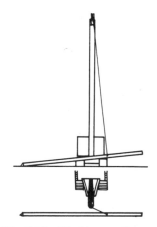

**Fig. 8.30** Pitching a pile

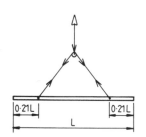

**Fig. 8.31** Transporting a pile

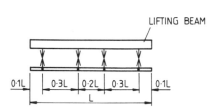

**Fig. 8.32** Transporting a long pile using a lifting beam

## SHEET PILES

Until fairly recently timber was almost universally used for shoring during excavation operations, but the escalation of prices for wood, coupled with the development of more rapid systems of erection, has forced engineers to turn to steel, and in particular to sheet piles. The modern sheet pile has several advantages, for instance:

   (i) high strength is combined with light weight
   (ii) the pile can be used many times over
   (iii) the pile can be driven much more quickly than a timber pile
   (iv) standard shapes and sizes are available to the designer
   (v) steel may be used for either permanent or temporary works

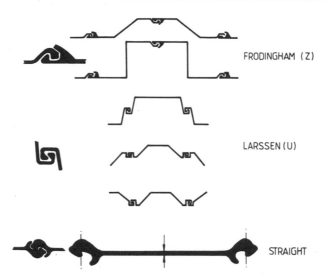

FRODINGHAM (Z)

LARSSEN (U)

STRAIGHT

**Fig. 8.33**  Sheet pile types

(vi) piles may be interlocked for rigidity and reasonable watertightness

(vii) piles can be extended by welding

(viii) long piles can be driven with heavy pile hammers

(ix) piles are easy to handle and store on site

(x) a range of special piles can be made for interlocking corners, junctions, etc.

(xi) the pile is strong along its length in bending, allowing internal bracing inside coffer-dams to be replaced with ground anchors.

Basically sheet piles are manufactured U-shape, Z-shape, and straight (Fig. 8.33).

Both Larssen U and Frodingham Z piles are equally suitable for most uses and, in general, contractors select the type most readily available. Applications include bulkheads, coffer-dams and retaining walls. Greater strength in bending can be obtained from strengthened piles e.g. composite construction (Fig. 8.34) or interlocking H piles. Straight web piles have little strength in bending and are mainly used for cellular coffer-dams, where interlocking piles are required to provide strength in tension. This type of structure is covered more fully in Chapter 9.

### Interlocks

The interlocks must be fairly loose to allow free sliding during driving. Consequently the joints cannot be made completely watertight. However, grease applied to the joint before driving helps to reduce friction and improve the driving, and also attracts fine particles into the joint.

Interlocks permit a deviation of only about 10° between adjacent

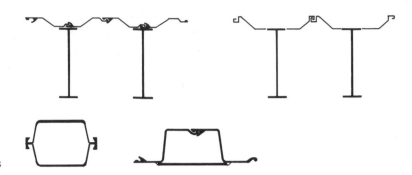

**Fig. 8.34** Composite piles

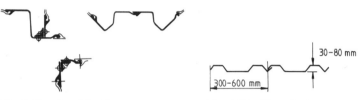

**Fig. 8.35** Special piles          **Fig. 8.36** Trench sheets

piles, hence changes in wall direction generally require specially pre-bent piles to produce a 'watertight' structure.

### Special piles

Special piles can be obtained for any angle up to 90°, (Fig. 8.35). Tee, cross and Y piles fabricated by riveting or welding sections together are also illustrated.

### Trench sheeting

The U pile is available with very shallow sections for use as trench sheeting (Fig. 8.36). Applications include supports to the sides of shallow trenches and excavations, temporary retaining walls along rivers and canals, etc. The piles are normally lap-jointed only, to aid speedy installation, but interlocking sections are available, and mild steel or high yield steel sections may be used.

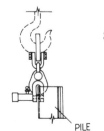

**Fig. 8.37** Lifting bracket for sheet piles

### 8.8 SHEET PILE-DRIVING

### Pitching and guiding

Sheet piling is driven in panels of 6 to 10 pairs of piles in order to maintain accuracy in both the vertical and horizontal directions. Each pile is usually supplied with a hole drilled near the top to which a 'quick-release' shackle (Fig. 8.37) can be attached. A crane is then normally used to lift and pitch the pile.

Forming the initial interlock can be troublesome, especially in windy conditions, but this problem can largely be overcome by attaching the lightweight aluminium guide bracket developed by the Dawson Company (Fig. 8.38). A sturdy frame made of heavy timbers (Fig. 8.39) acts to guide the pile and to counter wind forces. Walings should be provided at ground level with a second set as far away as practicable from the first to provide rigidity and directional restraint during driving.

### Pile-driving procedure

The piles are usually pitched in pairs to form a panel and the first and last pairs are partially driven first, which helps to prevent creep, as there is some play in the pile interlocks. The remaining piles are then driven down to the level of the top waling. The hammer is hung from a crane boom and usually a leader is required when using a diesel or single-acting hammer. This can be avoided with the double-acting hammer, where integral guide legs (Fig. 8.5) keep it on the pile and aid vertical driving. Sheet piling is quickly damaged if the hammer is allowed to lean over.

A panel of piles is often sufficiently stable on reaching the top waling for pile-driving to continue to ground level with this waling removed. For more accurate work, however, either hanging leaders or even a pile frame should be used for the final stage of ramming. When long runs of piles are involved it may be necessary to have several sets of guide frames to allow the following panels to be pitched and driven in stages as demonstrated in Fig. 8.40.

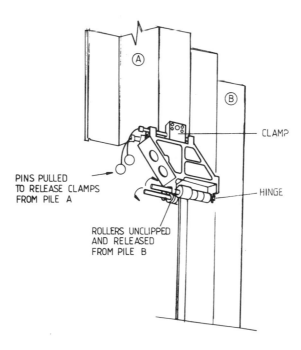

PINS PULLED
TO RELEASE CLAMPS
FROM PILE A

CLAMP

HINGE

ROLLERS UNCLIPPED
AND RELEASED
FROM PILE B

**Fig. 8.38** The Dawson sheet pile guide bracket

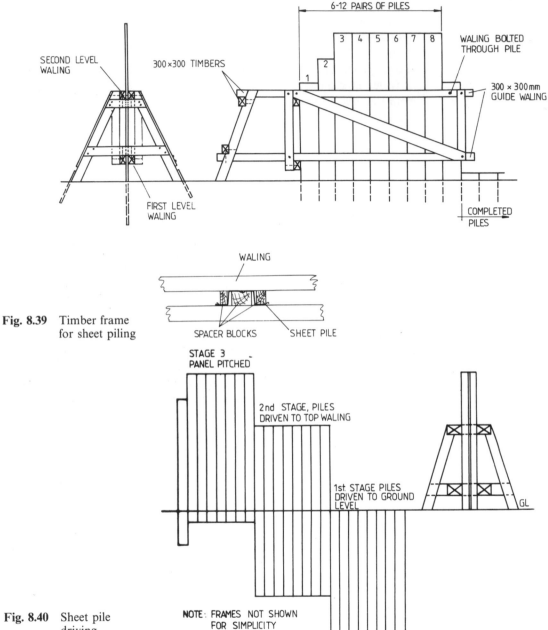

**Fig. 8.39** Timber frame for sheet piling

**Fig. 8.40** Sheet pile driving procedure

**NOTE**: FRAMES NOT SHOWN FOR SIMPLICITY

### Piling problems

(i) There is a tendency for the play in the interlock to cause leaning. This can be countered by placing the hammer slightly off centre or alternatively by pulling at the top of the pile (Fig. 8.41). If the leaning cannot be eliminated then the only recourse is to obtain

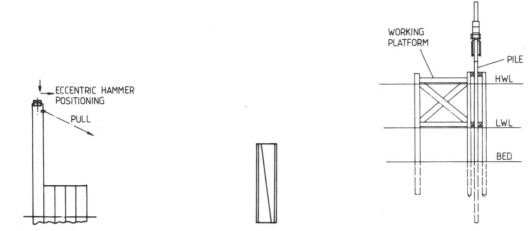

**Fig. 8.41**  Correcting off-line piles   **Fig. 8.42**  Special taper sheet piles   **Fig. 8.43**  Driving piles over water

special taper piles (Fig. 8.42) which may involve considerable procurement delays.

(ii)  Piles also tend to lean inwards during driving, and this can be countered by providing spacer blocks (Fig. 8.39) between the waling and pile face.

(iii)  When driving piles in soft material such as clay, previously driven adjacent piles might be dragged down. This can be prevented by cross-bolting the piles or bolting them to the waling. Overdriven piles should be jacked back into position.

(iv)  Limited head room may be overcome by driving short length piles and welding on extension pieces.

(v)  When driving piles below the water level, hammers can be equipped to work underwater.

(vi)  Piles often get damaged during driving, and also, because it is seldom possible to drive all the piles in a row to exactly the same level, sufficient pile should be left to allow burning off at 100 mm below the damaged parts.

### Piling over water

If a temporary platform can be erected to support a crane, then piling may be carried out using a rope-suspended hammer with the platform substructure acting as the guide frame for the panels of sheet piles (Fig. 8.43). Alternatively, where manoeuvrability is required, a pontoon-mounted pile frame or crane fitted with hanging leaders, incorporating a long dolly between the pile cap and hammer, may be preferred. Positioning must then be obtained by winching from temporary anchorages. Currents and tide unfortunately pose serious problems for this latter method.

## 8.9 PILING DATA

### Weight of hammer

$W$ = weight of hammer
$P$ = weight of pile

| | |
|---|---|
| Concrete piles | $W \simeq P$ |
| Timber piles | $W \simeq 2P$ |
| Steel tubes ⎫ | $W = 0.5P$ to $2.0P$ in dry sand |
| and ⎬ | $= 2.0P$ to $2.5P$ in saturated sand |
| Sheet piles ⎭ | $= 2.5P$ to $3.0P$ in clay |
| I-shaped piles | $W = P$ |

### Hammer/pile type selection

| | |
|---|---|
| Timber piles | – drop or single-acting hammer |
| Concrete piles | – drop or single-acting hammer with fall or stroke less than 0.5 m |
| Steel or sheet piles | – double-acting hammers giving a rapid blow rate. For silent driving, vibration or hydraulic methods may be adopted depending upon the type of soil. |

*Note*: These are the preferred types of hammer, but the demand for a very heavy hammer or for a diesel or hydraulic impact hammer giving high impact energy may be overriding factors with heavy piles and/or hard soils.

### Preferred hammer/soil type selection

| Light soils | Loose sands and gravels ⟶ Double-acting hammer (rapid blows needed) |
|---|---|
| | Soft clay ⟶ Drop or single-acting hammer |
| Medium soils | Medium dense sands Fine gravels → Double-acting hammer (rapid blows needed) |
| | Stiff clay ⟶ Drop or single-acting hammer |
| Hard soils | Dense sand ⟶ Double-acting hammer (rapid blows needed) |
| | Coarse gravel ⟶ Single-acting hammer or diesel hammer |
| | Very stiff clay ⟶ Drop or single-acting hammer |

*Note*: For heavy piles, the range of double-acting air hammers may be too small and single-acting hammers may be the only alternatives. Hydraulic impact hammers are alternatives to either drop or single-acting hammers in the above guidelines.

## 8.10  PRODUCTION RATES OF PILE-DRIVING

Many factors must be included, such as the time needed for setting up the pile frame or rigs, the skill of the work gang, weather losses, soil types, length of pile, type of pile hammer, etc., and it is therefore extremely hazardous to recommend typical driving rates. However, as a very general guide the following data may help in planning piling operations.

|  | | | *Single piles* (m/h) | *Sheet piles* (m²/h) |
|---|---|---|---|---|
| Light soils | ⎱ | using the appropriate hammer | 5–15 | 3–10 |
| Medium soils | ⎰ | | 3–9 | 2–6 |
| Heavy soils | | | 1.5–5 | 1–3 |

For sheet piling using the Taywood Pilemaster, up to 20 m²/h can be achieved in soft ground. Using vibratory methods, over 30 m²/h have often been recorded. Extracting piles with a compressed-air pile extractor requires about half the time required for driving.

### Piling gang

One ganger, one crane driver or winch operator, and two labourers to grease, prepare and handle piles, fix support brackets, etc. An additional labourer may be needed to assist with assembly when using a temporary timber guide frame. Transport should also be provided to bring the piles from the storage compound.

# 9

# SHORING SYSTEMS

## 9.1 INTRODUCTION

When carrying out excavation work on a congested construction site the sides of an excavation frequently require shoring (Fig. 9.1). When space is not at a premium, however, the sides are often left open and simply battered back (Fig. 9.2).

The choice of method depends upon technical factors, such as the depth of the foundation, the available space, the nature and permeability of the soil, the depth of the water table and economic considerations related to the period the excavation must be left open, the availability of labour, plant and materials and the construction programme. Table 9.1 provides an indication of the methods required in various ground conditions.

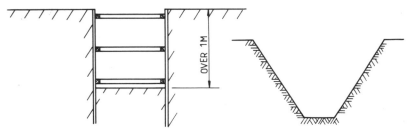

**Fig. 9.1**  Simple shoring method    **Fig. 9.2**  Open cut excavation

## 9.2 OPEN EXCAVATIONS (Fig. 9.3)

Slopes excavated in *dry sands* and *gravels* quickly find their natural angle of repose. The angle largely depends upon the degree of compaction and shape of the grains.

**Table 9.1** Guide to ground classification and suggested shoring method (based on original by the CITB)

| TRENCH DEPTH | UP TO 5FT | 5FT–15FT | OVER 15FT | SOUND ROCK | FISSURED ROCK | FIRM AND STIFF CLAYS | COMPACT GRAVELS SANDS | SLIGHTLY CEMENTED | FIRM PEAT | BELOW WATER TABLE GRAVELS SANDS | LOOSE | SOFT CLAYS AND SILTS | SOFT PEAT |
|---|---|---|---|---|---|---|---|---|---|---|---|---|---|
| SHALLOW | ░ | ▨ | ▨ | A | A | B-C | A | A | A | C | C | C | C |
| MEDIUM | | ▨ | ▨ | A | A | B-C | B | B | C | C | C | C | C |
| DEEP | | | ▨ | A | B | C | C | C | C | C | C | C | C |

A. NO SUPPORT MAY BE NECESSARY.   B. OPEN SHEETING.
C. CLOSE SHEETING OR BETTER

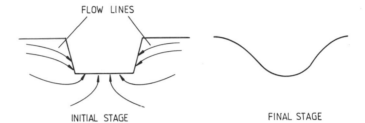

| | H:V |
|---|---|
| Solid rock, shale or cemented sand + gravel | 1:1 |
| Compacted angular gravels | 0.5:1 |
| Moist soil, drained clay or rubble | 1:1 |
| Rounded gravels and shingles | 1.25:1 |
| Dry sand | 1.5:1 |
| Dry soil | 1.75:1 |
| Well rounded loose sand, mixed gravel and sand | 2:1 |
| Wet sand | 2.5:1 |
| Wet clay | 3.5:1 |

Approximate angles of repose in soils without the presence of flowing ground water.

**Fig. 9.3** Suggested batters for open cut excavations

FLOW LINES

INITIAL STAGE              FINAL STAGE

**Fig. 9.4** Effect of ground water in open cut excavations

*Moist sands* tend to give a misleading picture as they will often stand almost vertically for a short time, but eventually the face either breaks up in lumps, or slumps to take up a natural resting position. Obviously, such soils are particularly dangerous, for example in pipe laying work, where unsuspecting workmen have frequently been trapped in collapsed trenches.

*Water-bearing sands* pose different problems. Here, the water flows into the excavation causing erosion of the toe of the slope, resulting in progressive collapse as shown in Fig. 9.4.

*Silty sands*, when dry, will stand almost vertically, especially if the silt is slightly cohesive. However, wet silts are extremely troublesome and suffer similar problems to water-bearing sands. Excavation in this material is extremely tedious as the soil flows into the excavation almost as quickly as it is removed, thus producing a very flat shape. Generally, open excavations in such material are impracticable.

*Cohesive soils* such as clay soil can theoretically be shown to stand vertically at a height dependent upon the cohesive strength. However, fissured clays, after excavation, tend to slide along these fissure planes. Even when fissures are not present, drying out on the surface causes shrinkage cracking, which later fills with water. Hydrostatic pressure then causes breaking away of sections of the clay face (Fig. 9.5).

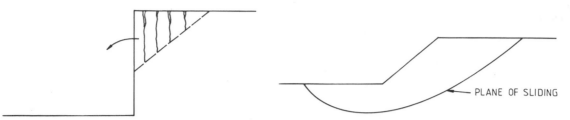

**Fig. 9.5** Fissuring of clay soils

**Fig. 9.6** Typical plane of sliding of a failed embankment in a clay soil

In most clays the safest action is to slope back the sides. The required section can be determined from slip circle calculations (Fig. 9.6).

### 9.3 CLOSED EXCAVATIONS

**Trenches**

SELF SUPPORT (Fig. 9.2)

Safety regulations do not require a trench to be lined for excavations shallower than 1.2 m; however, for greater depths a shoring method is recommended unless the sides are battered back.

PLANKS AND STRUTS

Planking and strutting was originally the accepted method of trench shoring. Figure 9.7 illustrates close boarding for applications in running soil and generally poor ground. Less planking may be possible in more stable conditions, for example when the boards are at about 2 m spacings the method is usually referred to as open timbering. Today, because of the high cost of timber and specialist skills required, other methods such as trench sheeting, lining panels and box systems are gaining popularity.

TRENCH SHEETS AND PROPS (Fig. 9.8)

The use of trench sheets is a very economical method of support. The installation procedure (Fig. 9.9) is similar to that for driving a panel of sheet piles.

The trench is first excavated to a depth of 0.5–1 m to steady the piles. A panel of sheets is pitched along opposite sides of the trench and propped apart. Pile-driving by means of a light double-acting hammer then proceeds to just below the next planned working level, excavation follows to this depth and further propping is installed. The process is repeated until the full depth of trench is obtained.

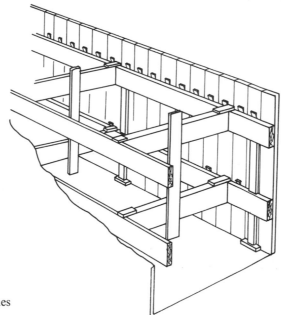

**Fig. 9.7**  Planking and
strutting for
timbered trenches

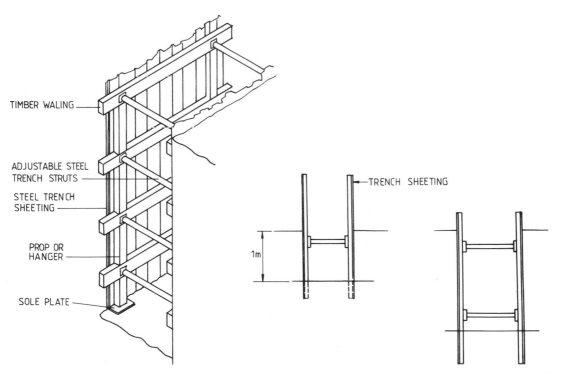

TIMBER WALING

ADJUSTABLE STEEL
TRENCH STRUTS

STEEL TRENCH
SHEETING

PROP OR
HANGER

SOLE PLATE

TRENCH SHEETING

1m

**Fig. 9.8**  Trench sheets and props        **Fig. 9.9**  Installing trench sheets

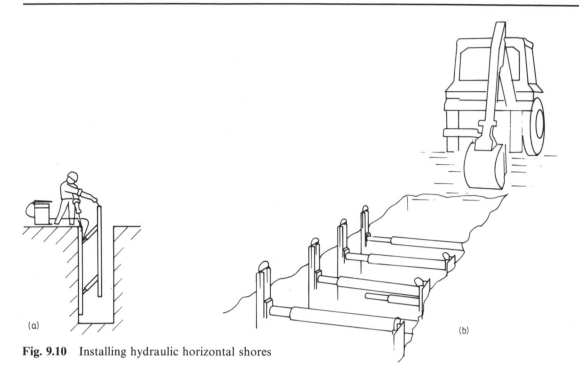

**Fig. 9.10**   Installing hydraulic horizontal shores

## TRENCH SHEETS AND HYDRAULIC VERTICAL SHORES (Fig. 9.10)

The system is primarily an alternative to open timbering and comprises integral hydraulically-operated props and support plates (Fig. 9.11).

Care must be taken during installation to ensure that the cylinders are perpendicular to the trench sides, otherwise on application of hydraulic pressure, one side of the frame may simply slide upwards. The hydraulic fluid is pumped in through flexible tubing from a hand pump operated at surface level. The whole system is very versatile and varying trench widths can be accommodated. The equipment is made in aluminium and is therefore light to handle.

In hard stable ground, such as soft rock, the frames may be spaced up to 2 m apart. In stiff clay and other stable cohesive soils, 1 m spacings may be sufficient. However, in sands and gravels 200–300 mm or less is often required. In all situations the ground should be free from excessive running water to avoid instability problems.

## TRENCH SHEETS AND HYDRAULIC HORIZONTAL SHORES (WALERS) (Fig. 9.12)

Horizontal shores are designed for use when close sheeting is required. Removing the sheets and backfilling is made much easier compared to a conventional propping method by the control over releasing the pressure in the cylinders. This allows the walers to be withdrawn from the trench in stages, thereby enabling backfilling to take place in safety.

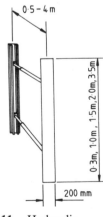

**Fig. 9.11**   Hydraulic vertical shore

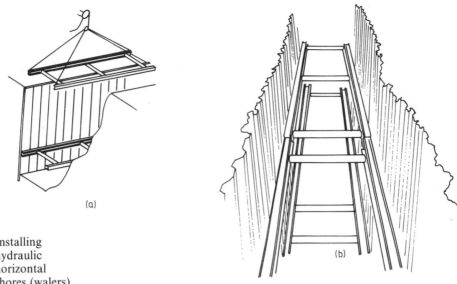

(a)

(b)

**Fig. 9.12** Installing hydraulic horizontal shores (walers)

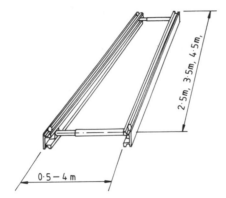

2·5m, 3·5m, 4·5m,

0·5 – 4 m

**Fig. 9.13** Hydraulic horizontal waler

The walers are fabricated in one piece aluminium units, which can be positioned from above (Fig. 9.13).

Lengths are available for 2–5 m, and units may be coupled together to line a full length of trench.

### DRAG BOXES (Fig. 9.14)

Drag boxes are used as safety boxes for operatives rather than trench supports. The method of operation requires the trench to be cut slightly wider than the box, which is then pulled forward into position by the excavator. This method is unsuitable when ground movement must be avoided, or where services cross the line of the trench. It is both a safe and quick procedure but requires a large excavator to handle the box.

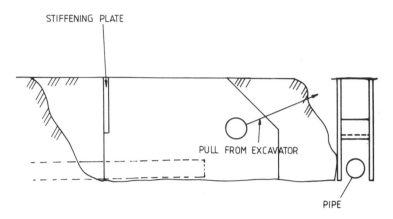

**Fig. 9.14**   Drag box

## TRENCH BOXES (Fig. 9.15)

For genuine trench shoring, trench boxes have recently appeared as an alternative to sheeting. The trench box is a modular system composed of two support walls separated by props. A method of installation is demonstrated in Fig. 9.16. Contractors have generally found that three boxes are sufficient to operate an efficient cycle of work – one box going down with the excavating, a second box already founded to provide protection for pipe installation and the third box coming up as backfilling proceeds. In order to avoid wedging action during installation it is normal practice to set the width at the toe 50 mm or so wider than that at the top of the box.

A large excavator capable of lifting is necessary to handle the boxes and achieve economic rates of digging and installation.

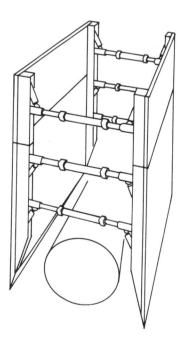

**Fig. 9.15**   Trench box

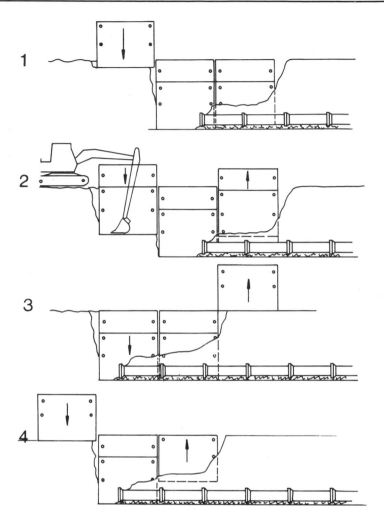

**Fig. 9.16**  Installing trench boxes

## SLIDE RAILS

The trench box system has been developed a stage further to incorporate slide posts driven ahead of the side plates to act as guides (Fig. 9.17). The manufacturers claim that this method overcomes the problems of withdrawal in granular soils.

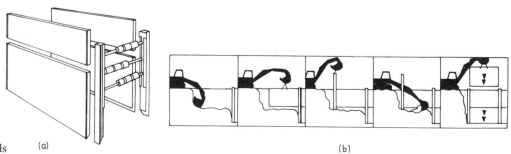

**Fig. 9.17**  Slide rails    (a)    (b)

Other specialist systems are available, but have been little used in the UK.

## GENERAL COMMENTS ON TRENCH SUPPORT

Trench sheets are relatively thin in section and therefore leave only a small void on extraction. However, bowing and buckling may take place and some ground movement is then unavoidable. Drag boxes provide adequate protection for the operative, but excessive ground movement is likely. Trench boxes and slide rails avoid most of the movement problems but extraction and ground services can cause severe difficulties.

The pressure of groundwater may cause problems for all the methods, particularly in running sands and silts subject to piping (see page 90). Production rates in such conditions might be only a fraction of normal, because material may continue to flow into the trench while excavation progresses (Fig. 9.18). In such circumstances a de-watering method should be installed.

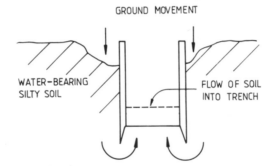

**Fig. 9.18**  Effect of flowing soil on a shored excavation

A summary of the merits of the various systems is given in Table 9.2.

## 9.4   OUTPUT DATA

Short has collected output data (Table 9.3) from a number of contracting firms. The units are expressed in $m^2/h$ of trench sides supported, rather than $m^3/h$ of trench excavated, because installation of the support is usually the slower activity.

### Recommended plant and labour selection (including pipe layers)

The choice of plant and gang size varies from contractor to contractor for any particular operation. In the example given below, two machines are assumed – one for excavating and the other for

**Table 9.2**   Comparison of trench shoring systems

| | |
|---|---|
| Slide rails | *Advantages*<br>(i) Operatives not required to enter unsupported trench<br>(ii) Provides a comparatively safe, clean and dry working area for pipe laying<br>(iii) Excavation and shoring in one operation<br>(iv) Requires a lower rated excavator than boxes<br>*Disadvantages*<br>(i) Capital outlay<br>(ii) Cross services are difficult to deal with<br>(iii) Installation becomes difficult if the system gets out of line |
| Trench boxes | *Advantages*<br>(i) Operatives not required to enter unsupported trench<br>(ii) Provides a safe, clean and dry working area for pipe laying<br>(iii) Increased output<br>(iv) Reduction in gang size<br>*Disadvantages*<br>(i) Large capital outlay<br>(ii) Boxes are heavy and require increased capacity of excavator<br>(iii) Cross services are difficult to deal with<br>(iv) There is a gap between adjacent boxes through which water and soil can flow; this is usually bridged by using a trench sheet |
| Drag boxes | *Advantages*<br>(i) Very high output rates<br>(ii) Operatives do not have to enter unsupported trench<br>*Disadvantages*<br>(i) Large excavator needed for pulling the drag box along, and a crane required for handling<br>(ii) Drag boxes have fixed dimensions<br>(iii) Large capital outlay<br>(iv) Cross services are a major problem |
| Trench sheets | *Advantages*<br>(i) The method is well known by planners and workmen alike – for many contractors it remains the 'only' solution to trench support<br>(ii) Trench sheets and props are easy to transport, require little storage space, and are available from numerous plant hire companies up and down the country<br>(iii) The method is versatile, varying trench widths and depths are easily accommodated<br>(iv) Differing ground conditions are catered for by varying the spacing of the sheets<br>(v) They are relatively cheap either to buy or hire<br>(vi) Cross services create few problems<br>*Disadvantages*<br>(i) It takes a considerable amount of time and labour to install and extract<br>(ii) The lowering of pipes through a mass of trench props is difficult<br>(iii) Workmen have to enter unsupported trenches to install and remove the supports |
| Hydraulic shores | *Advantages*<br>(i) Installation from above ground<br>(ii) Quick to install and remove<br>(iii) Fewer labourers needed in gang<br>(iv) Cross services are not a problem<br>*Disadvantages*<br>(i) More expensive than sheets and props<br>(ii) The systems require maintenance |

**Table 9.3** Output rate of various trench support systems in m²/h of lined trench

| System | Depth (m) | Strong | | Medium | | Loose | |
|---|---|---|---|---|---|---|---|
| | | **2–4** | **4–6** | **2–4** | **4–6** | **2–4** | **4–6** |
| Traditional trench sheeting | | 10 | 14 | 6 | 8 | 2.5 | 3.5 |
| Hydraulic walers and trench sheets | | * | * | * | 10 | * | * |
| Trench box | | * | 17 | 10 | 15 | * | 4.0 |
| Slide rails | | * | 6 | * | * | 2.5 | * |
| Drag box | | * | * | 8 | * | 3.5 | * |

\* No data were available

backfilling and lifting duties. At present, there is no standard available defining safe working load for excavators and therefore it is only possible to give approximate machine sizes to provide adequate safe lifting capacity.

There will of course be occasions when the recommendations given will not be valid, for example, when excavating in rock or for particularly shallow trenches. However, they will serve as a guide to new users of the systems.

## TRENCH SHEETS

One backhoe (with 0.50 m³ bucket) for excavation.
One backhoe (with 0.25 m³ bucket) for backfilling, placing pipes and pea gravel, and extracting trench sheets. It is common practice on many sites for the backhoe to spend 50% of its time on these tasks and the remainder on different jobs around the site. This is possible because of the machine's good manoeuvrability.
One ganger.
Five labourers.

*Note*: Sheet pile-driving equipment may be required in deep trenches (>3 m).

## TRENCH BOXES

There are two possible selections:

(i) One backhoe 0.5 m³ capacity for excavation.
One backhoe 0.5 m³ capacity (15–18 tonnes class) for backfilling, etc. plus extracting and placing the boxes. This machine is slightly underrated for the task and, therefore, the boxes have to be removed in individual

pieces (i.e. the top sections first and then the base unit). In granular material it has been known for a machine of this capacity to be unable to extract boxes because of the high ground pressures.

(ii) One backhoe 0.5 m³ capacity.

One backhoe (with 2.5 m³ bucket, 35–40 tonnes class) for back-filling etc. plus extracting and placing the boxes.

3 sets of trench boxes. Manufacturers recommend the use of four sets, but contractors find they can cope adequately with three.

One ganger.

Four labourers.

The second plant selection is recommended, although it is more expensive, since it does gives higher output and reduces the possibility of the boxes getting stuck.

## SLIDE RAILS

One backhoe 0.5 m³ capacity for excavation.

One backhoe 0.5 m³ capacity for backfilling etc., plus lifting and placing the slide rails.

Three sets of linings.

One ganger.

Four labourers.

## HYDRAULIC SHORES

One backhoe 0.5 m³ capacity for excavation.

One backhoe 0.25 m³ capacity (for 50% of the time).

One ganger.

Four labourers.

In strong ground conditions vertical aluminium shores are recommended, and in mixed or granular soils, trench sheets with horizontal aluminium waler frames. Contractors find that 3 m long frames provide enough working length. The number of layers of frames depends upon the depth of trench and the soil pressure.

*Note*:
 (i) With all systems, compaction plant and pumping equipment may be required.
 (ii) For trench boxes and linings, systems are at present available to provide supported excavations up to approximately 6 m deep and 5 m wide. This restriction is not applicable to sheet piling but careful design is required because of the considerable ground forces involved.

## 9.5 COFFER-DAMS

### Sheet piled coffer-dam

A coffer-dam is a 'watertight' structure which allows foundations to be constructed in the 'dry', for example bridge piers. Nowadays however the term also covers land-based operations and the coffer-dam is a common feature in foundation works. Sheet piling is a popular material used in the construction of medium to large coffer-dams, while trench sheeting or even timber boarding is more appropriate for shallow excavations (Fig. 9.19). A serious hindrance with these methods is the fairly cumbersome bracing required to hold the piles in position. However, modern developments in hydraulically-pressurised modular aluminium frames are proving quite successful where the sides do not exceed about 5 m, (Fig. 9.20). For more general applications involving irregular shapes etc. however, walings are normally fabricated from RSJs or similar. Indeed, individual sheet piles may also be strengthened in a composite construction.

An alternative and sometimes cheaper method is to prop from the base of the excavation as shown in Fig. 9.21 but obviously this can only be progressively carried out as the excavation is deepened, if excessive cantilevering is to be avoided.

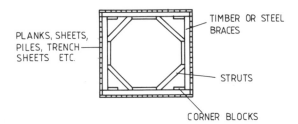

**Fig. 9.19** Timbered coffer-dam

Another alternative is to build a circular coffer-dam and form the walings in reinforced concrete, thereby utilising the advantage of ring compression as shown in Fig. 9.22.

Methods of sheet pile installation are discussed in Chapter 8.

### Design of sheet piled walls

The design of sheet pile walls and coffer-dams depends upon many factors, including the depth of excavation, water table, soil type, surcharge, propping the tying arrangement etc. To demonstrate all these variables is beyond the scope of this book and the reader is directed to the BSP Pocketbook.

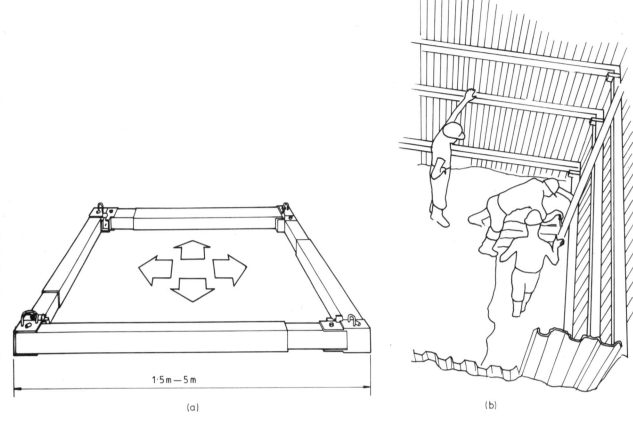

1·5 m — 5 m

(a)

(b)

**Fig. 9.20**    Piled coffer-dam using hydraulic horizontal walers

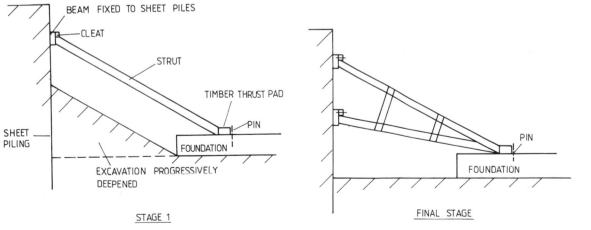

BEAM FIXED TO SHEET PILES

CLEAT

STRUT

TIMBER THRUST PAD

PIN

SHEET PILING

FOUNDATION

EXCAVATION PROGRESSIVELY DEEPENED

STAGE 1

PIN

FOUNDATION

FINAL STAGE

**Fig. 9.21**    Propped piled wall

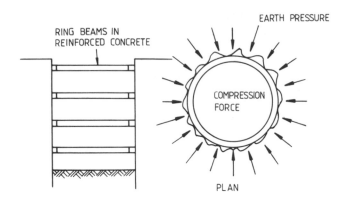

**Fig. 9.22** Coffer-damming using ring beams

## 9.6 REINFORCED CONCRETE PILED WALLING

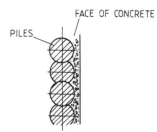

**Fig. 9.23** Forming a finished surface with reinforced concrete piled wall

In an attempt to reduce the cost of shoring, support systems have been designed as part of the permanent works. For example, the first methods involved the use of in situ concrete piles cast to form a continuous wall. Furthermore, the row of piles could be given a facing of reinforced concrete to improve the surface appearance (Fig. 9.23).

### Types of wall

(i) The simplest and cheapest method is the single row of tangentially touching piles (Fig. 9.24) (contiguous piles). The pile is constructed either by boring or by using a grabbing bucket. In poor or water-bearing ground, lining or bentonite slurry may be necessary to support the walls of the hole. Reinforcement is positioned and concrete tremied into place. Finally, the ground in front of the completed row of piles is excavated. A watertight connection between adjacent piles cannot be achieved, as in practice the positioning of each pile often leaves a 50–100 mm gap. A de-watering system may therefore be necessary.

**Fig. 9.24** Simple reinforced concrete piled walling

(ii) The above method can be improved to produce a closed joint by constructing alternate piles, followed by the intermediate ones as illustrated in Fig. 9.25. This is achieved by boring or cutting away part of the first phase piles while the concrete is still weak (i.e. less than 5 N/mm$^2$). The system is usually referred to as *Secant-pile* walling.

**Fig. 9.25** Secant concrete piled walling

## 9.7 CONTINUOUS DIAPHRAGM WALLS (ICOS WALLS)

This system developed from the technique of forming a coffer-dam from a row of bored and cast in situ piles sunk in close contact with one another.

## Method of construction (Fig. 9.26)

### STEP 1: CONSTRUCTION GUIDE WALLS

A trench is excavated about 1 m deep along the line of the wall, wide enough to accommodate the wall and to give enough clearance for the grab. The inner face only is shuttered.

GUIDE WALLS

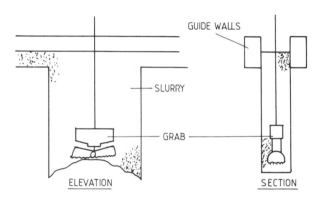

GUIDE WALLS

SLURRY

GRAB

ELEVATION

SECTION

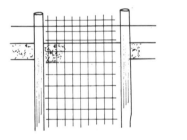

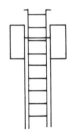

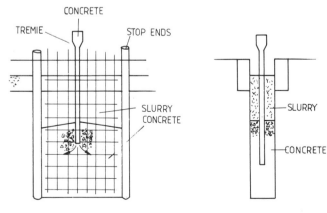

CONCRETE

TREMIE

STOP ENDS

SLURRY
CONCRETE

SLURRY

CONCRETE

**Fig. 9.26** Reinforced concrete diaphragm walling

## STEP 2: EXCAVATE FIRST AND SUBSEQUENT PANELS

(i) A slot is excavated to the requisite depth between the guide walls to form the first panel.
(ii) Slurry (3–10% powdered bentonite in water) is then pumped into the trench.

## STEP 3: PLACE CAGE OF REINFORCEMENT

(i) The trench is cleared of slurry/soil sediment at the base.
(ii) The ends of any abutting panel are scraped clean by the teeth of the excavator bucket.
(iii) Steel stop ends are inserted.
(iv) A cage of reinforcement fitted with spacer blocks is assembled.
(v) Reinforcement is lowered into the slurry.

## STEP 4: TREMIE CONCRETING

High slump concrete is finally placed in the trench using a tremie tube. The complete sequence of operations for subsequent panels is shown in Fig. 9.27a. The excavating grab shown, is of special design to suit the particular trench width and is generally hydraulically powered. A rotary excavator (Fig. 9.27b) however is proving increasingly popular and eliminates the cumbersome need to mechanically handle the excavated material plus improved accuracy.

*Conventional equipment*

| | |
|---|---|
| Construct panels in | 2–6 m lengths |
| Wall thickness | 0.5–1 m |
| Trench depth | Up to 35–40 m |
| Vertical accuracy | 1 in 80 |
| Wall production | 20–50 m$^2$/hr |

*Rotary equipment (Fig. 9.27b)*

| | |
|---|---|
| Rock strength | up to 100 N/mm$^2$ |
| Trench depth | up to 100 m |
| Machine size | approx. 80 kW |
| Rotary speed | 10–20 revs/min |
| Vertical accuracy | 1 in 250 |
| Excavation rate | up to 25 m$^3$/hr in soft material |

### Mixing and placing the slurry

The slurry is made from bentonite clay delivered to site as a dry powder and then mixed with water.

A typical systematic mixing set-up is shown in Fig. 9.28. The equipment consists of a mixing hopper with paddle wheels rotating at

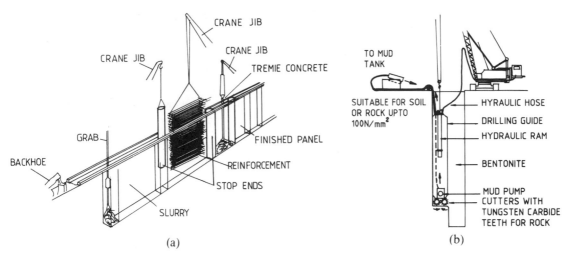

**Fig. 9.27**    (a) Installing procedure for diaphragm walling. (b) Diaphragm wall excavation equipment

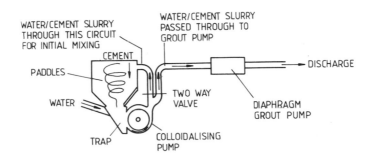

**Fig. 9.28**  Grout or bentonite mixer

about 250 rpm. The mix is passed through a centrifugal pump rotating at 1000–1500 rpm at the base of the drum and returned. The result is a rapid circulation of the mix through the hopper tank which produces the colloidalising effect.

The mixture is finally pumped to position with a piston pump. Because of the thixotropic properties of bentonite it may be necessary to place an agitating hopper between the mixer and delivery pumps when the material is not required continuously. Typical production data are given in Table 9.4.

**Table 9.4**   Data for low pressure grouting equipment

| | | | |
|---|---|---|---|
| Output (l/m) | 50 | 75 | 100 |
| Hopper capacity (l) | 10 | 10 | 10 |
| Pump discharge pressure* (bar) | 7 | 7 | 7 |
| Power required (kW) | 3 | 3 | 5 |

\* For high pressure grouting a piston pump capable of producing 70 bar may be required (see Chapter 7)

## BENTONITE PROPERTIES

The solids in the bentonite fill the pores of the soil to form a weak membrane. Because the density of the slurry is greater than the particle/water soil density, outward hydrostatic pressure forces the membrane against the sides of the excavation. Thus, support is provided and the hydrostatic head in the water table balanced.

The density of the fluid may be increased by raising the bentonite concentration as follows:

4% by weight bentonite – 1022 kg/m³ density is suitable in stiff clay
10% by weight bentonite – 1060 kg/m³ density is suitable in coarse soils.

## RE-CYCLING OF BENTONITE SLURRY

The bentonite slurry gets contaminated by soil particles during excavation which, if excessive, destroy its properties. However, to reduce costs, for example in a long section of walling, the displaced bentonite is led into a tank and allowed to settle. The top layer is then returned to the trench. Waste slurry is usually transported to a tip in tanker trucks for disposal.

### Testing the slurry

#### (i) VISCOSITY

A rough viscosity measurement may be taken on site using the Marsh Funnel (Fig. 9.29).

The time of outflow of 0.95 l of slurry is compared to that of water.

#### (ii) DENSITY

A given volume of slurry is weighed on a balance beam (Fig. 9.30). For a given bentonite concentration the difference between the density of fresh slurry and that measured on the balance will be due to the contamination by soil particles.

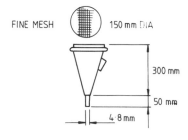

VISCOSITY

FINE MESH — 150 mm DIA

300 mm

50 mm

4·8 mm

MARSH FUNNEL

**Fig. 9.29** Viscosity test for bentonite

DENSITY

1   2   5   4

3

MUD BALANCE

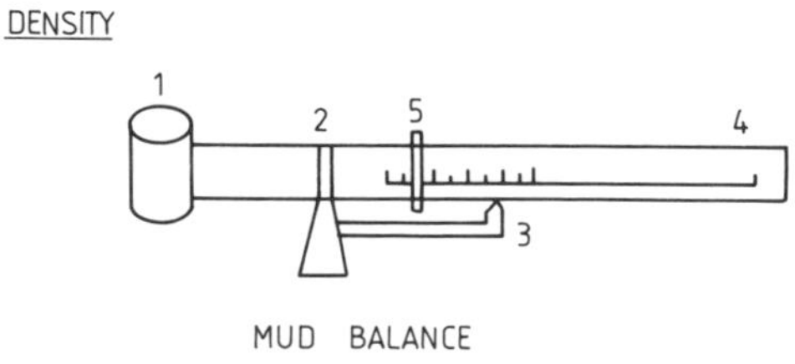

**Fig. 9.30** Density test for bentonite

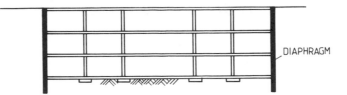

**Fig. 9.31** Diaphragm walling for a basement building

DIAPHRAGM

*Note*: Experience is at present the best guide to the most suitable bentonite concentration and the point at which slurry should be treated for re-use or disposed of.

## Applications

Bentonite diaphragm walling is particularly suitable when combinations of both temporary and permanent support are included in the design of the structure, for example

  (i)  base construction (Fig. 9.31)
 (ii)  pumping chambers ⎱ (Fig. 9.32)
(iii)  underground tanks ⎰
 (iv)  coastal defence works and river walls (Fig. 9.33)
  (v)  retaining walls (Fig. 9.34)

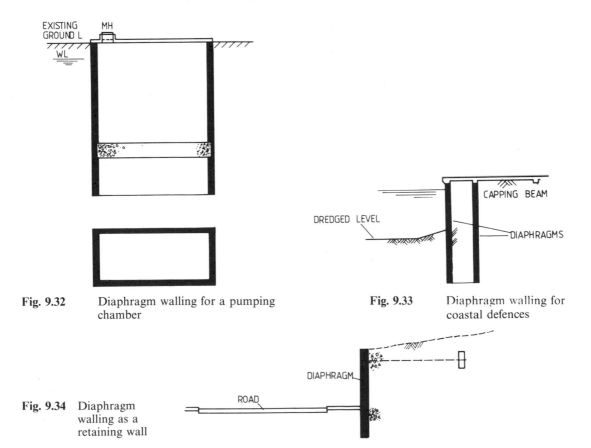

**Fig. 9.32**   Diaphragm walling for a pumping chamber

**Fig. 9.33**   Diaphragm walling for coastal defences

**Fig. 9.34**   Diaphragm walling as a retaining wall

### 9.8  INJECTED MEMBRANES

In water-bearing ground, a continuous impermeable membrane installed around the outside of a proposed excavation to provide a dry working area (Fig. 9.35) is sometimes a more economic solution than using sheet piling, diaphragms, etc. The method, originally developed by Études et Travaux Fondations of Toulouse, involves a series of H section piles. When approximately seven units are in place, the rearmost pile is withdrawn while the void beneath is injected with grout at 2–5 bar pressure through a 20 mm diameter pipe running along the face of the pile (Fig. 9.36). Extraction is usually performed with hydraulic rams, as the conventional hammer-type extractor method might disturb and damage the grouted membrane. The extracted pile is then transferred, redriven as the lead pile, and the process continued until the seal is completed.

The grout used is a clay/cement mix, e.g.

45% clay
35% water
20% cement

Approximately 250–300 kg of grout is required per m² of membrane. Output – 0.25 hours per m² of membrane may be required to drive and extract the piles, carry out the injection of grout, etc.

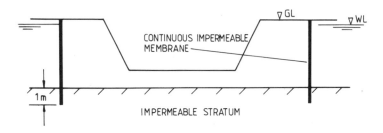

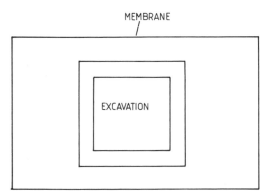

**Fig. 9.35**  Use of an injected membrane surrounding an excavation

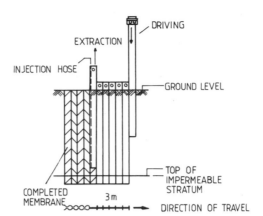

**Fig. 9.36** Installing injected membrane

## 9.9 GROUND FREEZING

Ground freezing can be used as a means of ground support, water cut-off or combinations of both, as illustrated by examples of a coffer-dam and tunnel heading shown in Fig. 9.37. Usually sheet piling, grouting, de-watering, compressed air, etc. would be first choices, but sheet piling, for example, is uneconomical for excavations deeper than about 20 m, especially where the wall area exceeds 200–300 m². Compressed air is only viable in heads of water less than 35 m and normally is not considered for depths in excess of 25 m. De-watering and grouting are only effective in a limited range of soil particle sizes. Such constraints are not usually imposed on the ground freezing process, but unfortunately the technique is expensive and slow to install.

### Methods of ground freezing

The basic principle of ground freezing involves removing heat energy from the surrounding water-bearing soil by passing a refrigerant down the core of a freeze pipe. Isotherms move out from the tube, and over a period of time a column of ice is built up. By spacing the tubes at

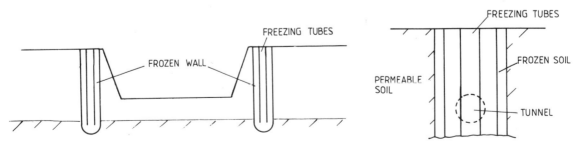

**Fig. 9.37**     Examples of ground freezing

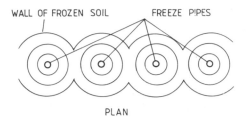

**Fig. 9.38**   Forming an iced wall

intervals, the columns of ice can be made to merge to form an enclosing wall (Fig. 9.38).

There are basically two freezing methods. (i) the two phase process and (ii) the single phase or direct process.

### (i) THE TWO PHASE PROCESS

This requires a primary refrigerant such as ammonia or freon to cool a secondary fluid such as glycol or calcium chloride brine. Typical plant is shown in Fig. 9.39. The brine is pumped down the inner core of a double wall pipe at 3–5 m/s and returned up the outer core to a chilling vessel.

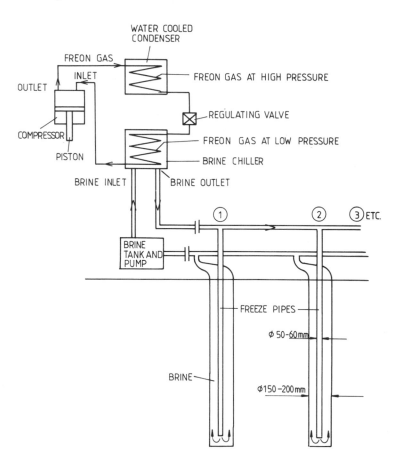

**Fig. 9.39**   Plant for the two phase freezing process

In the refrigeration plant a gas compressor reduces the volume of the freon gas, followed by cooling and condensing with circulating water. The resulting liquid is then fed through a regulating valve into the chilling coil, where the pressure is released, causing evaporation and cooling of the surrounding circulating brine. The gas is then represssurised in the compressor and the cycle continued.

Brine is a convenient fluid to use down to about $-40°C$ and requires a period of 3–10 weeks to complete the freezing process, depending upon the soil type and strength required.

### (ii) THE DIRECT PROCESS

This process uses liquid nitrogen which has a boiling point of $-196°C$. The fluid is simply passed down the central core of the freeze pipe and evaporated to atmosphere, passing up through the outer core (Fig. 9.40). This process does not involve refrigeration plant, but requires the liquid nitrogen to be stored in an insulated pressure vessel which can

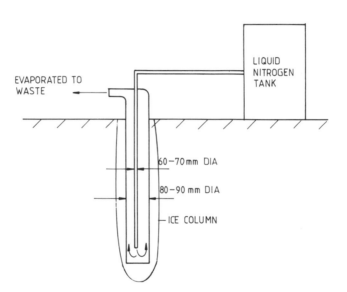

**Fig. 9.40**  Plant for the single phase freezing process

be periodically refilled from tanker trucks. Freezing by this method can be up to 50% more expensive than by the brine process, but is much quicker, as an ice wall can often be achieved in a week (approximately).

### Ground freezing principles

The freezing process passes through the following stages:

(i) cooling the soil particles and water to 0°C
(ii) removing the latent heat
(iii) reducing the soil and ice to the required temperature.

The heat removed from the ground when reducing the temperature from $t_1$ to $t_2$ is governed by the equation

$$Q = [\underset{\text{Specific heat}}{M \times S \times (t_2 - t_1)}] + [\underset{\text{Latent heat}}{M \times L \times (t_2 - 0°C)}]$$

where  $Q$ is the heat removed
$M$ is the mass of the body
$S$ is the specific heat of the body
$L$ is the latent heat of ice

Thus for soil and ice

$$Q = V_1 d_s S_s(t_1 - t_2) + V_2 d_w S_w(t_1 - 0°) \\ + V_2 d_i S_i(t_2 - 0°) + V_2 d_w L_i \quad (9.1)$$

where  $Q$ = heat removed in kcal
$V_1$ = solid volume of soil in m$^3$
$d_s$ = soil particle density $\simeq$ 2500 kg/m$^3$
$V_2$ = volume of water in the soil in m$^3$
$d_w$ = water density = 1000 kg/m$^3$
$S_s$ = specific heat of soil particles $\simeq$ 0.2 kcal/kg – °C
$S_w$ = specific heat of water $\simeq$ 1 kcal/kg – °C
$L_i$ = latent heat of ice = 80 kcal/kg
$d_i$ = density of ice = 900 kg/m$^3$
$S_i$ = specific heat of ice $\simeq$ 0.5 kcal/kg – °C
$t_2$ = final temperature of ice and soil in °C
$t_1$ = starting temperature (ambient) of water and soil in °C

*Note*:
$V$ = total volume of loose soil
$n$ = soil porosity
$V_1 = (1 - n)V$
$V_2 = V \times n$

Assuming that the process is only 50 to 70% efficient (because of heat losses) the size of refrigeration plant required is about 1.5 to 2.0 $\times$ $Q$.

From the above equation, the value of $Q$ needed to freeze 1 m$^3$ of earth is approximately 40 000–60 000 kcal, depending upon the difference between ambient and final temperature.

**Theoretical freezing time**

$$t = \frac{Q}{HA}$$

where  $Q$ is calculated from eqn. (9.1) in kcal
$H$ = heat removed by the freeze pipe, e.g. 200 mm diameter pipe removes 300–500 kcal/h – m$^2$ depending upon temperature
$A$ = surface area of pipe surrounded by soil in m$^2$.

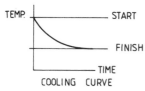

COOLING CURVE

**Fig. 9.41** Cooling plotted against time

According to Newton's Law of Cooling, the quantity of heat passing from a cooling liquid per unit of time is proportional to the temperature difference between the liquid and its surroundings (Fig. 9.41). Thus at the early stages the temperature difference between the soil/water and freeze pipe is large and the heat extraction will be high. Near the final temperature heat extraction will be relatively low. In practice, roughly only 30% of the initial rate of heat extraction is necessary to maintain the 'ice' column at its final temperature.

### Refrigeration plant rating

Refrigeration plant is rated in terms of the heat transfer rate for a particular differential between ambient and chilled brine. The unit chosen is usually tons of refrigeration $(T_R)$

whereby

$$1 T_R = 3.517 \, \text{kW or } 3.517 \times 4.187 \, \text{kcal/s}$$

or $\quad 1 T_R \simeq 12\,000 \, \text{Btu/h}$

Plant used for ground freezing typically ranges from 50–200 tons rating. (*Note*: 1 ton = 1.016 tonnes.)

### Disadvantages encountered with ground freezing

 (i) Salt water in the soil can make freezing difficult.
 (ii) Flowing water causes heat drain, e.g. water movement exceeding 1–2 m per day may render ground freezing impracticable. Thus tidal conditions, or adjacent de-watering may pose difficulties.
(iii) Water expands about 10% by volume on freezing, but in free draining soil this generally only displaces the surrounding water as the ice front advances and little ground heave results. However, where water is contained in fine gravel soils, ice lenses may develop and so cause heave (Fig. 9.42).

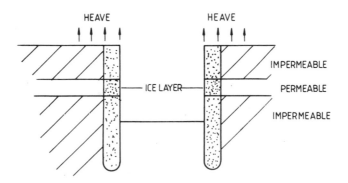

**Fig. 9.42** Ground heave caused by ground freezing

### Strength of frozen soil

The minimum 'ice-wall' thickness needed to resist bending and shear can be calculated by consideration of the active and passive forces to which the wall is subjected.

Approximate values of compressive strengths attainable in frozen soils are given in Table 9.5.

**Table 9.5**   Compressive strength of frozen soils (N/mm$^2$)

| | Final temperature | | |
|---|---|---|---|
| **Material** | $-10°C$ | $-15°C$ | $-20°C$ |
| Ice | 1 | 1.5 | 2 |
| Saturated clay | 5 | 7 | 9 |
| Saturated fine sand | 8 | 13 | 15 |
| Saturated medium sand | 8 | 13 | 20 |

## 9.10   ANCHORAGES

In order to provide a clear, unpropped working area in a coffer-dam, anchorage of the shoring system to the surrounding ground may be necessary for large and deep basements (Fig. 9.43).

The pull-out force that can be resisted varies markedly depending upon the soil type. The installation methods to deal with the particular ground conditions may be summarised as follows:

### Coarse sands and gravels

A hole is bored and lined, using a conventional rotary drilling rig. The steel anchor cable is located in position and cement grout pumped in under pressure. As the lining tube is withdrawn (Fig. 9.44) a bulb of grouted soil is produced around the bottom end of the cable. The resistance to pull-out is therefore highly dependent upon the effective grouted length ($L$) and the diameter of the anchor zone ($D$).

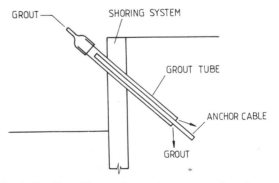

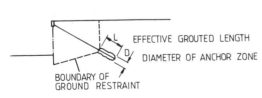

**Fig. 9.43**   Use of ground anchors

**Fig. 9.44**   Installing and grouting a ground anchor

### Fine to medium sands

Considerably lower pull-out resistance is obtained in these types of soils because the grains are too small to allow cement grout to penetrate the voids. The pull-out force is therefore largely controlled by the diameter of the borehole, although the pressure of the grout may cause some local compaction and so increase the anchor diameter.

### Clay

The cohesive strength which can be mobilised between the anchor and clay is fairly small. However, pull-out resistance may be increased by

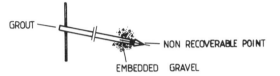

**Fig. 9.45** Use of a non-recoverable point grouting method

placing gravel in the boring along the anchor length. A lining tube fitted with a non-recoverable point (Fig. 9.45) is used percussively to cause the gravel to penetrate the clay wall of the borehole. The anchor cable is subsequently positioned and grouted in the normal way. Consequently the diameter of the anchored zone is increased.

Other methods of increasing the pull-out resistance involve under-reaming.

### Rock

The resistance to pull-out is dependent upon mobilising skin friction. A bored hole is drilled and cement grout pumped in to produce the fixed anchor length.

### Pull-out loads

These were generated from 100 mm diameter borings and fixed anchor length 4 m. Examples:

| | |
|---|---|
| Gravel | > 1000 kN |
| Medium sand | 400 kN |
| Clay | 600 kN |
| Chalk | 1000 kN |
| Rock | > 1000 kN |

*Note*: The load-carrying capacity should always be determined from a test anchor.

### Prestressed anchors (Fig. 9.46)

Prestressed anchors are mainly used for tying down permanent structures, such as a floating basement, retaining walls subject to

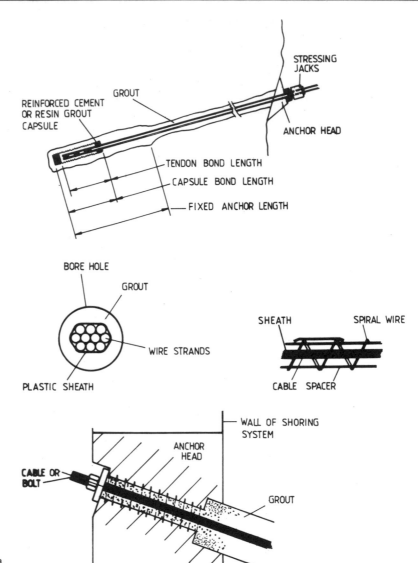

**Fig. 9.46** Prestressed ground anchor

**Fig. 9.47** Protecting a prestressed ground anchor

**Fig. 9.48** Anchorage for a prestressed anchor

imposed loading, etc. and also for improving stability, for example, of a rock face. The design principle aims to minimise movement of the structure when subject to surcharge and service loads.

Borehole diameters (200–250 mm) are a little larger than conventional anchors. The cable consists of bundles of high tensile steel strands greased and covered by a corrosion-resistant plastic sheath (Fig. 9.47). The fixed anchor length (which has the sheath covering removed) is cast into a 4–6 m long corrugated plastic tube, with an epoxy resin. The cable is then positioned in the borehole and grouted in the normal way. The exposed end of the cable is de-sheathed and post-tensioned against a permanent anchorage (Fig. 9.48).

## TENSION FORCE (EXAMPLE)

| | | |
|---|---|---|
| Diameter | 225 mm | |
| Fixed anchor length | 6 m | |

| | *gravel* | *clay* | *rock* |
|---|---|---|---|
| Tension force (kN) | 600 | 250 | $>600$ |

## LOSS OF TENSION FORCE

The tension in the cable may lessen with structural movement of the ground, for example up to 5% tension loss was recorded by Littlejohn over a 10 month period for anchors in London clay.

### Factors of safety (also see BS.DD81)

| | |
|---|---|
| Working tension in cable | $-0.6$ of breaking tension |
| Bond between anchor and soil | $-2$ for temporary works |
| | 3 for permanent works |
| Test load | $-1.3 \times$ working or design load |

## 9.11  DOUBLE-WALLED SHEET PILED DAM

Construction works to river banks, beaches, docks, etc. require some form of dam to provide temporary protection. Where space permits, a single row of sheet piles or similar supported by an earth embankment (Fig. 9.49) may suffice; in other situations a double row of conventional sheet piles, cross-tied, may be needed (Fig. 9.50).

### Design principles

The stability of double-walled structures may be designed on a similar principle to a gravity dam as illustrated in Fig. 9.51.

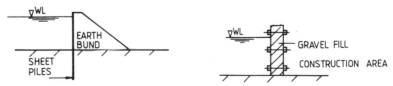

**Fig. 9.49**  Sheet piled wall protected with an earth bund

**Fig. 9.50**  Double-walled sheet piled dam

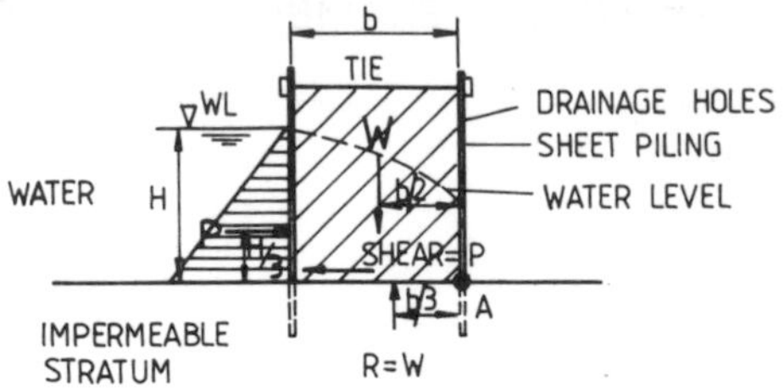

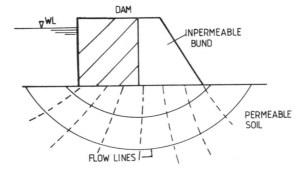

**Fig. 9.51**   Design principle of double-walling

Thus

$$P = \frac{\gamma_w \times H^2}{2} \qquad \gamma_w = \text{unit weight of water}$$

$$W = \gamma_s \times b \times H \qquad \gamma_s = \text{unit weight of fill/water}$$

Taking moments about $A$

$$\frac{Wb}{2} = \frac{PH}{3} + \frac{Rb}{3}$$

In order to keep $R$ within the middle $\frac{1}{3}$ of the base, a ratio of width ($b$) to height ($H$) for water-retaining structures of about 0.85 is required.

The structure must also be adequately designed to resist sliding, pile rise and vertical shear. This is beyond the scope of this book, but is adequately treated elsewhere, in Terzaghi (1945).

**Construction arrangement**

The dam is filled with permeable material such as sand, gravel or hardcore. If excessive seepage under the dam causes piping then an impermeable bund should be placed in front of the inner wall to increase the seepage path (Fig. 9.52) and thereby reduce the seepage head gradient. Adequate drainage holes may also be provided at various levels in the inner wall to reduce the internal hydrostatic pressure.

**Fig. 9.52**   Increasing the seepage path using an earth bund

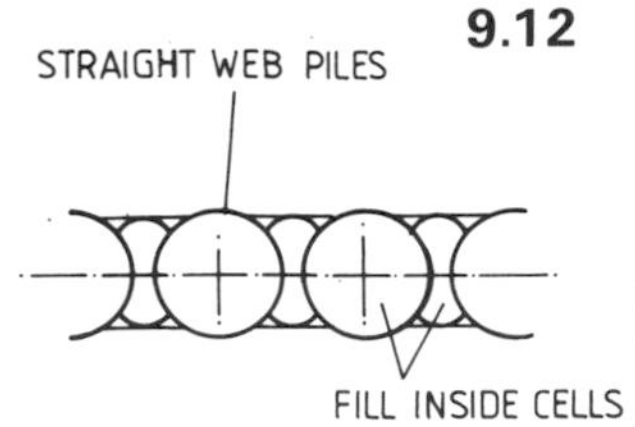

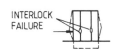

**Fig. 9.53** Cellular coffer-dam

STRAIGHT WEB PILES

FILL INSIDE CELLS

INTERLOCK FAILURE

**Fig. 9.54** Clutch separation causing failure of a cellular coffer-dam

## 9.12 CELLULAR COFFER-DAMS (Fig. 9.53)

Like the double-walled dam, the cellular structure may be designed as a gravity dam but the need for ties is avoided by taking the active pressure of the fill as ring tension in the piles. Special straight web steel piles are used, but great care is required during pile-driving to avoid straining the clutch, which could subsequently fail when placing the fill, causing complete collapse of the dam (Fig. 9.54). Seepage under the dam may be reduced by driving the piles deeper (see Chapter 6). The hydrostatic pressure inside the cell may be reduced by providing drainage holes on the inside of the coffer-dam (see Fig. 9.51).

Cells founded on clay soil are generally provided with a heavy bund on the inside face to improve stability and in particular to prevent sliding.

### Method of construction (Fig. 9.55)

A stable position, such as rail track, extended over completed cells, is necessary to facilitate accurate pile-driving. A full ring of straight web piles is pitched and driven around a rigid circular steel frame. If secondary cells are required then the special coupling piles (Fig. 9.56) must be correctly located during the pitching phase, and a further template used to obtain the correct shape for the cell.

A granular fill should be used to aid drainage and shear forces must be removed before any piles are withdrawn in order to avoid a major collapse.

## 9.13 OPEN CAISSONS

A caisson constructed in reinforced concrete may be considered as a coffer-dam during the construction phase but it ultimately forms part of the permanent structure – typically a deep foundation or bridge pier.

The caisson consists of an open cell sunk into position by excavating

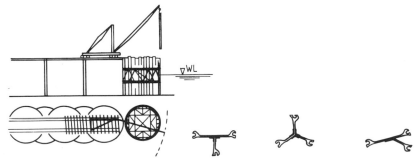

**Fig. 9.55** Installing method for a cellular coffer-dam

**Fig. 9.56** Special straight webbed coupling piles

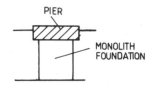

**Fig. 9.57**  Caisson used to
form a
foundation or
monolith

within each cell. It is finally sealed with mass concrete at the base and
ballasted to produce an independent foundation, as shown in Fig. 9.57.

### Applications

Open caissons may be:

(i)   used as foundation based (5–10 m deep), where the founding layer
      of soil or rock would be too hard for sheet piling to penetrate.
(ii)  used as coffer-dams or as shoring in highly porous soils which
      could not be de-watered and excavated in the normal way.
(iii) used as deep foundations (30–40 m deep).

They are unsuitable for sinking through soil containing large fragments
or where the skin friction between soil and caisson walls is excessive.
Also, caissons founded on an uneven or sloping stratum are likely to
slip from the final resting place.

### Caisson design and construction

(i)   The circular caisson offers less surface area for a given base area
      than a square or rectangular shape and is therefore favoured
      because of the lower skin friction. However, the shape of the
      permanent structure incorporating the caisson is often the deciding
      factor.
(ii)  The walls of the caisson should be constructed carefully. Rippled
      and non-parallel walls are likely to increase sinking resistance.
(iii) Verticality during sinking of the caisson can be controlled by
      choosing a multi-cell structure so that individual areas may be
      excavated in isolation (Fig. 9.58).
(iv)  The size and shape of the cells should be arranged for maximum
      open digging. Rectangular cells are preferred, especially in clay soil.
      There is a tendency to arch and wedge around the cutting edge if a
      circular form is selected.
(v)   The shape of the cutting edge should be varied to suit the soil
      type. The stiffer the soil, the steeper the angle of the cutting edge,
      but 35° from the vertical is typical (Fig. 9.59). The cutting edge
      should also be well reinforced to cope with high stress and
      therefore a steel shoe at the base is commonly incorporated.

**Fig. 9.58**  Multi-cell
reinforced
concrete
caisson

### Sinking theory

Consider the section of caisson wall and cutting edge shown in
Fig. 9.60.

The caisson comes to rest when the sinking force is overcome by skin
friction and ultimate bearing resistance.

Working in total stresses,

$$S = F_E + F_I + (CNc + P_o + P_w)A$$

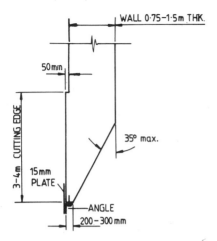

**Fig. 9.59**  Cutting shoe for a caisson

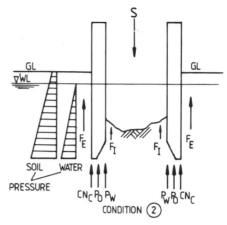

**Fig. 9.60**  Sinking principles of a caisson

where   $S$ = total weight of caisson and kentledge
   $F_E$ = external skin friction
   $F_I$ = internal skin friction
   $CN_c$ = net ultimate bearing resistance
   $P_o$ = effective overburden pressure
   $P_w$ = water pressure
   $A$ represents the area of monolith shoe.
   $CN_c$ is part of Terzaghi's equation for net ultimate bearing
       resistance
   $C$ = cohesive strength of the soil
   $N_c$ = approximate bearing capacity factor of 7 for $\phi = 0$ (for
       large $\phi$ values $N_c$ increases dramatically).
   $\phi$ = the internal angle of friction for the soil

The value of $C$ used in calculating $F_E$ and $F_I$ should be the remoulded shear strength, in order to reflect the effects of soil movement as the caisson sinks. However, practice has demonstrated that calculation of skin friction based on laboratory soil tests is unreliable. Terzaghi and Peck, however, have recorded loads needed to free stuck caissons and these have been interpreted to give values for skin friction as shown in Table 9.6.

Calculations should be made for several conditions, as shown in Fig. 9.61, including the initial settlement, intermediate points and when the caisson is constructed to its full height. In this manner the amount of kentledge, height of caisson above ground level, depth of undercutting, etc. can be determined for controlling the amount of sinking.

### Sinking and construction procedure (Fig. 9.62)

(i) Prepare a level bed and establish concrete pads where the toe of the walls is to be constructed.

**Table 9.6** Values of skin friction for caissons

| Type of soil | Skin friction (kN/m²) |
| --- | --- |
| Silt and soft clay | 7–28 |
| Very stiff clay | 50–200 |
| Loose sand | 12–35 |
| Dense sand | 35–67 |
| Dense gravel | 50–100 |

From K. Terzaghi and R. B. Peck (1967). *Soil Mechanics in Engineering Practice.* John Wiley & Sons, Inc. New York

**Fig. 9.61** Sinking procedure for a caisson

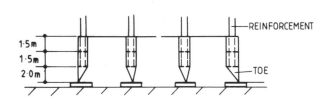

**Fig. 9.62** Excavating sequence for a multi-cell caisson

(ii) Construct the toe and walls up to about 5 m.

(iii) Break out the concrete pads and allow the caisson to settle. The amount of initial sinking will depend upon the strength of the soil.

(iv) Grab within the cells – this is the most commonly used method of excavating from the open cells. Grabbing must be done in a logical sequence to ensure that verticality of sinking is maintained. The grabs can either be mounted upon derricks or on mobile cranes. Sometimes other methods are appropriate, e.g. in sands, a jetting pipe often helps to loosen the soil particles, which are then lifted to the surface by an air-lift pump. This equipment (Fig. 9.63) contains compressed air which is injected into a riser pipe, and the aerated water, being of lighter density, flows up the riser pipe taking the loosened soil with it.

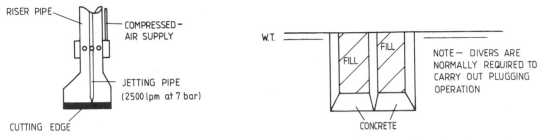

**Fig. 9.63**   Air-lift pump

**Fig. 9.64**   Plugging and hearting for a caisson

(v) The excavation is continued until about 1 lift of concrete (1.5 m) freeboard is remaining. Then the walls are extended. This in itself may result in significant sinking if the soil is weak. The height of the walls should be such that construction access is easy and the centre of gravity does not become too high to jeopardise the accuracy of sinking. About 6–7 m of freeboard is usual. Excavating within the cells is then continued.

(vi) On reaching founding level, open caissons are sealed (Fig. 9.64) by depositing a layer of concrete under water in the bottom of the wells (plugging). The wells are then pumped dry and further concrete or filling material is added to give dead weight (hearting).

### Control of verticality

(i) Add concrete (kentledge on one side or the other).
(ii) Jet under the cutting edge on the 'hanging' side.
(iii) Control the sequence of cell excavation.

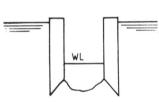

**Fig. 9.65**   Freeing a hanging caisson

### Releasing a 'hanging' caisson

Sometimes a caisson will meet a lens of material with a high friction coefficient, such that continued excavation in the cells causes no further sinking. If jetting or adding kentledge does not free the structure, then the only recourse is to reduce the natural level of water inside the cells (Fig. 9.65). The result is a reduction in the uplift pressure, thus, in effect, increasing the sinking force. In fine grained soils, however, caution should be exercised to avoid the possibility of a 'blow', i.e. piping, (see page 90) occurring, causing the caisson to settle and possibly tilt quite suddenly and significantly.

### Sinking aids

Introducing water, compressed air or bentonite slurry can reduce skin friction considerably. The material is introduced through pipes cast into the walls as shown in Fig. 9.66.

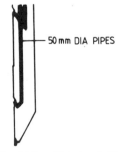

**Fig. 9.66**   Aids to sinking

### Output

Excavation is usually carried out from a rope-suspended grab of approximately 0.75–1.5 m³ struck capacity (Fig. 9.67). The operator is

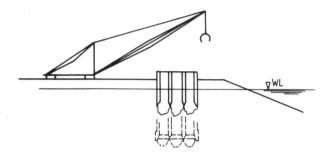

**Fig. 9.67** Grabbing method suitable for caisson sinking

working blind and digging through water and so normal output rates cannot be expected.

## EXAMPLES FOR 0.75 M³ GRABBING BUCKET

Easy dig     – 10 m³/h
Medium dig – 6 m³/h
Hard dig     – 4 m³/h

*Note*: These are bulked volumes and do not allow for stoppages or 'stuck' periods.

### Constructing a caisson from islands

Caissons for bridge piers and docks or lock walls are generally started from temporary islands as shown in Fig. 9.68. However, when deep water is involved a closed or box caisson is usually necessary.

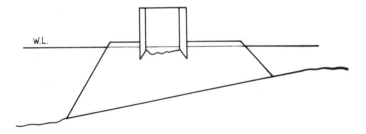

**Fig. 9.68** Constructing a caisson from an island

### 9.14 BOX CAISSONS

The use of open caissons for the construction of dock walls, harbour protection, breakers, watch towers, etc. is often impractical or uneconomic because of the temporary works necessary, e.g. islands or falsework. However, the problem can be avoided with the box caisson, which is constructed as a floating vessel on dry land or in a dry dock and subsequently floated into position and sunk onto its foundation.

The vessel can be of any desired shape, but a full structural and stability analysis must be undertaken at the launching, transporting and sinking phases, in order that planned and controlled procedures can take place.

### Foundations for box caissons

A gravel or crushed rock bed provides a suitable foundation if erosive currents are not present (Fig. 9.69). In fast flowing waters, such as rivers, a piled foundation is more appropriate (Fig. 9.70).

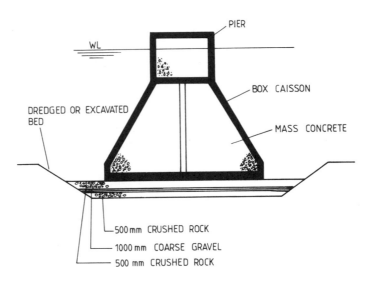

**Fig. 9.69** Caisson foundation where water currents are not excessive

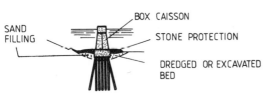

**Fig. 9.70** Piled caisson foundation in rapid flowing water

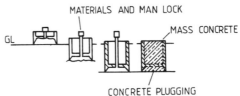

**Fig. 9.71** Sinking procedure for a pneumatic caisson

### 9.15 PNEUMATIC CAISSONS (see also compressed-air working in tunnels, Chapter 10)

A pneumatic caisson is a box made from steel or concrete constructed like a diving bell at its base. Compressed air is pumped into the bottom section, thereby excluding water, to allow workmen to excavate in a dry chamber. The structure is sunk in a similar fashion to the open caisson (Fig. 9.71) but men and materials must be passed through an air locking system (Fig. 9.72).

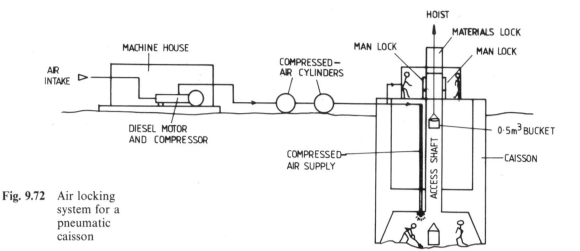

**Fig. 9.72** Air locking system for a pneumatic caisson

The pneumatic caisson is necessary where a foundation in open water must be founded well below the river or sea bed, or alternatively on dry land, where free-flowing soils might cause settlement of adjoining buildings, if an open caisson were chosen.

However, because men have to work in compressed air, the relatively short work shifts and high wages make the method expensive and, today, it is generally used only as a last resort.

### Sinking theory

The calculations required to determine stability during sinking and when in the final resting position are similar to those required for the open caisson, with the addition of the extra uplifting force in the compressed chamber (Fig. 9.73).

Safety regulations for working in compressed air limit pressures to about 3.5 bar, which is equivalent to about 35 m depth in water.

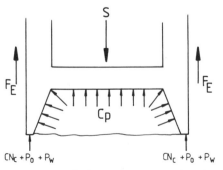

| | |
|---|---|
| $S$ | weight of caisson plus kentledge |
| $F_E$ | external skin friction |
| $C_p$ | uplift force from compressed-air |
| $CN_c$ | net ultimate bearing resistance |
| $P_o$ | effective overburden pressure |
| $P_w$ | water pressure |

**Fig. 9.73** Sinking principle of a pneumatic caisson

**Sinking control**

  (i) Excavate symmetrically.

 (ii) Maintain a berm around the inside cutting edge to reduce air leakage.

(iii) (a) In sands the inside pressure should be about 0.1 bar – higher than that required to maintain dry working conditions. Too much pressure may cause a sudden blowout and subsequent loss of pressure, causing material to flow into the chamber. In these circumstances, remove the men, flood the chamber and then re-pressurise.

    (b) In clays there is less air leakage, but attention must be given to the possibility of a sudden pressure drop when a layer of sand or gravel is encountered.

(iv) Excavate continuously wherever possible to allow continuous movement and avoid a build-up of static friction.

 (v) Have standby compressed-air plant and decompression facilities available.

(vi) The regulations governing work in compressed air are dealt with fully in the references and guidelines given in Chapter 10.

**Output**

An excavation rate by hand in easy material would be about 0.5 m³ per man hour. Up to ten workers in the digging gang is common, working the hours governed by the working in compressed-air regulations.

Typically, sinking rates vary between 0.05 and 0.3 m per day, depending upon soil type.

A topside gang is required to operate the compressor, hoist and lock and in addition it is common practice to have a fitter and a foreman.

# 10

# TUNNELLING

Tunnels are used for underground road and rail transport, for hydro and sewage conduits, drainage and many other services.

The methods adopted for the construction of tunnels vary from hand excavation to sophisticated power-driven tunnelling machines, depending upon the soil type, ground conditions, length of tunnel, etc. The techniques, however, may be roughly classified into six groupings:

  (i) Traditional methods (mainly for tunnelling in rock).
 (ii) New Austrian method.
(iii) Tunnelling with shields.
(iv) Tunnelling boring machines (TBMs).
 (v) Mini- and micro-tunnelling or boring.
(vi) Immersed tubes (including open cut).

## 10.1 TRADITIONAL METHODS

In short lengths of tunnel or where the diameter or shape is unsuitable for excavating machines, the traditional use of explosives to advance the heading is often favoured. Depending upon the stability of the rock the heading can be tackled by one of the following methods:

(a) advance the heading in full face without lining
(b) advance the heading in drifts followed by full face lining
(c) advance the heading in drifts followed by lining progressively in stages

classical methods (rarely used in modern times except for pilot tunnels)

### Full face heading without lining

Where the rock is stable and self-supporting, excavation of the heading in full face (Fig. 10.1(a)) is suitable for tunnels up to about 200 m² in cross-section. The procedure is to drill holes for explosives and

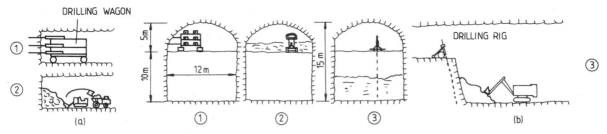

**Fig. 10.1**    Full face heading in rock without lining

advance the heading by blasting. The broken debris is loaded into trucks or onto a conveyor or transported by rail car away from the heading. Single or multiple drifter drills are commonly used to produce the holes for the explosive and these are discussed in detail in Chapter 3. The techniques of blasting are covered later in this chapter and in Chapter 4.

If the heading is large enough to be sub-divided and allow the use of excavating and materials handling plant and equipment, then progress can be increased by benching, as shown in Fig. 10.1(b).

When the cross-sectional area of the tunnel exceeds approximately 150–200 m², the tunnel roof and sides will often require propping, and work should then be carried out progressively from a series of subheadings or drifts.

### Classical methods

#### ADVANCE THE HEADING IN DRIFTS FOLLOWED BY FULL FACE PERMANENT LINING

The heading is usually opened out in segments, working from a small pilot tunnel driven the full length, which provides ventilation and a means of access for transporting materials and for carrying out temporary works such as grouting. If the rock is fairly stable and unlikely to suffer fracturing and bursting as the stresses in the rock are released, the drifts are sequentially advanced and temporarily supported until a complete ring of permanent lining can be put in place. Generally the English (Figs. 10.2 and 10.3) and Austrian (Fig. 10.4) methods follow this procedure.

In the English system, the top segment (1) may sometimes be driven the full length to improve ventilation. As the heading is enlarged, temporary support is given by timber props. The permanent lining is constructed when a full face has been excavated and the props are then removed. The face is progressively advanced in this manner.

The Austrian method is similar to the English system in that almost a complete ring of lining is positioned when a full circumference is available, but unlike the English method the segments of the heading are advanced in a series of drifts. As a consequence, the need to prop

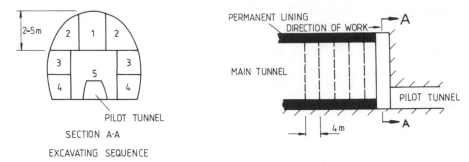

**Fig. 10.2** English method

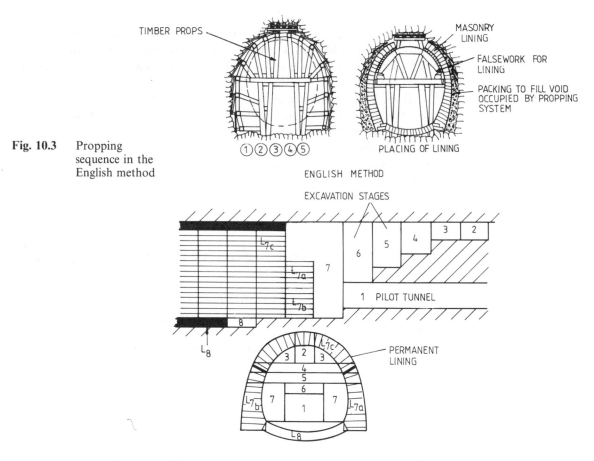

**Fig. 10.3** Propping sequence in the English method

**Fig. 10.4** Austrian method

the full face of the heading (thus causing complicated propping), as in the English method, is avoided. However, the Austrian method requires temporary support of the roof for a longer period, which is subject to repeated dismantling as the drifts are developed.

## ADVANCE THE HEADING IN DRIFTS FOLLOWED BY STAGED PERMANENT LINING

In brittle, fractured or soft rock, to avoid leaving the full face of the tunnel temporarily propped, the permanent lining is installed progressively as the drifts are advanced. The three classical techniques developed to deal with these conditions are the Belgian (Fig. 10.5), German (Fig. 10.6) and Italian (Fig. 10.7) methods.

In the Belgian method (Fig. 10.8) a sequence of drifts is used, such that the top half of the tunnel face is excavated and propped and the

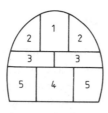

**Fig. 10.5** Belgian method    EXCAVATING SEQUENCE      LINING SEQUENCE

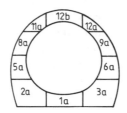

**Fig. 10.6** German method    EXCAVATING SEQUENCE      LINING ERECTION

**Fig. 10.7** Italian method    EXCAVATING SEQUENCE      LINING ERECTION

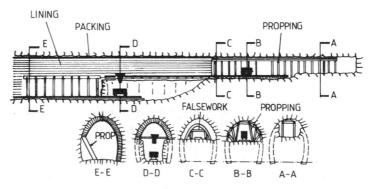

**Fig. 10.8** Propping sequence in the Belgian method

lining is then installed to form a flying arch. The remaining segments are subsequently excavated, working outwards from the centre.

In less stable rock, the German method is preferred. The core is left until last and provides a stable propping base. The side linings are erected next, followed by the roof lining. Like the Belgian method this sequence is performed in progressive stages.

When tunnelling in very unstable conditions, the Italian method is commonly adopted. Small drifts for subheadings are required, working from the base upwards.

## 10.2    NEW AUSTRIAN METHOD

Because of the high cost of timbering and the specialist skills required to install the temporary support, the classical methods of tunnelling have gradually been modified as technical improvements to support systems and permanent linings have been developed. The new temporary support techniques (which can also form part of the permanent lining) are simpler to install and include:

  (i) sprayed concrete
 (ii) rock bolting
(iii) stiffening ribs and liner plates
(iv) full lining

The New Austrian method (Fig. 10.9) is proving to be an economical sequence in rock-tunnelling and more recently in soft ground tunnelling when used in association with the new support methods. However, precise and continuous measurement of ground movements both in the tunnel and at surface level are essential in determining the selection of support method and timing of installation for stability (see Tables 10.1–10.3).

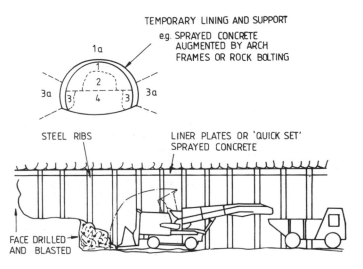

**Fig. 10.9**  New Austrian method

**Table 10.1** Guniting for temporary or permanent lining (Szechy)

| Type of material | Gunite thickness (mm) | Max. unbridged span (m) | Max. unbridged time | Comments |
|---|---|---|---|---|
| Firm | — | — | Non-specifiable | Support not required |
| Loosening over time | 20–30 | 3–4 | Up to $\frac{1}{2}$ year | Gunite only required in the arch roofing |
| Slightly friable | 30–50 | 2.5–3.5 | 1 week | Gunite only required in the arch roofing. Temporary propping needed |
| Friable | 50–70 | 1–1.5 | $\frac{1}{2}$ day | Temporary propping followed by mesh reinforced gunite |
| Very friable | 20–150 | 0.5–0.8 | $\frac{1}{2}$ h | Extensive temporary propping followed by mesh reinforced gunite |
| Immediate light ground pressure | 150–200 | Less than 0.5 | 1–2 min | Support from steel ribs, followed by mesh reinforced gunite |
| Immediate heavy ground pressure | — | 0.1–0.2 | A few seconds | Guniting not suitable |

**Table 10.2** Rock bolting for temporary tunnel support (Szechy)

| Type of material | Max. unbridged span (m) | Max. unbridged time | Rock bolt centres spacing (m) | Comments |
|---|---|---|---|---|
| Firm | — | Unspecifiable | Not needed | No support necessary |
| Loosening over time | 3–4 | Up to $\frac{1}{2}$ year | 2 | Rock bolting followed by wire mesh to support falling fragments |
| Slightly friable | 2.5–3.5 | 1 week | 1.5 | Rock bolting followed by wire mesh or gunited concrete |
| Friable | 1–1.5 | $\frac{1}{2}$ day | 1 | Rock bolting followed by mesh reinforced gunite up to 30 mm thick |
| Very friable | 0.5–0.8 | $\frac{1}{2}$ hour | 0.5 | Temporary props followed by rock bolting and reinforced gunited concrete |
| Immediate light ground pressure | Less than 0.5 | 1–2 min | Unsuitable | Rock bolting not appropriate |
| Immediate heavy ground pressure | 0.1–0.2 | A few seconds | Unsuitable | Rock bolting not appropriate |

### Sprayed concrete

It can be seen from Table 10.1 that guniting is an appropriate temporary lining and support for virtually all types of tunnelling. However, the method is most economical with full face working. In loose material, additional support using mesh or rock bolts or steel ribs may be required. Sprayed concrete of aggregate size up to 25 mm can be used, and with chemical additions strength, adhesion and setting time can be greatly improved. This type of lining may also serve the purpose of a permanent lining, depending upon the tunnel design.

A method of achieving good results is illustrated in Fig. 10.10.

**Table 10.3** Steel ribs and liner plates for temporary tunnel support (Szechy)

| Type of rock | Max. unbridged span (m) | Max. unbridged time | Comments |
|---|---|---|---|
| Friable | 1–1.5 | $\frac{1}{2}$ day | Rib support after full face excavation |
| Very friable | 0.5–0.8 | $\frac{1}{2}$ hour | Rib support and a ring of liner plates assembled in segments with subsequent grouting behind the liner plates |
| Immediate ground pressure | 0.1–0.2 | Less than 2 mins. | Rib support and a ring of liner plates assembled in segments, followed by an overlay of reinforced gunite |

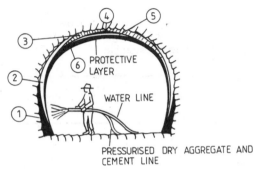

**Fig. 10.10** Sprayed concrete lining (Leins)

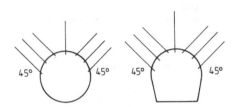

**Fig. 10.11** Rock bolting to support tunnel roofing and sides

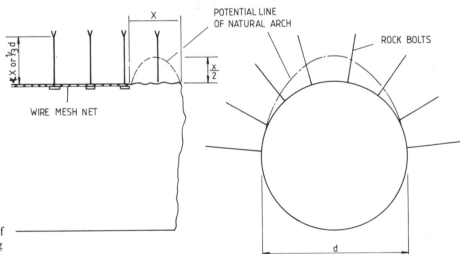

**Fig. 10.12** Principles of rock bolting method

## Rock bolting

Rock bolting, often applied in combination with sprayed concrete, can be used as a temporary or permanent support system as indicated in Fig. 10.11 and Table 10.2.

If the tunnel heading is advanced a distance $x$ as illustrated in Fig. 10.12 then unless the roof is supported, the fracture zone of rock will

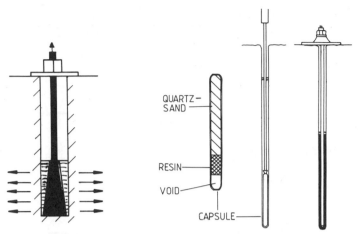

**Fig. 10.13**  Expanded-base rock bolt   **Fig. 10.14**  Grout-type rock bolt

begin to fall away and would ultimately settle in a natural parabolic arch, extending about $x/2$ beyond the perimeter of the circular tunnel. The rock bolts should therefore be longer than $x/2$ and generally not less than $\frac{1}{3}$ the tunnel diameter or width ($d$). The number of bolts and the spacing required to support a given area of tunnel roof and any compressive forces will depend upon the diameter, material and type of anchorage of the bolt.

## TYPES OF ANCHOR

Several proprietary systems are available involving a split or spreading device which presses against the walls of the drill hole and grips more tightly as the nut is turned (Fig. 10.13). Others rely on a chemical reaction between resin and quartz-sand. A capsule containing the two ingredients is introduced into the drill hole and mixed by the penetration and turning of the rock bolt (Fig. 10.14).

These methods use relatively small diameter bolts (15–30 mm). For more substantial rock bolting with post tensioning, a grouting technique is required (see Chapter 7).

### Stiffening ribs and liner plates

Table 10.3 illustrates that for very friable, shattered rock and certain categories of soft ground, where immediate heavy ground pressure is to be encountered, neither guniting nor rock bolting is likely to be a suitable means of providing the sole temporary support and in such conditions additional propping with steel ribs is required. When the material will stand unsupported long enough to install a complete unit or ring, then full face excavation is possible. Otherwise the face should be excavated in segments and parts of the rib temporarily propped until a complete unit can be formed (Fig. 10.15). In material which

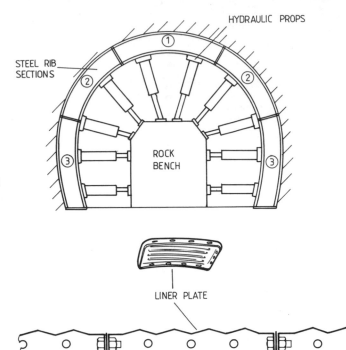

**Fig. 10.15** Propping very unstable rock with props and ribs

**Fig. 10.16** (*Top*) Liner plate
**Fig. 10.17** (*Bottom*) Combined liner plate and rib support

requires intermediate support between ribs or where immediate ground pressure is present during excavation, liner plates (Fig. 10.16) may be incorporated as shown in Fig. 10.17. The voids between the plates and tunnel surface may subsequently be grouted to distribute the loads uniformly (Fig. 10.18).

Additional strength may be obtained by overlaying the liner plates, ribs, etc. with reinforced gunited concrete as illustrated in Fig. 10.19. In more stable ground it is often possible to use the liner plates without the ribs, excavating a full face to the width of a liner plate.

## Full lining

When the permanent lining is designed in either cast iron or precast concrete units, as is occasionally the case, it can then perform a similar function to liner plates. The usual method is to build a ring of linings from the bottom upwards. It is therefore necessary to adopt a full face method of excavation. If heavy, the lining units can be positioned with special equipment (see Fig. 10.37).

The method may be used in conjunction with rock bolting to increase working space so that lining and excavation areas do not become too congested.

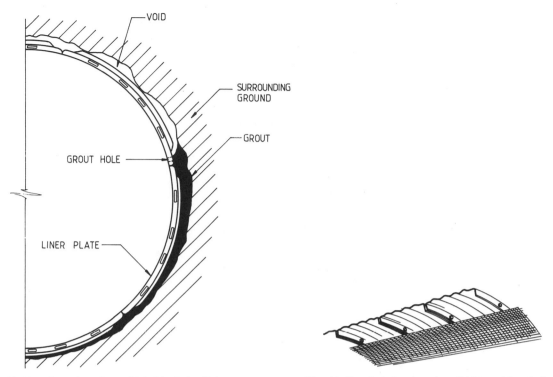

**Fig. 10.18**   Grouting the void behind the lining

**Fig. 10.19**   Reinforcing the ribbing with reinforced sprayed concrete

### Liner plate method

In relatively stable ground, where large inflows of water do not disturb the soil particles, full face excavation to a depth equal to the width of a lining panel can be carried out, followed by immediate extension of the lining ring as described earlier under the Full Lining method. Stiffening ribs may be included where ground conditions dictate. Unfortunately, this temporary lining system, which is very speedy, must remain in place whilst the permanent lining is erected, and is therefore irecoverable.

## 10.3   SHIELD TUNNELLING

Where the liner plate method is uneconomical or where the ground will not stand unsupported long enough to enable the permanent lining to be erected, then the shield technique is likely to be selected. The shield acts as a steel casing tube, which is pressed into the heading in front of the lining, thus providing protection whilst the face is excavated. This procedure is shown in Fig. 10.20. In order to reduce resistance to penetration, as much of the face as practicable is excavated in front of

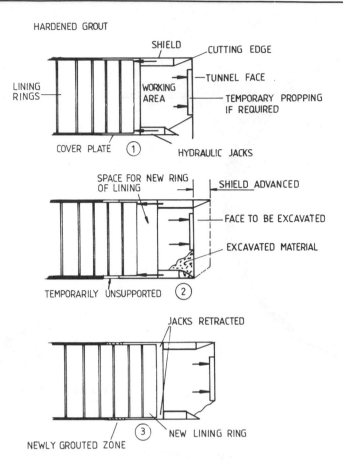

**Fig. 10.20**   Shield
tunnelling
method

the cutting edge. Hydraulic jacks at the rear use the lining as the source
of reaction to push the shield forward a distance equal to the width of
a ring of lining. The lining is subsequently extended under the
protection of the shield and the space left by the thickness of the shield
plating is finally filled by grout. (Good practice usually requires
grouting after the erection of each ring, to reduce the possibility of soil
movement.)

The excavation of the face is usually carried out manually with
power-assisted hand tools (Fig. 10.21) or, where the expense can be
justified, with mechanical excavating equipment such as a backhoe (Fig.
10.22).

Where blasting is required, for example in hard rock, the shield
method is usually avoided because of the technical problems associated
with steering. The tendency would be for the shield to wobble in a fully
excavated face. The shield method is suitable for tunnel diameters of
4–12 m. For the larger tunnels in this range staging is necessary to
carry out the excavation work (see Fig. 10.21) and to provide
foundation bases to prop the working face. Indeed, in a freely flowing
soil a special bulkhead shield is often required (Fig. 10.23).

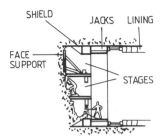

**Fig. 10.21**  Manual excavating method illustrating a shield with working platforms

**Fig. 10.22**  Excavating with mechanised equipment using the shield method

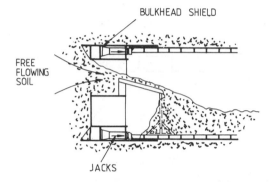

**Fig. 10.23**  Bulkhead shield for free flowing soils

### Thrusting forces required in the shield method

For a shield without stage levels, a thrust against the lining of up to 100 tonnes/m of shield circumference may be necessary. Where obstructions are encountered, considerably higher force could be needed. When the shield is fitted with staging levels, the required thrust is likely to be 300–400% more than the single unstaged shield.

## 10.4  TUNNEL BORING MACHINES (TBMs)

### Rock cutting machines (Fig. 10.24)

The demand for more economical tunnelling methods has gradually led to the development of rotary machines. Where uniform soil or rock is present and the length of the tunnel exceeds about 2 km, the high capital cost of the equipment can be justified by the increased production performance. However, if the nature of the ground is likely to change frequently, especially in soft ground where boulders, flowing sands, gravels, etc. might render the cutting head unsuitable for such a variety of material (the rotary machine requires a free standing face to attack), then a more traditional shield method would probably be chosen.

The rotary tunnelling machine combines a conventional shield with a rotary cutting head. The cutting head consists of 3 to 8 radial arms

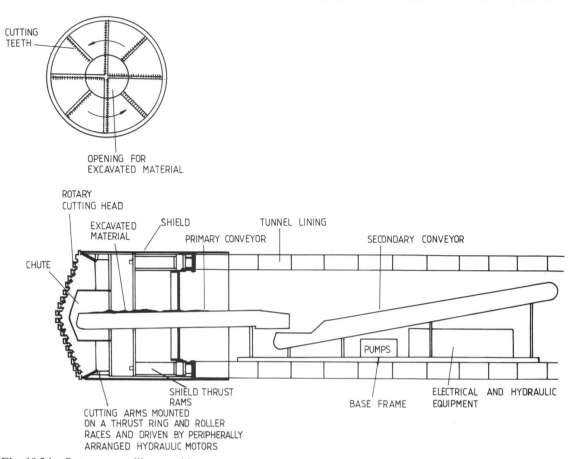

Fig. 10.24 Rotary tunnelling machine

fitted with chisels or discs and is rotated at 5–8 cycles per minute. The arms are mounted on a thrust ring and roller races and are driven by compact hydraulic motors arranged around the periphery. The force at the cutting edge is provided by the reaction of thrust rams against the lining. Excavated material is directed centrally through the cutting head and transported by conveyor to some convenient point away from the working area. A complete cycle of operations is shown in Fig. 10.25.

Fig. 10.25 Method of working with the rotary tunnelling machine in rock using a tail shield

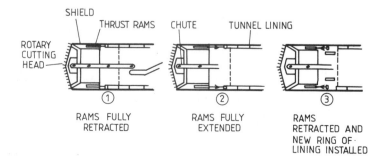

**Fig. 10.26** Method of working with the rotary tunnelling machine in stable rock

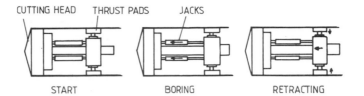

In hard and more stable rock the tail shield is not required. Also, the lining may be of much lighter construction and could not withstand the high jacking forces. The reaction must therefore be taken against the tunnel sides as indicated in the sequence shown in Fig. 10.26. Diameters ranging from 2–10 m are available. Thrusts of up to 2 kN/mm of cutting head diameter can be delivered for cutting through hard rock.

### Soft ground machines (Fig. 10.27)

Tunnelling in granular soils, such as water-bearing sands and gravels, usually presents severe difficulties to conventional shield-drive methods. For example the ground would have to be extensively grouted, dewatered, or the work carried out in compressed air. Also, the face must be propped to avoid collapse, thereby causing difficulties in excavating and in pushing the shield forward. However, the alternative of using a rotary shield requires a face that will stand vertically to allow the cutting action and spoil removal to take place efficiently. This problem led to the proposal of using bentonite slurry to provide the support and subsequently a machine was manufactured by R. L. Priestley Ltd. (Fig. 10.27). It is similar to a soft rock rotary-cutting machine, but the cutting head operates against the face in a chamber of circulated bentonite slurry. The mixture of spoil and slurry is continuously discharged through the centre of the cutting head into a sump and pumped away for disposal.

The presence of lumps of rock has in the past, posed severe difficulties for this kind of machine, requiring the isolated fragments to be laboriously removed by hand when the rotary head was stationary.

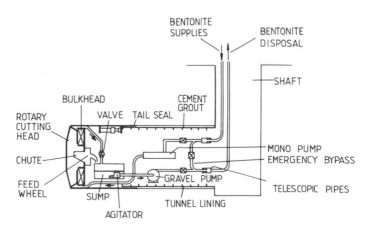

**Fig. 10.27** Soft-ground rotary tunnelling machine

**Table 10.4** Cutting heads for tunnelling machines

|  | Soft | Medium | Hard | Very hard |
|---|---|---|---|---|
|  | Shale<br>Clay<br>Limestone<br>Granular<br>  soils | Limestone<br>Sandstone<br>Sandy shales<br>Porphyries<br>Iron Ore | Silicon<br>Limestone<br>Dolomite<br>Granite<br>Iron Ore<br>Quartzite | Iron formation<br>Quartzite<br>Hard igneous rocks |
| Compressive<br>  strength<br>  $(N/mm^2)$ | up to 50 | 100 | 200 | 300 |
| Cutting<br>  heads | Teeth<br>  or picks | Roller<br>  discs | Roller<br>  buttons | Blasting with<br>  explosives |

More modern versions however, having a mixture of teeth and discs can now break up cobbles of softish rock ($60 \, N/mm^2$) with dimensions of 250 mm or so down to about 20 mm. The slurry type machine can operate effectively in up to 4 bar water head. Recently alternative forms of rotary machine have been successfully developed using the principle of Earth Pressure Balance (EPB). Material enters through slots and is compacted behind the cutting head to form a stable plug, which is progressively discharged through a gate in the bulkhead. The unit is best suited to silts and clays, as added water pressure in the chamber is often needed for water-bearing sands and gravels, and this then offers little advantage over the slurry type machine. Again upto about 4 bar pressure head is possible.

A gang of 5–6 workers is required to operate the machine, fix the lining, handle the spoil, etc.

### Selection of the cutting head

The methods available to deal with different rock hardnesses are shown in Table 10.4.

### (i) TEETH OR PICKS (Fig. 10.28)

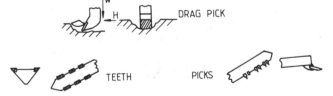

**Fig. 10.28** Teeth and picks for soft rock

Tungsten carbide tipped teeth are generally mounted on radial arms to form the cutting head which is either rotated or oscillated into the heading.

### (ii) ROLLER DISCS (Fig. 10.29)

These are arranged symmetrically on a strong base and a heavy force is applied to the cutting head, causing the rock to shatter and splinter. The alternative is roller cutters (Fig. 10.29), which produce slightly finer rock fragments.

**Fig. 10.29** Discs and cutters for medium rock

150 – 300mm | DISC CUTTER | ROLLER CUTTER

### (iii) ROLLER BUTTONS (Fig. 10.30)

Hardened metal buttons mounted on a conical roller are pressed with great force against the heading and rotated, causing fracturing of the rock surface.

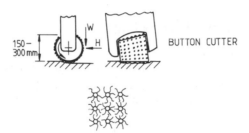

150 – 300 mm | BUTTON CUTTER

**Fig. 10.30** Roller buttons for hard rock

### (iv) BLASTING

At present, mechanical techniques are inadequate for excavating very hard rock and explosives must be used to loosen the material. It is then loaded with conventional machinery (Fig. 10.1). The use of explosives is discussed in Chapter 4.

## 10.5 BOOM CUTTER/LOADER FOR ROCK TUNNELLING

As an alternative to rotary machines or blasting, the boom cutter/loader can be used in rocks and hard soils with compressive strengths of up to 150 N/mm$^2$, i.e. medium/hard. However, drilling and blasting is likely to be more economical for rocks stronger than 50 N/mm$^2$. The machine (Fig. 10.31), can be mounted on rails or tracks and works against the face with a contact pressure of about 0.1 N/mm$^2$ using a series of rotating teeth or blades. The material is immediately transferred by conveyor to a convenient loading point. About 18 m$^3$/h of soft rock can be removed for example with a medium sized machine of about 100 kW power, and about 300 m$^2$/h has been achieved with very large equipment of 800 kW power.

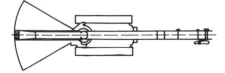

**Fig. 10.31**  Boom cutter/loader machine

## 10.6  PRODUCTION PERFORMANCE

### Rock tunnelling

(i) Explosives and modern propping systems (see Tables 10.1–10.3) – 1 to 7 m per 8 h shift.
(ii) Classical methods – 0.5 to 2 m per 8 h shift.
(iii) Tunnelling machines – Up to 15 m per 8 h shift, depending upon rock hardness and rate of lining erection.

### Soft ground tunnelling

(i) Tunnelling machines:
   (a) Up to 20 m per 8 h shift in uniform soil such as clay and very soft rock.
   (b) 2 to 5 m per 8 h shift in water-bearing granular soils.
(ii) Shields
   (a) hand digging
      (i) Uniform soil such as clay – 4 to 8 m per 8 h shift
      (ii) Poor ground – 1 to 4 m per 8 h shift
   (b) mechanical excavation, e.g. backhoe    Outputs similar to tunnelling machines.

*Notes*:
(i) The number of workers needed for excavating, and/or the machinery required to load material depend upon the working area available, e.g. 2 m diameter heading can accommodate two face miners.
(ii) In soft ground rotary machine tunnelling the rate of lining erection is often the limiting factor, e.g. in non-bolted segments, about 3 m/h of advance is likely to be the maximum rate.
(iii) It should also be noted that the wide range of output rates in rock tunnelling is dependent upon the amount of propping required, incoming water, etc.
(iv) The shield can take from 1 to 4 weeks to assemble and position in the heading, depend upon its size and associated mechanical equipment.

## 10.7 LINING ERECTION

A tunnel usually requires permanent lining which, wherever possible, is utilised for temporary support during advancement of the heading. In good practice, the lining is followed up a short distance from the excavation face. There are three common forms of lining used in tunnel design:

(i) Segmental
(ii) In situ reinforced concrete
(iii) Masonry

### (i) Segmental

(a) Cast iron units (Fig. 10.32) approximately 1–2 m long × 0.5–1 m wide when bolted together provide a strong lining which is able to resist both the heavy jacking forces from the shield and external loads imposed by the surrounding ground. By making the joints between butting flanges of lead wire or similar and using bituminous washers to cover bolt holes, the lining can be reasonably sealed against water penetration.

(b) Precast concrete segments (Fig. 10.33) have been introduced to avoid the high cost of using cast iron. Joints are made tongue and groove and caulked with rubberised bituminous strips. The segments may be bolted together in a similar way to cast iron units, but where uniform loads can be expected, bolting can be

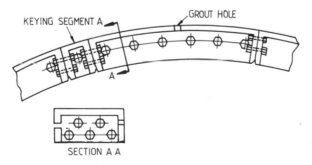

**Fig. 10.32** Cast iron permanent lining segment

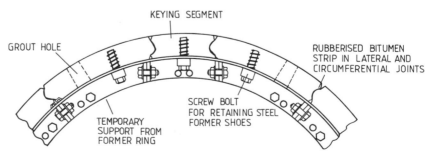

**Fig. 10.33** Precast concrete permanent lining segment

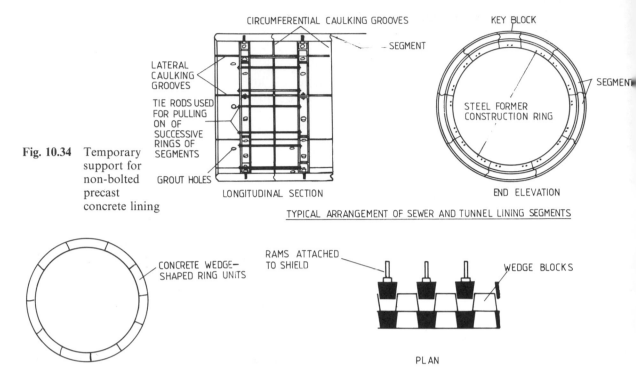

**Fig. 10.34** Temporary support for non-bolted precast concrete lining

**Fig. 10.35**   Wedge block system using precast concrete lining segment

eliminated, thereby providing some flexibility for adjustment to ground movements. However, as a result, the units must be temporarily supported until the keying section at the top can be placed in position (Fig. 10.34). This problem can be avoided by using a wedge block system shown in Fig. 10.35. Alternate blocks are positioned and are individually held in place by arms on the shield or rotary machine. The remaining blocks are subsequently pushed between the taper spaces, again by rams, to produce a tightly fitting ring. No bolting is required and installation rates of up to 3 m/h can be obtained with mechanical handling, and it is therefore a favoured method when using rotary tunnelling machines, where high rates of progress are needed to match the rate of advance of the machine.

Other materials have been tried, including steel; in particular, steel liner plates (see Fig. 10.16) provide an economic alternative where a heavy lining is not required, such as in rock tunnelling, or they can be incorporated into an in situ concrete permanent lining.

## METHOD OF ERECTION

In small diameter tunnels up to 2–2.5 m, segments can be manhandled (Fig. 10.36), but larger tunnels require special machinery (Fig. 10.37) to

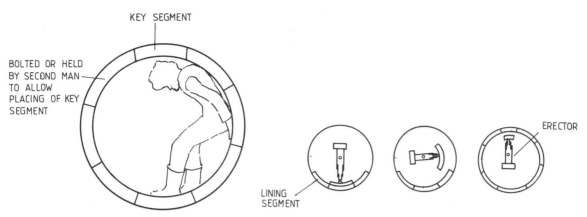

**Fig. 10.36**　Manual method of positioning lining segments

**Fig. 10.37**　Mechanical method of positioning lining segments

place the individual units. It is normal practice to work from the bottom, finishing with the key section at the top. The void between lining and tunnel wall is then grouted so that the ground pressures can be taken up by the ring of segments.

### (ii) In situ reinforced concrete

Frequently in rock tunnelling the roof is able to stay unsupported, or at least is stabilised with rock bolts or ribbing. For such tunnels an expensive cast iron lining is not required, as reinforced sprayed concrete may be adequate. If the lining requires greater strength or a preformed shape, then normal reinforced concrete can be cast behind specially designed travelling formwork and falsework (Fig. 10.38).

### (iii) Masonry

In earlier times, when using classical tunnelling methods, the complete lining was produced in masonry brickwork, which required temporary

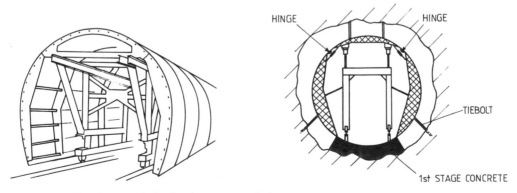

**Fig. 10.38**　Travelling formwork for in situ concrete lining

falsework. Today, such a lining is mostly used to provide a protective coating of other lining systems for tunnels exposed to corrosive elements.

## 10.8   STABILISATION METHODS

So far, problems caused by the ingress of water into the heading, or the changes encountered in soil texture and stability have been avoided. Such difficulties can generally be contained by the use of grouting or freezing techniques, as described in Chapters 7 and 9. A small diameter pilot tunnel driven prior to the main heading often enables lenses of water-bearing soil, rock fissures, etc. to be filled before being exposed by the full heading. However, where incoming water is likely to be a problem through much of the drive, the grouting may be too expensive and the use of compressed air should then be considered. In general, inflows of water in excess of 2–3 lps per m$^2$ of tunnel face would require containment.

## 10.9   COMPRESSED AIR FOR TUNNELS (INCLUDING CAISSONS AND DIVING WORK)

The pressure of water at the tunnel level can be counter-balanced by pressurising the working area, as shown in Fig. 10.39. However, in order to have sufficient head at the sole of the tunnel, excess pressure exists at the roof and some leakage is inevitable (Fig. 10.40), especially in granular soils. If the leakage is excessive, a 'blowout' may occur, causing a sudden rush of water into the heading. This possibility can be reduced by covering the ground surface above the line of the tunnel with a layer of clay (Fig. 10.41). In more severe cases extensive grouting

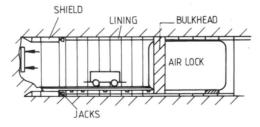

**Fig. 10.39**   Working in compressed-air

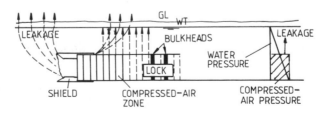

**Fig. 10.40**   Air leakage from a compressed-air zone

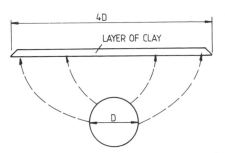

**Fig. 10.41** Method of reducing loss of compressed air

may be the only alternative and in any case ample spare compressed air supply should always be available. Hewett-Johanneson recommends provision of about 5 m³ free air per min per m² of face area for cohesive soils and 10 m³ per min per m² in granular soil. The compressed air in the working zone is built up against a bulkhead, through which both men and materials need access to the working area by means of an air lock. This consists of a small chamber, called a man-lock or decompression chamber (see Fig. 10.39) within which the air pressure can be reduced to or raised from atmospheric, depending upon whether personnel are entering or leaving. The procedures are known as compressing and decompressing, and must be carried out in strict sequence, otherwise serious injury to the miners will result.

In tunnelling schemes using compressed air, it is good practice to have two adjacent air-locks, one for personnel and one for materials. The latter does not require the elaborate decompressing process. However, space is often at a premium and only a single lock may be possible. In compressed-air working, pressures up to 50 psi (3.5 bar) above atmospheric have been used, but a more typical maximum is about 35 psi (2.5 bar). The compressing and decompressing procedures are laid down in 'The Work in Compressed-Air Special Regulations' (1958), which specifies decompressing times, lengths of working periods, medical checks and other requirements. However, medical criticism of these regulations has prompted the Medical Research Council to recommend new tables, known as the 'Blackpool Decompression Tables', after the first contract for which they were adopted.

Permission to use these new tables should be sought from the Health and Safety Executive.

Because of the rigorous procedures necessary for working in compressed air, the hours a workman can actually spend at the work face are fewer than those of a normal shift, although pay must be the same as for a normal length shift. Consequently this method is extremely expensive.

### Dangers of working in compressed air

The most visible dangers of working in compressed air are generally associated with inadequate decompression and fall into two categories – acute and chronic decompression sickness.

### ACUTE OR IMMEDIATE DECOMPRESSION SICKNESS

Type 1 – mild limb pain (the 'niggles')
  severe limb pain (the 'bends')
  skin mottling and irritation (the 'itches')

Type 2 – vomiting, with or without stomach pains
  vertigo (the 'staggers')
  tingling and numbness of the limbs
  paralysis or weakness of limbs
  choking
  severe headache
  visual defects
  angina, irregular pulse
  collapse, hypertension
  lung cysts
  coma
  death

### CHRONIC OR LONG-TERM DECOMPRESSION SICKNESS

 (i) Nervous and psychiatric forms, including paralysis
(ii) Disease to the bone joints

Medical evidence indicates that decompression sickness is primarily caused by nitrogen bubbles being trapped in the blood and tissues, following rapid decompression and that very slow decompression prevents this happening. However, an understanding of the medical problems is not complete at present and research is continuing.

## 10.10  TUNNEL VENTILATION

Air must be supplied to both workmen and machinery operating in the tunnel, so that the level of oxygen content does not fall below about 20%. A means of extracting carbon dioxide produced from exhaling, explosives, internal combustion in engines, etc. must also be provided.
  The following values are recommended:

Workmen      – 2 to 5 m$^3$ per min of fresh air per worker.
Diesel engines – 2 to 3 m$^3$ per min per kW of power generated.
Explosives    – Several formulae have been developed relating to the weight of explosives used, gases produced, minimum waiting time required after the explosion before the next explosion, number of workmen involved, etc., but a practical guide is:
  $Q = 10 \times$ cross-sectional area of tunnel in m$^2$
  $Q$ is in m$^3$ per min of fresh air.

## Methods of ventilation

Fresh air can be introduced to the work face by blowing the air along a pipeline of 300–500 mm diameter with its intake near to the tunnel entrance (Fig. 10.42). The displaced used air simply percolates back along the tunnel and is expelled at the entrance. Alternatively, in the reverse of this system, contaminated air is drawn away from the work face and fresh air pumped in at the entrance to the tunnel (Fig. 10.43). In very long tunnels intermediate shafts designed to provide permanent ventilation can be used to reduce the length of pipe runs and so limit the pressure drop caused by friction. A typical ventilation fan is shown in Fig. 10.44 with its corresponding characteristics in Fig. 10.45. Clearly, as the length of pipe run is increased, the power size of the blower must also be stepped up to achieve the desired delivery or extraction rate.

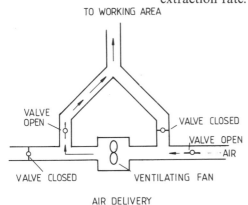

**Fig. 10.42**  Ventilation system using air delivery method

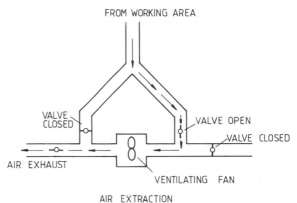

**Fig. 10.43**  Ventilation system using air extraction method

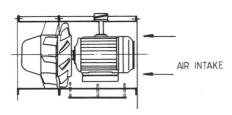

**Fig. 10.44**  Typical ventilation fan

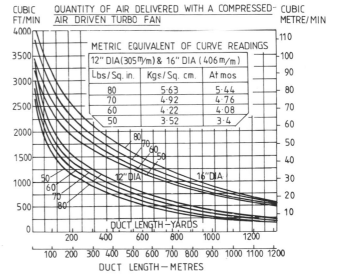

**Fig. 10.45**  Quantity of air delivered with a turbo fan

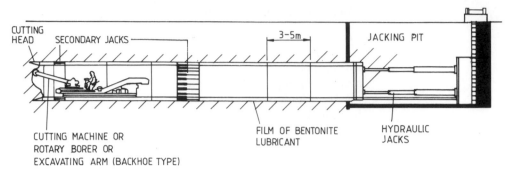

**Fig. 10.46**   Pipe-jacking method

## 10.11   MINI AND MICRO-TUNNELLING

The recent developments in pipe-jacking and auger boring are often referred to as mini-tunnelling for diameters 4.0 to 0.75 m and micro-tunnelling below this range.

### Pipe jacking

The demand for short lengths of medium diameter tunnel for sewers, subways, ducts, sleeves, etc. in a single drive of up to 500 m is increasingly being satisfied by pipe-jacking methods. Soft rock, soft/medium soil, water bearing granular materials, and even ground containing cobbles can all be tackled, using a system of concrete or steel pipes jacked (or pushed) into position from a pit.

### CONSTRUCTION PRINCIPLE

The pipe-jacking principle involves pushing sections of pipe with primary rams through the ground from a jacking pit. Excavation takes place behind a shield as the driving proceeds, performed manually by hand digging, or a backhoe for diameters down to around 1 m (Fig. 10.46). Smaller bores generally necessitate using rotary slurry cutting heads (Fig. 10.47) similar in concept to TBM's or an augering device, (Fig. 10.48), although both the latter systems are capable of dealing with larger diameters also.

### Rotary slurry machine (Fig. 10.47)

The first, i.e. leading pipe, is equipped with the rotary head and minor directional control achieved with secondary rams placed between it and the cutting head. Indeed as the length of drive increases and skin friction between the pipe and wall builds up, further rams may be necessary, located sufficiently close to overcome the driving resistance. However, ultimately the injection of lubricants may be unavoidable.

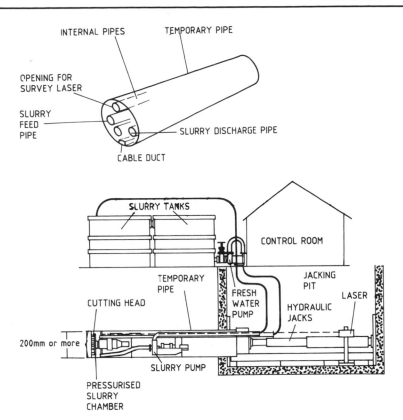

**Fig. 10.47** Pipe jacking using a rotary slurry boring machine

Producing a curved path is possible by inserting wedges between the pipes, and radii of 300 m have been successfully achieved on long drives of 500 m or so. The whole system uses a laser device located in the jacking pit to aid steering.

The latest rotary cutting heads can cope with cobbles up to about 250 mm and rock strengths of 60 N/mm² or so, by inducing a crushing action in the bulkhead chamber. Up to 4 bar water head can be accommodated.

### DETAILS

| | |
|---|---|
| Pipe diameter | 0.25 m to 4 m |
| Drive length | 500–600 m max. |
| Mains rams | 3000–3500 kN thrust each (10 kN/m² surface resistance) |
| Secondary rams | 500–800 kN thrust each |
| Accuracy | $\pm 20$ mm in 100 m |
| Load pipe, into breach | 30 mins |

PRODUCTION

(i) Hand digging by 2 miners      1 to 3 m per 8 hour shift.
(ii) Slurry rotary machine
     Uniform soils or soft rock      10 to 20 m per 8 hour shift.
     Water bearing granular soil      5 to 6 m per 8 hour shift.
       (10 m head)
     Slurry machines generally require 5 operatives:
       driver, desander, crane driver, pipeman, banksman.

### Auger boring

Auger boring equipment consists of a base frame located in a jacking pit, together with flight auger, cutting head and rams (Figs 10.48 and 10.49). Either steel sleeves or concrete pipes are thrust through the ground as the heading progresses. The technique can be used in soft ground to medium-hard rock (80 N/mm$^2$) when fitted with an appropriate cutting head and teeth/discs. Equipment specifications shown in Table 10.5 indicate the capabilities and approximate boring distances of machines from the jacking pit.

The larger units commonly separate the auger and cutting head, the former simply being used to move material from the face. The cutter in this case is designed to act independently, powered by individual motors. The hollow spine of the spiral flighted auger acts as the service duct.

With Earth Pressure Balance machines the cutting head allows soil to enter a bulkhead chamber, is compacted, and subsequently extruded onto the flighting. In this manner water bearing soils, up to about 4 bar lead can be accommodated without the need for complicated slurry power systems as on the rotary borer. Consequently the auger method is often favoured for simplicity.

DETAILS

Rotation speed    30–60 rpm
Accuracy         1° in 30 m
Diameters       Up to 1200 mm
Drive length     150 m max., limited by thrust needed to overcome
                friction
Output           Up to 10 m per 8 h shift in soft/medium uniform soils

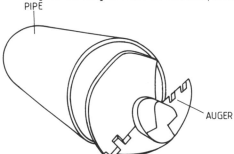

PIPE

AUGER

**Fig. 10.48**   Rotary boring method

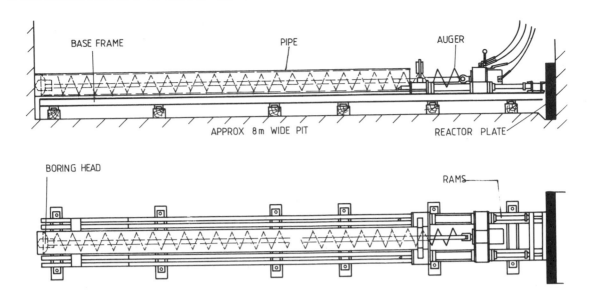

**Fig. 10.49**   Auger boring machine

**Table 10.5**   Boring machine specifications as distance in metres

| Thrust power | Auger diameter (mm) | | | | | | | | | | | | |
|---|---|---|---|---|---|---|---|---|---|---|---|---|---|
| | **80** | **150** | **200** | **300** | **400** | **500** | **600** | **750** | **900** | **1050** | **1200** | **1350** | **1500** |
| 46 kN 5 kW | 30 | 30 | 27 | 26 | 15 | | | | | | | | |
| 46 kN 7.5 kW | 36 | 30 | 26 | 21 | 15 | | | | | | | | |
| 130 kN 15 kW | 91 | 84 | 69 | 49 | 34 | 27 | 23 | | | | | | |
| 270 kN 15 kW | 91 | 91 | 91 | 84 | 58 | 44 | 35 | | | | | | |
| 270 kN 50 kW | | | 91 | 91 | 72 | 61 | 46 | 30 | 23 | | | | |
| 540 kN 50 kW | | | 91 | 91 | 91 | 85 | 67 | 46 | 33 | | | | |
| 560 kN 60 kW | | | | 91 | 91 | 80 | 73 | 56 | 44 | 38 | 33 | | |
| 1200 kN 60 kW | | | | 106 | 106 | 91 | 91 | 85 | 68 | 57 | 45 | | |
| 2700 kN 75 kW | | | | | | 152 | 152 | 152 | 114 | 144 | 99 | 76 | 61 |

*Note*: For both pipe jacking and auger boring, up to 1–2 weeks may be required to prepare and position the equipment in the jacking pit.

### 10.12  MICRO-SYSTEMS

#### Percussive boring

A new technique developed for installing pipes of 200 mm diameter and less has been developed. The method requires the pre-forming of a hole slightly larger in diameter than the pipe.

The equipment consists of a piston-driven chisel head (Fig. 10.50) operated by compressed air (minimum 6 bar). The head is lined up and started from a cradle (Fig. 10.51). Compressed air is then introduced and the device simply punches its way through the ground pulling new PVC piping along as progress is made. Obstructions such as bricks or stones are crushed or pushed aside (the head can be reversed out if necessary). The method is suitable for distances up to about 50–60 m but 20 m is more typical. Approximately 10 m/hour of progress, for example, is possible.

The latest developments with suitable attachments such as a cone slotted onto the pipe and mounted in front of the hammer end, have enabled pipes of 1500 mm diameter to be rammed into place. The technique basically allows soil to enter the casing which is subsequently extruded using compressed-air or water pressure after the full length of steel pipe has been installed.

Applications of the mole equipment have far exceeded their early uses and today are routinely adopted in bursting old drainage and sewer pipelines, fresh plastic pipe being pulled into position with the mole.

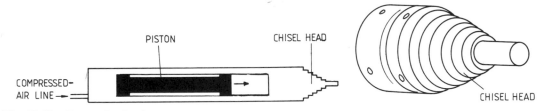

**Fig. 10.50**  Percussive boring

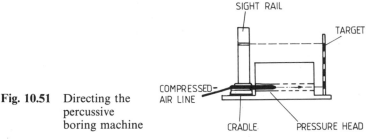

**Fig. 10.51**  Directing the percussive boring machine

**Fig. 10.52** Pulley and tackle to give extra thrust in percussive boring

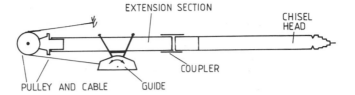

### Steerable micro-bore systems (Guided boring)

Micro-boring systems can now facilitate trenchless installation of plastic pipes in the range 25 to 300 mm diameter. Bores of 500 m or so can be achieved but in more normal practice about 200 m is the limit restricted by the pullback capacity of the rig itself.

The basic method (Fig. 10.53) involves a hydraulic or electric rotary motor and drilling rig capable of producing rotation of the drill string at up to 500 rpm. The rig is aligned at an angle of 15–20° to the surface and a jet of bentonite forced under pressure asymmetrically from the carbide tipped drill bit to produce both a cutting action and return fines along the bore to a small utility pit. Directional control is

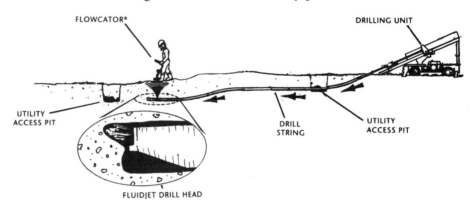

Step 2: Backreaming

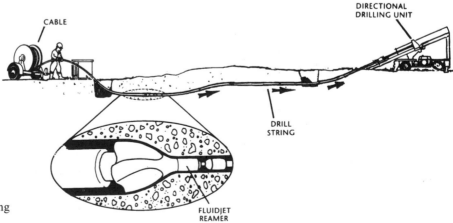

**Fig. 10.53** Steerable micro-boring method

achieved by means of a transmitter unit mounted in the drill head with the signal manually detected at the surface through a hand held antenna. Steering is then obtained by utilising the bias of the cutting jet to hollow out one side of the bore and so cause a change in direction. Accuracies of about 150 mm in the final location of the bore are practicable, indeed obstructions can be circumvented, and curves accommodated.

The bore usually takes place at depths in the range 0.5 to 10 m to form a curved pattern and is most suitable in clays, but has also been successfully used in sands and gravels.

Once the bore is complete as a small pilot hole, the cutting bit is replaced by a back-reaming head equipped with symmetrically positioned jets and a bore of up to 350 mm can be obtained. Any cable to be installed is usually pulled through simultaneously as the backreaming progresses.

## RIG INFORMATION

| | |
|---|---|
| Torque | 1 kNm |
| Feed and pullback forces | 60 kN |
| Rotary speed up to | 500 rpm |
| Rotary motor power up to | 100 kW |
| Drill rods | 3 m × 25 mm |
| Bentonite jet pressure | 35 N/mm$^2$ (5000 psi) |
| Drilling flow rate | 5–80 lpm |
| Operating weight | 1–1.5 tonnes |
| Rate of installation | up to 10 m per hr |

### Directional drilling

Larger drilling rigs are available up to about 250 kW power supplying 100 kN-m torque and 2500 kN pullback force, suitable for dealing with crossings, e.g. rivers up to 1000 m and pipe diameter of 700 mm or so. Down the hole instrumentation is used to locate the steering head. This larger scale activity is commonly called directional drilling and uses similar equipment to rotary drilling (see curved boreholes, p. 20).

## 10.13  IMMERSED TUBES

Units up to 100 m × 40 m are laid end to end to form the tunnel (Fig. 10.54). The procedure consists of:

(i) dredging the trench in which the tunnel is to be placed
(ii) preparing either a permanent screeded bed or piled foundation (Fig. 10.55), depending upon the soil, to support the tunnel sections
(iii) sinking the tunnel units
(iv) sealing the joints

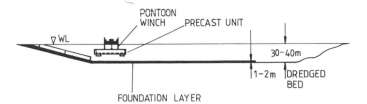

**Fig. 10.54** Principle of the immersed tube tunnel method

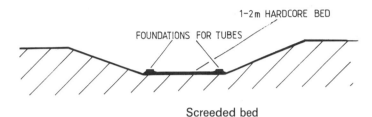

Screeded bed

**Fig. 10.55** Piled base and screeded bed for an immersed tube

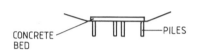

**Fig. 10.56** Pontoon-mounted winches for tube positioning

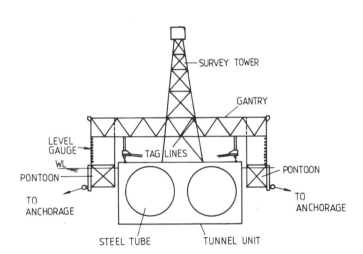

Reinforced concrete units are partially constructed on dry land, temporarily sealed at each end, launched into the water and built up to full dimension and towed into position. A unit is attached at each of the four corners to pontoon-mounted winches (Fig. 10.56). Ballasting is subsequently increased to cause a loss of buoyancy and the section is lowered into place on the foundation pads under winched control. The final position is checked by divers and adjusted if necessary. The joint between the segments is initially sealed with temporary rubber gaskets and subsequently sealed with concrete, placed with the aid of divers. To reduce the risk of movement, the space between the hard core bed and the underside of the unit should be filled with coarse sand (Fig. 10.57), pumped from barges.

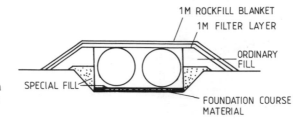

**Fig. 10.57** Protective covering for an immersed tube

An alternative method may be chosen, using steel tubing supported on a reinforced concrete keel and surrounded by a thin concrete shell to provide protection. The unit is sunk to position as described above, and the joint temporarily sealed with concrete on the outside. The space between the two diaphragms of the new and the previously placed unit is pumped out and the end diaphragms are removed, leaving only the seal at the other end of the newly placed tube. The units can then be welded together in the dry.

A final precaution for either method, particularly near heavy shipping, is the placing of a protective covering (see Fig. 10.57).

# EXCAVATING AND MATERIALS HANDLING EQUIPMENT

# 11

# INTRODUCTION

## 11.1 EARTHMOVING MACHINES

The history of mechanical equipment, particularly for excavating duties on the construction site is relatively recent. The canals of the eighteenth century were largely excavated by manual labour and the spoil transported away in horse-drawn carts. Progress was slow as a man could dig perhaps one cubic metre of earth in an hour. The railway builders of the middle 1800s saw the introduction of the first cumbersome steam-powered face shovels, which were capable of a few tens of cubic metres of production per hour. Today machines are available in many sizes to suit a multitude of different tasks and it has become almost unthinkable to carry out excavation work on even the smallest scale without mechanised assistance. However, although the tendency is towards the use of more and specialised machinery, earthmoving equipment falls roughly into only two categories: fixed-position and moving machines. The size of project, the topography, volume of earth to be removed and many other detailed factors influence the choice of type. In general, the moving machine is used for ground levelling and bulk earthmoving whilst the static-type machine is usually operated on specific tasks. *Stationary machines* (Fig. 11.1) include face shovels, backhoes, draglines and grabs. The excavator loosens the soil and loads without changing position, resulting in some loss of mobility when compared with scrapers and bucket loaders, but in compensation considerably more force may be applied at the excavation face.

**Fig. 11.1** Fixed-position excavating machines

FACE SHOVEL

BACKACTER

DRAGLINE/GRAB

**Fig. 11.2** Moving
excavating
machines

BULLDOZER   LOADER   SCRAPER   GRADER   TRENCHER

*Moving machines* (Fig. 11.2) include bulldozers, loaders, scrapers, graders and trenching machines. The excavated material is removed, transported and deposited in a cycle, which is a particularly useful feature when large volumes of earth need to be moved over rough terrain. For example, in both civil engineering and building work the initial construction site is often uneven and requires levelling. The most suitable plant for this purpose is generally selected from the group of moving machines. Specific excavation of the foundations, basement, drain trenches, etc., of the structure to be constructed can then be performed with a fixed-position excavator.

## 11.2   LIFTING PLANT

Lifting equipment has been used over the centuries, simple levers for example were known to the early civilisations. Ropes and pulleys have been applied with timber beams as a form of crane for most building work up to the present day. But like earthmoving machinery, steam power led to the development of heavier lifting devices, which have been slowly refined using electric motors, diesel engines, compressed air motors, etc. Today the choice of materials handling equipment is vast and includes all forms of crane, fork-lift truck, cableways, hoists, pumps and conveyors, as demonstrated in Fig. 11.3.

The selection of the appropriate device for a lifting application is rarely as obvious as that for earthmoving duties and many interrelated factors must be considered.

## 11.3   MODERN DEVELOPMENTS IN EQUIPMENT

The slow but steady improvement in construction equipment, especially earthmoving machines, and truck-mounted equipment, such as mobile cranes, relates to the introduction of the diesel engine and subsequently power-shift gears and the torque converter, whereafter the cumbersome and expensive steam engine lost its standing. More recently, the development of all-hydraulic transmissions, for example on backhoes, face shovels, bulldozers, and truck-mounted cranes, has virtually rendered redundant the manufacture of the traditional rope operated machines and not least the advances made in the structural design of large cranes now make lifts of 2000 tonnes possible. However, many of the old-type machines are still widely used, thus in the earlier chapters special attention is directed towards machine configuration in order

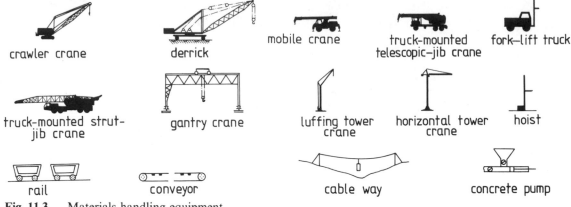

crawler crane     derrick     mobile crane     truck-mounted telescopic–jib crane     fork–lift truck

truck-mounted strut-jib crane     gantry crane     luffing tower crane     horizontal tower crane     hoist

rail     conveyor     cable way     concrete pump

**Fig. 11.3**     Materials handling equipment

that the earlier and later developments in construction equipment technology may be understood and compared. But before the reader is cast into details, it is first necessary to introduce a review of the power sources used with the various types of machine.

### Power sources

Figure 11.4 illustrates the range of torque available at the output shaft of a steam engine, internal combustion engine, gas turbine and electric motor. The graph for each is drawn to a common scale to aid comparison. It is seen that in general a reduction in rotation speed of the output shaft (i.e. lower engine revolutions per minute) leads to increased torque availability. The torque being the turning moment used to drive the transmission and attachments.

The available power may be calculated from the torque multiplied by the rate of turning of the output shaft. On the customary system the

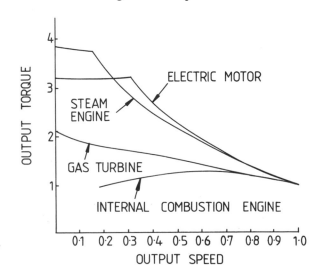

**Fig. 11.4**     Comparison of various forms of engine power

unit rate of working is 33 000 foot pounds per minute and is called the horse power, written hp. The equivalent metric unit is the watt (W), and 1 hp is 746 W.

It is obvious from the diagram that the maximum power and torque are not necessarily coincident at the same rate of engine revolutions.

An engine is usually described in terms of its maximum power output called the brake horse power (bhp).

### Types of engine

(a) *Steam* powered the early machines during the last century and represented immense improvements over animal power. Even today the high torque available from the steam engine is almost unsurpassed by other forms of power. Unfortunately, the engine and boiler are cumbersome and expensive to manufacture, and as a consequence steam engines have proved unpopular in recent times.

(b) *Diesel engines* were developed during the present century and have become popular only in the last forty years or so. The available torque is very much lower than from a steam engine, but with suitable gearing, considerable improvement is possible. However, the demands upon the driver as a result of continuously having to monitor the gearing are distracting and lead to reduced production.

Nevertheless, the diesel engine has gained much favour due to its compactness, low servicing and use of convenient fuel. Furthermore, diesel engines used with hydraulic power transmission fulfil both the power and manoeuvrability demands of an excavator. Rapid development is taking place in this direction.

(c) *Electric motors* show very favourable torque characteristics, but heavy expensive electric cable is required and hence only permanent, or semi-permanent fixed-in-place machines are powered in this way, e.g. tower cranes. Electric motors react very rapidly to torque requirement.

(d) *Diesel–electric motor*–a diesel engine drives an electric motor and thus no cables are needed. This method is expensive and only seldom used, e.g. large mobile cranes and excavators.

## 11.4  FACTORS AFFECTING EQUIPMENT SELECTION

Selecting the correct equipment for the job ideally forms part of the construction planning process and should be chosen for a particular task only after analysis of many interrelated factors. The important considerations for both earthmoving equipment and cranage include examination of:

(a) The function to be carried out: for example the choice of equipment in many instances will be dictated by the need to cope with combinations of horizontal, vertical and travelling movement. A crane can provide both horizontal and vertical movement, but is rather poor at manoeuvring, whereas most earthmoving machinery is mobile and will transport material very efficiently over long distances, but cannot be used for lifting.

(b) The capacity of the machine: the volume of material to be moved related to the time available in the construction programme is important. But for cranes and other lifting operations, the practical weight and size of the units as dictated by the design are most important factors.

(c) The method of operation: the distance and direction of travel, speed and frequency of movement, sequence of movement, state of the ground, etc., must all be taken into account.

(d) The limitations of the method: for many earthmoving operations the choice of method may be limited by the cost of temporary work facilities, for example, haul roads may not be permitted in a particular area, the establishing and dismantling costs of a crane or conveyor system may not be economic over the length of time required for the task. These are just two simple illustrations; in practice virtually every project presents different problems.

(e) The costs of the method: the most desirable method from a feasibility point of view may involve expensive and uneconomic units, which may not be available at the desired time without the incurrence of a high cost premium.

(f) The cost comparison with other methods: it is essential that all the alternatives be considered in the plan and costed.

(g) The possible modifications to the design of the project under consideration, in an attempt to accommodate the available materials handling method: for example a single large lift in a tower block, could perhaps be reduced in size. A smaller crane might then be installed. Although this procedure may appear obvious and straightforward, because of the one-off nature of most designs and construction projects the task of analysis is seldom simple. The choice of method is rarely clear cut – a compromise is the more usual result.

## PART ONE

# 12

# FIXED-POSITION EXCAVATING MACHINES

Fixed-position machines are divided into rope operated and hydraulically operated categories. The basic structural assembly of each is similar and both have a base frame and tracks or wheels supporting the digging arm and superstructure. Although both types are usually powered by a diesel engine, the mechanical parts and operating mechanisms of each are different.

A fixed-position rope operated excavator may be converted from a basic machine to operate as any one of the following types:

(a) face shovel;
(b) backacter (backhoe);
(c) grab;
(d) dragline.

Conversion to a crane, crane and fly-jib, pile driver with leaders, pile-boring machine, earth compactor, demolition tool, or rock breaker is normally a standard option.

In contrast, the hydraulically operated machines are generally manufactured for a specific application, e.g. backhoe, and are not convertible into other forms. Because of their lower maintenance costs and improved performance, increasingly hydraulics are replacing ropes and gears.

## 12.1 BASE FRAME ASSEMBLY

The base frame consists of a sturdy welded steel structure resting on the machine axles and supports the weight of the engine, counterweight, cab, controls, gears and other mechanical parts (Fig. 12.1). Today the rope operated machine is largely confined to the dragline, face shovel and grabbing crane and as such uses a system of mechanical gears or hydraulic motors to drive the tracks and attachments. The backhoe (backacter) is probably the most popular version of the fixed-position

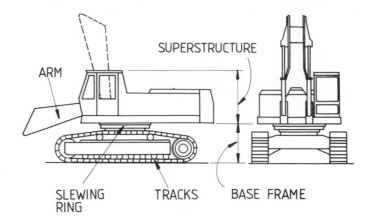

**Fig. 12.1** General arrangement of a tracked excavator

excavator on construction sites and is almost invariably of the all-hydraulic type. Both forms are powered by a diesel engine.

### Geared drive for wheels or tracks

Figure 12.2 shows a schematic arrangement of the transmission from the engine (1), to the tracks (6), rope drum (8) and slewing ring (11). One or more of the mechanisms may be selected by engaging appropriate gears (omitted from the diagram for simplification).

The engine power to drive the tracks and operate the bucket attachment on some makes of machine is directed through a torque converter to the main drive shaft (10), via a belt or chain drive (2) and speed regulation gearing (3). The torque converter is simply a fluid coupling, consisting of a series of pump blades set facing turbine blades, all housed in fluid. As the pump blades are rotated by the engine's drive shaft, a reciprocal rotation is caused in the turbine blades. Loading of the output shaft induces 'slipping' between the two sets of blades and consequent multiplication of the available output torque, as shown in Fig. 12.3. Thus a variable torque demand can be accommodated at constant engine speed. It is observed that the ratio of the output to input power is at a maximum when the output speed is approximately 0.6 of the input speed.

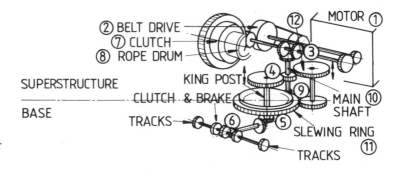

**Fig. 12.2** Schematic arrangement of transmission machinery for a tracked crane or excavator

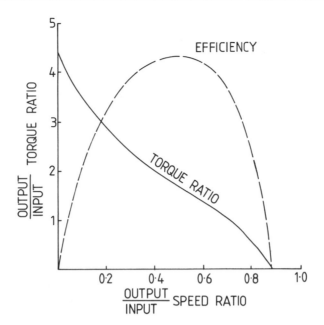

**Fig. 12.3** Performance of a typical torque converter

The power to drive the tracks is drawn off through the gears (4), (5) and (6) (Fig. 12.2) and the final connection is completed through a clutch and brake arrangement on each track. The machine is steered by disengaging one track while maintaining drive on the other.

The geared drive system is cumbersome and requires frequent maintenance.

### Hydraulic drive for wheels or tracks

Figure 12.4 shows the principles of a hydraulic transmission system. The engine (1), drives a hydraulic pump (3), via a series of speed-regulating gears (2), which in turn forces oil around a closed system to drive hydraulic displacement motors attached to each of the

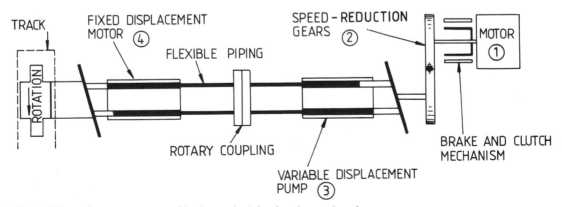

**Fig. 12.4** Schematic arrangement of hydrostatic drive for the tracks of a crane or excavator

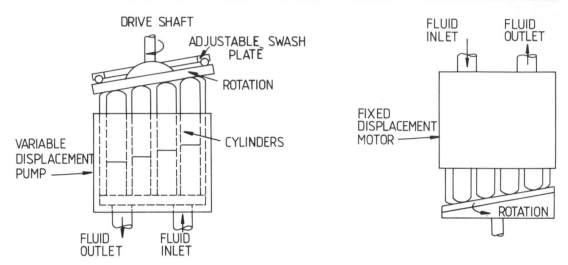

**Fig. 12.5**    Hydrostatic drive using a displacement pump and motor

tracks. A special rotary coupling, centred in the slewing ring, is needed to overcome the problem that the power source is mounted on the superstructure which must be able to rotate independently of the tracks and subframe, while maintaining the closed oil circuit.

A particular feature of displacement pump and motors (Fig. 12.5) is the variable torque output which can be obtained from a constant engine input speed. The pistons in the pump are activated against a movable swash plate, and the hydraulic pressure is converted to torque in the motor by pressing the pistons against a fixed inclined plate. The oil is thus pumped around the closed circuit and in turn converted to torque to drive the shaft in the motor. Clearly by altering the tilt of the plate in the pump the oil flow may be increased or reduced. Figure 12.6 shows a plot of fluid pressure against oil flow, thus as the torque requirement increases the flow of oil may be slowed down and the oil pressure increased to provide the required torque. The system is always in equilibrium, and is termed 'hydrostatic'.

This system is increasingly being preferred by manufacturers but is

**Fig. 12.6** Relationship of oil pressure and oil flow in a variable displacement pump

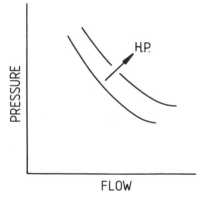

not universally installed on all makes of machine, and some still prefer the alternative of geared hydraulic pumps and motors.

## 12.2  SUPERSTRUCTURE CONSTRUCTION

The superstructure comprises the upper part of the machine, its weight is transferred to the base frame through the large slewing ring and carries the cab, engine, digging arm, bucket and counterweight. It is necessary for the superstructure to be able to turn through 360° to both dig and dump excavated material. With mechanical transmission (Fig. 12.2) slewing movement is achieved by locking in gears (9) and (12) to engage the slewing ring (11) mounted on the base frame. Alternatively, when using hydrostatic drive a separate hydraulic displacement motor engages the slewing ring (see Fig. 26.2).

## 12.3  ATTACHMENTS AND FITTINGS

### Rope operation

The digging attachment, usually a bucket, is operated and controlled by means of simple rope and drum arrangements. (A similar arrangement involving a second rope drum is also required to raise and lower the boom or arm on which the bucket is hung.) Power is supplied from a diesel engine (sometimes an electric motor on very large excavators), through a gear box, or more usually nowadays a torque converter. Band pulleys (2) transfer the power via the gear system (3) through a clutch (7) to the rope drum (8). Brakes are also needed to control the drum whenever the clutch is disengaged. The drum rope passes over a series of pulleys to the bucket. The force demanded at the excavation face is automatically regulated by the torque converter, therefore the engine may be set to run at constant speed and the need to change gear is eliminated. On the more modern units much of the mechanical gearing is replaced by hydraulic motors on each drum.

### Hydraulic operation

Rope operated machines, with the exception of draglines, grabs and a few of the very large face shovels, have largely given way to hydraulic operation. The system is similar to that used to power the tracks, but the hydraulic motor is replaced by hydraulic rams to manoeuvre the arm and bucket. The system is much simpler for the driver to operate and has resulted in neater and more compact cabs. The large levers used with rope operated machines are avoided, being replaced by finger controls.

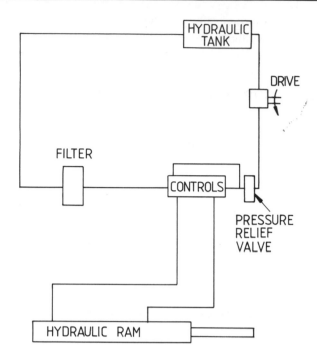

**Fig. 12.7** Hydraulic system to operate the arm and bucket on an excavator

## 12.4  METHODS OF EXCAVATION

### Face shovel

The excavator operates best from a flat prepared surface to work above the tracks against an excavation face. It serves only to loosen and load material and is mostly used in quarrying and road cuttings. Some leverage against the wall is possible, particularly with hydraulically controlled machines, enabling some faces to be tackled directly without the need for the soil to be loosened first.

### ROPE OPERATED FACE SHOVEL

The bucket is first positioned at the toe of the wall and pulled up the face, filling as it climbs. The upper part of the machine is then slewed to the dumping position, the trap opened, the bucket emptied and finally turned to its original position, ready for the next cycle.

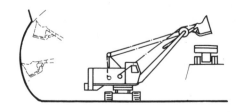

**Fig. 12.8**  Rope operated face shovel

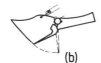

**Fig. 12.9**    Bucket
attachments for
a face shovel

*The bucket*

The bucket is controlled by means of ropes and pulleys, it is usually made of welded steel with bolted-on hardened teeth to improve penetration of the soil, and is usually fixed rigidly to the arm, as in (Fig. 12.9a). Adjustment is possible with some types of bucket to change the angle of dig for work in different types of soil, but this must be carried out manually when the machine is in a demobilised position (Fig. 12.9b). Clay soil presents problems and causes a reduction in output due to difficulty in releasing the material from the bucket.

## HYDRAULICALLY OPERATED FACE SHOVEL

Unlike the rope operated machine, the bucket is not rigidly attached to the arm, but connected to it by means of a knuckle joint and positioned with hydraulic rams. Thus the bucket is pressed into the soil with a regulated force and consequently considerably more control over the shape of the excavation face is possible than with ropes. A further advantage is that the bucket contents are discharged from the front, rather than relying upon a flap in the base of the bucket, as is the case with an operated excavator. This is a particularly useful facility when excavating clinging cohesive soils, like clay.

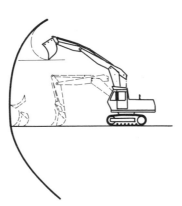

**Fig. 12.10**    Hydraulically
operated face
shovel

## METHOD OF WORKING WITH THE FACE SHOVEL

*Rope control.* The curve is radial (Fig. 12.11a) and the slope of cut is dependent upon the placing of the arm, positioning of the bucket and the power of the ripping force. *The hydraulic machine* may adopt many movements, it can work like a rope machine (curve b) or claw its way up the face by pressing and relieving as in (c). The teeth act like rippers causing the loosened material to fall into the bucket. Alternatively, the full power of the machine may be applied to force out rocks and boulders (d). The action may also be applied to strip the face in layers (e) or even perform the tasks of the loading shovel (f).

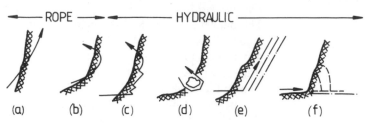

**Fig. 12.11** Excavation profiles using a face shovel

*Excavating techniques*

An excavation face is tackled either using the frontal or parallel approach. In the parallel method the machine follows parallel to the face of the cutting and can obviously only load the trucks from one side. The secondary face should be opened up to a depth of about $1\frac{1}{2}$ times the digging radius from the centre of rotation of the superstructure. The method is similar to block cutting shown in Fig. 19.8.

In the frontal approach the excavator simply cuts directly into the bank, until the front of the tracks have progressed up to the toe of the face. The excavator then reverses out, manoeuvres to overlap the previous area in readiness to repeat the procedure.

## COMPARISONS BETWEEN ROPE AND HYDRAULICALLY OPERATED FACE SHOVELS

|  | *Hydraulic* | *Rope* |
|---|---|---|
| 1. | Swivel bucket. | Fixed bucket. |
| 2. | Short arm action with potentially greater output and greater versatility. | Long arm action, causing a reduction in output. |
| 3. | Ripping power constant along the digging curve. | Full ripping power only possible at top of reach. |
| 4. | Light arrangement. | Heavy arm needed. |
| 5. | No emptying problems. | Cohesive soil causes emptying difficulties. |
| 6. | Medium-size machine, approximately 25 m³ max. bucket size. | Large machine possible, 40–50 m³ max. bucket size. |

### Backacter (backhoe)

The machine is used for excavating below the level of the tracks, e.g. trenches, basements, foundations and for other excavation work in confined situations and also increasingly as a small crane for handling duties in pipe-laying, installing trench sheets etc.

### ROPE OPERATED BACKHOE

The excavating efficiency and profile depends upon a combination of the weight of the bucket attachment and that of the arm to provide

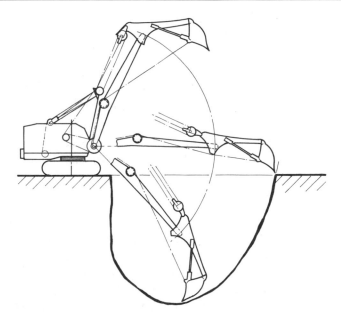

**Fig. 12.12** Rope operated backhoe

sufficient force to penetrate the soil. The bucket is connected to the arm by means of a knuckle joint which allows a motion similar to that of the human wrist.

## HYDRAULICALLY OPERATED BACKHOE

Both the arm and the bucket are independently controlled by means of separate hydraulic rams. Hence the shape of the excavation curve may be varied to suit the situation and allows the limits of the excavation to approach much nearer to the tracks of the machine. Additionally the variable force available with hydraulic powering gives the machine a much increased capability in harder soils. Buckets can also be fitted with a hydraulically operated 'thumb' to aid rock handling and similar duties.

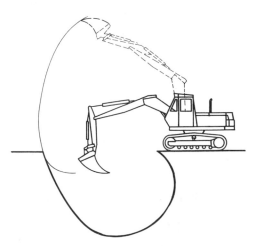

**Fig. 12.13** Hydraulically operated backhoe

Depths up to 8 m below the tracks may be excavated with some machines. However, the under-reaming indicated in Fig. 12.13 is the design limit. Naturally this could not be achieved in practice for fear of collapsing the sides of excavation.

## METHODS OF WORKING WITH THE BACKHOE

(a) When excavating large and deep pits, an operating plan should be sketched out to suit the removal of the spoil and accommodate any ramps needed for the trucks and dumpers. The material is best removed in strips working up and down the length of the excavation, to a depth well within the maximum digging reach of the machine.

(b) Small pits for column bases require careful excavation and obviously production will be reduced.

(c) Drain trenches should always be excavated working from a deep to a shallow section, especially in water-bearing ground. If a sump is located in the deep section, then water will flow away from the face of the excavation. It is usual to batter (i.e. cut) back the sides of the excavation faces to improve stability. A vertical face in most soils will quickly collapse.

(d) Lifting duties. Some National Standards permit excavators to carry out lifting duties but when a safe load indicator is not fitted, the permissible Safe Working Load (SWL) is only that stated by the manufacturer in the least stable condition. Check valves also need to be incorporated in the hydraulic circuitry in case of catastrophic failure, e.g. pipeline burst.

### Grabbing crane

## ROPE OPERATED GRAB

The grab is really a crane fitted with a grabbing bucket. The excavating efficiency is dependent upon the self-weight of the attachment and the machine is therefore limited to use in fairly loose soils.

With the rope operated machine the attachment comprises two half

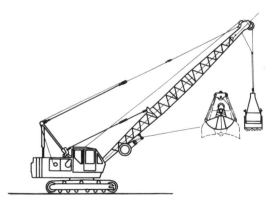

**Fig. 12.14**  Rope operated grab

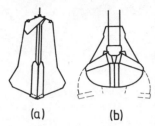

**(a)        (b)**

**Fig. 12.15** Grab and clamshell attachments

shells hinged at the centre, which may be of the grab type (Fig. 12.15a) or a clamshell (Fig. 12.15b). The grab is fitted with interlocking teeth to aid penetration of the soil for genuine excavating applications, whereas the clamshell has no teeth and is used only for stockpiling very loose materials such as sand. The whole unit is hung from a crane jib on a hoist rope, sometimes called the 'holding rope'. A separate rope is used to open and close the shells and a third rope is needed to alter the position of the boom. With three rope drums to control the arrangement is cumbersome and slow. The grabbing crane is only selected for excavating where other methods are not feasible; for example, the grab may be used in conjunction with a long boom to provide some advantage when long reaches are required, e.g. over water. The method is useful for excavating shafts and other deep confined excavations, large diameter wells and bored piles, diaphragm walls and in spoil heaps.

## HYDRAULICALLY OPERATED GRAB

Some modern developments now allow the grab to be opened and closed hydraulically, and even rotated. Clearly such improvements increase the operating efficiency and output especially when the grab is hung from a rigid arm as the teeth can then be forced into the earth to take larger bites. However, this technique reduces the loading efficiency as height of the grab may only be altered by lifting or raising the boom. Figure 12.16 shows a typical modern arrangement to try to overcome this difficulty.

## CHARACTERISTICS OF THE GRABBING CRANE

| | *Hydraulic grab* | *Rope grab* |
|---|---|---|
| 1. | Capable of spot digging. | Similar to hydraulic grab. |
| 2. | Excavating depth limit: approximately 14 m. | Excavating depth limited only by length of rope. |
| 3. | Excavating force directly related to hydraulic power available. | Excavating force very much dependent upon weight of grab. |
| 4. | Very limited reach. | Long reach, dependent upon jib length. |
| 5. | Difficulties in unloading when grab suspended from long arm extension. | Limited only by boom length. |

## Dragline

The dragline may only be used as a rope operated machine, powered through mechanical gearing or with hydraulic motors on the drums and like the grab is a crane with excavator attachments, but requiring three separate rope drums – a luffing winch, hoisting winch and drag winch.

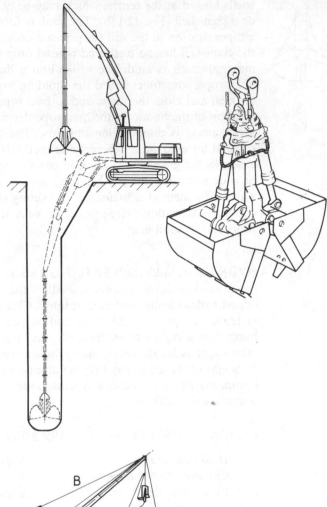

**Fig. 12.16** Hydraulically operated grab

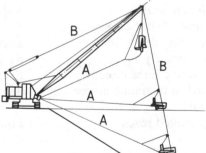

**Fig. 12.17** Dragline

The machine is especially suitable for excavating below the level of the tracks and may be used as an alternative to the backacter in many cases. This method of excavation presents the advantage that the bucket is cast out from a boom which itself may be long, so that excavating and dumping become possible at points wide apart. Unfortunately the same action is difficult to use in confined space. Applications include mainly excavation for drain trenches, large foundations and winning underwater deposits of gravel.

## METHOD OF WORKING WITH THE DRAGLINE

The bucket is controlled with two separate rope drums, the drag rope A and the hoisting rope B. The driver needs considerable skill to keep the two lines in unison: a rope which is given too much slack may result in tangling, catch in the works or damage the machine, etc. The excavating cycle is represented in the stages (1) to (4) in Fig. 12.18. The

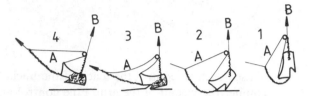

**Fig. 12.18** Dragline excavating cycle

cycle, for example, may start, say, in the 'having just dumped' position (1). The drag rope is slack and the centre of gravity of the bucket causes the bucket to hang downwards. The hoist rope is then played out and the drag rope slack is taken up until the cutting teeth begin to bite into the soil (2). The drag rope is then wound in and the bucket filled by pulling it through the material (3). The bucket, when full, is lifted by raising the hoist rope (4) and the material is emptied in position (1) by once again providing slack in the drag rope, and the cycle repeated. The excavating angle may be altered by changing the position of the drag chains' connections to the bucket.

## CHARACTERISTICS OF THE DRAGLINE

1. High output is possible, up to 90% of that of the face shovel in loose soils.
2. Long reach and depths below the tracks to roughly one-third the boom length may be excavated.
3. Excavation in water is possible with the use of a perforated bucket, but with much reduced efficiency and output because the driver is operating blindly and much of the material will be slurry. Roughly 20% of normal output is likely.
4. Only light to medium soil can be excavated.
5. It is difficult to cast the bucket to the exact position required and the method is therefore only appropriate for broad sweeping excavation work.
6. The system is very difficult to operate and requires the skills of a very experienced driver.

## VARIABLE COUNTERBALANCE BACKHOE (Fig. 12.19)

To increase the reach and digging depth of the backhoe, the manufacturers have incorporated some features of the dragline. A hydraulically controlled counterweight mounted on the superstructure

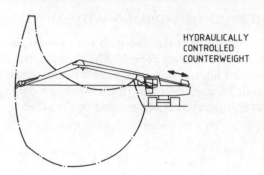

**Fig. 12.19**  Variable counterbalance excavator

is used both to control the reach of the bucket and simultaneously to counter the weight of the arm. Fine control and crowd force is provided through a hydraulic ram attached between the lower arm and bucket. The direct pull of the winch (on the dragline principle) draws the bucket through the material. An outreach of 20 m can be achieved with a 1.3 m³ bucket.

## MICRO, MINI AND MIDI EXCAVATORS

In recent years the trend towards small-scale sub-contracting of work has encouraged the use of miniaturised machines to replace manual excavation and materials-handing methods. The smallest version falls within the micro class of backhoe (Fig. 12.20a), weighs between ½ and 1 tonne and can be towed to site. Popular duties include excavation of shallow trenchwork, footings for small buildings and foundations. The mini machine (Fig. 12.20b) is slightly larger at about 3–5 tonnes, and is equipped with tracks to aid mobility around the work place. Duties are similar to a standard backhoe and other attachments such as hydraulic jack hammers can be fitted to increase versatility. The midi excavator is the next class in size, weighs 6–10 tonnes and mostly competes with the wheeled backhoe/loader bucket machine.

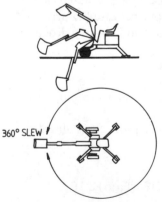

360° SLEW

**Fig. 12.20a**  Micro excavator

### General notes on fixed-position machines

(a) The rope operated machines described consist of a similar basic unit and may be quickly converted to other forms by adding the necessary attachments; for example, by changing from face shovel to backacter. For conversion to a crane, the rope and rope drums also require changing and this may take a day's work.

(b) The most common machines are in the range 18–40 tonnes basic weight (i.e. without attachments) and support excavating buckets of 0.25–1.5 m³ capacity. The output will vary depending upon the type of soil but typically a 1 m³ backhoe should on average be capable of excavating 75–100 m³/h, always provided that the machine is in good condition.

### 12.5 PRODUCTION INFORMATION

**Soil swell**

**Fig. 12.20b** Mini excavator

The production data given in Fig. 12.22 represent the output after bulking or swelling caused by disturbance when the excavator bucket is being filled. The in-place or bank volume represents the soil in the undisturbed condition. Thus

$$\text{bulked volume} = \text{in-place volume} \times \text{swell factor}$$

Bulking (swell) factors:

| | |
|---|---|
| Broken rock | 1.5–2.0 |
| Gravel | 1.0–1.1 |
| Clay | 1.25–1.4 |
| Sand | 1.0–1.3 |
| Common earth | 1.1–1.3 |

**Angle of swing**

The data are also based on excavating and depositing within an angle of swing of 90°. The output volumes should be multiplied by the following factors for other angles of swing:

| Angle of swing | 45° | 60° | 75° | 90° | 120° | 150° | 180° |
|---|---|---|---|---|---|---|---|
| Factor | 1.26 | 1.16 | 1.07 | 1.0 | 0.88 | 0.79 | 0.71 |

**Depth of dig/height of cut**

The actual output is related to the depth of dig for backhoes and draglines, and to the height of cut for the face shovel. The data in Fig. 12.22 are the production rates when the machine is operating at the optimum depth. Excavators working near the limits of the range will suffer considerably reduced output. Maximum output is achieved when the height/depth of face allows the bucket to be completely filled in a single movement from the bottom to the top of the face.

**Job conditions**

The production data also assume ideal conditions and uninterrupted working. For practical purposes some reduction in output is advisable. Example:

| | *Production loss (%)* |
|---|---|
| Weather conditions | 10 |
| Manoeuvring | 8 |
| Mechanical breakdowns | 5 |
| Operator efficiency | 7 |
| Waiting for other plant | 10 |
| | 40 |

**Fig. 12.21** Bucket rating

**Table 12.1(a)**   Tracked excavators – approximate specifications

**Rope operated machines**

| | | | | | | | | |
|---|---|---|---|---|---|---|---|---|
| Engine size | (kW) | (37) | (56) | (75) | (97) | (119) | (150) | (186) |
| | hp | 50 | 75 | 100 | 130 | 160 | 200 | 250 |
| Machine weight | tonnes | 15 | 22 | 30 | 50 | 60 | 75 | 90 |
| Fuel consumption | (litre/h) | (9) | (14) | (18) | (23) | (27) | (36) | (45) |
| | gal/h | 2 | 3 | 4 | 5 | 6 | 8 | 10 |
| *Backhoe* | | | | | | | | |
| Struck bucket capacity* | (m$^3$) | (0.38) | (0.57) | (0.76)–(0.96) | (1.14)–(1.34) | (1.53)–(1.91) | (2.68) | (3.25) |
| | yd$^3$ | $\frac{1}{2}$ | $\frac{3}{4}$ | $1$–$1\frac{1}{4}$ | $1\frac{1}{2}$–$1\frac{3}{4}$ | $2$–$2\frac{1}{2}$ | up to $3\frac{1}{2}$ | up to $4\frac{1}{4}$ |
| Max. digging depth | m | 5 | 6.5 | 7.5 | 8.0 | 9.0 | 9.0 | 9.5 |
| *Dragline†* | | | | | | | | |
| Struck bucket capacity* | (m$^3$) | — | (0.38)–(0.96) | (0.67)–(1.34) | (0.76)–(1.91) | (0.96)–(2.29) | (1.14)–(2.67) | (1.14)–(3.06) |
| | yd$^3$ | — | $\frac{1}{2}$–$1\frac{1}{4}$ | $\frac{7}{8}$–$1\frac{3}{4}$ | $1$–$2\frac{1}{2}$ | $1\frac{1}{4}$–$3$ | $1\frac{1}{2}$–$3\frac{1}{2}$ | $1\frac{1}{2}$–$4$ |
| Length of boom | m | — | 21 | 21 | 27 | 30 | 33 | 36 |
| *Face shovel* | | | | | | | | |
| Struck bucket capacity* | (m$^3$) | (0.38) | (0.57) | (0.76)–(0.96) | (1.14)–(1.34) | (1.53)–(1.91) | (2.68) | (3.25) |
| | yd$^3$ | $\frac{1}{2}$ | $\frac{3}{4}$ | $1$–$1\frac{1}{4}$ | $1\frac{1}{2}$–$1\frac{3}{4}$ | $2$–$2\frac{1}{2}$ | up to $3\frac{1}{2}$ | up to $4\frac{1}{4}$ |
| Max. cutting height | m | 7.0 | 7.0 | 7.0 | 7.5 | 8.5 | 10.0 | 10.5 |
| *Grab crane†* | | | | | | | | |
| Struck bucket capacity* | (m$^3$) | — | (0.57)–(2.29) | (0.76)–(2.29) | (0.76)–(3.82) | (0.76)–(3.82) | (0.76)–(3.82) | (1.34)–(5.34) |
| | yd$^3$ | — | $\frac{3}{4}$–$3$ | $1$–$3$ | $1$–$5$ | $1$–$5$ | $1$–$5$ | $1\frac{3}{4}$–$7$ |
| Length of boom | m | — | 21 | 21 | 27 | 30 | 33 | 36 |

**Hydraulic machines§**

| | | | | | | | | |
|---|---|---|---|---|---|---|---|---|
| Struck bucket capacity* | (m$^3$) | (0.19)–(0.48) | (0.57) | (0.76) | (1.25) | (1.67) | (2.1) | (2.5) |
| | yd$^3$ | $\frac{1}{4}$–$\frac{5}{8}$ | $\frac{3}{4}$ | 1 | $1\frac{1}{2}$ | 2 | $2\frac{1}{2}$ | 3 |
| Engine size | (kW) | (37) | (56) | (75) | (93) | (119) | (141) | (164) |
| | hp | 50 | 75 | 100 | 125 | 160 | 190 | 220 |
| Machine weight | tonnes | 12 | 15 | 18 | 22 | 30 | 36 | 45 |
| *Backhoe* | | | | | | | | |
| Max. digging depth | m | 5 | 6 | 7 | 8 | 8 | $8\frac{1}{2}$ | 9 |
| *Face shovel* | | | | | | | | |
| Max. cutting height | m | 5 | 6 | 7 | $7\frac{1}{2}$ | 8 | 8 | 9 |

\* Struck bucket capacity – material flush with strike-off plane. Heaped capacity – material stands inclined above strike-off plane (see Fig. 12.21). The Committee on European Construction Equipment (CECE) and the Society of Automotive Engineers (SAE) in the USA specify bucket sizes. CECE rates heaped bucket capacity on a 2:1 angle of repose while SAE stipulates 1:1. Note bucket sizes in the backhoe mode should be about 75% of face shovels for stability.

† The dragline and grab are operated like cranes and the maximum load is dependent upon the operating radius. Cranes are dealt with in Chapter 24.

§ The specification is very approximate for hydraulic machines, as engine power varies widely between manufacturers with similar bucket ratings. This is complicated further by the high hydraulic pressures used by some.

**Table 12.1(b)**   Very small hydraulic excavators

|  | Micro | Mini | Midi |
|---|---|---|---|
| Struck bucket capacity (m³) | 0.01–0.05 | 0.05–0.2 | 0.1–0.5 |
| Engine size (kW) | 10 | 20 | 45 |
| Machine weight (tonnes) | 1 | 3–5 | 6–10 |
| *Backhoe* | | | |
| Max digging depth (m) | 2 | 3 | 4 |
| *Face shovel* | | | |
| Max cutting height (m) | 3.5 | 5 | 6 |

**Table 12.1(c)**   Very large hydraulic excavators

| | | | |
|---|---|---|---|
| Struck bucket capacity (m³) | 4–8 | 8–12 | 16–30 |
| Engine size (kW) | 250 | 600 | 1700 |
| Machine weight (tonnes) | 80 | 150 | 500 |
| *Backhoe* | | | |
| Max digging depth (m) | 9 | 8 | 12 |
| *Face shovel* | | | |
| Max cutting height (m) | 12 | 15 | 20 |

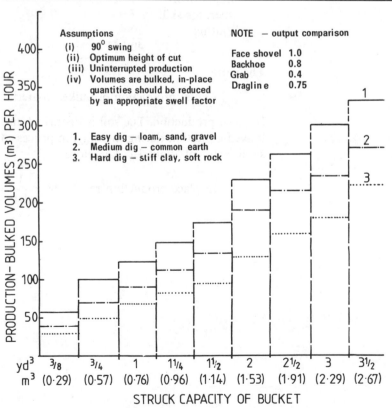

**Fig. 12.22**   Ideal output of the face shovel

## PROBLEM

Determine the hourly production of a $\frac{3}{4}$ yd$^3$ (0.57 m$^3$) capacity backhoe, excavating a foundation 3 m deep in common earth.

## SOLUTION

Ideal output for the equivalent face shovel in Fig. 12.22 for easy dig = 100 m$^3$/h.
Backhoe proportion of face shovel production = 0.80

Assume angle of swing is 90° therefore

Swing factor = 100.

Assume depth of dig allows optimum production, therefore

Adjusted production rate = 100 × 1.0 × 0.80 = 80 m$^3$/h.

Other factors affecting output:
Job conditions, losses caused by:

|  | % |
|---|---|
| Weather, say | 10 |
| Manoeuvring | 8 |
| Breakdowns | 5 |
| Operator skill | 7 |
| Waiting | 10 |
|  | 40 |

Therefore

Actual production of bulked material = 80 × 0.6 = 48 m$^3$/h

In-place production. The soil is disturbed and enters the buckets in a bulked condition during the digging process. Common earth has a swell factor of about 1.1. Therefore

In-place production rate = $48 \times \dfrac{1}{1.1} = 44$ m$^3$/h

# 13

# METHODS OF TRANSPORTING MATERIALS

The choice of system in transporting material from the loading point depends on many factors including:

(a) site conditions;
(b) volume of material to be moved;
(c) type of material;
(d) time available.

The choice is further complicated by considerations such as:

> Will transport be over metalled roads or not?
> Must the roads be kept clean?
> Management impositions, e.g. bonus targets, overtime, etc.

In terms of moving excavated earth, a rough guide in selecting an appropriate system is given in Fig. 13.1. The first six methods are

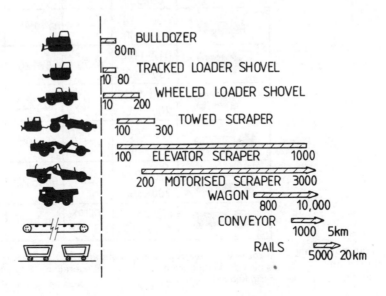

Fig. 13.1 Transport systems for earthmoving

discussed in other sections of the book, and this chapter will concentrate mostly on the means available for transport of materials over distances of several kilometres.

## 13.1 THE LORRY (TRUCK OR WAGON)

The wide range of trucks shown in Fig. 13.2 is now available to suit almost any type of work. For transporting building materials on public highways, conventional trucks such as types (1), (2), (5) or (6) are the most appropriate and a considerable choice of designs is available to suit the goods to be carried. Such lorries distribute the load more evenly on to the axles and are designed to limit the maximum axle load within legal requirements, which currently also stipulate the maximum vehicle laden weight and body length. Wagons used for transporting excavated earth are usually of the rear dump type (3), (4) and (7) and ideally should have four-wheel drive capability to overcome both difficult ground conditions and for travelling at relatively high speeds on smooth roads. The more popular trucks tend to be supported on two axles to improve manoeuvrability and reduce the turning circle but the load on the rear axle when laden is high compared to three-axle trucks. However, modern large volume tyres, twin mounted are capable of carrying these heavy loads and the vehicle is versatile in all but very soft going ground conditions. Three-axle wagons need only be used

| | GROSS VEHICLE HP (kW) | CAPACITY (m³) | NET WEIGHT EMPTY (tonnes) | PAYLOAD (tonnes) |
|---|---|---|---|---|
|  1 | (45-261) 60-350 | — | up to 10 | up to 16.26 * |
|  2 | (150-261) 200-350 | — | up to 20 | up to 24.38 * |
|  3 | (112-1194) 150-1600 | 6-70 | 6-120 | up to 150 |
|  4 | (150-2235) 200-3000 | 6-120 | 10-250 | up to 300 |
|  5 | (75-298) 100-400 | — | 4-10 | up to 16.26 * |
|  6 | (150-298) 200-400 | — | up to 20 | up to 38 * |
|  7 | (75-250) 100-325 | approx. 3-20 | up to 30 | up to 35 |

**Fig. 13.2** Transport using lorries and trucks

*Gross laden weight only, all others are carrying load. GLW is generally about 38 tonnes depending upon the country of operation – 8 tonnes or so is usually the permitted axle load

where it is necessary to limit the bearing load as on clays and fine soils. These trucks are more expensive than the two-axle version due to the difficulty of accommodating the large diameter wheels along a fairly short body length.

Trucks are described in terms of the total number of wheels and drive wheels. Thus a 4 × 4 truck has four wheels and four-wheel drive. An 8 × 4 truck has eight wheels, of which only four provide drive, the other four are free wheels. Sometimes wheels are mounted in pairs on the axle, to carry heavy loads. Thus although a 4 × 4 truck may perhaps have four wheels on the rear axle making a total of six tyres, the vehicle designation remains as 4 × 4. The modern rear dump truck is used for a variety of purposes from general site use up to the 300 tonnes versions in quarrying operations. The latter are frequently loaded by loader shovel, which tends to be more violent than other loading machines and requires the dump truck to be very robust.

## Transporting distances

Earthmoving trucks are mainly used to transport material over distances between 1 and 10 km but longer journeys are possible. Wagons are usually loaded by an excavator, but sometimes, in exceptional cases, by continuous equipment. After completion of loading there is a short delay while positioning the next truck for loading. Obviously the waiting time for the excavator is reduced as the wagon size increases. Conversely the smaller the wagon, the shorter the loading period and the corresponding waiting time for the truck, but the optimum wagon size is very difficult to determine and depends very much on site conditions. Many workers have attempted to represent the loading/discharge cycle in a mathematical form or simulation model, but as yet the information required for input is too diffuse to enable the users to rely with much confidence on the output information. In practice, the optimum wagon size seems to be between three and ten times the size of loading buckets, five being a good average ratio, but the final choice of truck size very much depends upon building up experience on a variety of differing types of contract, ground conditions, loading plant, etc.

## Haul roads

For economic transporting, haul roads should ideally be limited to a maximum gradient of 15%, preferably without bends and curves, have a hard surface and be continuously maintained in this condition by grading. The cost of a grader is easily recovered in the lower maintenance and repair costs of the trucks. When these savings are combined with the lower tyre wear and increased output resulting from the high speeds attainable, the grader proves a vital machine in the earthmoving operation.

### The decision to use large dump trucks

It is uneconomic to use very large dump trucks unless large volumes of earth are moved, but they can offer many advantages under certain conditions. For example, a reduced number of trucks means that fewer drivers are required, with a corresponding reduction in labour costs. The capacity of modern trucks now matches that of the largest loading machines, leading to less waiting time and thus a reduced operating cost. Large wagons are very robust and the exhaust system can also be arranged to provide heating and thereby prevent the soil clinging to the body. This is a particularly useful facility for carrying and emptying difficult materials.

Large rock fragments can be carried so obviating the expense of breaking the strata into small sizes.

Finally, some large trucks are now available in an articulated form which affords up to a 25% saving in manoeuvring time and furthermore helps when roads with sharp bends and obstruction have to be used. They are also longer and narrower than the rigid type, so providing savings in haul road widths. Being lighter, the power-to-weight ratio is better as well.

## 13.2 EXAMPLE OF TRUCK SELECTION

Trucks are loaded by a 1 m³ capacity tracked loader machine at the rate of 30 m³/h bulked material. The truck transports the material to a tip 3 km away. Select the size and number of trucks required.

Select truck size, say, loader bucket capacity × 5 = 5 m³ capacity truck.

|  | *Minutes* |
|---|---|
| Load time $= \frac{60}{30} \times 5$ ................................... | 10 |
| Cycle time to tip and back (average speed, 20 km/h) | 18 |
| Dumping, turning and accelerating .................... | 3 |
|  | 31 |

Truck 1    10    18    3

Truck 2    10    18    3

Truck 3    10    18    3

Truck 4    10    18    3

1 min waiting
for excavator

Thus three 5 m³ capacity trucks should be adopted.

*Note*: The analysis in practice is more complex, and an estimate of the quality of haul road, down time, etc., is necessary before the information assumed above can be obtained. In order to include the many variables which may affect the cycle time, Monte-Carlo simulation techniques have been developed. Models are available from the major equipment manufacturers, and operators concerned with haulage at long-term sites have also developed models to suit the particular situations. The principles, however, are fairly well established as demonstrated in the following example.

### Example of truck selection by simulation

Trucks arrive at an excavator from distribution points on a large earthmoving project. The arrival time intervals of the trucks are observed and yield the following results.

| Arrival time interval (min) | Frequency (%) |
|---|---|
| 2 | 10 |
| 3 | 15 |
| 4 | 30 |
| 5 | 25 |
| 6 | 20 |

The times taken to load the trucks, which are either 6 or 12 m$^3$ capacity, are fairly constant at 3 and 5 min, respectively, and both types are equally represented in the fleet.

If the excavator loads each of the trucks immediately it arrives, in the order that it arrives, calculate the total time that the excavator and trucks will be waiting in any one period of two hours selected at random.

### SOLUTION

The trucks arrive at the excavator at varying times in an indistinguishable pattern. However, the percentage distributions of the arrival times are known, therefore it is possible to simulate the arrival time interval of a truck by drawing a random number lying in the range represented by the observed distribution. Thus:

| Arrival time interval (min) | Frequency (%) | Random number |
|---|---|---|
| 2 | 10 | 00–09 |
| 3 | 15 | 10–24 |
| 4 | 30 | 25–54 |
| 5 | 25 | 55–79 |
| 6 | 20 | 80–99 |

Table 13.1 simulates the arrival of 50 trucks one after the other. The random numbers are selected from Table 13.2 and the type of truck arriving is determined by tossing a coin. Truck type A represents a 6 m$^3$ load. Truck type B represents a 12 m$^3$ load.

**Table 13.1** Simulated arrival times

| Random number | Arrival time (min) | Type of truck | Clock time of arrival (min) | |
|---|---|---|---|---|
| | | | (6 m³) | (12 m³) |
| 89 | 6 | A | 6 | |
| 29 | 4 | B | | 10 |
| 73 | 5 | A | 15 | |
| 50 | 4 | A | 19 | |
| 47 | 4 | B | | 23 |
| 51 | 4 | A | 27 | |
| 32 | 4 | B | | 31 |
| 58 | 5 | A | 36 | |
| 07 | 2 | A | 38 | |
| 49 | 4 | A | 42 | |
| 14 | 3 | A | 45 | |
| 02 | 2 | B | | 47 |
| 26 | 4 | B | | 51 |
| 97 | 6 | B | | 57 |
| 83 | 6 | A | 63 | |
| 47 | 4 | B | | 67 |
| 51 | 4 | A | 71 | |
| 55 | 5 | B | | 76 |
| 92 | 6 | B | | 82 |
| 07 | 2 | B | | 84 |
| 45 | 4 | A | 88 | |
| 85 | 6 | A | 94 | |
| 76 | 5 | A | 99 | |
| 15 | 3 | A | 102 | |
| 78 | 5 | B | | 107 |
| 68 | 5 | A | 112 | |
| 83 | 6 | B | | 118 |
| 33 | 4 | A | 122 | |
| 18 | 3 | B | | 125 |
| 04 | 2 | A | 127 | |
| 11 | 3 | B | | 130 |
| 98 | 6 | A | 136 | |
| 11 | 3 | B | | 139 |
| 18 | 3 | B | | 142 |
| 34 | 4 | B | | 146 |
| 84 | 6 | B | | 152 |
| 47 | 4 | B | | 156 |
| 18 | 3 | B | | 159 |
| 46 | 4 | A | 163 | |
| 71 | 5 | A | 168 | |
| 53 | 4 | B | | 172 |
| 88 | 6 | B | | 178 |
| 78 | 5 | A | 183 | |
| 30 | 4 | A | 187 | |
| 71 | 5 | A | 192 | |
| 35 | 4 | B | | 196 |
| 96 | 6 | A | 202 | |
| 78 | 5 | B | | 207 |
| 57 | 5 | A | 212 | |

**Table 13.2**  Random numbers

| | | | | | | | | | | | | | | | | | | | | | | | | |
|---|---|---|---|---|---|---|---|---|---|---|---|---|---|---|---|---|---|---|---|---|---|---|---|---|
| 29 | 97 | 38 | 28 | 97 | 54 | 95 | 94 | 54 | 79 | 93 | 88 | 1 | 82 | 40 | 62 | 93 | 78 | 8 | 88 | 64 | 58 | 31 | 6 | 45 |
| 78 | 26 | 21 | 3 | 10 | 14 | 30 | 18 | 18 | 84 | 4 | 11 | 62 | 16 | 70 | 2 | 16 | 7 | 31 | 97 | 79 | 56 | 98 | 79 | 39 |
| 75 | 49 | 97 | 87 | 79 | 32 | 66 | 57 | 89 | 56 | 81 | 70 | 53 | 83 | 21 | 25 | 26 | 56 | 55 | 56 | 34 | 59 | 74 | 21 | 76 |
| 94 | 25 | 80 | 13 | 50 | 67 | 95 | 11 | 78 | 42 | 25 | 91 | 82 | 74 | 30 | 9 | 21 | 90 | 26 | 44 | 23 | 81 | 74 | 51 | 76 |
| 54 | 40 | 59 | 35 | 35 | 47 | 55 | 55 | 73 | 50 | 85 | 38 | 61 | 49 | 8 | 99 | 14 | 2 | 17 | 94 | 61 | 95 | 25 | 85 | 66 |
| 35 | 76 | 40 | 98 | 88 | 45 | 58 | 65 | 12 | 18 | 7 | 43 | 36 | 28 | 10 | 7 | 84 | 16 | 0 | 49 | 32 | 19 | 66 | 87 | 12 |
| 37 | 87 | 64 | 2 | 11 | 8 | 2 | 20 | 20 | 6 | 12 | 16 | 95 | 60 | 69 | 81 | 76 | 75 | 88 | 69 | 95 | 13 | 76 | 70 | 19 |
| 11 | 60 | 36 | 32 | 48 | 57 | 96 | 71 | 43 | 65 | 62 | 37 | 78 | 51 | 87 | 3 | 64 | 45 | 33 | 47 | 17 | 3 | 29 | 15 | 33 |
| 61 | 57 | 28 | 33 | 96 | 48 | 50 | 86 | 85 | 83 | 1 | 53 | 45 | 70 | 19 | 24 | 48 | 10 | 90 | 99 | 68 | 89 | 58 | 52 | 94 |
| 84 | 10 | 80 | 38 | 90 | 52 | 99 | 85 | 52 | 49 | 66 | 63 | 69 | 12 | 32 | 37 | 32 | 28 | 96 | 59 | 78 | 62 | 25 | 95 | 84 |
| 44 | 14 | 10 | 13 | 1 | 66 | 69 | 55 | 12 | 1 | 20 | 94 | 22 | 93 | 90 | 16 | 82 | 64 | 28 | 47 | 20 | 71 | 67 | 21 | 93 |
| 92 | 3 | 96 | 16 | 65 | 42 | 68 | 71 | 56 | 75 | 17 | 72 | 63 | 80 | 81 | 3 | 42 | 50 | 28 | 92 | 45 | 44 | 13 | 45 | 21 |
| 13 | 55 | 9 | 80 | 31 | 51 | 35 | 96 | 32 | 71 | 20 | 22 | 80 | 42 | 18 | 57 | 5 | 5 | 20 | 1 | 80 | 8 | 15 | 42 | 15 |
| 38 | 43 | 40 | 25 | 46 | 88 | 84 | 87 | 75 | 2 | 39 | 99 | 3 | 76 | 76 | 62 | 88 | 97 | 89 | 7 | 97 | 15 | 70 | 27 | 27 |
| 39 | 9 | 51 | 67 | 63 | 4 | 76 | 76 | 86 | 10 | 50 | 28 | 97 | 9 | 87 | 65 | 13 | 33 | 28 | 96 | 12 | 27 | 31 | 88 | 48 |
| 89 | 29 | 73 | 50 | 47 | 51 | 32 | 58 | 7 | 49 | 14 | 2 | 26 | 97 | 83 | 47 | 51 | 55 | 92 | 7 | 45 | 85 | 76 | 15 | 78 |
| 68 | 83 | 33 | 18 | 4 | 11 | 98 | 11 | 18 | 34 | 84 | 47 | 18 | 46 | 71 | 53 | 88 | 78 | 30 | 71 | 35 | 96 | 78 | 57 | 19 |
| 42 | 44 | 69 | 11 | 31 | 49 | 98 | 33 | 76 | 12 | 81 | 30 | 83 | 3 | 87 | 5 | 88 | 21 | 78 | 89 | 94 | 32 | 37 | 83 | 74 |
| 48 | 73 | 61 | 15 | 45 | 74 | 41 | 27 | 26 | 48 | 22 | 46 | 86 | 27 | 12 | 21 | 28 | 38 | 56 | 48 | 86 | 18 | 39 | 39 | 19 |
| 89 | 99 | 62 | 81 | 98 | 40 | 33 | 62 | 77 | 60 | 85 | 37 | 15 | 69 | 76 | 71 | 38 | 27 | 32 | 40 | 53 | 36 | 4 | 51 | 54 |
| 83 | 15 | 51 | 17 | 86 | 77 | 66 | 84 | 50 | 84 | 44 | 96 | 92 | 78 | 37 | 24 | 49 | 35 | 54 | 43 | 78 | 50 | 40 | 32 | 56 |
| 17 | 25 | 7 | 90 | 90 | 70 | 48 | 70 | 69 | 23 | 8 | 85 | 8 | 95 | 84 | 53 | 85 | 55 | 11 | 93 | 41 | 6 | 66 | 72 | 39 |
| 23 | 76 | 93 | 41 | 27 | 30 | 77 | 61 | 72 | 74 | 81 | 13 | 73 | 21 | 99 | 1 | 47 | 52 | 44 | 19 | 51 | 25 | 29 | 43 | 54 |
| 76 | 57 | 8 | 22 | 23 | 26 | 22 | 70 | 63 | 70 | 6 | 6 | 59 | 75 | 92 | 86 | 60 | 50 | 87 | 81 | 36 | 80 | 83 | 43 | 17 |
| 96 | 79 | 7 | 87 | 51 | 5 | 17 | 61 | 43 | 13 | 64 | 77 | 45 | 7 | 55 | 68 | 20 | 0 | 17 | 23 | 64 | 83 | 61 | 76 | 37 |
| 68 | 23 | 26 | 10 | 82 | 97 | 77 | 2 | 89 | 33 | 70 | 46 | 23 | 45 | 83 | 99 | 55 | 95 | 4 | 41 | 89 | 33 | 49 | 89 | 86 |
| 19 | 93 | 65 | 67 | 40 | 81 | 96 | 44 | 68 | 47 | 78 | 3 | 18 | 58 | 1 | 48 | 19 | 2 | 34 | 49 | 99 | 56 | 54 | 71 | 65 |
| 34 | 76 | 58 | 86 | 21 | 86 | 84 | 70 | 40 | 23 | 89 | 26 | 42 | 62 | 69 | 10 | 63 | 32 | 80 | 30 | 18 | 12 | 75 | 34 | 81 |
| 26 | 45 | 91 | 80 | 51 | 26 | 64 | 71 | 6 | 49 | 96 | 57 | 56 | 49 | 81 | 91 | 77 | 95 | 44 | 51 | 61 | 34 | 89 | 73 | 78 |
| 98 | 26 | 56 | 88 | 66 | 51 | 69 | 71 | 48 | 14 | 72 | 40 | 57 | 32 | 23 | 54 | 36 | 66 | 29 | 10 | 99 | 4 | 41 | 86 | 60 |
| 53 | 60 | 78 | 66 | 81 | 67 | 45 | 56 | 64 | 78 | 19 | 79 | 10 | 2 | 55 | 61 | 32 | 20 | 20 | 49 | 11 | 89 | 54 | 64 | 96 |
| 0 | 99 | 16 | 71 | 84 | 95 | 51 | 9 | 72 | 70 | 81 | 23 | 33 | 89 | 62 | 20 | 78 | 77 | 64 | 37 | 52 | 39 | 88 | 16 | 92 |
| 24 | 43 | 49 | 79 | 28 | 16 | 10 | 94 | 49 | 35 | 95 | 98 | 59 | 27 | 70 | 55 | 34 | 62 | 91 | 88 | 56 | 71 | 79 | 75 | 31 |
| 30 | 14 | 32 | 60 | 8 | 33 | 73 | 62 | 89 | 63 | 65 | 18 | 15 | 22 | 39 | 29 | 89 | 8 | 58 | 11 | 83 | 66 | 4 | 15 | 74 |
| 39 | 58 | 83 | 63 | 94 | 73 | 84 | 48 | 95 | 17 | 79 | 74 | 78 | 39 | 10 | 38 | 35 | 75 | 74 | 70 | 69 | 54 | 82 | 75 | 50 |

Start →

A check on the distribution of arrival times simulated produces a reasonable fit, as shown below.

| Arrival time interval (min) | Observed frequency (%) | Simulated number of arrivals | Simulated frequency (%) |
|---|---|---|---|
| 2 | 10 | 4 | 8 |
| 3 | 15 | 8 | 16 |
| 4 | 30 | 17 | 34 |
| 5 | 25 | 11 | 22 |
| 6 | 20 | 10 | 20 |

It is now possible to simulate the whole loading process, since if the type of truck is known its loading time can also be determined. Table 13.3 shows the situation at the excavator over a two-hour period.

**Table 13.3** Simulation situation at the excavator

| Arrival type truck | Clock time of arrival (min) | Starts loading (min) | Loading time (min) | Completes loading (min) | Waiting time of excavator (min) | Waiting time of truck (min) |
|---|---|---|---|---|---|---|
| A | 06 | 06 | 3 | 09 | — | — |
| B | 10 | 10 | 5 | 15 | 1 | — |
| A | 15 | 15 | 3 | 18 | — | — |
| A | 19 | 19 | 3 | 22 | 1 | — |
| B | 23 | 23 | 5 | 28 | 1 | — |
| A | 27 | 28 | 3 | 31 | — | 1 |
| B | 31 | 31 | 5 | 36 | — | — |
| A | 36 | 36 | 3 | 39 | — | — |
| A | 38 | 39 | 3 | 42 | — | 1 |
| A | 42 | 42 | 3 | 45 | — | — |
| A | 45 | 45 | 3 | 48 | — | — |
| B | 47 | 48 | 5 | 53 | — | 1 |
| B | 51 | 53 | 5 | 58 | — | 2 |
| B | 57 | 58 | 5 | 63 | — | 1 |
| A | 63 | 63 | 3 | 66 | — | — |
| B | 67 | 67 | 5 | 72 | 1 | — |
| A | 71 | 72 | 3 | 75 | — | 1 |
| B | 76 | 76 | 5 | 81 | 1 | — |
| B | 82 | 82 | 5 | 87 | 1 | — |
| B | 84 | 87 | 5 | 92 | — | 3 |
| A | 88 | 92 | 3 | 95 | — | 4 |
| A | 94 | 95 | 3 | 98 | — | 1 |
| A | 99 | 99 | 3 | 102 | 1 | — |
| A | 102 | 102 | 3 | 105 | — | — |
| B | 107 | 107 | 5 | 112 | 2 | — |
| A | 112 | 112 | 3 | 115 | — | — |
| B | 118 | 118 | 5 | 123 | 3 | — |
| A | 122 | 123 | 3 | 126 | — | 1 |
| | | | | Total | 12 min | 16 min |

Waiting time of excavator = 12 min in 2 hours
Waiting time of trucks  = 16 min in 2 hours

The simulation should now be repeated several times, so that the results can be expressed with more confidence by the use of statistics. With the model established, and mounted on a computer, the user is in a position to experiment by changing the parameters, e.g. truck and excavator sizes, inter-arrival times to represent changes in the haul road, conditions etc.

# 14

# MOVING EXCAVATING MACHINES – THE BULLDOZER

The bulldozer shown in Fig. 14.2 is a very versatile machine and is used frequently for:

| | |
|---|---|
| Stripping top soil<br>Clearing vegetation | Fig. 14.1(a) |
| Shallow excavating | Fig. 14.1(b) |
| Pushing scrapers | see Chapter 15 |
| Maintaining haul roads | compare with the grader, Fig. 16.4 |
| Opening up pilot roads | Fig. 14.1(c) |
| Spreading and grading | Fig. 14.1(d) |
| Ripping | Fig. 14.11 |

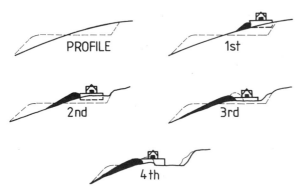

**Fig. 14.1(a)**  Stripping top soil and vegetation

**Fig. 14.1(b)**  Shallow excavation and stockpiling

**Fig. 14.1(c)**  Opening up pilot roads

**Fig. 14.1(d)**  Spreading and grading

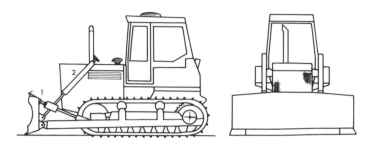

**Fig. 14.2**   Tracked dozer

The dozer is normally operated with a straight blade fixed perpendicular to the line of movement to push the material forward or angled to move the soil to the side.

## 14.1   STRUCTURAL ARRANGEMENT

The machine is assembled in two quite separate sections comprising, A, a base frame or robust welded construction to which are attached the mountings for the dozer blade, the drive sprockets and rollers for the tracks and supports for the upper body, and, B, the superstructure which carries the engine, transmission, hydraulics, cab and controls and is mounted on the base at three points 1, 2, 3 (Fig. 14.3). A heavy duty equaliser bar at point 3 separates the tracks and allows independent movement for better distribution of the machine's weight and improved traction when working on uneven ground.

### Transmission

Bulldozers are usually powered by a diesel engine with an output of about 60 hp (45 kW) for small machines, and 700 hp (530 kW) or more for the heavy duty dozer used for ripping operations. Sometimes a mechanical clutch is fitted between the engine and gears (Fig. 14.4) which even today provides the following advantages over other forms of transmission:

(a) Cost: the simple arrangements tends to result in a cheaper machine.
(b) The engine may be throttled down in order to produce more torque. This is a very useful facility when concentrating on forcing out large stones and rocks, but a manual gear change demands a high concentration from the driver in selecting the correct gear to prevent an engine stall during varying load conditions. Many drivers tend to reduce the efficiency by taking a lower gear than necessary. Today most modern bulldozers employ a torque converter and power-shift gears as shown in Fig. 14.4. The primary clutch is eliminated so dispensing with the need to disengage the engine during changing gear, as all the gear wheels are in connection. The gear requirement is altered by engaging the relevant secondary clutch available on each gear ratio. The torque

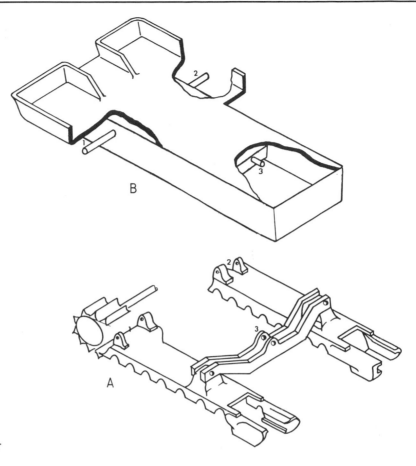

**Fig. 14.3**   Structural form
of tracked dozer

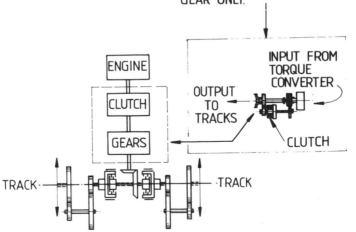

**Fig. 14.4**   Clutch and
gears for a small
tracked dozer

**Fig. 14.5(a)**   Dozer pivoted around one track

**Fig. 14.5(b)**   Dozer turned by slipping one track

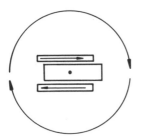

**Fig. 14.5(c)**   Dozer pivoted at centre of machine

converter provides these advantages compared with the mechanical clutch and gear box, e.g.

(i) shock loading when changing gears is eliminated;
(ii) full torque can be applied almost immediately as the engine works at constant power output. The engine speed will vary only marginally by about plus or minus 50 rev/min.

(c) The torque converter automatically adjusts the torque requirements in the selected gear to suit the working conditions, namely:

high torque requirement = low output speed
low torque requirement = high output speed.

(d) There is no primary clutch, so the driver can concentrate solely on the work. Such a machine may be effectively operated by a less skilled driver than is needed for a clutch and gearbox machine.

Some very modern machines are now manufactured with all-hydraulic transmission and control of the attachments. This system uses displacement pumps and motors as described earlier in Chapter 12. The machines have not as yet, however, proved robust enough to cope with heavy duty work and further development is taking place.

### Steering

Positioning is achieved by disengaging the track clutches, one for each track.

(a) For sharp turning, one track is used as the pivot point (Fig. 14.5a). The clutch to this track is disengaged and the brake applied, full drive is maintained on the other track.
(b) A broader turning circle is obtained by not completely disengaging the clutch on the pivot track as in (Fig. 14.5b). Unfortunately, both these methods tend to cause much clutch and track wear.
(c) A very small turning circle may be obtained on machines with the facility for allowing the tracks to rotate in opposite directions (Fig. 14.5c). Such a feature is not required for normal operation, but the extra cost is justified for work in confined areas, such as tunnelling.

## 14.2   THE WHEELED DOZER

The basic machine is similar in construction to the wheeled loader with the bucket attachment replaced by a bulldozer blade. Four-wheeled drive is required to give sufficient traction to perform the tasks of a bulldozer and a correspondingly larger engine than that needed for the tracked dozer is necessary. The increased self-weight in turn raises the bearing pressure at the point of contact between tyres and ground, causing compaction of the soil which may be undesirable in view of the fact that to be efficient a dozer requires loose material to push. Furthermore, the close centres of the two axles may cause stability

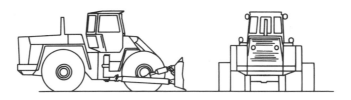

**Fig. 14.6**   Wheeled dozer

problems on uneven ground. Wheeled dozers are seldom used today for these reasons.

## 14.3   METHODS OF WORKING WITH THE BULLDOZER

### The blade

The dozer is fitted with a stiff welded steel blade and pushes material forward, transferring force from the tracks to the blade by means of two strong arm connections to the base frame (see Fig. 14.2). The blade is controlled with two sets of double hydraulic cylinders. One pair connected to the arm attachments (1) act to tilt the blade and in doing so control the depth of dig. The other pair of cylinders are attached to the upper frame (2) and serve to lift and lower the blade for digging effect.

The blade may be angled (Fig. 14.7) to deposit the material at one side along the push line which is a very useful facility when opening up pilot roads or backfilling trenches.

**Fig. 14.7**   Angled dozer blade

### STRAIGHT BLADE

Most blades are curved but the section perpendicular to line of push (Fig. 14.9a) is straight. The curvature causes the material to roll forwards (Fig. 14.8). The driver controls the depth of cut during the pushing action by feel. When the rear of the dozer is felt to rise, the blade is starting to dig in, the angle and depth of the blade are then 'tuned' to meet the resistance encountered. The volume of material that can be moved forward by the blade is dependent upon the blade size and shape. The maximum length of push should not exceed 100 m, as the machine is uneconomical for earthmoving over greater distances.

**Fig. 14.8**   Flow of material in front of a dozer blade

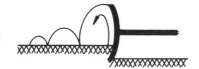

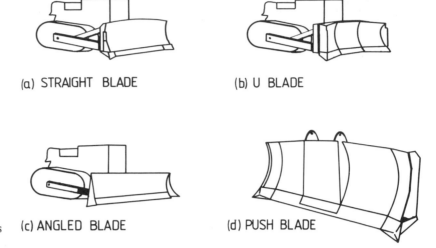

**Fig. 14.9**  Bulldozer blades     (a) STRAIGHT BLADE     (b) U BLADE     (c) ANGLED BLADE     (d) PUSH BLADE

### U-BLADE

The blade in cross section has a much deeper curvature approaching almost a U-shape (Fig. 14.9b), in addition the outer edges are angled slightly inwards. This type of blade will carry much larger volumes of material, but is limited to light, flowing-type soils.

### ANGLED BLADE

The blade in plan view is angled and casts the material to one side. The arrangement is used mainly for backfilling along a trench and for opening up pilot roads in hilly terrain and not as a general purpose earthmover as with the straight-ahead blade.

### PUSH BLADE

Push loading a scraper is one of the main functions of the bulldozer particularly for the larger machines (300 hp (223 kW) upwards). Modern motorised scrapers are only self-loading in very light soils and often must be assisted to load by a bulldozer. The blade (Fig. 14.9d), usually fitted with shock absorbers, effective at travelling speeds up to about 5 km/h, is of much stronger construction than the normal blade and is stiffened with a push plate.

Table 14.1 gives approximate data on the size of blade generally adopted with a given size of machine.

## 14.4  PRODUCTION DATA

The production information shown in Fig. 14.10 is ideal output assuming favourable conditions. The values should be adjusted to suit the circumstances of the construction site as explained in Chapter 12.

**Table 14.1** Bulldozer blades

| Engine size (hp) | 700 | 450 | 400 | 335 | 300 | 200 | 100–150 | 60–70 |
|---|---|---|---|---|---|---|---|---|
| (kW) | 520 | 340 | 298 | 250 | 224 | 149 | 75–112 | 45–52 |
| Machine weight (tonnes) | 85 | 50 | 45 | 35 | 25 | 20 | 10–14 | 5–8 |
| Blade length (m) | 5.5 | 5.0 | 4.5* | 4.0 | 4.5 | 3.5 | 3.5 | 3.0 |
| Blade height (m) | 1.8* | 1.8* | 1.8* | 1.5 | 1.5 | 1.2 | 1.0 | 0.8 |
| U-blade (m³) | 20 | 18 | 14 | 13 | 9 | 7 | 4 | 2 |

\* Not suitable for ground surface trimming because of size causing problems of control.

## 14.5  RIPPING ROCK

### Ripper

The principle of ripping hard soil and rock was known in the 1930s but only with the development of powerful bulldozers and robust tracks has the method become a practical proposition.

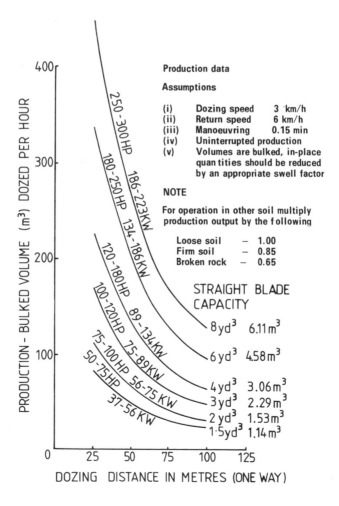

**Fig. 14.10**  Ideal output of the bulldozer

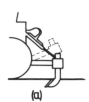

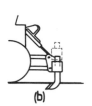

**Fig. 14.11**    Ripping methods

Until the development of the modern ripper, soil or rock which could not be excavated with traditional excavating equipment had to be blasted loose and into sizes which were easily disposable, but today large quantities of rock may be removed economically by ripping methods.

### Selection of ripping equipment

The ripping effect requires (a) penetration into the soil of a strong steel ripping tool, (b) a tractor with sufficient horse power to enable the ripping tool to advance the point through the material, (c) a heavy tractor to generate sufficient traction, (d) a strong robust tractor to take the strain. The ripping tool used is normally a radial arc type or parallel linkage type.

### RADIAL ARC TYPE

This is of simple construction (Fig. 14.11a) and requires little servicing, but because the ripping tool approaches the digging action along an arc, there is a tendency for the rear of the tractor to lift and thus diminish the available traction.

### PARALLEL LINKAGE TYPE

The ripping tool enters the ground upright, allowing the point to immediately dig in and thus pulls the tractor to the ground to increase traction. This method (Fig. 14.11b) is probably the better of the two systems, especially in tough materials. The latest versions can now be equipped with hydraulic vibrating mechanisms for improved penetration.

### SHAPE OF SHANK

The shank is made from high tensile steel usually fitted with a replaceable point. For hard compact material and laminated rock the straight shank (Fig. 14.12a) gives good performance. The curved shank (Fig. 14.12b) is more suited to ripping fractured rocks.

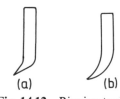

**Fig. 14.12**    Ripping tool shanks

### Ripping techniques

(a) The minimum tractor power required is 150 hp (112 kW), but for very tough work 400 hp (300 kW) or more may be needed.

(b) First gear should always be used, travelling at 1–2 km/h to obtain maximum torque from the torque converter.

(c) When conditions permit, it is advisable to rip with three teeth to reduce speeds to the least possible and minimise wear and tear not only to the ripping tool but also to the machine itself. The number of teeth is best chosen by trial and error but when working with

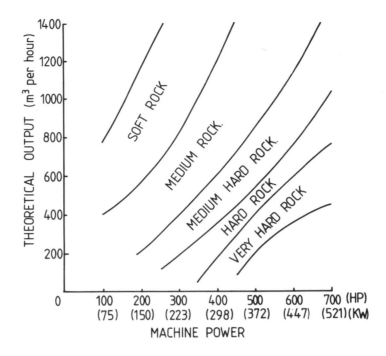

**Fig. 14.13** Theoretical output from a rippable rock

material which tends to break out in large slabs, generally only one point is necessary.

(d) Modern teeth can now rip to depths approaching one metre, the limiting depth depending very much on the type of ground. Usually it is desirable to rip as deep as possible.

(e) The distance between passes is again largely governed by the type of ground. As a guide, the following may be suitable:

hard material          1–1.5 m passes
light fractured material   2–2.5 m passes.

(f) The material should be removed either by scraper or loader.

(g) Output estimation is very difficult to assess and the only reliable method is to carry out site observation. For tendering purposes the figures shown in Fig. 14.13 which are based on records collected by Pohle at Aachen T.H., Germany, are very approximate guides.

(h) Ripping methods may offer up to 80% saving compared to the cost of using explosives.

## ROCKS WITH CHARACTERISTICS FAVOURABLE TO RIPPING METHODS

(a) Stratified rocks and soils.
(b) Rocks with fractures, faults and planes of weakness.
(c) Large-grained rock of brittle texture.
(d) Most shales, slates and mudstones.
(e) Soft rock.

## ROCKS WITH UNFAVOURABLE CHARACTERISTICS

(a) Non-crystalline rocks.

(b) Igneous rocks in general, but specifically, those with massive and uniform structures.

(c) Rocks without planes of weakness, and soils without bedding planes.

(d) Cohesive soils are also unsuitable for ripping.

It is now possible to obtain an indication of the rippability of the rock by seismological methods (see *Construction Planning, Equipment and Methods* by Peurifoy and Ledbetter).

### 14.6   OTHER ATTACHMENTS

### Towing winch

The larger machines often have a winch attachment at the rear, which serves for stump pulling, moving sheet piling and general skidding jobs.

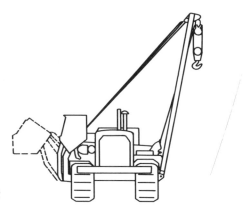

**Fig. 14.14**   Side boom attachment for a tracked dozer

### Side boom

Side booms have been developed mainly for use in the handling and laying of large diameter pipes. Boom lengths up to 6 m are available and loads up to 30 tonnes may be lifted with very large machines (400 hp (248 kW)). The modern machines have a hydraulically variable positionable counterweight, with independent boom and lift ropes.

### Marsh tracks

Bulldozers especially developed for operating on marshy ground have very wide tracks. Thus the contact area of the track is increased and bearing pressure is reduced:

|  |  |
|---|---|
| bearing pressure marsh tracks | approx. $0.03$ N/mm$^2$ |
| bearing pressure normal tracks | approx. $0.05$ N/mm$^2$ |
| bearing pressure under human foot | approx. $0.02$ N/mm$^2$. |

# 15

# MOVING EXCAVATING MACHINES – THE SCRAPER

## 15.1  INTRODUCTION

The bulldozer can be used effectively for moving earth over short distances up to 100 m. However, many projects, particularly road construction, necessitate a combined load, haul and discharge system at least up to a distance of 3 km. The situation calls for a robust excavator, capable of travelling over rough terrain to eliminate the need for the use of trucks and wagons on public roads. The scraper has been developed specifically to cater for this medium-distance haul. Essentially the earth is cut and loaded directly into the scraper box (or bowl), transported to the discharge area and finally spread in layers. The whole process takes place in a continuous cycle. The type of machine to be adopted depends upon the travelling distance; the tractor-pulled scraper is favoured for short hauls, while the motorised version is now almost universally preferred where the size of the project permits.

## 15.2  EXCAVATING ACTION

The scraper is a self-loading, transporting and spreading machine predominantly used for general levelling. It is usual to cut on a downward gradient to take full advantage of gravity. For cutting, the bowl is lowered and the apron (1) opened, forward movement of the machine directs the cutting edge (2) into the soil causing it to boil upwards into the bowl. Approximately 50–100 m of travel is required to fill the bowl. Excavation is carried out in layers of from 150 to 300 mm in depth, the levelling action is thus achieved as a gradual process. During the discharge stage ejection takes place whilst the unit

**Fig. 15.1**   Scraper action

is moving: again the height of the bowl is set to spread the material in a controlled layer and the soil is pushed out of the bowl with the aid of the ejector plate (3).

In general, tractor-pulled scraper capacities range from about 8 up to 30 m³ heaped capacity, motorised scrapers from 15 up to 50 m³ heaped capacity. The struck capacity is the volume of material contained in the bowl, when levelled off evenly with the top of the bowl. Some manufacturers quote the SAE (Society of Automotive Engineers, USA) heaped capacity of the bowl (see Fig. 12.21), which represents the volume of material contained in the bowl allowing for heaping at a slope of 1:1.

## 15.3   TOWED SCRAPER

The towed scraper comprises a tractor, frequently this will be a bulldozer, and a towed bowl supported on two axles running on four wheels. Large volume tyres are needed to cope with the heavy loads to be transported over rough uneven surfaces. A powerful bulldozer, capable of providing 300 hp (223 kW) or more, fitted with specially strong and deep webbed tracks, is needed to provide the necessary traction during the loading operation. The loading cycle takes up to 2 min, with about 90% of the bowl filled in the first minute, but because of the slow travelling speed to the discharge area, output is considerably reduced. In practice, hauls greater than 300 m make the method uneconomic relative to other means of removing the material. However, outputs of the order of 40 m³/h of struck material are achievable with, for example a 9 m³ heaped capacity scraper. The main advantage over the motorised system is the ability to load in heavy soils, to load on the upwards gradient, to manoeuvre in a small turning circle and to 'pump' load in patches of hard material.

ejector plate

**Fig. 15.2**   Towed scraper

## 15.4   MOTORISED SCRAPER

These machines were developed during the 1950s in an attempt to improve production on large earthmoving projects. Unlike the towed scraper, the engine is self-contained within the machine and power thus supplied directly to the wheels. The whole unit is supported on very large volume tyres.

The excavating and hauling action is carried out in a similar way to the towed scraper but frequently requires pushing assistance during the loading phase because of loss in traction when using wheels rather than tracks. However, haul speeds up to 60 km/h are possible with well-graded roads, yielding considerable improvements in output. For example, with the 20 m$^3$ heaped capacity scraper bowl output up to 150 m$^3$/h hauling over 1 km is possible.

The modern motorised scrapers are large earthmoving machines, ranging from 15 m$^3$ to 50 m$^3$ heaped capacity equipped with one or two large diesel engines capable of high power output, commonly up to 500 hp (373 kW) each. As with most modern excavation equipment the transmission is usually achieved through a torque converter and power-shift gears to the wheels. The gearbox often has from six to eight gear stages: the bottom four gears provide the high torque demanded during the loading operation and the top gears allow the high travelling speeds along the haul. Several types of scraper are available to suit different circumstances.

### Standard single-engine scraper

The standard scraper (Fig. 15.3) comprises the scraper bowl (1) mounted on a single rear axle (2). The front end is hitched to the drive axle (3), by means of a single arm called, by virtue of its shape, the *swan neck* (4). The bowl height is controlled through a pivot attachment to the swan neck and hydraulic cylinders (5). The standard scraper is not self-loading and requires a pusher bulldozer to provide the necessary traction.

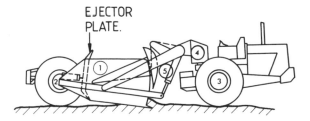

**Fig. 15.3**  Motorised scraper

### Double-engine scraper

The double-engine scraper (Fig. 15.4) is similar to the standard unit in both its construction and operating action. However, the rimpull required on soils with high coefficients of resistance (i.e. well rutted, soft, etc.) and/or high uphill gradients is often best achieved with four-wheel drive to utilise the total available grip (i.e. traction). A second engine is then usually located over the rear axle (Fig. 15.5).

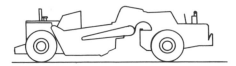

**Fig. 15.4** Double-engine scraper

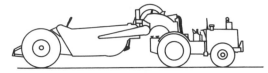

**Fig. 15.5** Four-sheeled tractor scraper

## 15.5 METHOD OF WORKING WITH THE MOTORISED SCRAPER

### Push loading

The method is used mainly with the standard scraper. The bulldozer is fitted with a very robust blade and mounted on to the dozer frame through shock absorbers to take up impact, while the scraper is stiffened at the rear to form a push block. The pusher is used only during the loading phase and the scraper moves off under its own power as soon as the bowl is fully loaded. Generally a pusher serves from three to five scrapers working as a team, so although in practice high production is possible, the high cost of the scraper and bulldozer team limits the use of the motorised system to fairly large projects. Several push load systems are in use and are shown in Fig. 15.6.

As a general rule:

$$\frac{\text{number of scrapers per pusher}} = \frac{\text{cycle duration of each scraper in minutes}}{\text{cycle duration of each pusher in minutes}}$$

### NUMBER OF RUBBER-TYRED MOTOR SCRAPERS SERVICED BY A BULLDOZER

It is assumed that a single dozer comprises the pusher set and that the average speed of a scraper is 30 km/h along a haul road. Two or more dozers per set may be required, depending upon the soil, size of scraper and dozer.

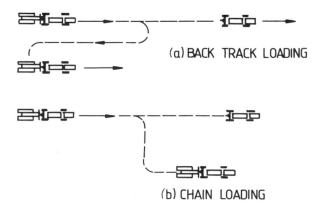

**Fig. 15.6** Push loading systems

(a) BACK TRACK LOADING

(b) CHAIN LOADING

| Haul distance (one way) (m) | Number of motor scrapers per pusher set |
|---|---|
| 100 | 2 |
| 200 | 2 |
| 300 | 2 |
| 600 | 3 |
| 900 | 4 |
| 1200 | 5 |
| 1500 | 6 |
| 1800 | 7 |
| 2100 | 8 |
| 2400 | 8 |
| 3000 | 10 |

### The push–pull method

This is an arrangement whereby two scrapers in turn assist each other to load without the need for a bulldozer pusher, however, a reasonable amount of coordination is necessary. The method is fairly effective in all but very heavy soils. Usually double-engine scrapers are used in tandem, whereby the rear scraper pushes the front scraper during loading and then the front machine pulls the rear scraper when it is loading. Clearly, the machines must be robust if they are to be operated in this way and modifications to the standard model include: strong swan neck and bowl hitch arms, a cushion push block and bail at the front end, and a hook and stiffened frame at the rear. A typical push–pull connection is shown in Fig. 15.7 in which the mechanism is controlled hydraulically from the driver's cab.

**Fig. 15.7** Push–pull scraper connection

### WORKING CYCLE

1. Scraper I ahead of scraper II after turning. Relative travelling speeds synchronised to 2–4 km/h.
2. Scraper I begins to load with the help of a push from scraper II.
3. Scraper II loads with the help of a pull from the fully laden scraper I.
4. Connection link between scraper I and scraper II is broken.

**Fig. 15.8** Push–pull scraper system

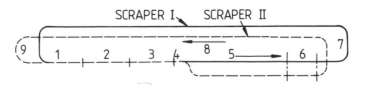

5. Scraper I accelerates away, and scraper II follows slightly off line.
6. Scrapers discharge.
7. Scrapers turn.
8. Scraper II takes the lead.
9. Scrapers turn to continue the loading operations at (1) with scraper I regaining the lead.

The size of the bulldozer to use as a pusher is related to the carrying capacity of the scraper and depends upon the hardness of the soil to be loaded. More than a single dozer may be needed to provide sufficient power, in which case the dozers are referred to as the 'pusher set'. Typically the power of the pusher set to match a standard scraper for the push loading method is shown in Table 15.1. Even the push–pull method often requires dozer assistance.

**Table 15.1**  Scraper and matching pusher dozers

| Heaped capacity of scraper in yd$^3$ (m$^3$) | Single-engine scraper hp (kW) | | Double-engine scraper hp (kW) Push–pull |
| --- | --- | --- | --- |
| | Standard scraper | Pusher dozer set | |
| 20(15.3) | 350(261) up to | 500(373) | 450(335) |
| 30(22.9) | 450(335) up to | 900(670) | 700(521) |
| 40(30.6) | 550(410) up to | 1200(894) | 950(707) |
| 50(38.3) | 550(410) up to | 1200(894) | 950(707) |

\* In the push–pull system two scraper units are used.

## 15.6  SELF-LOADING SCRAPER

The scrapers described above are, in general, not self-loading because of the high resistance encountered when cutting into the soil. In an attempt to overcome this problem, the apron is removed and replaced by an elevator which aids the loading action and helps to offset the power required for traction purposes. Such machines are self-loading in all but very hard or heavy soils.

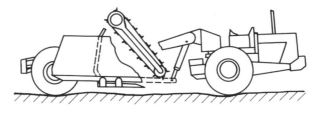

**Fig. 15.9**  Elevating scraper

## 15.7    GENERAL POINTS TO CONSIDER WHEN USING SCRAPERS

1. Keep haul roads broad enough to accommodate overtaking.
2. Ensure that haul roads are well maintained with graders and free from deep rutting and grooving and well drained.
3. Construct hauls with gradients less than 5% if possible, max 12%.
4. Plan haul road routes to avoid bottlenecks, short lengths of sharp gradient and curves.
5. Try to keep the rolling resistance as low as possible.
6. Towed scrapers are economic up to 300 m haul distance.
7. Single-engine standard scrapers are suitable for light soils, and low rolling resistances along the haul and require pusher assistance, economic up to about 3 km haul.
8. Push–pull scrapers – must be double-engined and robust. Capable of similar work described above under (7).
9. An elevating scraper is 10–15% heavier than the equivalent standard scraper and generates a greater rolling resistance. Purchase price is higher but as it is a self-loading machine it may be more economic than the standard scraper if push loaders are not needed on the project.
10. The single-engine scraper has lower fuel costs than either the double-engine or elevating scraper.
11. Push loading gives a shorter loading time than push–pull loading.
12. Cuttings – start along the outer edges of the projected cutting and work downhill towards the fill area. Maintain the formation in a convex shape to lean the scraper towards the sizes of the cutting and thereby encourage the maintenance of the full width of the cut. Continuously trim the cut sides with a grader (or a bulldozer as a second best), pushing the material to the scraper for removal. The sides should be cut in a series of steps by the scraper, as governed by the depth of soil layer attempted by the scraper.
13. Fillings – embankments are formed in layers working inwards from the outer limits. The formation is made concave to lean the scrapers towards the centre and so reduce the possibility of the scraper slipping over the edges.

## 15.8    PRODUCTION DATA

Ideal outputs from the towed and motorised scrapers are given in Figs. 15.10 and 15.11 respectively. In practice, production may vary considerably from these recommendations and should be adjusted accordingly as typified by the example on output estimation in Chapter 12.

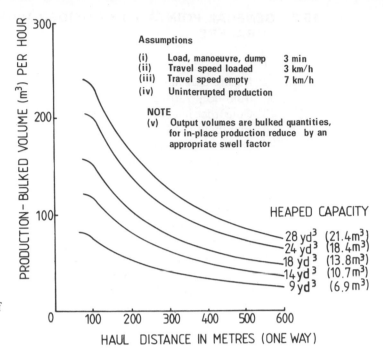

**Fig. 15.10** Ideal output of the towed scraper

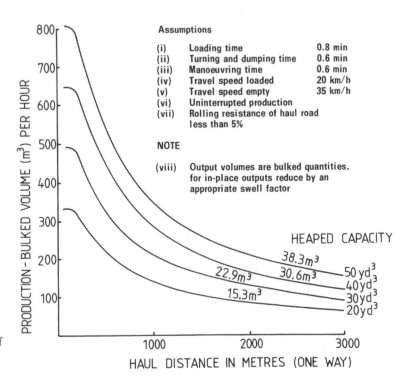

**Fig. 15.11** Ideal output of the motorised scraper

### 15.9    HAUL ROADS

**Resistance to movement**

The haul road is the length of track connecting the load and discharge areas on an excavation site. The haul road can quickly become rutted and holed with the constant pounding of the earthmoving plant. The production data given in Figs. 15.10 and 15.11 assume a good haul road with a rolling resistance of about 2%. Rubber-tyred scrapers are particularly affected as the haul road deteriorates. For example, as the rolling resistance approaches 10%, output from a scraper could be halved, because the machine is forced to travel at a much lower speed and far more manoeuvring is involved. The rolling resistance is expressed as a percentage of the combined weight of the excavating machine and any self-contained load. Clearly the machine must supply sufficient power to the wheels to overcome rolling resistance. Table 15.2 gives approximate values of percentage rolling resistance for rubber tyres and also tracks operating on different soil types.

**Table 15.2**   Rolling resistance

| | Rolling resistance (%) | |
| Ground surface condition | Rubber tyres | Tracks |
|---|---|---|
| Concrete | 1.5 | 0 |
| Fine gravel/sand | 2.0 | 0 |
| Loose gravel | 10.0 | 5 |
| Loose sand | 10.0 | 5 |
| Soft mud | 16.0 | 7 |
| Deeply rutted loam | 16.0 | 7 |
| Loose loam | 4.0 | 3 |
| Firm loam | 2.0 | 1 |
| Packed snow | 2.5 | 0 |

### 15.10    RIMPULL AND TRACTION

The maximum speed attainable by a moving machine naturally depends on the rolling resistance of the ground, however, the grip (traction) between the ground surface and the wheels or tracks must be sufficient to prevent slippage and therefore must exceed the supplied rimpull, if the vehicle is to move at all.

The *supplied* rimpull is the tractive force applied between the tyres of the drive wheels and ground surface, measured in newtons (N) and is determined by the power of the engine and influenced by the efficiency of the transmission, the gearing ratios and wheel/track arrangements, and is further affected by the influence of altitude and air temperature on the engine power.

The *usable* rimpull, i.e. traction, however, depends upon the coefficient of traction for rubber tyres or tracks running on particular soils. The coefficient of traction is defined as the factor by which the load (i.e. downward force) on the drive wheels or tracks may be multiplied before slipping occurs. This value is the maximum *usable* rimpull, which clearly must be greater than the supplied rimpull or the wheels will spin.

### Example

A rubber-tyred motorised scraper has a capacity of 21/30 yd$^3$ (16/23 m$^3$) (struck/heaped). The machine is operated on firm loam up a gradient of 6% at an altitude of 900 m where the air temperature during the day is 30 °C. The rated power of the engine is 450 hp (335 kW). The gross laden weight of the unit is 72 000 kg, made up of 36 000 kg self-weight and 36 000 kg payload. The total load is distributed with 50% on each axle. Determine:

(a) if the wheels will slip;
(b) if sufficient rimpull is available given the travelling speeds for the following gear ratios:

| | |
|---|---|
| 1st gear | 5 km/h |
| 2nd gear | 12 km/h |
| 3rd gear | 18 km/h |
| top gear | 40 km/h. |

### INFORMATION

Approximate coefficients of traction are given in Table 15.3. For wet materials the coefficients are difficult to determine but cohesive materials are especially affected and the coefficient of traction could be reduced by up to 50%.

**Table 15.3**  Approximate coefficients of traction

| Type of surface (dry) | Rubber tyres | Tracks |
|---|---|---|
| Smooth concrete | 0.8–1.0 | 0.3–0.6 |
| Clay | 0.5–0.8 | 0.6–0.9 |
| Firm sand/gravel | 0.3–0.8 | 0.7–0.9 |
| Loose sand | 0.1–0.2 | 0.3–0.5 |
| Loose gravel | 0.2–0.4 | 0.4–0.7 |
| Packed snow | 0.1–0.4 | 0.2–0.6 |
| Firm loam | 0.4–0.8 | 0.6–1.0 |
| Loose loam | 0.4–0.6 | 0.7–1.0 |

SOLUTION

(a) (i) From Table 15.3 the coefficient of traction for firm loam is, say, 0.6. The gross laden weight of the unit is 72 000 kg of which 36 000 kg is supported on the drive axles. Therefore the maximum usable rimpull is $36\,000 \times 0.6 \times gN$, where $g$ is the force of gravity, $9.81$ m/s$^2$, which reduces to 212 kN.

(ii) Maximum power supplied from the engine $= 450 \times 0.746 = 335$ kW. Typically for this size of machine, the efficiency in transferring engine power into rimpull is about 80% and supplied rimpull (in kN)

$$= \frac{3.6 \times \text{engine power (kW)} \times \text{efficiency}}{\text{speed (km/h)}}$$

$$\text{Supplied rimpull: in 1st gear} = \frac{3.6 \times 335 \times 0.8}{5} = 193 \text{ kN}$$

$$\text{Supplied rimpull: in 2nd gear} = \frac{3.6 \times 335 \times 0.8}{12} = 80 \text{ kN}$$

$$\text{Spllied rimpull: in 3rd gear} = \frac{3.6 \times 335 \times 0.8}{18} = 54 \text{ kN}$$

$$\text{Supplied rimpull: in top gear} = \frac{3.6 \times 335 \times 0.8}{40} = 24 \text{ kN}$$

(iii) *Approximate reduction in engine horse power for altitude* – depends upon the manufacturer's engine performance specification but for a four-cycle diesel engine the losses are approximately:

*Altitude*
0–300 m ....................no loss in performance
300 m and above ..........reduces available rimpull by 3% per 300 m.

(iv) *Approximate reduction/increase in engine power for temperature*

| Temperature (°C) | −15° | 0° | 15° | 30° | 45° |
|---|---|---|---|---|---|
| Reduction (%) | +6 | +3 | 0 | −3 | −5 |

Therefore total reduction of rimpull $= 6\% + 3\% = 9\%$.
Available rimpulls corrected for temperature and altitude:

1st gear = 176 kN
2nd gear = 73 kN
3rd gear = 49 kN
top gear = 22 kN.

Supplied rimpull in first gear 176 kN. This is less than usable rimpull of 212 kN (see (i)), therefore the wheels will not slip.

(b) (i) Rolling resistance from Table 15.2 for firm loam = 2%.
Supplied rimpull required to overcome rolling resistance

$$= \text{total weight of unit} \times \text{rolling resistance}$$

$$= 72\,000 \times \frac{2}{100} \times \frac{9.81}{1000} = 14 \text{ kN}$$

(ii) Supplied rimpull required to overcome grade resistance

$$= 72\,000 \times \frac{6}{100} \times \frac{9.81}{1000} = 42 \text{ kN}$$

Therefore total supplied rimpull required = 56 kN.

The scraper will generate sufficient rimpull in second gear to climb the grade and thus if the unit is to be operated at heaped capacity the maximum speed along the haul is 12 km/h. However, top gear may be engaged where there is no gradient. Should it be necessary to operate the machine on loading in first gear, on an upward gradient, then clearly the available rimpull is 176 − 56 = 120 kN. In effect this value is the *draw bar pull*, and is the available pull which could be exerted on a load that is being towed, for example a box scraper attached to a crawler dozer.

*Note*: The reader should now repeat the calculation in the unloaded condition.

### Example of a machine selection exercise

MASS HAUL

Bulk earthworks operations on a road construction contract involve forming cuttings and embankments. All earthmoving is to be carried out by tractor-pulled scrapers. Tips and borrow pits are available as required.

Table 15.4 gives quantities computed from longitudinal sections, and cross sections. Bulking is 10% after recompaction. Costs are as follows:

(i) dig and form banks up to 200 m haul, £0.30 per m³;
(ii) dig and form banks between 200 m and 300 m, £0.80 per m³;
(iii) dig and form banks between 300 m and 400 m, £1.60 per m³;
(iv) cart to tip, £0.50 per m³;
(v) bring from borrow pit, £0.75 per m³.

SOLUTION

The cumulative volumes are plotted graphically on Fig. 15.12. With the aid of dividers the 200 m, 300 m and 400 m hauls are marked on the diagram and the respective volumes of earth moved by the scraper

**Table 15.4**

| Length along road (m) | Volume (m³) | | | Cumulative volume (m³) |
|---|---|---|---|---|
| | Cut | Fill | Fill $\times \frac{10}{11}$ | |
| 0 | 0 | | | 0 |
| 200 | 40 000 | | | 40 000 |
| 400 | 70 000 | | | 110 000 |
| 600 | 30 000 | | | 140 000 |
| 800 | | 20 000 | 18 200 | 121 800 |
| 1000 | | 40 000 | 36 400 | 85 400 |
| 1200 | | 50 000 | 45 500 | 39 900 |
| 1400 | | 20 000 | 18 200 | 21 700 |
| 1600 | 40 000 | | | 61 700 |
| 1800 | 40 000 | | | 101 700 |
| 2000 | | 20 000 | 18 200 | 83 500 |

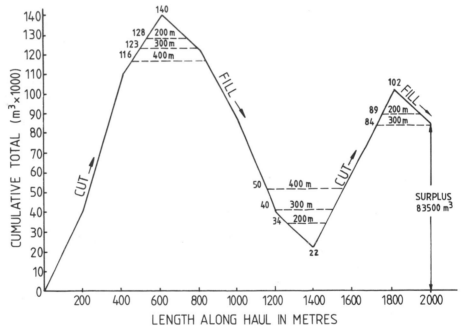

**Fig. 15.12** Mass haul diagram

are obtained by subtracting values from the gradient turning points. The volumes of cut and fill are read off the graph between the respective values established between the haul limits.

The cost for each combination of haul, cut and fill are calculated as follows:

*200 m haul*

| | | | | |
|---|---|---|---|---|
| Hauled = | 37 000 m³ | cost = | 37 000 × 0.3 = | £11 100 |
| Fill = | 94 000 + 5500 | cost = | 99 500 × 0.75 = | £74 625 |
| Cut = | 128 000 + 55 000 | cost = | 183 000 × 0.5 = | £91 500 |
| | | | | £177 225 |

*300 m haul*

(ii) Hauled 300 m = 53 000 m³
 (i) Hauled 200 m = 37 000 m³
                    16 000 m³

Fill =  83 500 m³     (i) cost =  37 000 × 0.3  =  £11 100
Cut = 167 000 m³     (ii) cost =  16 000 × 0.8  =  £12 800
                          cost =  83 500 × 0.75 =  £62 625
                          cost = 167 000 × 0.5  =  £83 500
                                                   £170 025

*400 m haul*

Hauled 400 m = 70 500 m³
Hauled 300 m = 53 000 m³
Hauled 200 m = 37 000 m³

Fill =  66 000 m³     (i) cost =  37 000 × 0.3  =  £11 100
Cut = 149 500 m³     (ii) cost =  16 000 × 0.8  =  £12 800
                    (iii) cost =  17 500 × 1.6  =  £28 000
                          cost =  66 000 × 0.75 =  £49 500
                          cost = 149 500 × 0.5  =  £74 750
                                                   £176 150

The cost of each respective haul distance is plotted on Fig. 15.13, where it is seen that the economic haul distance is approximately 300 m.

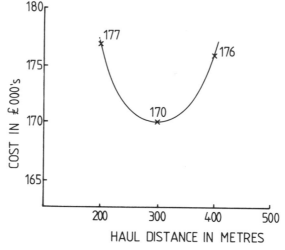

**Fig. 15.13**  Economic haul distance = 300 m (i.e. scrapers are only used up to 300 m hauls; outside this limit material is transported to tip or brought in from borrow pits)

# 16

# MOVING EXCAVATING MACHINES – THE GRADER

## 16.1  INTRODUCTION

Many earthmoving projects require the final ground to be accurately finished, so that the surface is smooth and level without undulations and ridges. Although a skilful driver using a bulldozer can in many instances achieve adequate results, the grader (Fig. 16.1) has been specifically developed for trimming the subgrade, sub-base surface on roads and road cuttings and banks, for smoothing off the walls on

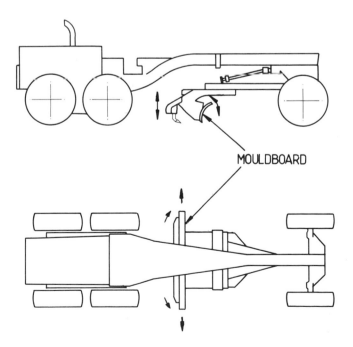

**Fig. 16.1**   Grader

earth-fill dams and maintaining haul roads. It is self-transporting and supported on two or three axles. The large road programmes over the past few years have demanded heavier and more robust machines and the three-axle configuration has proved necessary to carry the large engine and provide more traction and is today the more popular type.

## 16.2  MECHANICAL ARRANGEMENT

The engine, transmission, driver's cab and controls are supported on the rear frame. Modern machines allow the driver to work in the seated position who until fairly recently had to stand in order to see clearly to operate the mouldboard.

The front wheels support a long bridging beam from which the mouldboard is hung. On some types the beam is pivot-jointed to the rear frame to allow a reduced turning circle, increased manoeuvrability and to permit offset axle grading. On others the connection is rigid and steering is only possible through the front axle. This design allows the wheels to: (a) lean (Fig. 16.2), up to about 15° each side of vertical to resist side loads, for example, when angling the mouldboard, and (b) to operate at different levels (Fig. 16.3) while shaping banks, making ditch cuts, forming crossfalls and so on. Combination of the two facilities means that direction may be held without the need for excessive concentration by the driver on steering, so freeing his attention for direction of the mouldboard.

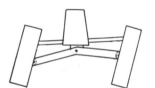

**Fig. 16.2**   Leaning to resist side load

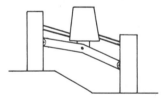

**Fig. 16.3**   Tilting axle

### The mouldboard

The mouldboard on modern machines is operated hydraulically from the driver's cab, and although it works similarly to the bulldozer blade, it is hung between the axles, and the magnifying effect of uneven and stony ground is very much reduced, as demonstrated in Fig. 16.4. The pitch of the mouldboard is adjusted to suit the work. It is tilted back when cutting and forwards when spreading. The more upright the mouldboard, the greater the mixing and rolling which will be given to the material being spread.

**Fig. 16.4**   Comparison of action between the grader and bulldozer

 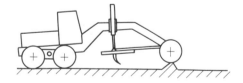

### Grader blade length and corresponding engine power

| | Engine | Blade length | Machine weight |
|---|---|---|---|
| hp | (kW) | (m) | (kg) |
| 35–50 | (26–37) | 2.5 | 2 500 |
| 50–70 | (37–52) | 3.0 | 3 500 |

| | | | |
|---|---|---|---|
| 70–90 | (52–67) | 3.0–3.5 | 6 500 |
| 90–120 | (67–89) | 3.0–3.5 | 8 000 |
| 120–140 | (89–104) | 3.5–4.0 | 10 000 |
| 140–160 | (104–119) | 3.5–4.0 | 13 000 |
| 160–180 | (119–134) | 3.5–4.0 | 14 000 |
| 180–300 | (134–224) | 4.0–4.5 | 25 000 |

### 16.3 METHODS OF WORKING WITH THE GRADER

The blade can be used in several positions for:

(a) Levelling and trimming on the horizontal, with the mouldboard central or swung out either to the left or right (Fig. 16.5a). If the mouldboard is set at an angle on plan, the material will roll off the blade to form a windrow. However, with the blade at right angles to the line of movement, then only spreading or trimming is obtained.

(b) Levelling and trimming to the slope and vertical face (Fig. 16.5b).

(c) Forming ditches (Fig. 16.5c). The mouldboard is angled both on plan and in the vertical and set such that the blade just protrudes beyond the outside line of the wheels nearest to the ditch to be shaped. A windrow is formed along the top of the ditch. The ditch is deepened gradually, trimming off a layer at a time, keeping the nearside wheels in the ditch.

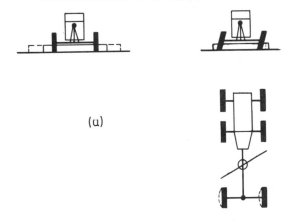

(a)

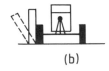

(b)

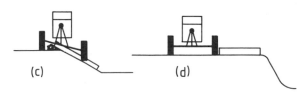

(c)          (d)

**Fig. 16.5** Methods of working with the grader

(d) Backfilling along trenches (Fig. 16.5d). The action is similar to that for producing a windrow.

### Controlling the blade

Without guidance the blade trims to the plane dictated by the unevenness of the ground, but because the mouldboard can be positioned by means of hydraulic cylinders, the driver is able to superimpose some measure of control, independent of the posture taken by the wheels. However, in order to obtain very smooth and even surfaces it is necessary to use some form of predetermined levelling device. Frequently for road works, wires are set along the direction line and sensors attached to the mouldboard adjust the blade height automatically. As an alternative device, a laser beam activates photoelectric cells which control the hydraulic adjusters fitted between the blade and the frame of the grader.

**Fig. 16.6** Wire and sensors to control grading levels

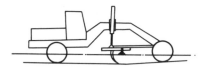

### 16.4   EXTRA ATTACHMENTS

Many graders also carry a scarifier mounted just in front of the mouldboard, it is raised and lower hydraulically and used to break up the soil to facilitate an easier grading action. A ripping tool may sometimes be mounted at the rear end and a conventional bulldozer blade at the front.

### Recommended operating speeds

|                                        | km/h   |
| -------------------------------------- | ------ |
| Grading site roads                     | 4–9    |
| Scarifying (e.g. soil stabilisation)   | 8–18   |
| Forming ditches                        | 4–8    |
| Spreading                              | 4–10   |
| Trimming and levelling                 | 9–40   |
| Snow ploughing                         | 8–20   |
| Self-transporting                      | 10–40  |

The demands on the grader clearly are very varied, so robust engines up to 300 hp (223 kW) are common. Hydraulic control of the attachments is standard on the latest models.

# 17

# MOVING EXCAVATING MACHINES – THE LOADER SHOVEL

The loader shovel, sometimes called the tractor shovel or front end loader, exists in two forms; with tracks (Fig. 17.1) or wheels (Fig. 17.2). It is a machine which serves the purpose of both the fixed-position excavator and transporter over short distances of perhaps 10–20 m.

It is the ability of the machine to cope with duties usually associated with the face shovel and also to overlap with the functions of the bulldozer which has caused the rapid increase in numbers in plant fleets over recent years.

The choice between tracks or wheels is fairly straightforward. The tracked loader is a genuine excavator, whereas the wheeled version is more suited to stockpiling and digging in loose soils. The tracked machine can apply more traction and will cope with tougher

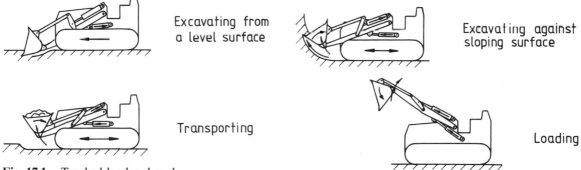

Excavating from a level surface

Excavating against sloping surface

Transporting

Loading

**Fig. 17.1**   Tracked loader shovel

**Fig. 17.2**   Wheeled loader shovel

conditions. The tracked loader is typically available with bucket sizes of up to 9 yd³ (7 m³), whilst buckets over 40 yd³ (30 m³) are possible, especially on the larger wheeled machines. The tracked loader is of course limited by its slow travel speed and becomes uneconomic when the distance from the excavation or discharge points exceeds about 80 m. The wheeled loader being faster is viable up to 200 m or so, and is able to travel on tarmacadam roads without causing damage and is certainly more mobile and manoeuvrable than the equivalent tracked machine. However, tracks are more stable on soft and rutted surfaces, which helps to improve the production performance.

## 17.1 MECHANICAL ARRANGEMENT OF THE TRACKED LOADER

Tracked loaders are almost identical in appearance to bulldozers, the engine, gearing and hydraulics are all similar (Fig. 17.3) but the tracks are positioned more to the front to provide counterweight when loading and to distribute the bearing pressure more evenly when the bucket is fully loaded. Unlike the bulldozer, however, independent and differential movement of the tracks is impossible.

The bucket attachment is controlled by hydraulic rams and an automatic cut-out device disengages the hoist control lever when the bucket reaches a predetermined height. A similar device is sometimes fitted on the tilt rams to keep the bucket horizontal when tipping.

## 17.2 MECHANICAL ARRANGEMENT OF THE WHEELED LOADER

Structurally the wheeled loader is quite different from the tracked machine and has in fact been developed from the farm tractor into a

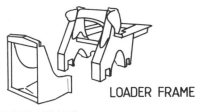

LOADER FRAME

FRONT FRAME

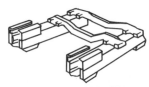

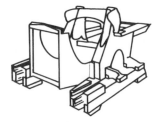

**Fig. 17.3** Construction of the tracked loader

TRACK FRAMES JOINED BY RIGID BARS

FRAME ASSEMBLY.

**Fig. 17.4**  Pivot connection

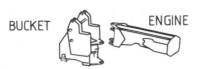

form which today consists of a pivoted frame with the engine mounted over the rear wheels (Fig. 17.4) and the cab over either the front or rear frame dependent upon the manufacturer's preference. The pivot arrangement gives the loader very good manoeuvring capabilities, facilitating the front frame to turn in plan up to 40° either side from the forward position (Fig. 17.2) to work in a relatively small turning circle.

Power is supplied from a diesel engine through a torque converter, and power-shift gears to drive the wheels; four-wheel drive is common, necessitating all wheels to be the same size. The machine, however, can be operated as two-wheel drive only. The rear-wheel drive loader is better at digging while front-wheel drive gives better traction when carrying a full bucket. On most machines the front axle is fixed, whilst the rear axle may oscillate up to $\pm 15°$ from horizontal (Fig. 17.5), i.e. a total of 30° to incorporate differential movement between the wheels and allow more effective action on uneven ground.

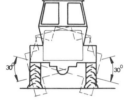

**Fig. 17.5**  Rear axle pivot action

### Modern developments in loader design

Some very modern machines have complete hydrostatic drive, whereby displacement pumps, powered direct from the engine, drive displacement motors on the wheels or tracks. Furthermore, a few of the wheeled machines have a hydraulic rather than a pivot connection between the two frames to give articulation in the vertical plane for both the rear and front axles.

## 17.3  METHODS OF WORKING WITH THE LOADER

The size of the machine depends upon the particular circumstances of the job. Both the tracked and wheeled types are able to carry out similar operations and are basically used for loading loose material or for excavating at the level of the wheels or tracks in fairly loose soils. By careful positioning of ramps the excavator may be used to form large pits and other deep excavations. The machine is also very suitable for stripping soil in thin layers – acting like a bulldozer.

Modern machines are quite robust and can compete favourably with face shovels, especially in quarries.

### Loading methods

(a) *V-loading* (Fig. 17.6a) is a very efficient method and is most frequently adopted when the wagons are able to take up the required positions.

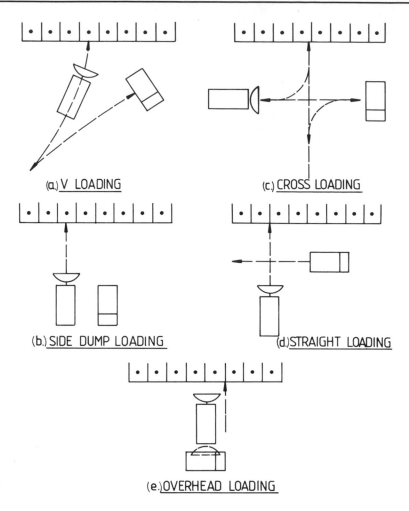

**Fig. 17.6** Loader movements

(b) *Side dump loading* (Fig. 17.6b) requires no turning of the loader and is used mainly for backfilling ditches or close quarter loading. It is not commonly used in general construction work.

(c) *Cross loading* (Fig. 17.6c) requires far more manoeuvring than the V-method and is slightly less efficient.

(d) *Straight loading* (Fig. 17.6d) is used when it is possible to place the wagon along the line of the excavation, the method appears in principle to be very effective. However, the need to co-ordinate two machines is a little cumbersome and therefore seldom used.

(e) *Overhead loading* (Fig. 17.6c) is used mainly in tunnelling work.

**The loading bucket**

(a) *The general purpose bucket* (Fig. 17.7a) is made from thick high-strength, wear-resilient plate. The cutting edge extends forward, and the addition of steel teeth gives an improved penetration.

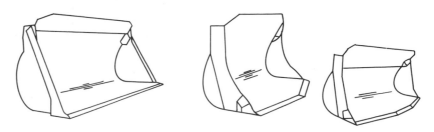

**Fig. 17.7** Loader buckets

(a) GENERAL PURPOSE BUCKET. (b) ROCK BUCKET (c) ROCK BUCKET

(b) *The rock bucket* is made of special heat-treated alloy steel with high strength and wear resistance. The bucket has either a V-cutting edge (Fig. 17.7b) or a modified V-edge (Fig. 17.7c) and cut-back sides to aid penetration.

(c) *The side dump bucket* (Fig. 17.8) dumps to the left or to the right side, providing considerable advantages in confined areas and for backfilling along trenches.

(d) *The multipurpose bucket:* a four-in-one bucket (Fig. 17.9) was marketed originally by the Drott Company and since then any machine fitted with this type of bucket has tended to be called a 'Drott'. The bucket action is hydraulically controlled and used in one of four forms; a dozer blade, shovel, grab or clamshell, scraper. The most popular size is about 1 m$^3$ for use with small and medium-sized bulldozers and wheeled or tracked loaders. It is this bucket attachment which has provided the loader with most of its advantages over alternative types of excavator.

(e) Other attachments include log grapples, fore-lift tool carrier, brush, etc.

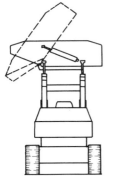

**Fig. 17.8** Side dump bucket

**Fig. 17.9** Four-in-one bucket

DOZER      SHOVEL      CLAMSHELL      SCRAPER

## 17.4  WHEELED BACKHOE/BUCKET LOADER

Many construction sites require excavation for manholes, drain trenches, etc., and the loading of loose material into wagons to be carted away to tip. Much of this work, particularly for building-type projects, is on a small scale and takes place spasmodically. The backacter-type loader shovel adequately fills such demands. The machine comprises a rigid frame (no pivoting is possible), a loader bucket and backhoe. Only the rear wheels are powered, so the front wheels do not need to be of the same size. The engine and transmission are in general similar to those of other loading units and operate with a similar loading action, however, the driver usually moves to a separate seat and control panel to operate the backhoe which is mounted on a horizontal bar for digging wide trenches. The backhoe can turn through 90° in plan and excavate approximately 5 m below the wheels. The machine must be stabilised by jacks in this mode, otherwise too much movement and strain on the axles will be generated from the voluminous rear tyres.

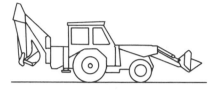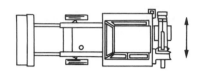

**Fig. 17.10** Wheeled backacter/bucket loader

## 17.5 SKID-STEER LOADER

The machine is wheel mounted and very robust. Hydrostatic drive facilitates quick forward and backward changes of direction to provide excellent manoeuvrability in confined spaces such as basement excavation and inside coffer-dam enclosures. Typically the weight range of models varies from about 1000 kg up to 6500 kg equipped with buckets of from 0.3 to 2.5 m³ capacity, respectively.

## 17.6 LOADERS – POPULAR RANGE

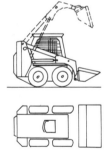

**Fig. 17.11** Skid-steer loader

| Bucket heaped capacity | | Engine | |
|---|---|---|---|
| yd³ | (m³) | hp | (kW) |
| 1–1½ | (0.76–1.14) | 60–80 | (45–60) |
| 1½–2¼ | (1.14–1.72) | 80–100 | (60–75) |
| 2¼–3 | (1.72–2.30) | 100–150 | (75–112) |
| 3–4 | (2.30–3.06) | 150–220 | (112–164) |
| 4–5½ | (3.06–4.21) | 220–320 | (164–238) |
| 5½–7 | (4.21–5.36) | 320–400 | (238–298) |
| 7–9 | (5.25–6.75) | 400–700 | (298–525) |

## 17.7 PRODUCTION DATA

The production information given in Figs. 17.12 and 17.13 is for ideal conditions and should be adjusted for a particular construction site, as explained in Chapter 12.

### Example of machine selection

Select suitable plant to excavate one million cubic metres of firm silty clay for the foundations of a modern power station.

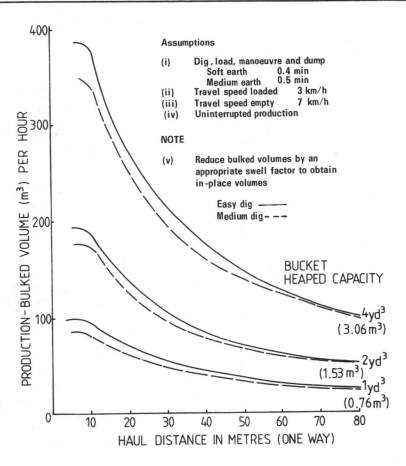

**Fig. 17.12** Ideal output of the tracked loader shovel

## INFORMATION

(a) All material may be directly loaded into dump trucks and ramped up to ground level.
(b) Average haul for excavator is 30 m.
(c) Swell factor is 1.2.
(d) Excavating season, March to October inclusive.
(e) Tip located on site.

## SOLUTION

In the dry season trucks may approach the excavation area on firm ground, therefore a tracked vehicle is selected, because of its improved traction capabilities over wheeled loaders.

Output from a 4 yd³ (3.06 m³) heaped capacity tracked loader given in Fig. 17.12 is 200 m³/h. This is a bulked production rate.

$$\text{In-place output} = \frac{\text{bulked output}}{\text{swell factor}} = \frac{200}{1.2} = 167 \text{ m}^3/\text{h}$$

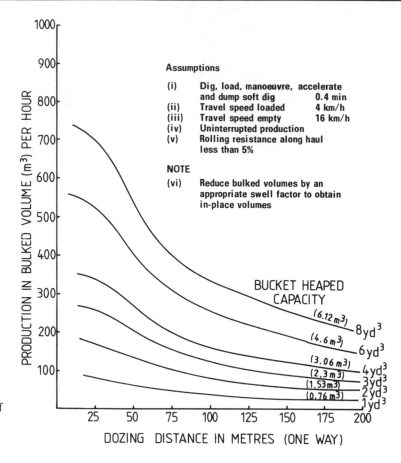

**Assumptions**

| | | |
|---|---|---|
| (i) | Dig, load, manoeuvre, accelerate and dump soft dig | 0.4 min |
| (ii) | Travel speed loaded | 4 km/h |
| (iii) | Travel speed empty | 16 km/h |
| (iv) | Uninterrupted production | |
| (v) | Rolling resistance along haul less than 5% | |

**NOTE**

(vi)   Reduce bulked volumes by an appropriate swell factor to obtain in-place volumes

BUCKET HEAPED CAPACITY

(6.12 m³) 8 yd³
(4.6 m³) 6 yd³
(3.06 m³) 4 yd³
(2.3 m³) 3 yd³
(1.53 m³) 2 yd³
(0.76 m³) 1 yd³

PRODUCTION IN BULKED VOLUME (m³) PER HOUR

DOZING DISTANCE IN METRES (ONE WAY)

**Fig. 17.13**  Ideal output of the wheeled loader shovel

Potential losses (%):

| | |
|---|---|
| Weather | 10 |
| Breakdown | 5 |
| Operator efficiency | 7 |
| Waiting | 10 |
| | 32 |

$$\text{Realisable in-place production rate} = 167 \times \frac{68}{100} = 113 \text{ m}^3/\text{h}$$

$$\text{Time required} = \frac{1\,000\,000}{113} = 8850 \text{ h}$$

Muck-shifting season, approximately 30 weeks with an average 50 hour working week = 1500 h.

$$\text{Number of machines required} = \frac{8850}{1500} = 5.9, \text{ say 6 excavators}$$

Truck size, say 5 × loader bucket capacity = 5 × 4 = 20 yd³ (15.3 m³).

# 18

# TYRES

Tyres are required to spread the machine weight on to the ground surface, to convert the rimpull at the points of contact with the ground surface into traction and to aid and improve the steering capabilities of the machine.

## 18.1 TYRE CONSTRUCTION

The design and manufacture of tyres has improved slowly during the past 10 years with new versions to match the improved performance of the present generation of machine only gradually appearing from the major manufacturers. Currently there are two principal methods of tyre construction; *cross-ply* and *radial ply*, both types are mounted on earthmoving machines (but not mixed together). A special *beadless* tyre is now being introduced on some loader machines.

### Cross-ply tyre

The cross-ply tyre is shown in Fig. 18.1 and consists of:

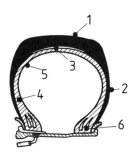

**Fig. 18.1** Cross-ply tyre

1. Tread – this is the outer wearing surface which makes contact with the ground and gives protection to the inner layers. The tread is an important part of the tyre, since the rate of wear and friction coefficient are influenced by the quality of the rubber and tread pattern.
2. Side walls – these are a protective layer of rubber covering the tyre plies.
3. Belts – these provide an intermediate layer of protection, usually confined to the tread area. They wrap around the tyre circumference.
4. Body plies – consist of alternating layers of synthetic material which cross each other at a 45° angle. They form the main carcase of the tyre and provide its strength.

5. Inner liner – this is an inner lining which seals in the air.
6. Beads – these consist of bands of steel wire integrated with the body plies to anchor the tyres to the wheel rim. When the tyre is inflated, a good air seal is maintained.

### Radial tyre

The radial tyre is constructed in a different manner from the cross-ply tyre. The beads are made from a single bundle of steel wires, while the tyre carcase is a single layer of steel cables fixed radially from bead to bead to form an arch shape. A more flexible and wider bodied tyre is the result.

## 18.2 TYRE SIZE

The tyre size is usually stated, for example as 40–39, where the first number is the cross section width of the tyre and the rim diameter is the second number in inches. The tyre may also be designated by its aspect ratio, which is the ratio of the tyre cross section height to width. Aspect ratios are: standard tyre 1.0, wide tyre = 0.85 and low profile tyre = 0.65.

Thus a tyre designated 65R40–39 represents a low profile, radial tyre.

## 18.3 PLY-RATING

This is an index defining the strength of the tyre, and is directly related to the type of material forming the body plies. The ply-rating for a given tyre size may vary depending upon the working requirements.

For example for a tyre size 18–25 in:

| Ply-rating | Permissible load on tyre (kg) | Tyre pressure N/mm² | (atm) |
|---|---|---|---|
| 12 | 4675 | 0.2 | (2.0) |
| 16 | 5675 | 0.275 | (2.75) |
| 20 | 6475 | 0.3 | (3.0) |
| 24 | 7275 | 0.425 | (4.25) |
| 28 | 8000 | 0.5 | (5.0) |

## 18.4 TYRE SELECTION

Generally, the larger the tyre the greater the machine output because large diameter tyres require lower tyre pressures and therefore present a larger contact area to the ground surface, so reducing rolling resistance and thus reducing tyre slippage and wear and tear. As a consequence treads may be shallower for the same tyre life, with a corresponding further reduction of rolling resistance. The introduction of the wide-bodied tyre, further emphasises these advantages. However, increased tyre diameter lowers the rimpull.

Presently, radial tyres are preferred for hauling operations because the heat build up tends to be less than with cross-plies and they have a slightly lower rolling resistance.

Finally, the lower the ply-rating the lower the heat build up. Use deep treads for better grip on soft ground and durability on rock or hard surfaces. Smooth or ribbed tyres are recommended for light flowing soils.

## 18.5    TYRE LIFE

The tyre life is affected by the tyre size and type, heat build up, travel surface, tyre pressure and load, and is further influenced by the operator's skill and vehicle maintenance. In particular, the tyre should be matched with the load to be carried and the correct tyre pressure maintained. As a further precaution manufacturers have introduced a system to relate heat build up and tyre safety known as the ton – mile/h ratings (=mean tyre load × average speed).

The mean tyre load is the weight on the tyre when the machine is fully loaded plus the weight on the tyre when the machine is empty, divided by two.

The average speed is calculated from the total distance travelled in kilometres during the work day, divided by the length of the work day in hours.

When operating scrapers, for example, the cost of tyres can amount to up to 30% of the total owning and operating costs of the machine. It is therefore important to observe the manufacturer's advice.

### A guide to tyre life

| Machine | Operating conditions | | |
|---|---|---|---|
| | Favourable (h) | Average (h) | Unfavourable (h) |
| Scrapers | | | |
|     Twin engine | 4000 | 3000 | 2500 |
|     Single engine, tractor | 4000 | 3000 | 2500 |
|     Single engine, box | 5000 | 3500 | 2500 |
| Loaders | 3500 | 2500 | 1500 |
| Graders | 5000 | 3000 | 2000 |
| Trucks | 4000 | 3000 | 2000 |

The tyre life on loaders can be improved by fitting chains.

# 19

# SPECIALISED EXCAVATING MACHINES

The rapid build up of the oil and gas industries during the past 40 years has called for pipelines in the distribution network to be buried below ground. As a result machines have been developed to improve the efficiency of excavating long sections of trench. In conjunction with this development, there has been the need for machines to cut canals and work vast mineral deposits.

The equipment for performing these duties most effectively is either the chained-bucket or rotary-bucket excavator. Both are expensive, and technically sophisticated machines, which require relatively large amounts of capital investment compared to most other types of earthmoving equipment.

## 19.1 CHAINED-BUCKET EXCAVATOR

The machine is very efficient for excavating shallow trenches in fairly loose soils, and the example shown in Fig. 19.1 is a simple type for use on small scale operations. Large and more sophisticated machines are available for more substantial work, such as canal construction.

### Method of working

The soil is excavated along the direction of travel by means of an endless chain of buckets mounted on a boom attachment fixed to the main frame. The contents are discharged on to a conveyor belt integral with the machine and then deposited into some other means of spoil disposal. Wear and tear tends to be rapid in abrasive soils as the drag force puts considerable tension into the chain, demanding almost continuous maintenance, with the associated high costs. Power is supplied either by a diesel or electric engine, the latter source being used mainly for large units. The range of sizes on the market is now sufficient to meet most demands, from the small unit of 20 m³/h output up to the giants of 300 m³/h. However, the length of trench must be long in order to justify the large capital purchase and would only be

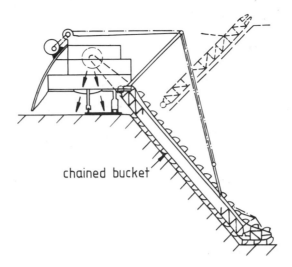

**Fig. 19.1** Chained-bucket trenching machine

possible if much work of a similar character could be obtained in future contracts. Furthermore, the type of soil very much influences the output capability of the machine and heavy clay soils are almost impossible to excavate using this method. Thus if the nature of the soil is likely to change frequently over a run of trench excavation, the use of other traditional excavating methods on some sections may be necessary and the machine's economic advantage becomes seriously eroded.

## 19.2  ROTARY-BUCKET EXCAVATOR

Only on extremely large excavation works, for example opencast ore mining, does it become economic to use the large-wheeled bucket excavator. However, the rapid developments now taking place in design may result in the smaller types being operated for more general construction work. Currently machines with outputs of the order of 100 m³/h up to 20000 m³/h are available.

A large-diameter wheel placed at the end of a long boom is rotated in the excavation face. The wheel consists of six to eight buckets arranged in a regular pattern around the rim, the diameter of which may vary from 2 m up to 20 m depending upon the size of the machine, buckets on the smaller units may be quite small of perhaps 10 litre capacity, increasing to 4000 litre on the giant versions. The wheel is rotated at 10–30 cycles/min and about 25% of the circumference comes into contact with the excavation face.

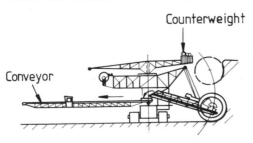

**Fig. 19.2** Rotary-bucket excavating machine

### 19.3 BUCKET SHAPE

**The cellular wheel**

The cellular wheel comprises an integral series of bucket cells through which excavated material is discharged on to a conveyor belt. Some difficulty may be experienced in dislodging cohesive soils from the cells, the only recourse then is to run the wheel at a lower speed to reduce the centrifugal force, a correspondingly lower output is obtained.

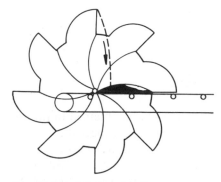

**Fig. 19.3** Cellular wheel

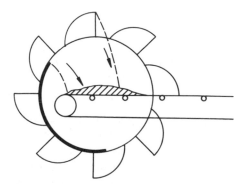

**Fig. 19.4** Bucket wheel

**The bucket wheel**

The bucket wheel consists of single cells mounted on the rim, through which the soil falls as each bucket reaches its highest position on the rotation cycle. A ring plate prevents discharge taking place earlier. The arrangement is much lighter than that of the cellular wheel and also less liable to clog up in cohesive soils.

### 19.4 CUTTING ACTION

(a) Forward cutting (Fig. 19.5) – the cells are eased into contact with the soil starting at (1) gradually deepening as the wheel rotates into

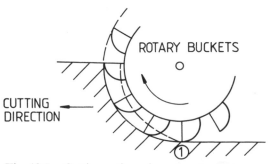

**Fig. 19.5** Cutting action of a rotary-bucket excavator

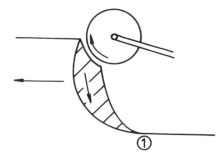

**Fig. 19.6** Downward cutting

the face. The upward rotation reacts against the weight of the machine and so provides an ideal method for tackling heavy soils.

(b) Downward cutting (Fig. 19.6) – this time the cutting action finishes at the toe of the cutting (1) and a much cleaner surface is obtained. The applied force between cells and soil face is dependent upon the weight of the machine, little reaction can be gained from the soil in comparison with the forward cutting action.

### Block and line cutting

To excavate a face deeper than the diameter of the wheel, it is usual to remove the material in stages (Fig. 19.7). In this way perhaps 20% of the potential output is lost, due to the forwards and backwards shunting action demanded of the machine. The alternative of leaving a deep under-reamed face would be rather unsafe.

The block cutting system shown in Fig. 19.8 or the alternative line cutting method in Fig. 19.9 have been developed to deal with a deep excavation face.

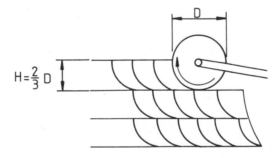

**Fig. 19.7** Cutting a deep excavation face

**Fig. 19.8** Block cutting method

**Fig. 19.9** Line cutting method

### Production data

The output of the machine very much depends upon the type of soil and the quantity which is available for continuous excavation. For estimating purposes the following figures as used by Pohle at Aachen. T.H., Germany, give a guide:

| Wheel diameter (m) | 4 | 6 | 8 | 20 |
|---|---|---|---|---|
| Bucket capacity (m$^3$) | 0.2 | 0.3 | 0.5 | 5 |
| Soil – light (m$^3$/h) | 400 | 800 | 1600 | 16 000 |
| Soil – medium (m$^3$/h) | 300 | 600 | 1200 | 12 000 |
| Soil – heavy (m$^3$/h) | 200 | 400 | 800 | 8 000 |

# 20

# DREDGES

Dredging is a highly specialised excavating method used predominantly for deepening rivers, estuaries and coastal areas for improved navigation; land reclamation as typified by the work of the Dutch in the Polder regions of Holland; the winning of sands and aggregates from river and sea beds; and specialist excavation on civil engineering works where the dredging barge can approach the site. The dredging action may be: (a) of the suction form which stirs up material and pumps the mixture on to nearby land, into barges or is stored on board, or alternatively (b) mechanical whereby the material is physically excavated and loaded into tugs and barges. Several different types of dredge are operated within these two broad classifications each with features to provide an advantage in a particular dredging situation.

## 20.1 SUCTION DREDGE

The suction dredge is basically a floating platform equipped with a powerful centrifugal pump and suction pipe. The material is first loosened by jetting in front of the suction pipe entrance using special high-pressure pumps, the resulting spoil and water mixture is then drawn up the suction pipe and pumped into barges or directly to shore. The exact output depends upon the length of pipe, suction pressure and friction loss. The pumped mixture consists of between 1:6 and 1:10 soil to water, and, on some of the larger vessels, over $10\,000$ m$^3$/h may be pumped.

The method is most suited for operation in sands for land reclamation use. Where the site is too distant for the capacity of the pumps on the dredge the spoil is discharged into specially equipped barges then pumped away over distances of 10 km or more in large-diameter pipes. Positioning is achieved by winching from anchors.

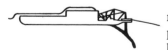

**Fig. 20.1**  Suction dredge

## 20.2    TRAILER SUCTION HOPPER DREDGE

**Fig. 20.2** Self-propelled hopper dredge

The hopper dredge is self-propelled and stores the excavated material in hoppers built into the hull and discharges through valves at the bottom of the vessel.

Suction pipes are mounted along the sides and may be pulled up for transporting. During dredging the boat is propelled forward at about 1 m/s (4 km/h) drawing the spoil up the large-diameter suction pipe through a draghead (Fig. 20.3) which, being heavy and wide, presses into the bed resulting in a stirred up mixture of spoil and water which is then discharged into the hopper.

The largest types are capable of dredging to depths of 30 m and will store up to 10 000 m$^3$ of spoil on board and work in waters with wave heights up to 2.5 m.

The output depends upon the soil type but dragheads are available for use in fine and coarse sands, clay and mud. Naturally, the volume of soil dredged will vary since the spoil is a mixture of soil and water, but modern pumps are capable of filling the 10 000 m$^3$ hopper in 1 hour. The hopper dredge is a self-contained seagoing ship and can sail to almost any dredging location. It is used mainly in fine loose soils for improving waterway schemes.

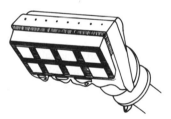

**Fig. 20.3** Draghead

## 20.3    CUTTING SUCTION DREDGE

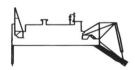

**Fig. 20.4** Cutterhead dredge

**Fig. 20.5** Cutterhead

Where the effect of normal suction pressures or jetting is insufficient to loosen the soil, a mechanical method, such as the cutterhead dredge, is preferred. The equipment consists of a rotating head (Fig. 20.5) fixed to the end of a stiff jib on which the drive motor and suction pipe are also carried. The cutter operates by rotating the arm in an arc and the dredged materials are pumped ashore along a floating pipeline. Depths up to 15 m can be worked: the method is suitable for cutting soft to medium hard rock but is also capable of operating in clays, sands and granular deposits.

The cutting dredge is generally not self-propelled since the vessel needs to be held in position whilst working and this is achieved with two piles at the rear of the barge. Gradual positioning adjustment may be effected in conjunction with a system of anchors and ropes (Fig. 20.6) and the barge is 'walked' forward by raising one pile at a time.

### 20.4 BUCKET DREDGE

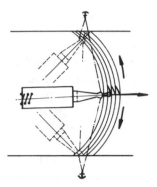

**Fig. 20.6** Walking the barge

The bucket dredge is appropriate for most types of soil but is best suited to medium sand, gravel, mud or clay, and loose rock fragments. Although the modern high-powered suction pumps and especially the cutting head dredge are proving more effective, the bucket dredge is probably still preferred where working in restricted space is required, for example, along quay walls and in dock systems. The method is also principally used for mining ores and precious metals.

The operation of the dredge consists of an endless chain of buckets continuously passing through the body of the barge to the working surface and is therefore only suitable in quiet waters. The buckets hold 50–1000 litre each but as a guide 50% of the contents is likely to be water, and output may range between 15 and 300 m³. The barge is winched into position, can work to depths of 30 m, and discharges down a shoot into mud barges moored alongside. The chain and buckets are subjected to intense wear and the consequent maintenance costs are high.

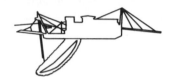

**Fig. 20.7** Bucket-ladder dredge

### 20.5 DIPPER DREDGE

**Fig. 20.8** Dipper dredge

The dipper dredge is a heavy-duty excavator operating with a similar action to the face shovel. The arm can turn through 180° and discharge to barges. The vessel is supported by two front piles, and a single pivoted rear pile to provide forward movement.

The main advantage is the powerful crowding action, so that the bucket can be forced into the underwater materials, which allows excavation in very compact and oversize materials such as rock, sandstone and coral, without the need for blasting. Depths up to 15 m may be worked and output of over 100 m³/h is achievable with a $1\frac{1}{2}$ yd³ (1.14 m³) bucket.

### 20.6 CLAMSHELL DREDGE

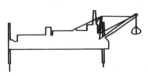

**Fig. 20.9** Clamshell dredge

The clamshell dredge is little more than a pontoon-mounted grab crane and is used mainly for dredging in confined areas such as wharfs and breakwaters. The method is only suitable for use in loose soils and rocks.

## 20.7 GENERAL POINTS ON THE CHOICE OF DREDGING METHOD

(a) *Pure suction* is suitable in granular soil, such as medium sands with a particle diameter 0.2–0.6 mm which flow fairly easily.

(b) *Suction dragging* is suitable for use in most types of soil, but is most appropriate in naturally occurring fine sands and light loams, for example, 0.2 mm particle diameter and less.

(c) *Suction cutting* is best operated in clays, marl, gravel and soft rock. The clays in particular lend themselves well to transport to shore along pipes because of their ability to suspend in water.

# 21

# EARTH COMPACTING EQUIPMENT

Fill for the construction of dams, airports, roads, backfilling foundations and trenches is normally loose and bulked after the excavation process and must be compacted to prevent distortion, settlement and softening when in place. The attainable compaction, measured by the change in density, depends very much upon the type of material and moisture content, and whilst heavy rolling will significantly reduce the voids in granular soils, cohesive materials will consolidate only very slowly under a sustained pressure. In order to obtain satisfactory compaction in the vast range of soils to be met in practice, several different types of compacting plant have been developed, and are classified into those imparting static weight, impact, vibration or kneading.

## 21.1 STATIC WEIGHT ROLLERS

These rollers rely purely on self-weight to achieve compaction.

### Smooth wheel rollers

**Fig. 21.1** Smooth wheel roller (pulled type)

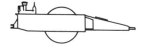 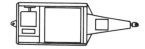

### SINGLE-AXLE ROLLER

The single-axle roller consists of a frame and smooth steel cylinder ballasted with sand or water to increase self-weight. As the roller is pulled forwards a wave of soil is pushed up (Fig. 21.2) in the direction of movement and with successive passes over the ground surface the soil gradually compacts. The effect is more exaggerated with the pulled

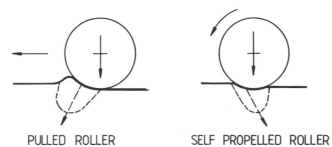

**Fig. 21.2**   Rolling actions

PULLED ROLLER          SELF PROPELLED ROLLER

as compared to the self-propelled roller, and the latter is preferred in so far as the soil is disturbed less to produce a more even surface. However, for both methods the wave form will be larger as the weight of the roller is increased and the diameter decreased, the relative magnitude being related to composition of the soil. Such rollers may be used on most types of granular soils but are fairly inefficient when compared to similar machines fitted with a vibratory mechanism.

## THREE-WHEELED ROLLER

This consists of a wide front roller and two narrower rear rollers and is predominantly used for compacting bituminous materials on road surfacing operations. Most three-wheel and also some tandem rollers are steered by the front drum with the drive provided through the rear rollers, usually set to overlap the track of the front roller by about 150 mm to prevent ridge formations. Differential gearing on the rear axle helps the machine to accommodate bends and curves without causing ruckling of the surface. In general, for all smooth wheel static rollers a thin layer near the surface is compacted very effectively, but unfortunately the soil below about 150 mm deep remains virtually untouched.

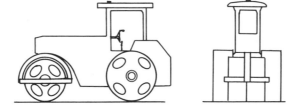

**Fig. 21.3**   Three-wheel
roller

## RUBBER-TYRE ROLLER

It was observed that the constant pounding of traffic on a bituminous road surface caused grooving and further compaction of the road base and subgrade. On the basis of the evidence the rubber-tyre roller was developed. Today the method is mostly used for rolling base courses on roads and fill for large earthworks involving soil of a loamy texture. The machine is usually self-propelled and with ballasting, using either water, sand or pig iron, the weight may be increased by a factor of about 2.

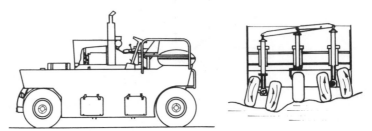

**Fig. 21.4** Rubber-tyre roller

## ROLLING METHOD

In order to avoid ruckling of the surface when operating in curves and bends the wheels must have independent couplings combined with a swivel action to distribute the weight on each tyre evenly on undulating surfaces. Generally construction is with two axles comprising a total of seven tyres (three front and four rear) or nine tyres (four front and five rear), so arranged that the paths of the rear overlap those of the front to prevent ridges. A major advantage lies in the ability to control the ground contact pressure by:

(a) altering the weight of the machine;
(b) increasing the number of wheels;
(c) increasing the tyre width;
(d) changing the contact area of the tyre by altering the tyre pressure.

Manufacturers usually supply a table of ground contact pressures for given combinations of the above factors. Consideration of the bulb stresses under the tyres (Fig. 21.5) indicates:

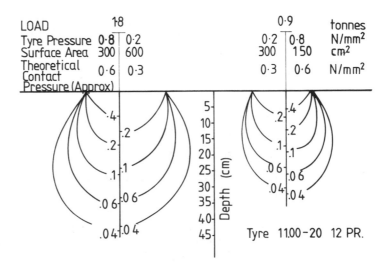

**Fig. 21.5** Bearing pressures under rubber tyres

(a) the load on the tyre influences the depth to which compaction is possible;

(b) the tyre pressure and the tyre load are important in achieving high levels of compaction near the surface, but the tyre pressure becomes less important as the depth increases.

### COMPACTING GUIDELINES

(a) Loose lifts in sands and loamy soils according to Pohle of Aachen T.H., Germany, are:

        up to 300 mm layers (lifts) – tyre loads 1.5–1.7 tonnes
        up to 500 mm layers (lifts) – tyre loads 2.0–2.5 tonnes
        up to 700 mm layers (lifts) – tyre loads 4.0–4.5 tonnes

(b) Blacktop layers:

        up to 80 mm layers (lifts) – tyre loads 1.5 tonnes
        up to 130 mm layers (lifts) – tyre loads 2.5 tonnes
        up to 200 mm layers (lifts) – tyre loads 4.0 tonnes

When compacting earth, from four to eight passes are usually sufficient while for rolling blacktop, from four to six passes are recommended. Travelling first forwards and then backwards along the road at speeds up to 20 km/h produces the best results.

### 21.2  SHEEP'S-FOOT ROLLER

The sheep's-foot roller has projecting feet mounted on the surface, so-called because in the earlier models they resembled the shape of a sheep's foot. Roughly ten to twenty such feet are arranged around the rim at between 100 and 200 mm centres along the axis; several different shapes of foot are available to suit varying soil types. In general the method is preferred when attempting to compact highly cohesive materials. The small surface area of each foot transmits high pressure so kneading the soil particles together, and with repeated passes, the feet gradually 'climb' out of the fill. Practice has demonstrated that satisfactory compaction in a layer of fill equal to the length of a sheep's foot may be achieved.

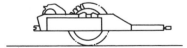
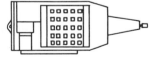

**Fig. 21.6**  Sheep's-foot roller

### 21.3  DYNAMIC COMPACTION

Such methods rely on either vibration or impacting kinetic energy.

### Vibrating rollers

By setting the rim of a roller into oscillation, vibrations are transmitted to the soil particles and improve compaction. In comparison a much heavier static weight roller is necessary to achieve similar results. Today the sheep's-foot and smooth wheel rollers are constructed with a vibration unit, the frequency of which may be varied to render the method appropriate on most types of soil, including compacting broken rock.

## VIBRATION MECHANISM

Several systems are adopted by the various manufacturers, but most use the principle of the rotating eccentric weight, installed inside the roller drum (Figs. 21.7a and 21.7b). Unlike the static weight machine the weight may not be increased by ballasting the drum but only by the addition of external weights which are usually hung from the support frame.

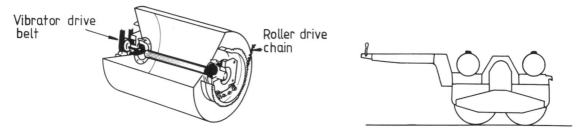

**Fig. 21.7(a)**   Roller vibration system

**Fig. 21.7(b)**   Tandem roller (self-propelled)

## VIBRATING PLATES

Two eccentric weights are placed either side of the centre of gravity of the plate and rotated out of phase in opposite directions. The resulting combination of forces impacts upon the soil surface to cause compaction, however, the horizontal components sometimes combine to act in the same direction when the vertical components are acting upwards allowing the operator to advance horizontally. By pivoting the whole block the movement may be set either for continuous forwards or backwards movement. The method is mainly used on fairly small patches and walked by the operator, it is more suited to granular materials.

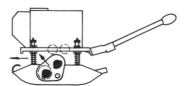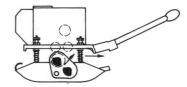

**Fig. 21.8**   Vibrating plate compactor

Typically to achieve adequate compaction in cohesive materials a unit capable of delivering 1400 kg/m$^2$ is appropriate reducing to about 900 kg/m$^2$ for uniformly graded granular material. Approximately four to six passes may be necessary with layers up to 200 mm deep.

### Impact plates

Kinetic energy is utilised by raising a heavy weight and allowing it to fall on to the surface. The common form is the hand-operated explosion stamper used in trenches and around small foundations and effective to approximately 500 mm depth.

**Fig. 21.9**   Impact stamper

## 21.4   OTHER TYPES OF COMPACTION EQUIPMENT

### Freefall hammer

Again kinetic energy is used by dropping a heavy flat weight from the jib of a crane. In this way it is sometimes possible to compact layers of soil several metres below the surface.

### High-speed compactor

Modern earthmoving projects such as motorways, earth dams and airports, involve large earthmoving machines and correspondingly rolling equipment must match the high production achieved by these methods. The introduction of the high-speed roller is proving very effective for this purpose. It comprises a pin-jointed frame for steering

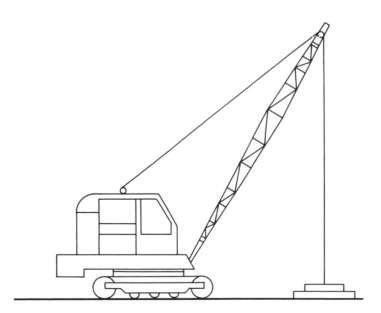

**Fig. 21.10**   Freefall
                 hammer

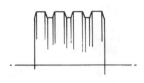

**Fig. 21.11** High-speed compactor

and is driven on four polygonal wheels, each of the front of which has from four to five polygon segments with the rear segments arranged to run between the furrows of the front wheels, so assuring complete coverage during the pass. The compacting action combines static weight, impact and kneading, and provides considerable advantages over other types of roller on massive areas involving cohesive and plastic soils, silts, gravel and hardcore. Adequate density results are possible with three to five passes with working speeds of 10–16 km/h.

### Approximate compaction factors

|  | In-place volume | Bulked volume | Compacted volume |
|---|---|---|---|
| Broken rock | 1.0 | 1.5–2.0 | 1.3–1.4 |
| Gravel | 1.0 | 1.0–1.1 | 0.8–1.0 |
| Clay | 1.0 | 1.25–1.4 | 0.8–1.0 |
| Sand | 1.0 | 1.0–1.3 | 0.9–1.0 |
| Common earth | 1.0 | 1.1–1.3 | 0.8–0.9 |

Thus compacted volume = in-place volume × compaction factor, assuming of course that the material was compacted from a bulked condition.

### Soil classification

| Soil description | Condition | Particle diameter (mm) | In place (bank) density (tonnes/m³) |
|---|---|---|---|
| Sand | Dry | 0.05–2.0 | 1.4–1.6 |
|  | Damp |  | 1.6–1.8 |
|  | Wet |  | 1.8–1.9 |
| Silt and clay | Dry | 0.001–0.05 | 1.6–1.8 |
|  | Wet |  | 1.8–2.1 |
| Broken rock | — | 50.0 up | 1.5–1.7 |
| Loam/ Common earth | Dry | Varies | 1.5–1.6 |
|  | Wet |  | 1.6–1.7 |
| Gravel | — | 2.0–5.0 | 1.7–1.9 |

**Table 21.1** Comparison of roller applications

| Type | Self-weight (tonnes) | Ballasted weight (tonnes) | Drum or wheel diameter (mm) | Horse power, hp (kW) | Working speed | Frequency (Hz) | Loose lift (mm) | Remarks |
|---|---|---|---|---|---|---|---|---|
| Vibration | 0.3–0.7 | — | up to 1000 | approx. 5 (3.7) | up to 3 km/h | up to 80 | 100–200 | Trench reinstatement, small scale works and compacting bituminous surfaces |
| Vibration (including seated driver) | 0.5–13.0 | — | up to 1500 | 10–200 (7.5–150) | up to 10 km/h | up to 80 | 200–1000 | Small hand-steered version similar to above, larger types with seated driver for large works, e.g. sub-bases and bearing courses on roads |
| Vibration | 4.0–15.0 | — | up to 2500 | 6–200 (4–150) | Towed | 20–30 | 300–1000 | Tractor-drawn suitable on large scale works, road subgrades, earth dams, heavy foundations, stone crushing |
| Static | 4.0–15.0 | up to 20.0 | up to 2500 | Towed | Towed | — | up to 300 | |
| Static | 8.0–14.0 | up to 20.0 | up to 2000 | 50–100 (37–75) | up to 10 km/h | — | up to 300 | Mainly used for road surfacing operations |
| Rubber-tyred static | 7.0–14.0 | up to 30.0 | — | 100–200 (75–150) | up to 20 km/h | — | up to 700 | Suitable for loamy soils on major construction, e.g. road bases, earth dams |
| Vibrating plate | 0.1–1.0 | — | — | 2–10 | up to 25 m/min | 10–80 | up to 700 | Used on gravel and sand, very manoeuvrable, ideal for small scale work in awkward situations |
| Stamper | approx. 0.1 | — | — | approx. 3 (2) | — | 60–80 blows per min | up to 500 | Very small scale works, compacting foundations, trenches, etc. |

## PART TWO

**Notes:**

(i) Where safe working load (SWL) tables are given, these are intended only as guidelines for use in equipment selection exercises as the permitted lifting capabilities of machines vary depending upon the regulations operating in different countries. Typically in this book SWL is quoted as 75% of the tipping load, but up to 85% is allowable under some National Standards.

(ii) Cranes designed for grabbing or magnet use, the calculated load is increased by 25%, i.e. safe working load is 80% of that for cranes.

(iii) The use of terms such as winch and hoist, jib and boom, derricking and luffing are freely interchanged to have the same meaning.

# SIMPLE LIFTING MECHANISMS

## 22.1  INTRODUCTION

The choice of lifting equipment is vast, ranging from simple ropes and pulleys, to large cranes and the trend is towards ever more sophisticated lifting devices. For example over 3000 tonnes capacity crawler cranes are currently in the design stage. However, even the large crane relies upon the principles of hoisting tackle, i.e. lifting rope, pulley blocks and a winch. This basic arrangement has been used successfully to design temporary works-lifting mechanisms for many tasks, including the erection of bridges, high-rise structures and industrial buildings. It is only during the past century that the modern crane has extended the application of such devices.

## 22.2  PRINCIPLES OF HOISTING TACKLE (ROPES AND PULLEYS)

Figure 22.1 shows a single line fixed pulley block, clearly $P$ and $W$ are only equal if there is no friction. When the block is free to rotate on an axis, the friction is a small component of the total forces and $P$ and $W$ can be assumed equal. Such a system, called 'single part' or 'fall', is commonly used for reeving on cranes.

**Fig. 22.1**  Single line
pulley block

**Fig. 22.2**  Blocks and
tackle

The load-carrying capacity of the hoist tackle may be increased by
using two or more pulleys as in Fig. 22.2. In the arrangement shown, if
$n$ is the number of ropes leading from the lower to the upper blocks,
then neglecting friction, each rope supports $W/n$. Thus $P = W/n$. It is
obvious, however, that considerably more rope must be wound in to
raise $W$, compared with the single line system. The velocity ratio of the
system, defined as the ratio of the distance moved by $P$ to the distance
moved by $W$ is calculated as follows:

$$\text{Work done by } P = \text{work done by } W$$
$$PH = Wh$$

therefore    $$\frac{H}{h} = \frac{W}{P} = \frac{W}{W/n} = n$$

In practice the pulley blocks at the top and bottom are mounted on
a single axis.

The reeving loads are governed by BS 1757:1964, which covers all
the possible reeving arrangements from single up to eight and above
falls.

Figure 22.3 shows the methods of reeving for two, four, six and eight
falls. The rope follows the numbers on the pulleys beginning with the
lead line over the top block.

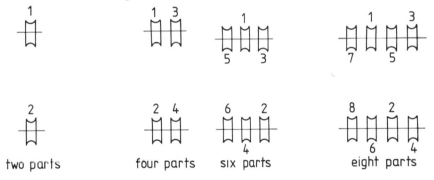

**Fig. 22.3**  Reeving
methods

two parts          four parts          six parts          eight parts

## 22.3  ROPES

6 × 37 Cable

**Fig. 22.4**  Wire-rope
configuration

Cranes use wire ropes which comprise strands made up of individual
wires, of which there are several configurations as typified by the
example shown in Fig. 22.4. The load carrying capacity of the rope
depends upon its diameter and the number of wires and is
manufactured with a safety factor of approximately five.

The three common lays are ordinary lay, Lang's lay and non-rotating
lay.

### Ordinary lay

In the right-hand lay method the wire spirals to the left and the strands
to the right (Fig. 22.5). In the left-hand lay the arrangement is *vice*

**Fig. 22.5** Ordinary right-hand lay

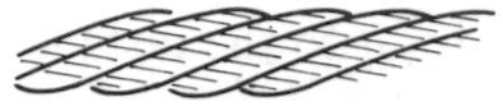

**Fig. 22.6** Ordinary left-hand lay

**Fig. 22.7** Right-hand Lang's lay

**Fig. 22.8** Left-hand Lang's lay

*versa* (Fig. 22.6). These types of lay are useful as slings, but tend to wear quickly if operated as hoist ropes, since only the crown wires are in contact with the pulley.

### Lang's lay

Lang's lay has both the wires and strands spiralling in the same direction (Figs 22.7 and 22.8), but obviously both ends must be secured to prevent twisting. This lay has better wearing properties.

### Non-rotating lay

For most hoisting work with single part reeving the rope must avoid twisting. This is achieved by using a double rope construction, e.g. the inner rope can be in right-handed Lang's lay and the outer strands in left-handed ordinary lay. This method produces a non-rotating rope.

The size of the rope may be designed to suit the load requirements but for a medium size crane, e.g. 30 tonnes lifting capacity, an 18–20 mm diameter rope with up to six part reeving is usual.

## 22.4 WINCHES

All cranes and most lifting tasks with a block and tackle require a winch and rope drum to provide the lifting force. The winch may be powered by compressed air, electric motor, hydraulic motor or a diesel engine.

For portability around the construction site, the electric or compressed-air winch is favoured. A typical arrangement of an electric winch is shown in Fig. 22.9. The hoist speed of most winches may be varied to accommodate the line load. For example, with four-part reeving the described electric winch of 100 hp (74.6 kW) can handle 4000 kg at speeds up to 100 m/min, whereas at the full 20 000 kg load, the maximum operating speed is about 20 m/min.

Within these working ranges, both the lowering and raising speeds may be finely controlled, particularly on the modern independently powered electric or hydraulic winches. For the traditional diesel powered all-mechanical transmission winches, this is achieved by gear selection in conjunction with a brake and clutch arrangement. The hoisting speed may be further adjusted by altering the engine speed with the throttle.

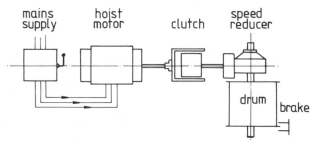

**Fig. 22.9** Typical arrangement of an electric winch

# 23

# CRANES – SHEAR LEGS AND DERRICKS

## 23.1  INTRODUCTION

The applications of the simple block and tackle are somewhat, restricted by the limitation of reach. Thus cranes of various forms have been designed to combine both horizontal and vertical movement, so providing a three-dimensional capability.

The earliest devices for use in civil engineering work were shear legs and later the derrick. Even today shear legs are widely used as a site-made temporary works crane, designed for special tasks peculiar to the particular duties called for on the project.

## 23.2  SHEAR LEGS

Where very heavy lifts are needed, then shear legs offer a very simple and inexpensive method. The equipment comprises a winch, and block and tackle suspended from a pin-jointed frame fabricated from steel tubing or RSJs, etc., stabilised by guy ropes. Loads may only be raised and lowered, horizontal movement is not available. However, for river work, shear legs mounted upon a barge or pontoon can offer three-dimensional movement. Several bridges have been erected using

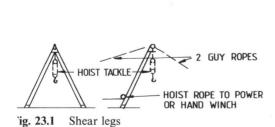

Fig. 23.1   Shear legs

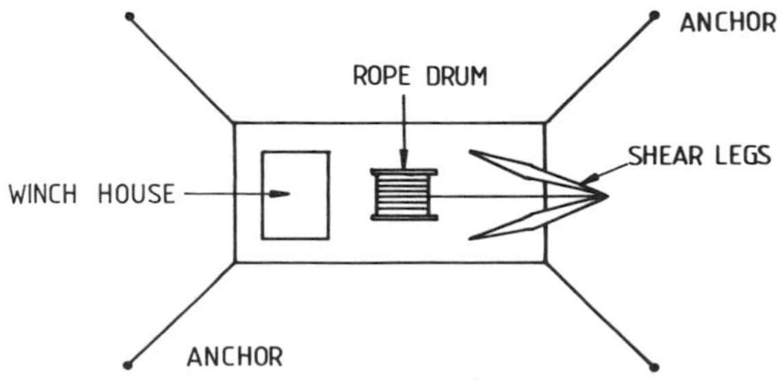

Fig. 23.2   Shear legs mounted on a pontoon

this technique, but great skill in manoeuvring the vessel is required, and generally movement is only obtained by winching from anchorages located at convenient positions.

### Design principles

Resolving horizontally

$$T \cos \alpha = R \cos \theta \tag{1}$$

Resolving vertically

$$T \sin \alpha + W = R \sin \theta \tag{2}$$

Thus $T$ or $R$ is calculated by solving equations (1) and (2). It is usual to apply a factor of safety of at least three to the forces in the members when selecting the actual material requirements.

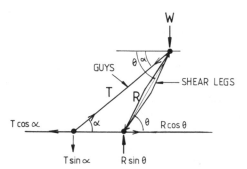

**Fig. 23.3** Forces acting on shear legs

## 23.3 GUYED DERRICK

The guyed derrick consists of a single boom and mast. The mast stands vertically and is guyed to anchorages in a similar fashion to shear legs. The arrangement allows both luffing (changing of radius) and slewing (turning), but lifting is usually only attempted under a guy rope. It is possible to arrange for full 360° slewing, which makes for a very versatile crane and was at one time widely used where lifting facilities were required over a long period, and could justify the setting-up costs, e.g. the erection of steel-framed structures.

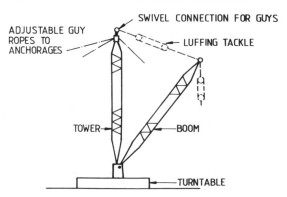

**Fig. 23.4** Guyed derrick

### 23.4 DERRICK

The Scotch or stiff-leg derrick is more commonly chosen in preference to either the shear legs or guyed derrick, and is used for heavy lifting over long and high reaches. The derrick is extensively used for steelwork erection, especially in heavy plant construction, such as power stations and process plants, although with the continued development of the tower crane it is gradually losing its advantages.

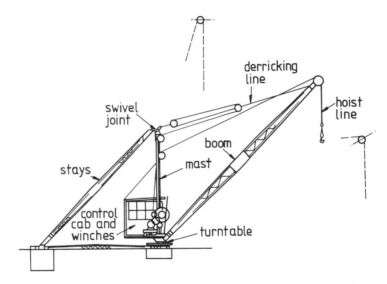

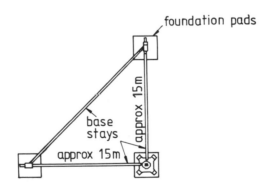

**Fig. 23.5** Derrick

### Construction

The derrick consists of a vertical mast, usually made of steel plate, supported by two sloping fixed legs. The whole arrangement is seated upon a triangular frame of lattice construction, with the centre mast free to rotate on bearings at its top and bottom supports. The boom, attached to the base of the mast may be rotated through 270°, between the mast stays, and is capable of hoisting, slewing and luffing. The stays are relatively short compared to the length of the boom, and heavy ballasting is required at the base plates of the mast and stays. As a general rule the weight of kentledge at each ballasting point should be about four times the maximum lifting capacity of the crane. It is therefore essential that the unit is firmly supported on well prepared foundations. A common method of providing suitable foundations is to mount the derrick on bogies at the base apex points. The bogies themselves are supported on rails and sleepers, to provide the extra dimension of mobility.

The working height of the derrick may be increased by means of lattice towers, called gabbards, placed under the base apex points.

### Design principles

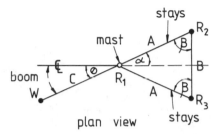

**Fig. 23.6** Forces acting on a derrick

(a) Taking moments about a line through $R_2 R_3$

$$W \times (C \cos \theta + A \cos \alpha) = R_1 \times A \cos \alpha$$

$$R_1 = W \frac{(C \cos \theta + A \cos \alpha)}{A \cos \alpha} \qquad (3)$$

Therefore when $\theta$ increases to 90°, $R_1$ is equal to $W$.

(b) Taking moments about a line through $R_1 R_2$

$$R_3 \times B \sin \beta = W \times C \sin(\alpha \mp \theta)$$

$$R_3 = W \times \frac{C \sin(\alpha \mp \theta)}{B \sin \beta} \qquad (4)$$

(c) Taking moments about a line through $R_1 R_3$

$$R_2 \times B \sin \beta = W \times C \sin(\alpha \mp \theta)$$

$$R_2 = W \times \frac{C \sin(\alpha \mp \theta)}{B \sin \beta} \qquad (5)$$

It can be seen that as $\theta$ changes the reactions also change.

### Method of erection

Complete erection, including commissioning, takes about 40 hours using three men and crane assistance. Dismantling can be achieved in about half the erection time. The following procedure is recommended:

(a) prepare the foundation bases and position the base frame;
(b) place the mast sole plate on the frame;
(c) erect the crane mast and temporarily guy to anchors;
(d) erect the stays in turn and secure with holding bolts;

(e)  load on ballast;
(f)  connect boom to mast;
(g)  fit gears, winches, ropes, pulleys, etc.;
(h)  raise the boom.

### Operating the derrick

The derrick is provided with two rope drums, one for derricking (i.e. luffing) and one for hoisting. Both are driven through gearing by an electric or hydraulic motor, steam or diesel engine. The more common form is electric, when separate motors are provided for slewing, hoisting and luffing. The hoisting facility is usually available with a gear change, the fast speed for light loads, and the slow speed for heavy loads. Some types of derrick have a third drum for opening a clamshell bucket when used as a grab.

### Derrick crane characteristics

The derrick is designed as a pin-jointed lattice frame structure to lift a certain load, called the design capacity, and is typically available with maximum load capacities up to 200 tonnes. Table 23.1 shows the corresponding maximum and minimum radius of operation for the more common sizes. It can be seen in Fig. 23.7 that between the minimum possible radius ($x$) and maximum permitted radius ($y$) the load is limited to the design capacity load. For loads less than the design capacity the permitted operating radius ($z$) may be slightly increased. The corresponding values are shown in Table 23.2. The working ranges for various lengths of boom are given in Fig. 23.8.

**Table 23.1**  Maximum operating radius at design capacity of derrick

| Boom length (m) | Max. ($y$) radius (m) | Design capacity (tonnes) | | | | | | | | | | | |
| | | 3 | 5 | 7 | 10 | 15 | 20 | 25 | 30 | 35 | 40 | 55 | |
|---|---|---|---|---|---|---|---|---|---|---|---|---|---|
| 36 | 27 | 8 | 8.5 | 8.75 | 9 | 9 | 9.25 | 9.25 | 10 | 11 | 11 | 11 | Min. operating |
| 46 | 36 | 9 | 9 | 9.5 | 9.5 | 10 | 10.5 | 11 | 11 | 11 | 11 | 12 | radius ($x$) (m) |

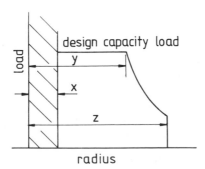

**Fig. 23.7**  Load–radius diagram for the derrick

**Table 23.2** Permissible load at maximum operating radius of derrick (*z*)

| Boom length (m) | Radius (z)(m) | Design capacity (tonnes) | | | | | | | | | | | |
|---|---|---|---|---|---|---|---|---|---|---|---|---|---|
| | | 3 | 5 | 7 | 10 | 15 | 20 | 25 | 30 | 35 | 40 | 55 | |
| 36 | 35 | 2.5 | 2.5 | 3 | 4.25 | 5 | 5.75 | 7.5 | 7.5 | 8.75 | 10 | 14 | Load (tonnes) |
| 46 | 42.5 | 2.5 | 2.5 | 3 | 4.25 | 5 | 5.75 | 7.5 | 7.5 | 8.75 | 10 | 14 | at radius (z)(m) |

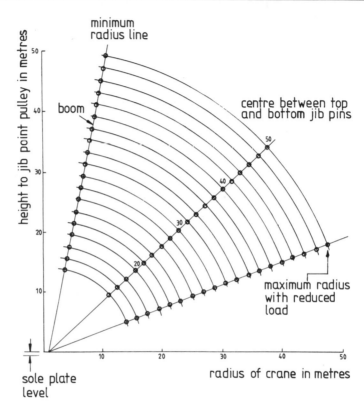

**Fig. 23.8** Boom height–radius diagram for the derrick

### Derrick characteristics

| | Up to 7 tonnes capacity | 10 tonnes capacity and over |
|---|---|---|
| Hoisting speed – lifting design capacity (m/min) | 30–35 | 10–15 |
| Hoisting speed – lifting light load (m/min) | 70 | 20–30 |
| Derricking (luffing) speed (m/min) | 30 | 12–15 |
| Slewing speed (rev/min) | 1 | 0.3 |
| Hoist motor (kW) | 50 | 50 |
| Slewing motor (kW) | 10 | 30 |
| Derricking motor (kW) | 40 | 50 |
| Travelling speed (m/min) | 40 | 10 |
| Travelling motor (kW) | 20 | 60 |

*Note*: All are electric motors.

### Example of derrick selection

A derrick crane is used to construct three concrete monoliths
20 m × 10 m on plan, for the widening of the docks entrance shown in
Fig. 23.9. The monoliths are constructed in 1.5 m lifts and a lift of
reinforcing steel protrudes above the top of the shuttering. A maximum
of four lifts of concrete is allowed above ground level as shown. The
crane is to be used (i) for all materials handling including a 764 litre
(1 yd³) concrete skip, (ii) for excavating by grabbing action inside the
monoliths with a 1000 litre (33 ft³), 15° CECE rating, capacity grab,
(iii) for lifting a complete cell of formwork weighing 3 tonnes.

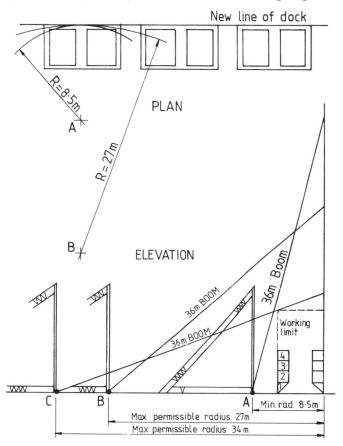

**Fig. 23.9** Example layout
of derrick crane

### PROBLEM

Select a suitable size derrick crane.

### SOLUTION

(a) *Choosing the correct position*

Figure 23.9 shows that given the smallest available boom of 36 m
and allowing two clear 1½ m lifts between the top of the

reinforcement and the underside of the boom, the derrick can reach the four sides of the monolith when set up at its minimum operating radius of 8.5 m (position A on the section diagram), or alternatively at the crane's maximum radius (position C). This latter position however, is beyond the maximum permissible operating radius for the design capacity of the derrick. Clearly, therefore, the derrick can provide clearance at the maximum load permitted radius of 27 m (position B) and thus may be established in any position between points A and B. The top corners of each monolith, however, are outside the reach of the boom at all placings. Therefore the derrick should be rail mounted to cover all three monoliths.

(b) *Choosing the appropriate crane capacity*
Possible loads are:

   (i) Shuttering – 3 tonnes.
  (ii) Concrete skip + concrete

$$
\begin{array}{lrl}
1 \text{ yd}^3 \text{ skip} & = & 500 \text{ kg} \\
1 \text{ yd}^3 \text{ concrete} & = & \underline{1800 \text{ kg}} \\
& & 2300 \text{ kg}
\end{array}
$$

 (iii) Grab + contents

$$
\begin{array}{lll}
1000 \text{ litre grab} & = 1350 \text{ kg} & \text{(see Table 24.4)} \\
\text{moist earth} & = 1875 \text{ kg} & \text{(heaped) (see Table 24.2)} \\
& \phantom{=} 3225 \text{ kg}
\end{array}
$$

 (iv) Hook block, etc. = 775 kg.
Max. possible load = 3225 + 775 = 4000 kg.
Include an extra 25% for surcharge, thus max.
load = 4000 × 1.25 = 5000 kg.
*A 5 tonnes capacity derrick is needed.*

(c) *Ballast*
The ballast required on each foundation is approximately 4 × derrick capacity, i.e. 20 tonnes.

# 24

# CRAWLER-MOUNTED CRANE

## 24.1 INTRODUCTION

On many construction sites a crane is needed to lift small to medium loads, such as concrete skips, reinforcement, formwork, etc. Flexibility of movement around the site is at a premium, thus where the works is spread over a wide area beyond the reach of tower cranes and derricks, a crane capable of operating on unprepared surfaces is demanded. Rubber-tyred mobile cranes are excellent for lifting on level firm surfaces, but on many sites the ground conditions are so bad that these models would become bogged down and unable to work easily and efficiently. In conditions such as these the crawler-mounted crane is the most advantageous model to use. This is because the weight of the crane is spread over a large bearing under wide and long tracks.

The crawler crane has the further advantage that conversion from crane to grabbing cranc or dragline for excavation purposes is readily achieved.

## 24.2 CRANE CONSTRUCTION

The crane is built in three sections; the base frame, superstructure and boom, the whole unit being powered by a diesel engine.

### The base frame

The base frame is made from a welded steel channel to which the two machine axles are attached, and supports the weight of the engine, gearing and winches, controls, cab, boom and counterweight.

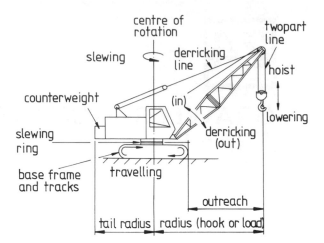

**Fig. 24.1** Crawler-mounted strut-boom crane

### The superstructure

The superstructure consists of a revolving frame sitting on a large turntable mounted on the base frame to give 360° slewing. The engine, gears, winches and counterweight are all mounted on this part of the unit.

The machine relies on a mechanical or hydraulic transmission system and two rope drums with independent brakes and clutches facilitate gravity lowering of the load. The front drum near the boom is the hoist winch and serves to raise and lower the crane hook and load, while the rear drum is used for luffing the boom.

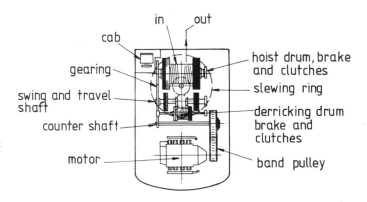

**Fig. 24.2** Gearing and winching layout (mechanical-type) transmission)

**Fig. 24.3** Simple head sheave

### The boom

The basic boom is assembled in two sections each constructed from four high tensile steel rectangular hollow sections braced with round tube lacing members. The sections are pin-connected and the top section incorporates a head sheave (Fig. 24.3) for light loads of up to

10 tonnes or a hammerhead boom point (Fig. 24.4) for heavy lifts. Intermediate sections may be inserted to extend the length of the boom.

### Hoisting tackle

The hoist rope between the head sheave and hook block may be arranged with one, two, three, four or more falls to accommodate the load being raised as described in Chapter 22. A single part tackle naturally permits fast hoisting, but the weight of the load must be kept within the permissible tensile strength of the cable.

**Fig. 24.4**   Hammerhead boom point

### Fly-jib

The straight boom suitably extended is ideal for general lifting duties but because of the inclined angle, obstructions often restrict load positioning. To overcome this problem a fly-jib may be attached to the boom point as shown in Fig. 24.5.

The fly-jib is of similar construction to the main boom and is available in various lengths depending on the duties and capacity of the crane mounted in line with the main boom to act as a simple extension or more customarily at 30° offset to provide increased operating radius. If the crane is supplied with a third drum it is feasible to use this as the hoist winch with the fly-jib, leaving the main tackle on the boom for heavy lifts. Fly-jibs are designed for load lifting purposes only and are not suitable for grabbing crane or dragline operations.

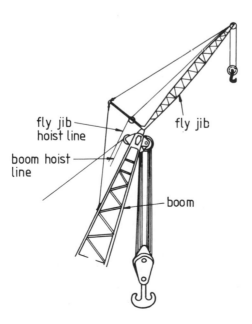

**Fig. 24.5**   Fly-jib

### Working loads

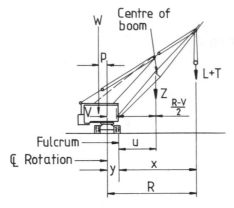

**Fig. 24.6**   Forces acting on the strut-boom crane

It is preferable to operate the crane on a flat well-prepared surface. The tipping load may be determined by using the following formula.

$$L = \frac{W(p + y) - Zu}{x} - T$$

where

$L$ = tipping load of the crane;
$W$ = weight of the machine without the boom (but including counterweights);
$Z$ = weight of boom;
$T$ = weight of head sheave;
$R$ = radius to load from centre of slewing ring;
$y$ = fulcrum distance;
$p$ = centre of gravity of machine without boom to centre line of slewing ring;
$v$ = connection point of boom to centre line of slewing ring;
$x$ = $R - y$;

$$u = \frac{(R - v)}{2} + v - y.$$

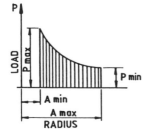

**Fig. 24.7**   Load–radius diagram for the strut-boom crane

Therefore safe working load $(P) = L -$ margin for safety.

The shape of the load-radius diagram is shown in Fig. 24.7. The weight of the hook block, together with any slings, etc., should be included when selecting a crane of suitable lifting capacity. If a fly-jib is attached but not in use then the safe working load should be reduced in accordance with the manufacturer's recommendations. Approximate load reductions are:

| Fly-jib length (m) | Load reduction (kg) |
|---|---|
| 6 | 850 |
| 9 | 1000 |
| 12 | 1200 |
| 18 | 1500 |
| 24 | 2100 |
| 30 | 2600 |
| 36 | 3000 |

## CRANE RADIUS DIAGRAM

The precise range of boom lengths and the corresponding reaches and radii vary with individual manufacturers, but the dimensions shown in Fig. 24.8 are typical of the makes of machine that are available.

### Crane capacities

The lifting ability at a given radius varies slightly with the particular crane manufacturer, with capacity being designated in terms of the maximum load at the minimum operating radius. The most popular sizes are in the range 15 to 120 tonnes but cranes with capacities of

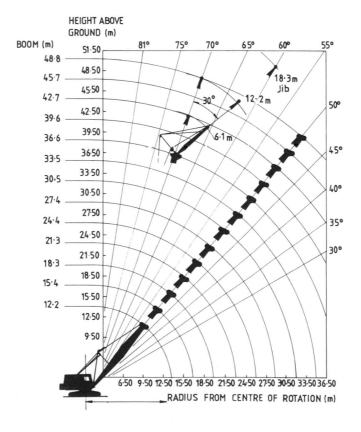

**Fig. 24.8** Boom height–radius diagram for strut-boom crawler crane: (a) Crane supported on tracks only: (b) Crane on tracks and ringer system

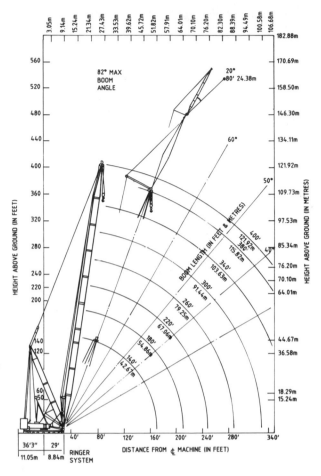

**Fig. 24.8(b)**

3000 tonnes or so are available. These larger units often have a ringer configuration (Fig. 24.9), which as a rough rule-of-thumb doubles the SWL compared to the conventional tracked mounted version.

### Transport and assembly

The crane is not intended to be self-transporting and will travel at little more than walking speed, and must be moved from depot to site on a low-loader truck. This choice is therefore not a practical option when lifting capacity is required for only a short period of hours or even a few days as the time taken to load, transport, unload and prepare for work may require a full day or more.

## 24.3 CONVERSION OF STRUT-BOOM CRANE TO GRABBING CRANE OR DRAGLINE (SEE CHAPTER 12)

Conversion to a grabbing crane or dragline can be readily obtained by changing the size of the hoist rope and hoist drum. The derricking

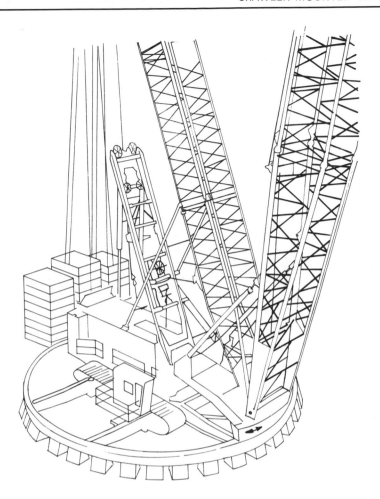

**Fig. 24.9** Ringer configuration

drum is retained and an additional drum (i.e. third drum) included to control the bucket. Larger and wider tracks are often fitted to provide greater stability.

### Working ranges of the dragline and grab crane

The dragline, like the crane, is usually designated with safe working loads based on a fixed percentage of the tipping load, depending on the country of operation, with the grabbing crane (and magnet cranes) normally being 80% of crane values.

The boom size may be varied to suit the required operating radius and lifting height as shown in Fig. 24.10. As a rule-of-thumb the dragline is capable of digging to a depth below its tracks of roughly from one-third to half the length of the boom. Furthermore, the throw of the bucket beyond the radius of the boom may be of similar length, depending upon the skill of the operator.

The safe working loads of cranes shown in Table 24.1 can be used to calculate the dragline working range, but should be reduced by 20% for

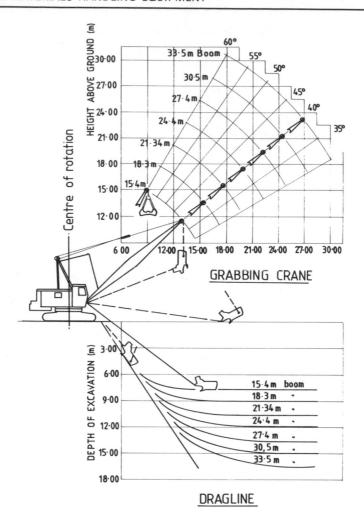

**Fig. 24.10** Working ranges of the dragline and grabbing crane

**Table 24.1** Guidelines to safe working loads of crawler-mounted strut-boom cranes.

**(a) Main Boom Rating: 15 Tonnes Crane**

| Radius (m) | Boom (m) | | | |
|---|---|---|---|---|
| | 10.7 | 15.2 | 21.3 | 27.4 |
| 3.0 | 15.00 | | | |
| 4.5 | 8.6 | | | |
| 6.0 | 5.6 | 5.5 | 5.4 | |
| 7.5 | 4.2 | 4.1 | 3.9 | 3.7 |
| 9.0 | 3.2 | 3.1 | 3.0 | 2.9 |
| 10.5 | 2.7 | 2.5 | 2.5 | 2.4 |
| 12.0 | | 2.1 | 2.0 | 1.9 |
| 15.0 | | 1.5 | 1.4 | 1.4 |
| 18.0 | | | 1.0 | 0.9 |
| 21.0 | | | | 0.8 |
| 24.0 | | | | 0.7 |

**(b) Main Boom Rating: 20 Tonnes Crane**

| Radius (m) | Boom (m) | | | |
|---|---|---|---|---|
| | 10.7 | 15.2 | 21.3 | 27.4 |
| 3.0 | 20.3 | | | |
| 4.5 | 10.6 | | | |
| 6.0 | 6.6 | 6.4 | 6.4 | |
| 7.5 | 4.9 | 4.7 | 4.6 | 4.5 |
| 9.0 | 3.9 | 3.7 | 3.6 | 3.4 |
| 10.5 | 3.3 | 3.0 | 2.9 | 2.8 |
| 12.0 | | 2.5 | 2.9 | 2.2 |
| 15.0 | | 1.8 | 1.7 | 1.5 |
| 18.0 | | | 1.2 | 1.1 |
| 21.0 | | | | 0.8 |
| 24.0 | | | | 0.5 |

**(c) Main Boom Rating: 35 Tonnes Crane**

| Radius (m) | Boom (m) | | | | | |
|---|---|---|---|---|---|---|
| | 12.2 | 15.2 | 18.3 | 21.3 | 24.4 | 27.4 |
| 3.5 | 35.0 | | | | | |
| 6.0 | 17.0 | 16.9 | 16.8 | 16.7 | 16.2 | |
| 8.0 | 11.4 | 11.2 | 11.1 | 11.1 | 10.9 | 10.8 |
| 10.0 | 8.4 | 8.2 | 8.1 | 8.1 | 7.9 | 7.8 |
| 12.0 | 6.5 | 6.4 | 6.2 | 6.2 | 6.0 | 5.9 |
| 14.0 | 5.4 | 5.1 | 5.0 | 4.9 | 4.7 | 4.6 |
| 16.0 | | | | 4.0 | 3.8 | 3.7 |
| 18.0 | | | | 3.3 | 3.1 | 3.0 |
| 20.0 | | | | 2.7 | 2.6 | 2.5 |
| 22.0 | | | | | 2.1 | 2.0 |
| 24.0 | | | | | | 1.7 |
| 26.0 | | | | | | 1.4 |

**(d) Main Boom Rating: 67 Tonnes Crane**

| Radius (m) | Boom (m) | | | | | |
|---|---|---|---|---|---|---|
| | 15.2 | 21.3 | 27.4 | 33.5 | 39.6 | 45.7 |
| 3.5 | 67.0 | | | | | |
| 6.0 | 33.0 | 31.9 | 30.0 | | | |
| 8.0 | 21.5 | 20.9 | 20.3 | 19.4 | 17.5 | |
| 10.0 | 15.5 | 15.1 | 14.6 | 14.2 | 13.9 | 12.3 |
| 12.0 | 12.2 | 11.8 | 11.5 | 11.2 | 11.1 | 10.4 |
| 16.0 | 8.1 | 7.5 | 7.3 | 6.9 | 6.7 | 6.3 |
| 20.0 | | 5.3 | 4.8 | 4.6 | 4.3 | 4.1 |
| 24.0 | | | 3.8 | 3.4 | 3.1 | 3.0 |
| 28.0 | | | | 2.7 | 2.2 | 2.1 |
| 32.0 | | | | | 1.5 | 1.4 |
| 36.0 | | | | | 0.9 | 0.8 |

### (e) Main Boom Rating: 91 Tonnes Crane

| Radius (m) | Boom (m) | | | |
|---|---|---|---|---|
| | **18.3** | **27.4** | **36.6** | **45.7** |
| 4.5 | 91.0 | | | |
| 6.0 | 82.8 | 63.8 | | |
| 8.0 | 42.9 | 42.1 | 41.1 | |
| 10.0 | 32.3 | 31.5 | 30.6 | 30.0 |
| 12.0 | 25.4 | 24.6 | 23.7 | 23.1 |
| 16.0 | 17.1 | 16.3 | 15.4 | 14.7 |
| 20.0 | | 11.7 | 10.8 | 10.1 |
| 24.0 | | 8.8 | 7.8 | 7.2 |
| 28.0 | | | 5.8 | 5.2 |
| 32.0 | | | 4.3 | 3.7 |
| 36.0 | | | | 2.6 |
| 42.0 | | | | 1.3 |

### (f) Main Boom Rating: 110 Tonnes Crane

| Radius (m) | Boom (m) | | | | | |
|---|---|---|---|---|---|---|
| | **18.3** | **27.4** | **36.6** | **45.7** | **57.9** | **71.6** |
| 4.5 | 110.0 | | | | | |
| 6.0 | 77.6 | 22.0 | | | | |
| 8.0 | 59.9 | 59.3 | 51.6 | | | |
| 10.0 | 46.8 | 46.2 | 45.7 | 35.2 | | |
| 12.0 | 37.1 | 37.5 | 37.1 | 35.1 | 21.4 | |
| 16.0 | | 25.4 | 24.9 | 24.7 | 20.1 | 12.8 |
| 20.0 | | 18.9 | 18.5 | 18.3 | 17.6 | 12.3 |
| 24.0 | | 14.9 | 14.4 | 14.2 | 13.6 | 11.8 |
| 28.0 | | | 11.7 | 11.5 | 10.9 | 10.2 |
| 32.0 | | | 9.7 | 9.5 | 8.8 | 8.2 |
| 36.0 | | | | 8.0 | 7.4 | 6.7 |
| 42.0 | | | | 6.3 | 5.7 | 4.8 |
| 48.0 | | | | | 4.4 | 3.7 |
| 52.0 | | | | | 3.8 | 3.1 |
| 58.0 | | | | | | 2.2 |

### (g) Main Boom Rating: 217 Tonnes Crane

| Radius (m) | Boom (m) | | | | | |
|---|---|---|---|---|---|---|
| | **18.0** | **27.0** | **36.0** | **45.0** | **57.0** | **63.0** |
| 4.5 | 217.8 | | | | | |
| 6.0 | 180.0 | | | | | |
| 8.0 | 122.5 | 122.0 | | | | |
| 10.0 | 87.0 | 86.6 | 86.3 | | | |
| 12.0 | 67.1 | 66.7 | 66.3 | 65.7 | | |
| 16.0 | 45.4 | 44.9 | 44.5 | 44.0 | 43.0 | 42.8 |
| 20.0 | | 33.3 | 32.9 | 32.4 | 31.6 | 31.2 |
| 24.0 | | 26.0 | 25.7 | 25.1 | 24.3 | 23.9 |
| 28.0 | | | 20.7 | 20.2 | 19.4 | 19.0 |
| 32.0 | | | 17.1 | 16.6 | 15.8 | 15.4 |
| 42.0 | | | | | 10.9 | 9.5 |
| 48.0 | | | | | 7.6 | 7.2 |
| 56.0 | | | | | | 5.0 |

### (h) Main Boom Rating: 270 Tonnes Crane

| Radius (m) | Boom (m) | | | | | | | |
|---|---|---|---|---|---|---|---|---|
| | 21.34 | 27.43 | 36.58 | 45.72 | 51.82 | 57.91 | 64.01 | 73.15 |
| 5.6 | 270.0 | | | | | | | |
| 7.0 | 212.0 | | | | | | | |
| 9.0 | 148.4 | 148.2 | | | | | | |
| 12.0 | 94.2 | 94.0 | 93.5 | 93.2 | | | | |
| 16.0 | 62.3 | 62.00 | 61.4 | 61.0 | 60.5 | 60.2 | 59.9 | 59.4 |
| 20.0 | 45.7 | 45.4 | 44.8 | 44.5 | 44.1 | 43.8 | 43.4 | 42.9 |
| 30.0 | | | 25.4 | 25.0 | 24.6 | 24.3 | 24.0 | 23.5 |
| 40.0 | | | | 16.1 | 15.6 | 15.2 | 14.9 | 14.4 |
| 45.0 | | | | | 10.7 | 10.3 | 9.8 | 9.2 |
| 50.0 | | | | | | | 6.8 | 6.1 |
| 55.0 | | | | | | | | 5.0 |

Crawler crane.

### (i) Main Boom Rating: 550 Tonnes Crane

| Radius (m) | Boom (m) | | | | | | | | | | | | |
|---|---|---|---|---|---|---|---|---|---|---|---|---|---|
| | 22.86 | 30.48 | 38.1 | 45.72 | 53.34 | 60.96 | 58.58 | 76.2 | 83.82 | 91.44 | 99.06 | 106.68 | 114.3 |
| 5.4 | 544 | | | | | | | | | | | | |
| 7.0 | 442 | 440 | | | | | | | | | | | |
| 8.8 | 362 | 359 | 357 | 342 | | | | | | | | | |
| 12.2 | 229 | 228 | 228 | 225 | 230 | 230 | 227 | 224 | | | | | |
| 21.3 | 100 | 99 | 99 | 97 | 102 | 102 | 98 | 98 | 98 | 98 | 97 | 96 | 94 |
| 30.5 | | 60 | 60 | 57 | 60 | 61 | 60 | 59 | 58 | 58 | 57 | 56 | 54 |
| 39.6 | | | | 38 | 43 | 42 | 41 | 40 | 41 | 39 | 38 | 37 | 35 |
| 48.8 | | | | | 32 | 31 | 30 | 29 | 28 | 27 | 26 | 25 | 24 |
| 57.9 | | | | | | 23 | 22 | 21 | 20 | 19 | 18 | 17 | 17 |
| 67.1 | | | | | | | 17 | 16 | 15 | 15 | 14 | 13 | 11 |
| 77.7 | | | | | | | | | 11 | 10 | 9 | 8 | 6 |

Crawler crane.

### (j) Main Boom Rating: 680 Tonnes Crane with 18 m dia. ringer

| Radius (m) | Boom (m) | | | | | | |
|---|---|---|---|---|---|---|---|
| | 42.7 | 54.9 | 67.1 | 79.2 | 91.4 | 103.6 | 121.9 |
| 21.3 | 680 | 676 | 503 | 391 | | | |
| 30.5 | 370 | 381 | 379 | 353 | 273 | 217 | 149 |
| 39.6 | 232 | 261 | 258 | 256 | 247 | 197 | 134 |
| 48.8 | | | 193 | 190 | 188 | 176 | 118 |
| 57.9 | | | 147 | 149 | 147 | 144 | 102 |
| 67.1 | | | 107 | 118 | 115 | 116 | 88 |
| 77.7 | | | | 85 | 92 | 92 | 57 |
| 82.3 | | | | | 81 | 85 | 40 |
| 86.9 | | | | | 70 | 75 | — |
| 91.4 | | | | | 61 | 55 | — |

Crawler crane.

**(k) Main Boom Rating: 910 Tonnes Crane**

|  | Boom (m) | | | | | | | |
|---|---|---|---|---|---|---|---|---|
| Radius (m) | 47.2 | 62.5 | 77.7 | 92.9 | 108.2 | 123.4 | 138.6 | 153.9 |
| 13.7 | 907 | | | | | | | |
| 21.3 | 500 | 495 | 492 | 476 | 415 | | | |
| 30.5 | 295 | 290 | 287 | 284 | 281 | 252 | 224 | 161 |
| 39.6 | 205 | 200 | 196 | 193 | 190 | 185 | 170 | 128 |
| 48.8 | 153 | 148 | 144 | 141 | 138 | 133 | 129 | 99 |
| 57.9 | | 115 | 111 | 108 | 105 | 98 | 96 | 75 |
| 67.1 | | | 88 | 85 | 82 | 77 | 73 | 53 |
| 77.7 | | | 71 | 65 | 62 | 57 | 53 | 43 |
| 82.3 | | | | 59 | 56 | 50 | 46 | 41 |
| 86.9 | | | | 52 | 49 | 44 | 40 | 37 |
| 91.4 | | | | 48 | 44 | 35 | 35 | 32 |
| 93.5 | | | | | 38 | 32 | 29 | 25 |
| 103.6 | | | | | 33 | 27 | 26 | 20 |

grab cranes. Also, in practice, the working load may be further restricted in order to maintain an adequate factor of safety on the roping system, which will be in single fall reeving. The weight of the bucket must be included in the weight of the load to be handled, as recommended in Tables 24.2, 24.3, 24.4 and Fig. 24.11.

**Table 24.2** Approximate densities of soils

| Material | Density, lb/yd$^3$ (kg/m$^3$) |
|---|---|
| Earth – moist | 2500 (1490) |
| Sand – dry | 2700 (1600) |
| Sand – wet | 3300 (1960) |
| Gravel | 2900 (1720) |
| Loose stone | 2700 (1600) |
| Clay – wet | 3000 (1780) |
| Coal | 1350 (800) |

**Table 24.3** Dragline bucket data

| Capacity (yd$^3$) | 5 | $4\frac{1}{2}$ | 4 | $3\frac{1}{2}$ | 3 | $2\frac{1}{2}$ | 2 | $1\frac{3}{4}$ | $1\frac{1}{2}$ | $1\frac{1}{4}$ |
|---|---|---|---|---|---|---|---|---|---|---|
| (m$^3$) | 3.82 | 3.44 | 3.06 | 2.670 | 2.29 | 1.91 | 1.53 | 1.35 | 1.15 | 0.96 |
| Weight (lb) | 9200 | 7700 | 7000 | 6400 | 5700 | 4700 | 4250 | 3300 | 2900 | 2300 |
| empty (kg) | 4175 | 3495 | 3175 | 2905 | 2585 | 2130 | 1925 | 1495 | 1315 | 1040 |
| Length (m) | 6.8 | 6.2 | 6.1 | 5.9 | 5.8 | 5.5 | 5.5 | 4.8 | 4.7 | 4.4 |

**Table 24.4**  Grabbing crane – medium-weight grabs

| Capacity* (ft³) | 100/80 | 90/72 | 80/64 | 71/57 | 63/51 | 50/40 | 44/35 |
|---|---|---|---|---|---|---|---|
| (m³) | 2.75/2.25 | 2.5/2.0 | 2.25/1.75 | 2.0/1.6 | 1.75/1.50 | 1.4/1.1 | 1.25/1.00 |
| Weight (lb) | 6550 | 5200 | 5100 | 3900 | 3850 | 3050 | 2950 |
| empty (kg) | 2975 | 2400 | 2350 | 1800 | 1750 | 1375 | 1350 |
| Length (m) | 4.4 | 4.2 | 4.0 | 3.8 | 3.8 | 3.5 | 3.4 |

\* Capacities given are heaped at 15° (CECE rating).

Heaped  15°(CECE Rating)

**Fig. 24.11**  Dragline and grabbing buckets

Dragline bucket          Clamshell bucket

## Ground pressure under the tracks

Pressure is given in N/mm².

| | Crane capacity | |
|---|---|---|
| | 30 tonnes | 80 tonnes |
| Crawlers, standard tracks | 0.06 | 0.113 |
| Short crawlers, wide tracks | 0.06 | 0.098 |
| Long crawlers, standard tracks | 0.055 | 0.076 |
| Long crawlers, wide tracks | 0.047 | 0.067 |

The pressure from a human foot is 0.02 N/mm².

## Lift crane data

| | |
|---|---|
| Hoisting speed (single fall line) | approx. 40–50 m/min |
| Derricking (max. to min. radius) | approx. 50–100 s |
| Slewing speed | approx. 2 rev/min |
| Travelling speed | $2\frac{1}{2}$–3 km/h |
| Max. gradients when travelling: loaded | 1 in 16 |
| no load | 1 in 5 |

## Example of crawler crane selection

A crawler crane is used to lift a load of 5 tonnes (including hook block and tackle) from a temporary road (which acts as a flat and well-founded surface), to the centre of a reinforced concrete tank, as shown in Fig. 24.12. The crane must provide $\frac{1}{2}$ m gap between the underside of the boom and the edge of the tank. Select an appropriate crane to perform the task.

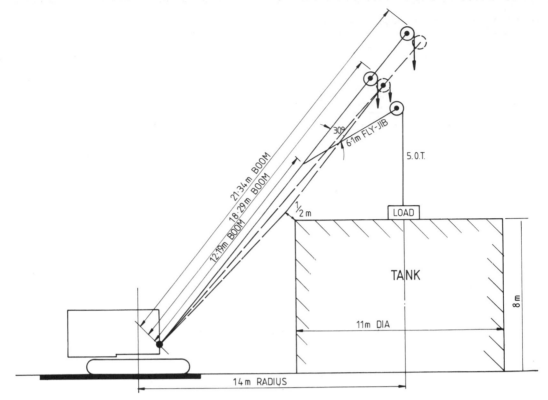

**Fig. 24.12** Example of crawler crane in use

SOLUTION

(a) The crane is operated from the temporary road and with the gap between boom and tank maintained at the minimum distance of $\frac{1}{2}$ m a crane with a 21.34 m long boom is required. The alternative smallest next size of 18.29 m will not provide the necessary reach of 14 m radius from the centre line of rotation of the machine to the lifting hook.

Reference to Table 24.1 reveals that the smallest size of crane capable of lifting 5 tonnes at 14 m radius is the 35 tonnes capacity crane. However, this may only be achieved with an 18.29 m boom, which is too short. The next size of crane would therefore be required.

A cheaper alternative is to attach a 6.1 m fly-jib to a 12.19 m boom on the 35 tonnes capacity crane, (the lifting capacity of the crane is equivalent to the 18.29 m boom). The sketch shows that in this configuration the desired operating radius of 14 m is achieved with some loss of operating height (which is not critical in this case). Thus a load of 5 tonnes may be lifted. It will be noticed that there is sufficient clear height between boom and tank to raise the load clear of the tank rim when slewing the load.

(b) The crane may be used for lifting with the main boom and sheave, with the fly-jib attached and out of action. The safe working load of the crane therefore must be reduced from the values shown in Table 24.1 by 850 kg. Thus the maximum 5 tonnes load at an operating radius of 14 m with a 12.19 m boom is reduced to 4.15 tonnes.

(c) The crane may also be re-roped and operated as a dragline or grab crane. (Note – Only from the main boom, not the fly-jib.)

  (i) *Grab crane* – the lift crane safe working load is reduced by 20% for grabbing or clamshell purposes.

    *Small bucket* – 1250 m³ heaped capacity bucket plus dry sand = 1350 kg self-weight + (1600 kg/m³ × 1.25 m³) = 3350 kg. From Table 24.1 for a 35 tonnes capacity crane, the permissible operating radius is about 14 m with a 24.40 m boom, i.e. permissible load is 4.7 × 0.8 = 3.76 tonnes, which is greater than actual load of 3.35 tonnes. (Note – from Fig. 24.10 jib angle is at the upper limit thus ruling out the longer 27.4 m boom.)

    *Large bucket* – 2750 m³ heaped capacity bucket plus dry sand = 2975 kg self-weight + (2.75 m³ × 1600 kg/m³) = 7375 kg. From Table 24.1 for a 35 tonnes capacity crane the maximum operating radius is about 9 m with an 18.2 m boom, i.e. interpolating the permissible load 9.5 × 0.8 = 7.6 tonnes, which is greater than actual load.

    Note – from Fig. 24.10 jib angle is near the limit.

 (ii) *Dragline* – the lift crane safe working loads are directly applicable for dragline selection.

    Thus using dragline data from Table 24.3 a $1\frac{1}{4}$ yd³ (0.96 m³) capacity bucket + dry sand weight = 1040 + (0.96 × 1600) = 2576 kg. From Table 24.1 for a 35 tonnes capacity crane, the permissible operating radius with a boom of suitable length, say 21.34 m, is 2.7 tonnes at 20 m. Assuming the bucket is cast one-third of the distance of the boom length beyond the boom, the boom could be set at an operating radius between 12 and 20 m, to keep within the permitted working range given in Fig. 24.10.

    A final check should be included to ascertain if the rope strength is sufficient to carry the working load.

# SELF-PROPELLED CRANE ON RUBBER-TYRED WHEELS

## 25.1 INTRODUCTION

Since the end of the Second World War considerable growth in the use of cranes on wheels has taken place. The reasons for this are many but undoubtedly an important factor has been the demand for specialist one-off lifting facilities, effectively facilitated by the greater mobility afforded to cranes with the development of the diesel engine, efficient gear boxes and more recently by the introduction of telescopic booms – all aided by the crane rental market, which has allowed higher utilisation of specialist equipment than could otherwise have been achieved by individual construction companies owning their own plant.

Self-propelled cranes divide roughly into two classifications: the strut-boom type, but with the crawler tracks replaced by rubber-tyred wheels, and the mobile telescopic-boom vehicles. Whereas the mobile version is capable of travelling at 30–40 km/h on public highways, the 'converted crawler' crane achieves perhaps 8–10 km/h. The market trend seems to be moving away from the latter and they may, in the near future, cease to be manufactured.

## 25.2 STRUT-BOOM CRANE

The crane is built in three sections similar to the crawler crane and comprises the base frame, superstructure and boom, the whole unit being powered by a diesel engine.

The base frame consists of a welded steel chassis and power is transferred to the wheels via a king-post gear passing through the turntable to a differential gearbox on the drive axle. The other axle is used for steering. The chassis usually has two-wheel drive, but for use on bad ground some machines are available with four-wheel drive.

The boom and winching arrangements are similar to the crawler crane, with all-mechanical transmission on the older versions and

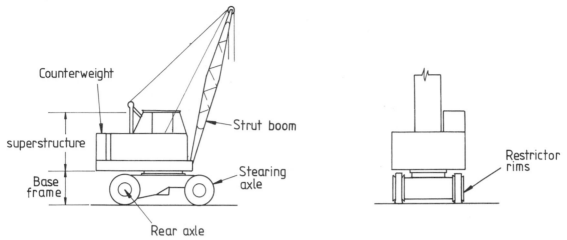

**Fig. 25.1**  Self-propelled strut-boom crane

hydraulic motors on newer models. The machine may also be used in the grabbing mode, with suitable rearrangement of the drum sizes and ropes.

## Uses and operation

The crane is operated on fairly hard ground such as temporary hardcore roads in stockyards, in scrapyards (where it is usually fitted with a magnet attachment), and for lifting and transporting relatively light loads over short distances of a few tens of metres.

The load is transferred to the ground directly through the large rubber tyres, but some cranes' wheels are made with restrictor rims. There are two large-diameter heavy gauge steel discs mounted one each side of the rubber-tyred wheels. In effect they act as tyre stabilisers by restricting the flexing of the tyre and so dampen down the tendency of the machine to bounce when lifting and transporting. The restrictors also have a lipped rim which considerably increases the bearing area under load, particularly when operating from hard surfaces such as concrete.

## Outriggers

To increase the operating range of the crane, outriggers are incorporated into the base frame on the larger versions, as shown in Fig. 25.2. They are housed in heavy steel compartments and are extended and retracted by a winding mechanism. The effective width of the supporting base is thereby extended and any differential level of the ground can be taken out by adjustment of the pads attached to the extremities. The whole unit, including wheels, is raised clear of the ground when lifting, all the load being transferred through to the

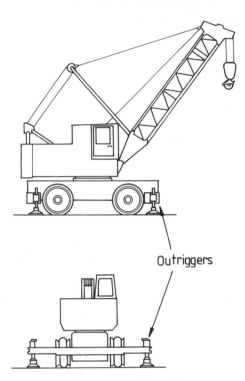

**Fig. 25.2** Outriggers for improved stability

outriggers. This arrangement is called the *blocked* position. It is essential that the ground is well prepared and a solid foundation is available particularly wherever heavy lifts are undertaken. The crane is of course stationary in this configuration and is only used in this way for maximum lifting.

### Crane capacity

The crane size is defined by the maximum load at the minimum operating radius. Lifting capacities *free* on the wheels, i.e. not *blocked*, range from small cranes of 5 tonnes up to 15 tonnes. Typical capacity–boom–radius data are shown in Table 25.1 for free-on-wheels and blocked-on-outriggers arrangements. It is apparent from these data that the load–radius diagram for wheeled machines is similar in shape to that of crawler cranes (Fig. 24.7), but suitable only for light lifting duties compared to the range of crawler cranes.

### Crane characteristics

| | |
|---|---|
| Hoisting speed (single fall line) | 40–60 m/min |
| Derricking (max. to min. radius) | 20–40 s |
| Slewing speed | 3 rev/min |
| Travelling speed (unladen) | 10 km/h |
| Max. gradients: max. load | 1 in 16 |
| no load | 1 in 8 |

| Crane capacity (free-on-wheels) (tonnes) | 5 | 10 | 15 |
|---|---|---|---|
| Engine power, kW (hp) | 30 (40) | 37 (50) | 52 (70) |
| Machine self-weight (tonnes) | 14 | 18 | 28 |

## 25.3  CANTILEVER-BOOM CRANE

### Comparison of strut- and cantilever-booms

A strut-boom offers the advantage of high lifting capacity when placing heavy loads at a wide radius, but being pin-jointed to the crane fairly close to the ground restricts the ability to lift and travel with wide loads when height restrictions are in force, for example, stockyard work where access into buildings is required. A cantilever-boom, however, is pivoted at a much higher position on the superstructure and so provides greater clearance, thereby facilitating the handling of bulky loads as shown in Fig. 25.4.

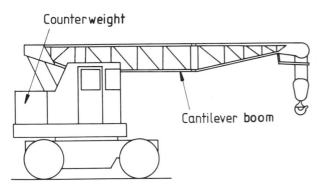
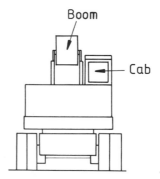

**Fig. 25.3**    Self-propelled cantilever-boom crane

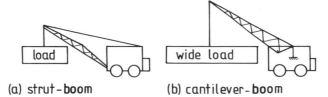

**Fig. 25.4** Comparison between strut- and cantilever-boom cranes

(a) strut-boom        (b) cantilever-boom

### Design principles of the cantilever-boom

The forces in a strut-boom are compressive, the tension component of the load being transferred to the superstructure through the derricking ropes. In contrast a cantilever-boom has a wide base tapering to the sheave support and is pin-jointed to the superstructure at the underside of its lower end and restrained by the derricking rope running over a sheave along the top side as shown in Fig. 25.5. The top member lattice frame is thus in tension and the bottom member in compression.

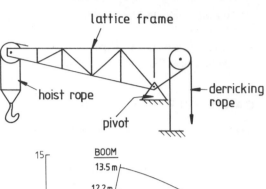

**Fig. 25.5**  Cantilever-boom
principles

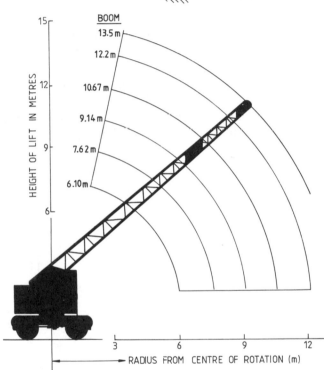

**Fig. 25.6**  Boom radius–
height diagram
for the
cantilever-boom
crane

## Crane characteristics

The choice of cantilever crane is limited within the range 5–15 tonnes
free-on-wheels capacity.

### SIX TONNES CAPACITY CRANE (EXAMPLE ONLY)

Max. safe working load ⎱ free-        6 tonnes at 3.0 m radius on 9.14 m
Max. radius and safe      ⎰ on-            jib
  working load            ⎰ wheels    850 kg at 13 m radius on 15.2 m jib
On outriggers                          Not available
Max. hoisting speed (single fall       40 m/min
  line)
Slewing speed                           3 rev/min
Derricking (max. to min. radius)    25 s

| Travelling speed (unladen) | 7 km/h |
|---|---|
| Weight of crane | 16 tonnes |
| Engine horse power | 48 kW (66 hp) |

## ELEVEN TONNES CAPACITY CRANE

The lifting characteristics are given in Table 25.1. The weight of the crane is about 20 tonnes and is designed only for lifting *free*-on-wheels.

**Table 25.1**   Guidelines to safe working loads of strut-boom cranes on rubber-tyred wheels

**(a) 7 Tonnes Crane FREE ONLY**

| Radius (m) | Boom (m) | | | | |
|---|---|---|---|---|---|
| | 7.6 FREE | 10.7 FREE | 13.7 FREE | 16.8 FREE | 19.8 FREE |
| 3.0 | 6.0 | | | | |
| 3.5 | 5.3 | 5.0 | | | |
| 4.0 | 4.2 | 4.0 | 3.5 | | |
| 5.0 | 3.2 | 2.8 | 2.7 | 2.3 | |
| 6.0 | 2.5 | 2.1 | 2.1 | 1.8 | 1.7 |
| 7.0 | 1.7 | 1.6 | 1.6 | 1.5 | 1.3 |
| 8.0 | 1.4 | 1.4 | 1.4 | 1.2 | 1.1 |
| 9.0 | | 1.1 | 1.1 | 1.0 | 0.9 |
| 10.0 | | 1.0 | 1.0 | 0.9 | 0.7 |
| 11.5 | | | 0.9 | 0.8 | 0.6 |
| 13.0 | | | 0.7 | 0.6 | 0.5 |
| 14.5 | | | | 0.4 | 0.3 |

**(b) 11 Tonnes Crane FREE ONLY**

| Radius (m) | Boom (m) | | |
|---|---|---|---|
| | 6.1 FREE | 9.1 FREE | 12.2 FREE |
| 3.0 | 11.0 | 11.0 | |
| 3.5 | 9.2 | 9.0 | |
| 4.0 | 7.5 | 7.4 | 6.2 |
| 5.0 | 5.5 | 5.4 | 5.2 |
| 6.0 | 4.3 | 4.2 | 4.1 |
| 7.0 | | 3.4 | 3.3 |
| 8.0 | | 2.9 | 2.8 |
| 9.0 | | 2.5 | 2.4 |
| 10.0 | | | 2.0 |
| 11.0 | | | 1.6 |

Cantilever crane.

**(c) 12.5 Tonnes Crane**

| Radius (m) | Boom (m) | | | | |
|---|---|---|---|---|---|
| | 9.1 BLKD | 12.2 BLKD | 15.2 BLKD | 18.3 BLKD | 21.3 BLKD |
| 3.0 | 12.5 | | | | |
| 3.5 | 10.6 | 10.2 | | | |
| 4.0 | 9.4 | 8.9 | 8.5 | | |
| 5.0 | 7.4 | 6.8 | 6.6 | 6.0 | |
| 6.0 | 5.9 | 5.0 | 4.7 | 4.5 | 4.3 |
| 7.0 | 4.3 | 4.0 | 3.8 | 3.6 | 3.5 |
| 8.0 | 3.5 | 3.3 | 3.1 | 3.0 | 2.9 |
| 9.0 | 3.0 | 2.9 | 2.7 | 2.6 | 2.5 |
| 10.0 | | 2.5 | 2.4 | 2.3 | 2.1 |
| 11.5 | | | 1.9 | 1.8 | 1.7 |
| 13.0 | | | 1.7 | 1.6 | 1.4 |
| 14.5 | | | | | 1.1 |

**(d) 12.5 Tonnes Crane**

| Radius (m) | Boom (m) | | | |
|---|---|---|---|---|
| | 9.1 FREE | 12.2 FREE | 15.2 FREE | 18.3 FREE |
| 3.0 | 7.0 | | | |
| 3.5 | 6.3 | 5.8 | | |
| 4.0 | 5.4 | 5.2 | 5.1 | |
| 5.0 | 4.4 | 4.0 | 3.9 | 3.8 |
| 6.0 | 3.3 | 3.1 | 3.0 | 2.9 |
| 7.0 | 2.7 | 2.6 | 2.5 | 2.3 |
| 8.0 | 2.2 | 2.1 | 2.0 | 1.9 |
| 9.0 | 1.8 | 1.7 | 1.7 | 1.6 |
| 10.0 | | 1.5 | 1.4 | 1.3 |
| 11.5 | | | 1.2 | 1.1 |
| 13.0 | | | 1.0 | 0.9 |
| 14.5 | | | | 0.7 |

(*continued*)

Table 25.1 (*continued*)

### (e) 15 Tonnes Crane

| Radius (m) | Boom (m) | | | | | | |
|---|---|---|---|---|---|---|---|
| | 9.1 BLKD | 12.2 BLKD | 15.2 BLKD | 18.3 BLKD | 21.3 BLKD | 24.4 BLKD | 27.4 BLKD |
| 3.0 | 15.0 | | | | | | |
| 3.5 | 13.2 | 12.6 | | | | | |
| 4.0 | 12.0 | 11.6 | 11.6 | | | | |
| 5.0 | 9.6 | 9.0 | 9.0 | 8.5 | 8.3 | | |
| 6.0 | 7.3 | 7.2 | 7.2 | 7.1 | 7.1 | 7.2 | |
| 7.0 | 6.0 | 6.0 | 6.0 | 5.9 | 5.9 | 5.9 | |
| 8.0 | 4.9 | 4.9 | 4.9 | 4.8 | 4.8 | 4.8 | |
| 9.0 | 4.1 | 4.0 | 4.0 | 3.9 | 3.9 | 3.8 | 3.6 |
| 10.0 | | 3.5 | 3.5 | 3.4 | 3.4 | 3.3 | 3.2 |
| 11.5 | | 2.9 | 2.9 | 2.8 | 2.8 | 2.7 | 2.6 |
| 13.0 | | | 2.6 | 2.5 | 2.5 | 2.4 | 2.3 |
| 14.5 | | | | 2.0 | 2.0 | 1.9 | 1.8 |
| 16.0 | | | | 1.6 | 1.6 | 1.6 | 1.5 |
| 17.5 | | | | | 1.4 | 1.3 | 1.2 |
| 19.0 | | | | | | 1.1 | 1.1 |
| 20.5 | | | | | | | 0.9 |

### (f) 15 Tonnes Crane

| Radius (m) | Boom (m) | | | | | |
|---|---|---|---|---|---|---|
| | 9.1 FREE | 12.2 FREE | 15.2 FREE | 18.3 FREE | 21.3 FREE | 24.4 FREE |
| 3.0 | 7.0 | | | | | |
| 3.5 | 6.3 | 6.1 | | | | |
| 4.0 | 5.6 | 5.4 | 5.3 | | | |
| 5.0 | 4.4 | 4.3 | 4.2 | 4.1 | | |
| 6.0 | 3.5 | 3.4 | 3.3 | 3.2 | 3.0 | |
| 7.0 | 2.8 | 2.7 | 2.6 | 2.5 | 2.4 | 2.3 |
| 8.0 | 2.3 | 2.2 | 2.1 | 2.0 | 1.9 | 1.8 |
| 9.0 | 2.0 | 1.9 | 1.8 | 1.7 | 1.6 | 1.5 |
| 10.0 | | 1.7 | 1.6 | 1.5 | 1.4 | 1.3 |
| 11.5 | | 1.4 | 1.3 | 1.2 | 1.1 | 1.0 |
| 13.0 | | | 1.1 | 1.0 | 0.9 | 0.8 |
| 14.5 | | | 0.9 | 0.8 | 0.7 | 0.6 |
| 16.0 | | | | | | 0.4 |

### (g) 32 Tonnes Crane

| Radius (m) | Boom (m) | | | | | | |
|---|---|---|---|---|---|---|---|
| | 7.6 BLKD | 12.2 BLKD | 15.2 BLKD | 19.8 BLKD | 22.9 BLKD | 27.4 BLKD | 30.5 BLKD |
| 3.0 | 32.0 | | | | | | |
| 3.5 | 31.0 | | | | | | |
| 4.0 | 28.0 | 26.0 | | | | | |

**Table 25.1** (*continued*)

### (g) 32 Tonnes Crane

| Radius (m) | Boom (m) | | | | | | |
|---|---|---|---|---|---|---|---|
| | 7.6 BLKD | 12.2 BLKD | 15.2 BLKD | 19.8 BLKD | 22.9 BLKD | 27.4 BLKD | 30.5 BLKD |
| 5.0 | 23.0 | 22.0 | 21.0 | 18.0 | | | |
| 6.0 | 19.0 | 18.0 | 18.0 | 17.0 | | | |
| 7.0 | 15.0 | 15.0 | 15.0 | 14.0 | 13.5 | | |
| 8.0 | | 12.0 | 11.5 | 11.3 | 11.2 | 10.0 | 8.8 |
| 10.0 | | 8.0 | 8.0 | 7.8 | 7.8 | 7.7 | 7.4 |
| 12.0 | | | 5.8 | 5.8 | 5.8 | 5.8 | 5.7 |
| 14.0 | | | 4.8 | 4.7 | 4.6 | 4.5 | 4.5 |
| 16.0 | | | | 3.9 | 3.8 | 3.7 | 3.7 |
| 18.0 | | | | 3.3 | 3.2 | 3.2 | 3.1 |
| 20.0 | | | | | 2.8 | 2.7 | 2.7 |
| 24.0 | | | | | | 2.0 | 1.9 |
| 28.0 | | | | | | | 1.5 |

### (h) 32 Tonnes Crane FREE ONLY

| Radius (m) | Boom (m) | | | | | | |
|---|---|---|---|---|---|---|---|
| | 7.6 FREE | 12.2 FREE | 15.2 FREE | 19.8 FREE | 22.9 FREE | 27.4 FREE | 30.5 FREE |
| 3.0 | 15.5 | | | | | | |
| 3.5 | 13.8 | | | | | | |
| 4.0 | 12.3 | 12.1 | | | | | |
| 5.0 | 9.8 | 9.7 | 9.7 | 9.7 | | | |
| 6.0 | 7.9 | 7.9 | 7.8 | 7.7 | | | |
| 7.0 | 6.5 | 6.4 | 6.3 | 6.3 | 6.2 | 5.5 | |
| 8.0 | | 5.3 | 5.3 | 5.2 | 5.2 | 5.1 | 5.0 |
| 10.0 | | 4.0 | 3.9 | 3.9 | 3.8 | 3.7 | 3.7 |
| 12.0 | | 3.2 | 3.1 | 3.0 | 3.0 | 2.9 | 2.8 |
| 14.0 | | | 2.5 | 2.4 | 2.4 | 2.3 | 2.2 |
| 16.0 | | | | 2.0 | 1.9 | 1.8 | 1.8 |
| 18.0 | | | | 1.7 | 1.6 | 1.5 | 1.4 |
| 20.0 | | | | | 1.3 | 1.2 | 1.2 |
| 24.0 | | | | | | 0.9 | 0.8 |

# 26

# MOBILE TELESCOPIC-BOOM CRANE

Since the early 1960s telescopic booms have been steadily increasing their share of the crane market at the expense of the self-propelled strut-boom model which has hitherto been very economical and efficient for recurrent usage on specific tasks, e.g. stockyard work. The mobile telescopic version is suitable for similar tasks but is more versatile and is capable of travelling at 20–30 km/h quickly moving from site to site in a particular area. The telescoping action is very flexible and various one-off lifts can be easily accommodated.

Unfortunately, the crane is expensive and as yet has not entirely replaced the self-propelled type for the duties required on the small-to-medium size site. In the long term, however, it must be considered as a serious competitor, particularly the specific versions developed to cope with rough terrain work, when the duties now fulfilled by the crawler crane, such as placing concrete, etc., may also be threatened.

## 26.1    CRANE CONSTRUCTION

### Chassis

The chassis consists of two welded steel rectangular hollow side beams connected at each end by similar hollow boxes. A diesel engine is mounted at the rear of the vehicle and drives a rigidly mounted rear axle through a torque converter and power-shift gear box (see Chapter 12, Fixed-position Excavating Machines, for details).

On some vehicles the rear as well as the front wheels have independent steering as shown in Fig. 26.1. Steering on the latest models is controlled by hydraulic cylinders attached between the chassis and wheel hubs.

Two- or four-wheel drive actions are available, four-wheel drive is virtually standard on models made for rough terrain duties on

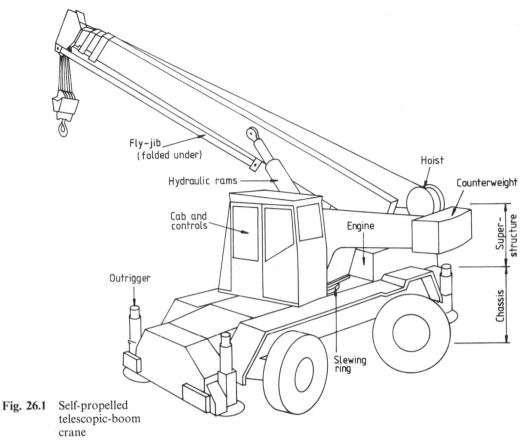

**Fig. 26.1** Self-propelled telescopic-boom crane

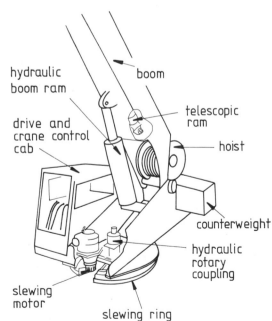

**Fig. 26.2** Arrangement of superstructure

construction sites. The superstructure comprises the drive cab and controls, telescopic boom, hoist and counterweight. The whole unit sits on a turntable mounted on the chassis and thus 360° slewing is available. An independent hydraulic motor is used to induce the slewing motion. The hydraulic pressure is produced from a pump located near the engine on the base frame and thus a special rotary coupling located at the centre of the slewing ring is required to deliver hydraulic fluid to both the hoist and slewing motors.

### The boom

The boom is designed on the cantilever principle shown in Fig. 26.3. Whereas the self-weight of a strut-type boom is carried by the derricking ropes, the cantilever version is designed to support its own weight in addition to the load, the overturning moment being balanced by the counterweight with a suitable factor of safety. The boom is pin-jointed at its base to the super-structure and derricking is controlled by one or two hydraulic rams.

The boom itself consists of three or four sections, one sliding within the other. The telescoping action is provided by a hydraulic ram and a multiplying chain arrangement, as shown in Fig. 26.4.

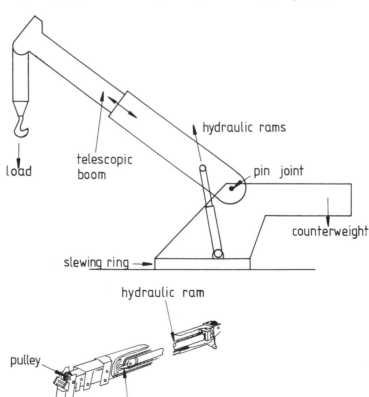

**Fig. 26.3** Principles of the cantilever–telescopic boom

**Fig. 26.4** Section through a telescopic boom

This configuration causes each section to be telescoped simultaneously and so ensures that the boom is maintained in a tapered shape to match the bending moment caused by the load.

## BOOM DESIGN

The restrictions on the design of the telescopic boom are those imposed by the self-weight and weight of the hydraulic telescoping rams which reduce the payload compared to the strut-jib. However, developments in metal strengths are gradually permitting the use of larger booms, and so increasing the operating range of the crane. Most manufacturers adopt a trapezoidal boom cross section made from high tensile alloy steel plate and sometimes the final jib section is of the lattice construction to reduce the weight of the boom.

### Hoist operation

The hoist is operated by a hydraulic motor driven from the main hydraulic pump. The hoisting and lowering speed may be varied for precision control of the load.

### Steering

Operation of the crane is carried out from a single cab mounted on the super-structure. The steering may be arranged to suit the duties required as follows:

(a) As a general purpose mobile unit crane is either two-wheel drive from the rear axle (Fig. 26.5a) or alternatively four-wheel drive for improved traction (Fig. 26.5b) and is steered by the front wheels. Maximum travelling speed is about 50 km/h.

(b) To improve manoeuvrability, particularly for factory use such as stacking and warehousing duties, some versions of the crane have four-wheel drive and four-wheel steer (Fig. 26.5d) which enables crab steering for diagonal movements (Fig. 26.5c). Maximum travelling speed is 30–35 km/h.

(c) For heavy duty work, models equipped with the facilities described under (b) have a strengthened chassis and are provided with large diameter tyres for improved grip and flotation, and have increased ground clearance. These are called 'rough terrain' cranes and range up to about 150 tonnes maximum capacity. They are specifically designed to cope with the bad ground conditions found on construction sites. The tendency has been to try to use these cranes as genuine self-travelling cranes on the public highways to increase the utilisation factor. But because they are not genuine on-off highway cranes, and due to the heavy maintenance requirements, they have not yet proved popular enough to supplant the crawler crane for difficult site work or the all-terrain

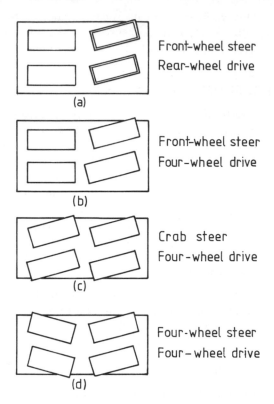

Front-wheel steer
Rear-wheel drive

(a)

Front-wheel steer
Four-wheel drive

(b)

Crab steer
Four-wheel drive

(c)

Four-wheel steer
Four-wheel drive

(d)

**Fig. 26.5** Steering alternatives

model for on-off highway duties. Maximum travelling speed is less than 30 km/h.

### Outriggers

The operating configuration may be unblocked on the tyres, but the lifting capacity is greatly improved when used with outriggers. Those shown in Fig. 26.1 can be hydraulically operated and independently controlled from the driver's cab. They can be quickly set to provide a wide and stable base.

### Fly-jib

The main boom may be provided with the operational facility of a fly-jib which when not in use is stored on the side or underneath the boom. For lifting duties the jib is swung into position by means of the hoist rope and the guy ropes are attached. It may be used in line with the main boom or offset up to about 25°. A fly-jib option is usually available only on the large lifting capacity cranes of 10 tonnes and more – lengths up to about 8 m are available. On the larger rough terrain versions of the crane the fly-jib may also be telescoped (Fig. 26.7) to improve the crane's reach for speedy applications.

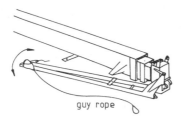

Side folding fly-jib

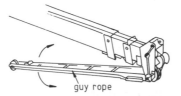

**Fig. 26.6** 'Quick assembly' fly-jib types

Under folding fly-jib

### Crane chararacteristics

The smaller versions (up to 10–12 tonnes capacity) tend to be called 'mobile cranes', selected for stockyard work, whilst the larger and more robust machines are likely to be labelled 'rough terrain' cranes, being more appropriate for site work. The shape of the load–radius diagram for these cranes is shown in Fig. 26.8 and the load–radius data for individual cranes are given in Table 27.1.

### Examples

FOUR TO TEN TONNES CAPACITY CRANES OPERATING AT MINIMUM RADIUS (i.e. SMALL CRANES)

**Fig. 26.7** Telescopic fly-jib

| | |
|---|---|
| Min. lifting radius | 2.5 m |
| Engine size | 100 hp (74 kW) |
| Machine weight | 12–15 tonnes |
| Max. hoisting speed (single fall line) | 10 m/min on older models, 50–60 m/min on the latest models |
| Derricking (max. to min.) | 10 s |
| Slewing speed | 2 rev/min |
| Travelling speed (unladen) | 30 km/h |
| Turning radius | 6 m |
| Road: gradient: loaded | 1 in 6 |
| no load | 1 in 4 |
| Min. boom length | 4–5 m |
| Max. boom length | 6–8 m |
| Overall height | approx. 2.75 m |

**Fig. 26.8** Load–radius diagram for a telescopic-boom crane

| | |
|---|---|
| Overall width | approx. 2.5 m |
| Overall length | approx. 6–7 m |

The lifting capacity at maximum radius with the boom fully extended operating at an angle of 30° with the horizontal is about 1 tonne.

## POPULAR ROUGH TERRAIN CRANES

| | |
|---|---|
| Safe working load at minimum radius on outriggers | 15–40 tonnes |
| Engine size | 150–200 hp (112–150 kW) |
| Machine weight | 20–40 tonnes |
| Max. hoisting speed (single fall line) | up to 120 m/min |
| Derricking (max. to min.) | 25 s |
| Slewing speed | 3 rev/min |
| Travelling speed | 30 km/h |
| Telescoping: three-part jib | 20 s |
| four-part jib | 40 s |
| Turning radius | 10 m |
| Road gradient: unladen | 1 in 25 |
| Boom length: four-part | 20 m |
| three-part | 14 m |
| Overall height | 3 m |
| Overall width | 2.5 m |
| Overall length | 8 m |

*Note*: The safe working load in the *blocked* position is more than halved when lifting *free*-on-wheels only.

It may be observed by reference to tables in Chapter 24 that telescopic mobile cranes as currently manufactured have a significantly lower lifting range of performance than the comparable size of strut-boom crawler models routinely used in placing concrete, handling reinforcement and bulky form-work panels. Thus when conversion to an excavating machine is considered as just one of the options available with the crawler crane, it is readily apparent that the wheeled crane is not particularly attractive for semi-permanent use on the construction site. Although this does not detract from the advantages of mobility between sites for short-term hire, and the large and more robust rough terrain of about 18 to 30 tonnes capacity suits the needs of a section of the market demand for cranage.

# 27

# TRUCK-MOUNTED TELESCOPIC BOOM-CRANE

While self-propelled and mobile cranes are suitable for on-site applications in stockyards, warehousing, dockyards, etc., modern construction sites often require cranes to provide medium-to-heavy lifting capacity over high and wide reaches. For example, placing precast concrete floor decks in high-rise construction, mechanical equipment in power station boiler houses, placing bridge deck beams, etc. Frequently the vehicle is only required for a short period of perhaps hours and fast travel between sites then becomes of paramount importance for economic viability. The obvious solution to the problem was development of the conventional truck or lorry to support a lifting unit, the first model being produced about fifty years ago, since then they have gradually become more efficient and reliable, with improvements made to the diesel engine and transmissions and more recently by the introduction of the telescopic boom.

The great advantage of the truck-mounted telescopic version is that travel at normal lorry speeds on the public highways is achievable and when on site takes only a few minutes to prepare for the lifting operations. Such a facility considerably reduces the hire cost, thus making the crane very competitive indeed.

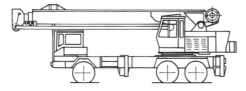

**Fig. 27.1** Truck-mounted telescopic boom-crane

Cranes with a telescopic boom are available from about 10 to 800 tonnes capacity. However, vehicles less than 120 tonnes or so are increasingly being labelled 'all-terrain', and only the upper end of the range are true truck-mounted cranes. The all-terrain version (Fig. 27.2) generally has all-wheel drive, with more robust construction and

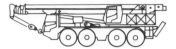

**Fig. 27.2**  All-terrain crane

greater stability better suited to the temporary road surfaces found on typical construction sites. The latest models also have crab steering capabilities and improved speeds for highway travel. The boom on the all-terrain machine is usually shorter than the equivalent truck-mounted version to reduce overhang in the travelling mode, the chassis being much shorter. Finally, all-terrain models of 180 tonnes capacity are now appearing on the market and the dividing line between these and truck-mounted units will inevitably be pushed to higher capacities.

Beyond these sizes telescopic-boom technology is currently inadequate and the conventional lattice boom must be used.

## 27.1  CONSTRUCTION OF THE VEHICLE

### Chassis

The basic carrier comprises a chassis constructed from two universal beams with integral outrigger boxes. The chassis supports the power units, transmission, cab, boom, counterweight and hoists.

The vehicle has the appearance of a conventional truck and is described in terms of the total number of wheels and drive wheels, for example the truck in Fig. 27.1 is designated 6 × 2 wheel drive. The number of axles required depends upon the travelling weight of the truck as the legal restrictions specifying the permissible axle load vary depending upon the country, typically 8–12 tonnes.

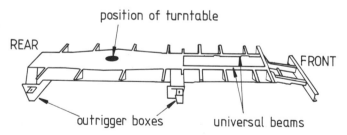

**Fig. 27.3**  Chassis arrangement

### Superstructure

The upper part of the vehicle is similar in construction to mobile telescopic models, except that the counterweight is usually placed at a lower position to improve travelling stability. The whole unit is seated on a turntable to provide 360° slewing. On the smaller vehicles the crane-operating cab and drive cab are combined and a single diesel engine powers all mechanical parts, whereas larger versions have separate cabs and power sources for travelling and lifting duties. In this latter situation the hydraulic pumps and motors are all located on the

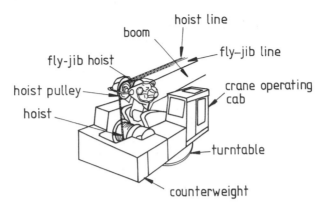

**Fig. 27.4** Arrangement of superstructure

superstructure and the hydraulic rotary coupling at the centre of the slewing ring is eliminated. The operating controls are also positioned on the superstructure to give the driver a clear view of the load and provide at least 180° of vision without having to look back, as is necessary when operating from the driver's cab.

### Crane characteristics

| | |
|---|---|
| Max. hoisting speed (single fall line) | approx. 70–150 m/min |
| Slewing speed | 3 rev/min |
| Derricking (max. to min. radius) | up to 1 min |
| Telescoping: in | 10 m/min |
| out | 20 m/min |
| Travelling speed | up to 70 km/h |

### Travelling on the highway

It is essential that the boom is fully retracted and well secured in the horizontal position, as shown in Figs. 27.1 and 27.2 to avoid danger of damage to bridge decks.

### Crane capacity

Mobile rough-terrain, all-terrain and truck-mounted telescopic boom cranes are designed in accordance with the following criteria:

(a) All capacities are limited to a proportion of the tipping load.
(b) With a fly-jib attached and extended but not working, the load on the main boom should be reduced, e.g. if length of fly-jib is 6–10 m, working load reduction is about 380 kg.
(c) Boom sections should be extended equally.
(d) The crane may be operated blocked on outriggers or free-on-wheels, with a corresponding reduction in capacity for the latter.
(e) The weight of the hook block, together with any slings or other lifting tackle, must be included in the working loads to arrive at a suitable size of crane.

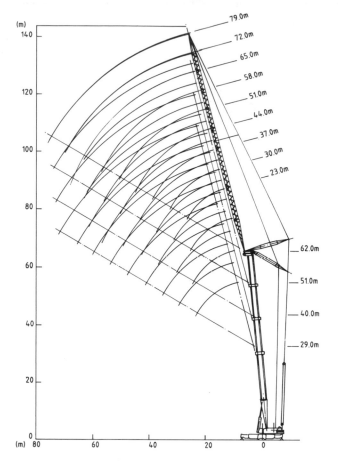

**Fig. 27.5** Boom radius–height diagram for truck/all-terrain cranes

## LOAD–RADIUS–BOOM DATA

The data given in Table 27.1 are general and should only be used for approximate crane selection purposes as the capacity of similarly designated cranes varies from make to make.

**Table 27.1** Guidelines to safe working loads of mobile, rough, all-terrain and truck-mounted telescopic-jib cranes.
(Note: Boom extensions quoted are maximum permitted values at the stated radius.)

**(a) 6 Tonnes Crane Mobile**

| Radius (m) | Boom (m) | | | |
| | **6.6** | | **8.7** | |
| | **BLKD** | **FREE** | **BLKD** | **FREE** |
|---|---|---|---|---|
| 2.5 | 6.0 | | | |
| 2.75 | 5.5 | 5.25 | | |
| 3.0 | 5.0 | 4.5 | 4.85 | |
| 3.5 | 4.4 | 3.75 | 4.25 | 3.75 |
| 4.0 | 3.9 | 3.2 | 3.75 | 3.2 |
| 4.5 | 3.4 | 2.6 | 3.25 | 2.6 |

**(b) 12 Tonnes Crane Mobile**

| Radius (m) | Boom (m) | | | |
| | **7.92** | | **18.3** | |
| | **BLKD** | **FREE** | **BLKD** | **FREE** |
|---|---|---|---|---|
| 3.0 | 12.0 | 5.25 | 10.0 | |
| 3.5 | 10.5 | 4.5 | 8.6 | |
| 4.0 | 9.1 | 4.0 | 7.4 | 4.0 |
| 5.0 | 6.5 | 2.7 | 5.5 | 2.7 |
| 6.0 | 4.0 | 1.8 | 4.3 | 1.8 |
| 7.0 | | 1.1 | 3.5 | 1.1 |

**Table 27.1** (*continued*)

**(a) 6 Tonnes Crane Mobile**

| Radius (m) | Boom (m) | | | |
|---|---|---|---|---|
| | 6.6 BLKD | FREE | 8.7 BLKD | FREE |
| 5.0 | 3.0 | 2.2 | 2.8 | 2.2 |
| 5.5 | 2.75 | 1.9 | 2.6 | 1.9 |
| 6.0 | | | 2.3 | 1.5 |
| 6.5 | | | 2.1 | 1.35 |
| 7.0 | | | 1.9 | 1.2 |
| 7.5 | | | 1.8 | 1.1 |

**(b) 12 Tonnes Crane Mobile**

| Radius (m) | Boom (m) | | | |
|---|---|---|---|---|
| | 7.92 BLKD | FREE | 18.3 BLKD | FREE |
| 8.0 | | 0.6 | 3.0 | 0.6 |
| 9.0 | | | 2.4 | 0.4 |
| 10.0 | | | 1.9 | |
| 11.5 | | | 1.5 | |
| 13.0 | | | 1.1 | |
| 14.5 | | | 0.75 | |

**(c) 18 Tonnes Crane Rough-Terrain**

| Radius (m) | Boom (m) | | | | | | |
|---|---|---|---|---|---|---|---|
| | 7.3 BLKD | 9.0 BLKD | 11.0 BLKD | 12.2 BLKD | 14.6 BLKD | 16.5 BLKD | 18.3 BLKD |
| 3.0 | 18.3 | 15.6 | 14.8 | 14.0 | | | |
| 3.5 | 14.8 | 14.8 | 14.2 | 13.7 | 12.9 | | |
| 4.0 | 13.4 | 13.4 | 13.1 | 12.9 | 12.0 | | |
| 4.5 | 11.9 | 11.9 | 11.9 | 11.9 | 11.0 | 10.1 | 9.7 |
| 6.0 | 9.1 | 9.1 | 9.1 | 9.1 | 9.1 | 8.6 | 8.0 |
| 7.5 | | 5.9 | 5.9 | 5.9 | 5.9 | 5.8 | 5.8 |
| 9.0 | | | 4.2 | 4.2 | 4.2 | 4.2 | 4.2 |
| 10.5 | | | | 3.3 | 3.3 | 3.3 | 3.3 |
| 12.0 | | | | | 2.5 | 2.5 | 2.5 |
| 13.5 | | | | | | 1.8 | 1.8 |
| 15.0 | | | | | | 1.6 | 1.6 |
| 16.5 | | | | | | | 1.3 |

**(d) 22 Tonnes Crane All-Terrain**

| Radius (m) | Boom (m) | | | | | | | |
|---|---|---|---|---|---|---|---|---|
| | 9.75 BLKD | FREE | 13.4 BLKD | FREE | 17.1 BLKD | FREE | 24.4 BLKD | FREE |
| 3.0 | 22.0 | 7.0 | 19.6 | 6.7 | | | | |
| 3.5 | 18.6 | 5.8 | 16.9 | 5.2 | | | | |
| 4.0 | 16.0 | 4.7 | 15.1 | 4.3 | | | | |
| 4.5 | 13.5 | 3.5 | 13.4 | 3.4 | 12.0 | 3.2 | | |
| 6.0 | 9.8 | 2.2 | 9.5 | 2.0 | 8.6 | 1.8 | 6.5 | 1.1 |
| 7.5 | 6.5 | 1.2 | 6.5 | 1.1 | 6.0 | 0.9 | 5.2 | 0.7 |
| 9.0 | | | 4.5 | 0.6 | 4.5 | 0.5 | 4.0 | 0.3 |
| 12.0 | | | 2.5 | | 2.5 | | 2.3 | |
| 15.0 | | | | | 1.6 | | 1.4 | |
| 18.0 | | | | | | | 0.9 | |
| 21.0 | | | | | | | 0.7 | |

(*continued*)

**Table 27.1** (*continued*)

**(e) 30 Tonnes Crane Rough-Terrain**

| Radius (m) | Boom (m) | | | | | |
|---|---|---|---|---|---|---|
| | 9.75 BLKD | 13.25 BLKD | 16.76 BLKD | 20.37 BLKD | 22.73 BLKD | 30.48 BLKD |
| 3.0 | 30.0 | 17.5 | 16.4 | | | |
| 3.5 | 26.0 | 17.5 | 16.4 | | | |
| 4.5 | 20.0 | 17.5 | 15.0 | 12.0 | | |
| 6.0 | 16.0 | 14.5 | 13.0 | 10.5 | 8.5 | |
| 7.5 | 14.0 | 12.5 | 11.5 | 9.2 | 7.2 | 5.0 |
| 9.0 | | 10.3 | 10.0 | 8.2 | 6.2 | 4.5 |
| 12.0 | | | 5.5 | 5.6 | 4.7 | 3.5 |
| 15.0 | | | | 3.5 | 3.6 | 2.7 |
| 18.0 | | | | 2.2 | 2.3 | 2.3 |
| 21.0 | | | | | 1.3 | 2.0 |
| 24.0 | | | | | | 1.5 |
| 27.0 | | | | | | 0.9 |
| 30.0 | | | | | | |

**(f) 40 Tonnes Crane All-Terrain**

| Radius (m) | Boom (m) | | | | | | |
|---|---|---|---|---|---|---|---|
| | 9.75 BLKD | 11.6 BLKD | 13.4 BLKD | 15.2 BLKD | 18.9 BLKD | 23.8 BLKD | 30.8* BLKD |
| 3.5 | 41.0 | 35.0 | 30.0 | 27.0 | | | 2.5 |
| 4.5 | 32.4 | 28.8 | 26.4 | 23.9 | | | 1.7 |
| 6.0 | 24.0 | 24.0 | 22.8 | 20.8 | 16.8 | | |
| 7.5 | 15.4 | 15.4 | 15.4 | 15.4 | 14.5 | 11.2 | 7.8 |
| 9.0 | | 10.6 | 10.6 | 10.6 | 10.6 | 10.0 | 6.3 |
| 12.0 | | | | 5.8 | 5.8 | 5.8 | 4.3 |
| 15.0 | | | | | 3.8 | 3.8 | 3.2 |
| 18.0 | | | | | | 2.5 | 2.1 |
| 21.0 | | | | | | 1.7 | 1.6 |
| 24.0 | | | | | | | 1.1 |
| 27.0 | | | | | | | 0.7 |
| 30.0 | | | | | | | |

**(g) 60 Tonnes Crane All-Terrain**

| Radius (m) | Boom (m) | | | | | | | |
|---|---|---|---|---|---|---|---|---|
| | 10.4 BLKD | 13.4 BLKD | 17.1 BLKD | 20.1 BLKD | 23.5 BLKD | 27.1 BLKD | 29.9 BLKD | 36.2* BLKD |
| 3.0 | 59.0 | 49.0 | | | | | | |
| 3.5 | 54.0 | 46.0 | 39.0 | 32.0 | | | | |
| 4.5 | 45.0 | 42.0 | 35.0 | 30.0 | 26.0 | | | |
| 6.0 | 36.0 | 36.0 | 30.0 | 24.0 | 22.0 | 21.0 | 16.0 | 7.5 |
| 7.5 | 25.0 | 26.0 | 25.0 | 20.0 | 18.0 | 16.0 | 13.0 | 7.0 |
| 9.0 | 18.0 | 18.0 | 18.0 | 17.0 | 15.0 | 14.0 | 11.0 | 6.0 |
| 12.0 | | | 10.0 | 10.0 | 10.0 | 9.0 | 7.0 | 5.0 |

**Table 27.1** (*continued*)

## (g) 60 Tonnes Crane All-Terrain

| Radius (m) | Boom (m) | | | | | | | |
|---|---|---|---|---|---|---|---|---|
| | 10.4 BLKD | 13.4 BLKD | 17.1 BLKD | 20.1 BLKD | 23.5 BLKD | 27.1 BLKD | 29.9 BLKD | 36.2* BLKD |
| 15.0 | | | 6.0 | 6.0 | 6.0 | 6.0 | 6.0 | 4.0 |
| 18.0 | | | | | 4.0 | 4.0 | 4.0 | 3.5 |
| 21.0 | | | | | | 3.0 | 3.0 | 2.8 |
| 24.0 | | | | | | 2.0 | 2.0 | 2.2 |
| 27.0 | | | | | | | 1.2 | 1.8 |
| 30.0 | | | | | | | | |
| 33.0 | | | | | | | | |

## (h) 90 Tonnes Crane All-Terrain

| Radius (m) | Boom (m) | | | | | | |
|---|---|---|---|---|---|---|---|
| | 13.4 BLKD | 18.2 BLKD | 23.2 BLKD | 28.0 BLKD | 32.9 BLKD | 42.7* BLKD | 52.4* BLKD |
| 3.0 | 90.0 | | | | | | |
| 3.5 | 80.0 | 53.0 | 48.0 | | | | |
| 4.5 | 69.0 | 51.0 | 44.0 | | | | |
| 6.0 | 52.0 | 42.0 | 37.0 | 33.0 | 29.0 | | |
| 7.5 | 39.0 | 35.0 | 31.0 | 26.0 | 23.0 | | |
| 9.0 | 28.0 | 28.0 | 27.0 | 22.0 | 19.0 | 19.0 | |
| 12.0 | | 17.0 | 17.0 | 16.0 | 14.0 | 14.0 | 10.0 |
| 15.0 | | 11.0 | 11.0 | 11.0 | 10.0 | 10.0 | 8.0 |
| 18.0 | | 7.0 | 7.0 | 7.0 | 7.0 | 7.5 | 6.5 |
| 21.0 | | | 5.0 | 5.0 | 5.0 | 6.0 | 5.5 |
| 24.0 | | | | 3.0 | 3.0 | 4.0 | 4.5 |
| 27.0 | | | | | 2.0 | 3.0 | 3.5 |
| 30.0 | | | | | | 2.5 | 2.8 |
| 33.0 | | | | | | 1.6 | 2.2 |
| 36.0 | | | | | | 1.0 | 1.7 |
| 39.0 | | | | | | 0.5 | 1.1 |
| 42.0 | | | | | | | 0.6 |

## (i) 112 Tonnes Crane All-Terrain

| Radius (m) | Boom (m) | | | | | | |
|---|---|---|---|---|---|---|---|
| | 13.9 BLKD | 21.3 BLKD | 28.6 BLKD | 36.0 BLKD | 43.0 BLKD | 52.8* BLKD | 62.6* BLKD |
| 3.5 | 112.0 | 53.0 | | | | | |
| 4.5 | 89.0 | 52.0 | | | | | |
| 6.0 | 67.0 | 45.0 | 40.0 | | | | |
| 7.5 | 53.0 | 40.0 | 34.0 | 31.0 | | | |
| 9.0 | 42.0 | 35.0 | 30.0 | 27.0 | 24.0 | | |
| 12.0 | | 26.0 | 23.0 | 20.0 | 18.0 | 16.0 | |
| 15.0 | | 18.0 | 18.0 | 16.0 | 13.0 | 13.0 | 9.0 |

(*continued*)

**Table 27.1** (*continued*)

**(i) 112 Tonnes Crane All-Terrain**

| | Boom (m) | | | | | | |
|---|---|---|---|---|---|---|---|
| Radius (m) | 13.9 BLKD | 21.3 BLKD | 28.6 BLKD | 36.0 BLKD | 43.0 BLKD | 52.8* BLKD | 62.6* BLKD |
| 18.0 | | 12.0 | 12.0 | 12.0 | 11.0 | 10.5 | 7.5 |
| 21.0 | | | 8.0 | 8.0 | 8.0 | 8.5 | 6.5 |
| 24.0 | | | 6.0 | 6.0 | 6.0 | 6.8 | 5.7 |
| 27.0 | | | | 4.5 | 4.5 | 5.3 | 5.1 |
| 30.0 | | | | 3.0 | 3.0 | 4.5 | 4.4 |
| 33.0 | | | | 2.0 | 2.0 | 3.3 | 3.6 |
| 36.0 | | | | | 1.0 | 2.4 | 2.9 |
| 39.0 | | | | | | 1.8 | 2.2 |
| 42.0 | | | | | | 1.1 | 1.6 |
| 45.0 | | | | | | 0.6 | 1.1 |

**(j) 150 Tonnes Crane Truck-Mounted**

| | Boom (m) | | | | | |
|---|---|---|---|---|---|---|
| Radius (m) | 15.24 | 21.34 | 27.43 | 33.53 | 39.62 | 45.72 |
| 4.0 | 150.0 | | | | | |
| 6.0 | 100.0 | 98.0 | | | | |
| 8.0 | 73.9 | 73.3 | 72.7 | | | |
| 10.0 | 56.2 | 55.9 | 55.700 | 55.4 | 52.0 | |
| 14.0 | 33.9 | 33.7 | 33.3 | 33.1 | 32.9 | |
| 20.0 | | 20.4 | 20.000 | 19.8 | 19.5 | 32.7 |
| 24.0 | | | 15.6 | 15.2 | 15.0 | 19.3 |
| 30.0 | | | | 11.0 | 10.6 | 14.3 |
| 35.0 | | | | | 8.4 | 10.4 |
| 40.0 | | | | | | 7.9 |
| | | | | | | 6.5 |

**(k) 350 Tonnes Crane Truck-Mounted**

| | Boom (m) | | | | | |
|---|---|---|---|---|---|---|
| Radius (m) | 16.75 | 28.75 | 40.0 | 52.0 | 72* | 86* |
| 3 | 350.0 | | | | | |
| 4.5 | 251.0 | 150.0 | | | | |
| 7 | 163.0 | 117.0 | 80.0 | | | |
| 9 | 125.0 | 99.0 | 69.0 | 45.0 | | |
| 11 | 99.0 | 85.0 | 60.0 | 44.5 | 25.5 | |
| 13 | 79.0 | 74.0 | 52.5 | 41.0 | 25.4 | |
| 18 | | 47.0 | 39.0 | 33.0 | 21.5 | 9.0 |
| 24 | | 28.0 | 29.0 | 25.5 | 17.7 | 8.4 |
| 28 | | | 22.0 | 21.5 | 15.7 | 7.6 |
| 34 | | | 14.5 | 16.6 | 13.1 | 6.4 |
| 40 | | | | 11.8 | 10.8 | 5.4 |
| 46 | | | | 8.5 | 8.6 | 4.5 |

**Table 27.1** (*continued*)

### (k) 350 Tonnes Crane Truck-Mounted

| Radius (m) | Boom (m) | | | | | |
|---|---|---|---|---|---|---|
| | 16.75 | 28.75 | 40.0 | 52.0 | 72* | 86* |
| 54 | | | | | 5.0 | 3.5 |
| 64 | | | | | | 2.5 |

*Note: This is the largest crane able to carry its telescopic boom in the road travelling configuration.*
\* Including fly-jib.

### (l) 500 Tonnes Crane Truck-Mounted

| Radius (m) | Boom (m) | | | | 57.0 Fly-jib (m) | | |
|---|---|---|---|---|---|---|---|
| | 18.6 | 31.4 | 44.2 | 57.0 | 24 | 48 | 72 |
| 3 | 500.0 | | | | | | |
| 4 | 380.0 | | | | | | |
| 6 | 288.0 | 190.0 | | | | | |
| 8 | 230.0 | 189.0 | 140.0 | | | | |
| 10 | 177.0 | 158.0 | 137.5 | 100.0 | | | |
| 15 | 115.0 | 105.0 | 101.0 | 80.5 | | | |
| 18 | | 82.0 | 87.0 | 71.5 | 45.0 | | |
| 22 | | 60.0 | 65.0 | 60.5 | 44.0 | | |
| 25 | | 49.0 | 53.0 | 53.0 | 41.5 | | |
| 30 | | | 38.5 | 43.5 | 39.0 | 22.0 | |
| 36 | | | 26.5 | 31.5 | | 22.0 | |
| 40 | | | 20.5 | 25.0 | | 22.0 | 8.8 |
| 46 | | | | 18.0 | | 21.7 | 8.4 |
| 50 | | | | 15.0 | | 20.2 | 8.2 |
| 60 | | | | | | | 7.8 |
| 70 | | | | | | | 7.0 |

### (m) 800 Tonnes Crane Truck-Mounted

| Radius (m) | Boom (m) | | | 29 Fly-jib (m) | | | 62 Fly-jib (m) | | |
|---|---|---|---|---|---|---|---|---|---|
| | 18 | 40 | 62 | 23 | 51 | 93 | 23 | 51 | 72 |
| 4 | 800 | | | | | | | | |
| 6 | 500 | | | | | | | | |
| 7 | 420 | 300 | | | | | | | |
| 8 | 360 | 260 | | | | | | | |
| 9 | 320 | 235 | 175 | | | | | | |
| 10 | 290 | 215 | 165 | | | | | | |
| 14 | 200 | 160 | 130 | 120 | | | | | |
| 18 | | 125 | 102 | 117 | | | 67 | | |
| 22 | | 105 | 82 | 104 | 65 | | 64 | | (*continued*) |

**Table 27.1** (*continued*)

**(m) 800 Tonnes Crane Truck-Mounted**

| Radius (m) | Boom (m) | | | 29 Fly-jib (m) | | | 62 Fly-jib (m) | | |
|---|---|---|---|---|---|---|---|---|---|
| | 18 | 40 | 62 | 23 | 51 | 93 | 23 | 51 | 72 |
| 26 | | 85 | 70 | 92 | 63 | | 59 | 37 | |
| 30 | | 62 | 60 | | 59 | | 55 | 35 | |
| 34 | | 48 | 53 | | 54 | 16 | | 33 | 21 |
| 40 | | | 37 | | 47 | 14 | | 28 | 19 |
| 48 | | | 23 | | 38 | 13 | | 26 | 19 |
| 60 | | | 10 | | | 11 | | | 16 |
| 72 | | | | | | 9 | | | 12 |
| 84 | | | | | | 8 | | | |

## Example of truck-mounted telescopic crane selection

Select a truck-mounted crane as an alternative to the crawler crane for lifting 5 tonnes to the centre of the concrete tank described in the example on page 337.

### SOLUTION

(a) It is seen in Fig. 27.6 that to place the load a boom length of at least 18.9 m is required. Table 27.1 shows that a 40 tonnes capacity

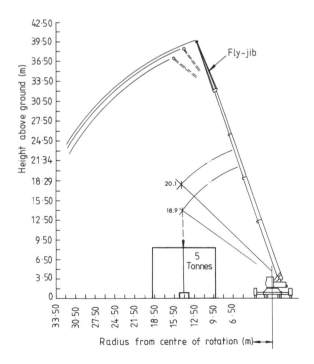

**Fig. 27.6** Example of truck-mounted crane in use

crane (which is sufficient in the case of a crawler crane) is slightly undersize. Interpolation between the 12 m and 15 m radii given in the table indicates that at 14 m radius the crane will lift about 4.5 tonnes. The next available size crane in the tables is 60 tonnes capacity, interpolating this will lift approximately 7.3 tonnes at 14 m radius.

(b) If the load were very wide and deep, say, half the diameter of the tank and 5 m deep, then to clear the tank rim whilst slewing would require a 20.1 m boom.

# 28

# TRUCK-MOUNTED STRUT-BOOM CRANE

Before the development of the telescoping action, the strut-boom was the predominant type of truck-mounted crane. Today, however, these cranes are limited to very specialist duties for lifts of 400 tonnes and more and are currently available with capacities up to 1200 tonnes. Designs are in hand up to 2000 tonnes.

## 28.1  CRANE CONSTRUCTION

The carrier consists of a stiff chassis mounted upon two or more axles depending upon the travelling weight. The chassis incorporates a diesel engine, the transmission and outriggers. The engine is used to power the vehicle for transporting only. The lifting section of the crane is a conventional strut-boom crane superstructure without the crawler tracks, which sits on a turntable mounted on the chassis. It comprises the counterweight, hoist and derricking drums, slewing mechanism, boom and operating controls, and an independent diesel engine to power the crane parts which like most other modern machines use independent hydraulic motors to provide precision hoisting and lowering control.

### Boom assembly

This type of crane is frequently selected because of its ability to provide very specific heavy lifting capacity, economically over a short period. To do this effectively, the crane should ideally be ready and prepared for use almost immediately upon arrival at the construction site. The strut-boom unfortunately, however, cannot travel in this ready form, and must be folded and securely held. Consequently much setting up time is needed upon arrival, for example 4 hours preparation time would not be untypical. For use with the basic length boom, i.e. top and bottom sections only, the equipment packs down neatly as shown

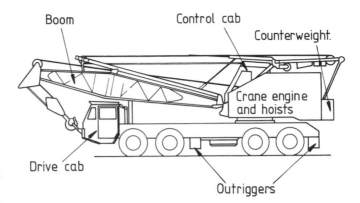

**Fig. 28.1** Truck-mounted strut-boom crane arranged for travelling on a highway

in Fig. 28.1. But extensions beyond this basic length require the sections to be laid end-to-end on an area of level ground before erection. The assembly then proceeds as shown in Fig. 28.2.

These very large capacity cranes often require two carriers; one for the crane (i.e. drive cab, engine, turntable, counterweights, hoist, etc.) and one to transport the boom and fly-jib. The mechanisms for slewing, hoisting and derricking of this size crane are usually driven by electric motors powered from the main diesel engine. Extra outriggers and kentledge are also necessary (Fig. 28.3) and the setting up time is much increased. The size of the load to be placed, however, is usually so specified that this can be taken into account in the construction programme.

**Crane characteristics**

| | |
|---|---|
| Travelling speed | up to 75 km/h |
| Slewing speed: up to 500 tonnes | 0.5–1.0 rev/min |
| 500–1000 tonnes | 0.1–0.3 rev/min |
| Derricking (min. to max.) | Varies, e.g. 3 min for 300 tonnes crane with 15 m boom |
| Max. hoisting speed (single fall line) | 60–120 m/min |
| Optional features | Operation as a tower crane |

## LOAD–RADIUS–BOOM CAPACITY

The weight of the snatchblock, slings and other handling devices are significant and must, of course, be added to the load. The crane may be operated with a fly-jib and can be used either blocked-on-outriggers or free-on-wheels. Table 28.1 shows the lifting capacities of 450 and 800 tonnes capacity cranes, respectively.

It will be noticed that these machines have very high lifting ability and are particularly suited for heavy duties, such as in offshore work, boiler installation in power stations, etc. The number of such cranes in the country are few and lifts must be planned well in advance to ensure availability of a crane of suitable capacity.

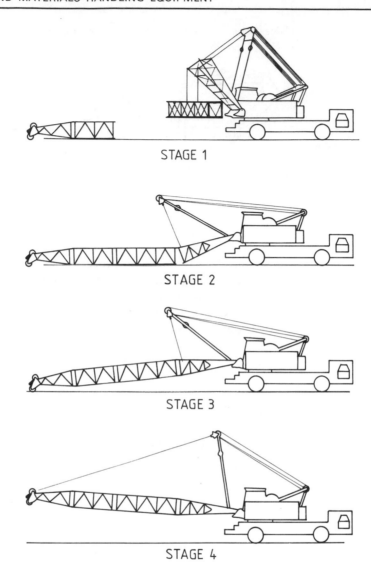

STAGE 1

STAGE 2

STAGE 3

**Fig. 28.2** Assembly of strut-boom

STAGE 4

**Fig. 28.3** Large truck-mounted crane (chassis not shown)

**Table 28.1** Guidelines to safe working loads of truck-mounted strut-boom cranes

**(a) Main Boom Rating: 450 Tonnes**

| Radius (m) | Boom (m) | | | | | | |
|---|---|---|---|---|---|---|---|
| | 16 | 30 | 44 | 58 | 72 | 86 | 100 |
| 5 | 450 | | | | | | |
| 7 | 350 | | | | | | |
| 9 | 295 | 292 | | | | | |
| 12 | 210 | 205 | 200 | 192 | | | |
| 14 | 176 | 174 | 170 | 160 | 125 | | |
| 18 | | 131 | 127 | 124 | 100 | 77 | 54 |
| 24 | | 95 | 92 | 90 | 75 | 59 | 31 |
| 30 | | 75 | 72 | 69 | 61 | 47 | 24 |
| 40 | | | 51 | 47 | 44 | 38 | 18 |
| 56 | | | | 27 | 24 | 20 | 10 |
| 68 | | | | | 16 | 14 | 4 |
| 80 | | | | | | 7 | 1 |

**(b) Main Boom Rating: 800–1200 Tonnes**

| Radius (m) | Boom (m) | | | | | | |
|---|---|---|---|---|---|---|---|
| | 23 | 35 | 47 | 59 | 71 | 93 | 95 | 113 |
| 5 | 800 | | | | | | | |
| 7 | 650 | | | | | | | |
| 9 | 529 | 526 | | | | | | |
| 10 | 478 | 475(1200) | 470 | | | | | |
| 12 | 401 | 397(1000) | 393 | 363 | | | | |
| 18 | 268 | 265(645) | 261 | 258 | 205 | 184 | | |
| 22 | 216 | 214(600) | 210 | 207 | 176 | 165 | 139 | 93 |
| 26 | | 178(504) | 174 | 171 | 152 | 144 | 126 | 85 |
| 32 | | 135(410) | 131 | 128 | 126 | 116 | 108 | 75 |
| 40 | | | 94 | 91 | 89 | 86 | 84 | 62 |
| 56 | | | | 54 | 51 | 48 | 46 | 41 |
| 64 | | | | | 40 | 37 | 34 | 30 |
| 76 | | | | | | 26 | 23 | 18 |
| 84 | | | | | | | 17 | 12 |
| 88 | | | | | | | | 10 |

Note: Figures in brackets for 35 m boom are with alternative counterweight.

# 29

# TOWER CRANES

On travels far and wide the number of cranes in use is quite striking, with the skyline of the major cities often cluttered by new high-rise constructions. The tower crane, however, is also very suitable for low-rise work concentrated within a limited area where access by crawler or other mobile cranes is restricted.

The main advantage of the tower crane is that the jib or boom is supported at the top of a tall tower which may be set at a sufficient height to clear any obstructions. This configuration allows the crane to stand very close to, or even in, the structure under construction. In this way a relatively short boom provides more reach in comparison with other types of crane. Thus for high-rise buildings the tower crane is often the cheapest form of device.

There are two versions, the horizontal-boom and the luffing-boom, both powered electrically with 350–415 V supply.

## 29.1  LUFFING-BOOM

This arrangement consists of a lattice-framed vertically standing mast mounted on a sturdy turntable. The boom is hinged near to the top of the mast and the luffing line, attached to the extremity, is passed over a pulley at the top of the tower then down to the kentledge to counterbalance the weight of the boom. The kentledge is generally located on the turntable but some manufacturers prefer to have both the turntable and kentledge at the top of the mast as shown in Fig. 29.2 so that all the moving parts are above the structure under construction. The whole unit is placed on a well-prepared foundation. The crane is electrically powered and electric motors are used to drive the hoist and luffing winches and for slewing. The main advantage of the luffing-boom tower crane is that the boom can be raised clear of nearby obstructions when slewing.

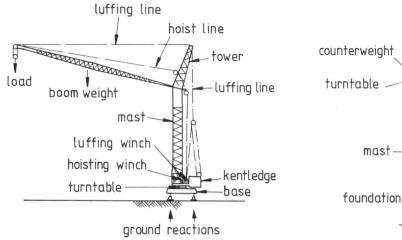

**Fig. 29.1**  Luffing-boom tower crane (turntable at base of mast)

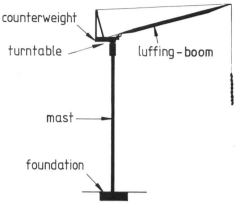

**Fig. 29.2**  Luffing-boom tower crane, alternative kentledge at top of mast

## 29.2    HORIZONTAL BOOM

The crane comprises a vertical standing lattice-framed central mast, which supports a horizontal boom in two parts, the larger section being used for lifting, and carries a 'trolley' or 'saddle' travelling on guides along the length of the boom (Fig. 29.4). Thus the radius is changed by moving the trolley and not by luffing the boom. On the opposite side of the mast a shorter boom supports a kentledge block and serves as a counterbalance. The resistance to overturning when lifting (and from wind pressures) is transferred through the tower to a heavy foundation base. Like luffing-boom models, the tower crane must be designed to resist torsion from side loads acting on the boom, e.g. a swinging load, wind, etc. It is usual for crane operation to be suspended when the wind gust speed exceeds 61 km/h (38 mph) and the boom is left to swing freely on the turntable to reduce the torsional effect.

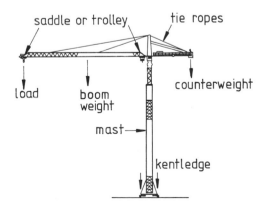

**Fig. 29.3**  Horizontal or saddle-boom tower crane

4 FALLS

The crane has 360° slewing capability and the turntable is commonly mounted at the top of the mast. This configuration usually involves transporting the crane in sections with an associated slow assembly on site, but rapid development in self-erecting cranes (Fig. 29.5) is taking place. The whole unit is supported on a strong foundation connected to the base of the mast with the main counterweight also located in this position. The centre of gravity is thus brought nearer to the ground to improve the crane's stability.

Electric power provides the drive for the slewing and hoist motors.

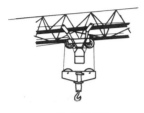

2 FALLS

**Fig. 29.4**  Saddle or trolley for horizontal boom crane

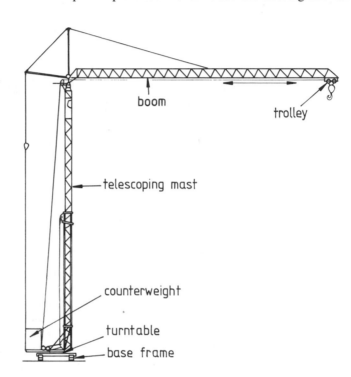

**Fig. 29.5**  Self-erecting saddle-boom tower crane

## 29.3  LOAD–RADIUS DIAGRAM FOR THE LUFFING-BOOM CRANE

The tipping load of the luffing boom is calculated from the following principles (excluding side loads).

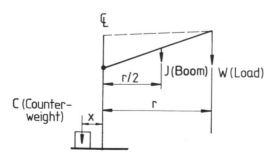

**Fig. 29.6**  Forces acting on a luffing-boom tower crane

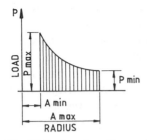

**Fig. 29.7** Load–radius diagram for the luffing-boom tower crane

Taking moments about the centre of rotation

$$Wr + J\frac{r}{2} = Cx$$

$$W = \frac{Cx}{r} - \frac{J}{2} \qquad (29.1)$$

Safe working load $(P) = W -$ margin for safety.

Equation (29.1) if plotted with varied $r$ draws the curve shown in Fig. 29.7.

### 29.4    LOAD–RADIUS DIAGRAM FOR THE SADDLE-BOOM CRANE

Taking moments about the centre of rotation

$$Wr + Jl = C_1 x + C_2 y$$

$$W = \frac{C_1 x + C_2 y - Jl}{r} \qquad (29.2)$$

Safe working load $(P) = W -$ margin for safety.

Equation (29.2) if plotted with varied $r$ produces Fig. 29.9, similar in shape to Fig. 29.7.

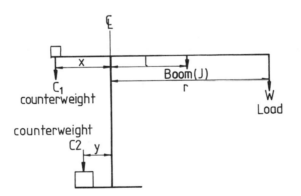

**Fig. 29.8** Forces acting on a saddle-boom tower crane

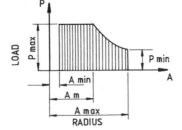

**Fig. 29.9** Load–radius diagram for the saddle-boom tower crane

For safety, the luffing-type crane operating radius of the boom is restricted to $A$ min $\simeq 0.25\ A$ max. For the saddle-jib the minimum operating radius $A$ min is limited to approximately 0.05–0.2 $A$ max. In general the load–radius diagram is further restricted as shown in Fig. 29.9 for loads lifted near to the mast, and $A$ m $\simeq 0.25$–0.4 $A$ max. $P$ min $\simeq 0.1$–0.2$P$. The saddle-boom crane is often quoted in terms of the load moment capacity, which is calculated by multiplying the load by the radius. Usually the maximum load moment capacity occurs at $A$ min and the minimum load moment at $A$ max.

Tower cranes are now available to suit many situations and the data shown in Tables 29.1, 29.2, 29.3, respectively for luffing-booms, saddle-booms and self-erecting models are typical of the free-standing capabilities with the crane operating with the longest possible boom.

**Table 29.1** Luffing-boom load–radius data

| | | | | | | | | |
|---|---|---|---|---|---|---|---|---|
| Max. radius ⎱ (m) | 18 | 26 | 30 | 36 | 42 | 50 | 56 | 75 |
| Lifting capacity ⎰ (tonnes) | 0.75 | 1.2 | 0.6 | 1.0 | 1.4 | 1.8 | 2.0 | 5.6 |
| Max. lifting capacity ⎱ (tonnes) | 1.2 | 5 | 5 | 11 | 12 | 21 | 16 | 45 |
| Radius ⎰ (m) | 7 | 6 | 7 | 6 | 8.5 | 8.5 | 13 | 19 |
| Height to jib pivot (m) | 18 | 27 | 29 | 35 | 46 | 52 | 58 | 76 |
| Max. hook height (m) | 27 | 49 | 52 | 61 | 75 | 90 | 106 | 140 |
| Track gauge (m) | 2.35 | 2.8 | 3.8–4.0 | 4.4–4.5 | 5.0 | 6.3 | 6.3–7.1 | 10.0 |
| Power supply (kW) | 20 | 40 | 50 | 70 | 80 | 120 | 150 | 260 |

**Table 29.2** Saddle-boom tower crane load–radius data

| | | | | | | | | |
|---|---|---|---|---|---|---|---|---|
| Max. jib radius (m) | 36 | 40 | 45 | 50 | 60 | 70 | 70 | 80 | 80 |
| Capacity (tonnes) | 1.0 | 1.5 | 2.5 | 2.9 | 3.6 | 5.0 | 12.2 | 14.5 | 22.8 |
| Max. capacity ⎱ (tonnes) | 3.0 | 8.0 | 10.0 | 12.0 | 20.0 | 20.0 | 50.0 | 64.0 | 64.0 |
| Radius ⎰ (m) | 14.4 | 10.6 | 14.0 | 14.6 | 16.3 | 22.4 | 20.0 | 23.8 | 34.8 |
| Max. load moment (mt) | 43 | 85 | 140 | 175 | 326 | 448 | 1000 | 1523 | 2238 |
| Min. radius (m) | 7.0 | 4.0 | 3.0 | 3.0 | 3.0 | 3.9 | 3.9 | 4.0 | 4.0 |
| Height under hook (m) | 30 | 38 | 44 | 47 | 66 | 80 | 90 | 140 | 105 |
| Track gauge (m) | 2.8 | 4.5 | 4.5 | 4.5 | 8.0 | 8.0 | 10.0 | 15.0 | 15.0 |
| Power supply (kW) | 40 | 40 | 60 | 100 | 150 | 150 | 170 | 200 | 200 |

**Table 29.3**    Self-erecting saddle-boom tower crane load–radius data

| | | | | | | | | | | | | |
|---|---|---|---|---|---|---|---|---|---|---|---|---|
| Max. jib radius) (m) | | 11 | 16 | 18 | 20 | 25 | 30 | 35 | 35 | 40 | 45 | 50 |
| Capacity          (tonnes) | | 0.3 | 0.65 | 0.75 | 0.8 | 1.0 | 1.0 | 1.0 | 3.0 | 1.5 | 1.75 | 2.0 |
| Max. capacity   (tonnes) | | 0.45 | 1.5 | 1.5 | 1.7 | 3.0 | 4.0 | 6.0 | 8.0 | 8.0 | 8.0 | 10.0 |
| Radius            (m) | | 7.8 | 8.2 | 10.3 | 11.42 | 10.3 | 9.4 | 8.8 | 14.9 | 10.7 | 13.4 | 14.0 |
| Max. load moment (mt) | | 3.5 | 12.3 | 15.5 | 19.4 | 30.9 | 37.6 | 52.8 | 119.2 | 85.6 | 104.8 | 140.0 |
| Min. radius (m) | | 3 | 3 | 3 | 3 | 3 | 3 | 3 | 3 | 3 | 3 | 4 |
| Track/wheel gauge (m) | | 2.0 | 2.32 | 2.8 | 3.2 | 2.8 | 3.8 | 4.5 | 6.0 | 5.0 | 5.0 | 6.0 |
| Power supply* (kW) | | 16 | 16 | 16 | 22 | 20 | 25 | 40 | 35 | 50 | 50 | 60 |
| Max. hook height (m) | | 10.6 | 16.0 | 18.0 | 18 | 20 | 20.0 | 32.8 | 29.3 | 32.8 | 37.8 | 32.8 |

* Self-contained generator available on some models.

Note: Available (1) static on outriggers; (2) travelling on rail track; (3) on crawlers.

## 29.5    TYPES OF TOWER CRANE (BOTH LUFFING- AND SADDLE-BOOM TYPES)

### Free-standing crane

The free-standing crane, stands self-supporting on a well prepared foundation. The mast is bolted to a strong steel cruciform base (Fig. 29.10). Ballast consisting of blocks of concrete, iron or gravel, is placed on the base to provide a counterbalance.

Free-standing cranes are available with up to 100 m clear distance between the hook and ground level. To increase the height further the crane must be tied to the structure under construction.

The free-standing arrangement is commonly adopted and includes both the saddle- and luffing-boom types. It is usually preferable to position the crane outside the building, as this makes for easier erection and dismantling and also avoids the costs of leaving out and making good parts of the structure.

With the fixed-position base the crane boom must be capable of reaching all parts of the job and so tends to be used for building construction and particularly high-rise construction where regular shape structures are involved as typified in Fig. 29.11.

### Rail-mounted cranes

An alternative and popular foundation is the rail-mounted base. With this arrangement the crane can travel with the load on the hook on a specially prepared track, which may be straight or curved.

The base frame is made from sturdy welded steel plate. For straight track the frame is of simple construction mounted on four wheels as shown in Fig. 29.13. For curved track (Fig. 29.14) the frame is more

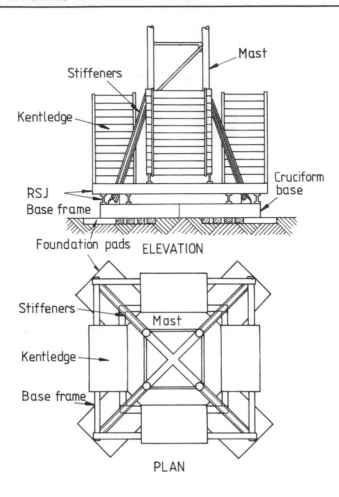

**Fig. 29.10** Free-standing tower crane base layout

**Fig. 29.11** Free-standing tower crane applications

sophisticated and the arms and wheels are pivoted to accommodate the bends in the track. Usually two wheels are provided under each support. The crane can be used to construct both high and long buildings and is able to travel to material stockpiles, concrete batching plants, etc., located in close proximity to the structure.

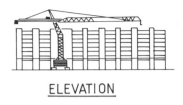

ELEVATION

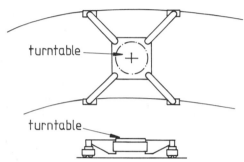

**Fig. 29.12**   Rail-mounted tower crane

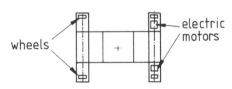

**Fig. 29.13**   Tower crane base for straight rail track

**Fig. 29.14**   Tower crane base for curved rail track

There are some disadvantages compared with the free-standing tower crane, as follows:

(a) Track is expensive to purchase and lay.
(b) The site must be firm and the crane can only be used on a level track, rails out of level by only a few millimetres render the mast unstable.
(c) The crane's travel motors can accommodate gradients only up to about 1 in 200.
(d) The track must be kept clear at all times, which may entail supervision, especially on a busy site, and so add to the costs.
(e) The track may occupy a valuable area of the construction site, which cannot be used for storage of materials.

Rail-mounted cranes are electrically powered with separate three-phase a.c. motors, requiring a high voltage supply of 350–415 V. The travel wheels are driven through gears to enable the crane to move as fast as is safe with the particular load being carried. Sometimes the crane is provided with a special braking system to allow fine movement when precision positioning the load.

### Crawler-mounted tower cranes

There has been a limited demand for tower cranes with greater mobility than is provided by rails. For example, crawler-mounted tower cranes

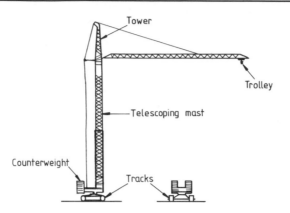

**Fig. 29.15** Crawler-mounted tower crane

(Fig. 29.15) have been successfully used to pour the concrete foundations and floors, and lift out the waste materials for the construction of housing and low-rise dwellings. For economic application, however, duties would need to include the placing of all materials plus positioning any units such as precast floors and walls. Such cranes, however, are relatively small – up to 1500 kg capacity at 40 m radius. The crane is brought to site on a low loader.

### Tied-in tower crane

The free-standing tower crane supported on a ballasted base or rails can be used up to heights of about 100 m. Above this the crane must be tied in to the structure under construction at points recommended by the manufacturers of the crane, to coincide with suitable positions on the building, such as the floors of a high-rise dwelling. The tie frame is of special design usually in the form of a lattice frame which spreads out from the mast and is attached to the structure to resist forces from all horizontal directions. By careful selection of the fixing positions the operating height may be extended up to 200 m or more. The base may be ballasted as in Fig. 29.10 but frequently, where the site conditions allow, a concrete foundation block is cast into the ground as shown in Fig. 29.17. The mast is then simply bolted down. Most of the overturning moment is taken by the structure rather than the foundation, resulting in a lighter and neater base than either rail or ballast types, with a consequent increase in working space on a congested site.

### Truck-mounted tower crane

The strut-boom truck-mounted crane of the lattice type may be assembled to operate as a luffing tower crane as an optional feature on the large cranes of 100 tonnes capacity and greater. The fly-jib is replaced by a much longer luffing boom which operates with the safe working load at 75% of the tipping load. The crane is suitable for short duration one-off tasks where heavy loads need to be lifted into high wide structures. Lifting data are given in Table 28.1.

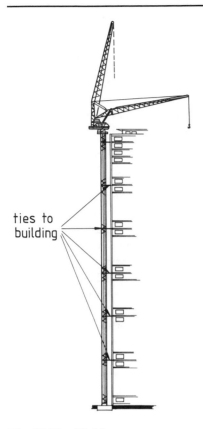

ties to
building

**Fig. 29.16**   Tied-in tower crane

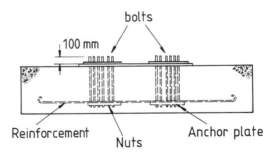

bolts

100 mm

Reinforcement      Nuts      Anchor plate

**Fig. 29.17**   Base for a tied-in tower crane

**Fig. 29.18**   Truck-mounted tower crane

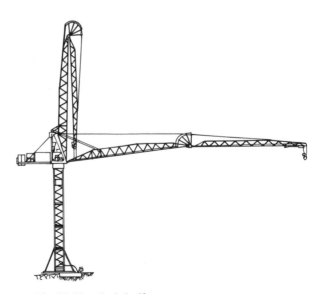

**Fig. 29.19**   Jack-knife tower crane

### Jack-knife crane

The jack-knife crane is designed to work in extremely tight quarters, for example, work between two high buildings. A free-standing horizontal boom crane would have to free sail over the top of such structures and thus might pose an unnecessarily expensive solution to the problem. Capacities are available for all the common site duties and permit about 2 tonnes to be raised at 30 m radius.

### Climbing crane

Where external site space is at a premium and the shape of the building allows, a tower crane may be located inside the building. Frequently the lift shaft in high-rise structures serves as a convenient position to locate the mast. Generally the tied-in crane referred to earlier is used in this situation but where the client permits, a climbing crane may be cheaper, as the foundation and most of the mast is dispensed with. The horizontal and vertical thrusts must therefore be taken by the structure itself.

Initially the mast is mounted on a fixed base which is usually part of the foundations of the structure. The crane then proceeds to erect the building around itself up to at least the second floor. It is then secured by special collars at each floor level. Thereafter it climbs without support away from the original foundations base. The two principal methods of raising the mast are by winches or hydraulic jacks.

**Fig. 29.20**   Climbing tower crane

### WINCH RAISING

A line (F) is attached to the collar (H) which is firmly fixed to the structure. The line is passed under two pulleys (E) fixed to the base of the mast and over a third pulley (G) connected to the pulley (D). A second line (J) from a winch (A) mounted at the base of the mast is passed over the pulley (C) which is hung from the top of the mast section, it is led down and under pulley (D) and finally attached firmly to the support at (C). The winch winds in rope (J) and raises the

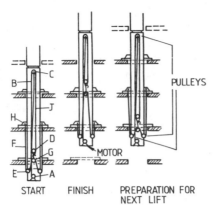

**Fig. 29.21** Raising a climbing tower crane by winching

START    FINISH    PREPARATION FOR NEXT LIFT

mast, lifting it clear of the bottom collar. The collar is then firmly clamped and the lifting tackle is positioned ready for the next lift.

## HYDRAULIC RAISING

The lifting procedure starts with the mast in position (A) supported on a climbing ladder (1), hung from a bearing plate (6). The collar (2) is firmly clamped to the crane mast, with the spring claws (5) engaged on rungs (4). A sliding collar (3) is linked to collar (2) by means of the hydraulic jack (7). To raise the mast, the claws on collar (3) are positioned to engage ladder rung (8). The jack is extended and thereby lifts the crane from position (A) to position (B). The jack is then withdrawn raising collar (3) in the process.

Dismantling requires the boom and mast to be taken apart in sections and lowered to the ground by winches.

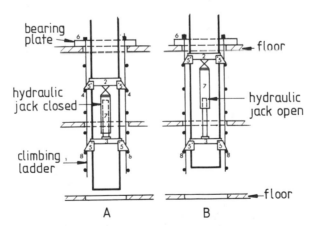

**Fig. 29.22** Raising a climbing tower crane by hydraulic jacking

## 29.6  ERECTION OF TOWER CRANES

### Turntable at the top of the mast

An example of a winch-assisted method of erection is shown in Fig. 29.24 and 29.23(a). For some makes the erection procedure is lengthy

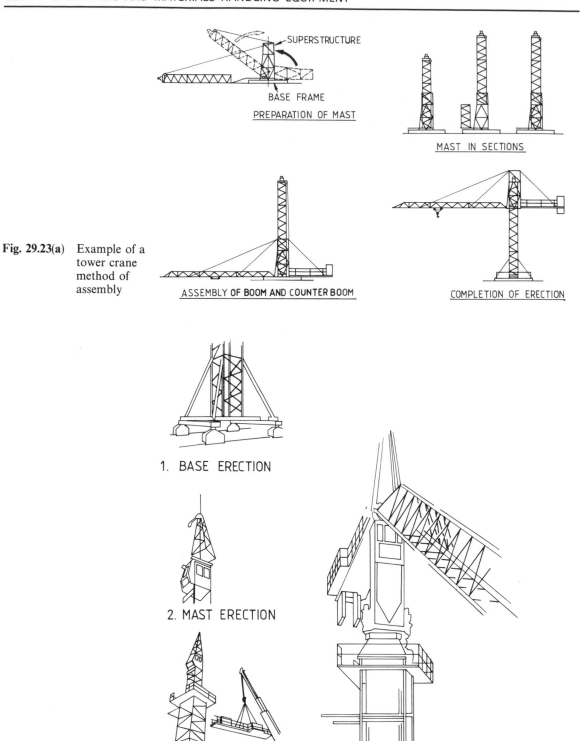

**Fig. 29.23(a)** Example of a tower crane method of assembly

SUPERSTRUCTURE

BASE FRAME

PREPARATION OF MAST

MAST IN SECTIONS

ASSEMBLY OF BOOM AND COUNTER BOOM

COMPLETION OF ERECTION

1. BASE ERECTION

2. MAST ERECTION

3. COUNTERWEIGHT BOOM ERECTION

4. MAIN BOOM ERECTION

**Fig. 29.23(b)** Erection of a tower crane

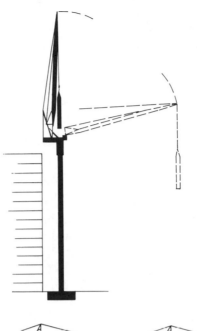

**Fig. 29.24** Extending the height of a tower crane from the top

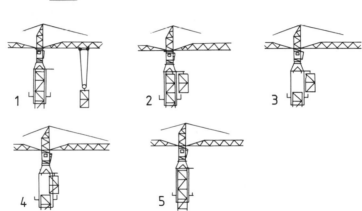

**Fig. 29.25** Extending the height of a tower crane through the side

and usually requires the assistance of a separate crane, usually a truck-mounted telescoping boom-type mobile (Fig. 29.23(b)). All the parts are brought to site on trucks and the crane is put together in sections as follows:

Stage 1: The base is prepared and the first section of mast is lifted into position.

Stage 2: The climbing cage and turntable are next positioned over the mast.

Stage 3: The counterweight and arm are fixed in place. (Saddle-boom type only.)

Stage 4: The boom is attached and finally self-hoisted or lifted by crane into position.

The crane height is further increased either by introducing mast sections at the top of the mast as shown in Fig. 29.24 or alternatively through the side as in Fig. 29.25.

**Fig. 29.26** Transporting a tower crane on the highway

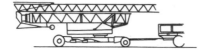

**Fig. 29.27** Self-erecting tower crane sequences

## Turntable at the base of the mast

Cranes of both the luffing- and saddle-boom types can be made self-erecting by designing the mast on a telescoping principle. The operating height, however, is limited to the design size of the crane.

The whole unit usually arrives in the collapsed arrangement towed by an articulated wagon (Fig. 29.26) or even as a completely mobile truck-mounted crane (Fig. 29.18).

Using built-in winches on the crane the tower and jib are unfolded as shown in Fig. 29.27.

## Crane assembly time

None-self-erecting cranes require 1 day for small units and 2 days for large units with crane assistance. Without crane assistance, using only winches, up to 4 days with three men may be required. Self-erecting cranes require from $\frac{1}{2}$ to 1 day. Laying of a 10 m length of rail track takes from 1 to 2 days. The ratio of erection to dismantling time is about 60:40.

## Tower crane characteristics

|  |  |  | Crane size | | |
|---|---|---|---|---|---|
|  |  |  | *small* | *medium* | *large* |
| Max. hoist speed | | | | | |
| (two part line): raise | $V_R$ | m/min | 30.0 | 55–90 | 150 |
| lower | $V_L$ | m/min | 80 | 100 | 150 |
| Slewing speed | $U$ | rev/min | 1.0 | 0.8 | 0.6 |
| Trolley speed | $V_s$ | m/min | 40 | 30 | 2 |
| Travelling speed (on rails) | $V_T$ | m/min | 40 | 30 | 0 |
| Luffing | $t_D$ | s | 20 | 50 | 90 |

## Example: cycle time for a luffing-boom tower crane (see Figs 29.12 and 29.28)

1. Hook on load            $t_{O_1} = 1$ min
2. Raise load from ground level    $t_{R_1} = (h + h_s)/V_R$

3. Slewing $(\beta_1 + \beta_2)$       $t_{S_1} = [(\beta_1 + \beta_2)/360] \times U$
           $(\beta_3)$         $t_{S_2} = [\beta_3/360] \times U$
4. Travelling along $S$ (on tracks)   $t_{T_1} = S/V_T$
5. Derricking jib into position      $t_D$
6. Lowering load $(h_S)$            $t_{L_1} = h_S/V_L$
7. Unhook load                  $t_{O_2} = 1$ min
8. Raise hook $(h_s)$             $t_{R_2} = h_S/V_R$
9. Slewing $(\beta_3)$              $t_{S_3} = [\beta_3/360] \times U$
10. Derricking (as 5)         $t_{D_2}$
11. Lowering hook $(h + h_S)$    $t_{L_2} = (h + h_S)/V_L$
12. Travelling along $S$        $t_{T_s} = S/V_T$
13. Slewing $(\beta_1 + \beta_2)$     $t_{S_4} = [(\beta_1 + \beta_2)/360] \times U$

The bar programme is shown in Fig. 29.28.

**Fig. 29.28**  Tower crane cycle time

The number of cycles per hour $(L_0)$ is given by

$$L_0 = \frac{60}{\Sigma t_0 + \Sigma t}$$

This is the theoretical output which in practice may be reduced by other factors such as weather stoppages, breakdowns, etc. These may add up to total losses of 30%. The experience and skill of the driver are also important, for example the efficiency factor may be 0.95, 0.85 or 0.75 for good, average and poor drivers, respectively. Thus calling the production losses factor $e_1$ and the driver efficiency factor $e_2$

$$\text{expected output} = \frac{60e_1e_2}{\Sigma t_0 + \Sigma t} \text{ cycles per hour}$$

Typically $\frac{1}{2}$ yd$^3$ skip of concrete can be placed at the rate of about 7 m$^3$/h and the corresponding rate for a 1 yd$^3$ skip is 12 m$^3$/h at heights up to 25 m.

### Tower crane exercise (1)

A tower crane required to concrete a 100 m long retaining wall is positioned as shown in Fig. 29.29 such that the maximum length of boom is 20 m. The unit is mounted upon rails which run parallel and 20 m distant from the retaining wall. The crane data are as follows:

| | |
|---|---|
| Raising speed | 30 m/min |
| Lowering speed | 60 m/min |
| Slewing speed | 1 rev/min |
| Travelling speed | 40 m/min |
| Derricking speed | 40 s |

The operating efficiency of the crane is 85%. The raising and lowering height of the load from ground level is 20 m. The times to fill and empty the skip are:

| | |
|---|---|
| Fill | 30 s |
| Empty | 30 s |
| Position over wall | 30 s |

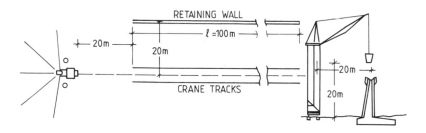

**Fig. 29.29**   Example of a tower crane in use

## PROBLEM

(a) How many skips per hour can the crane handle?
(b) How far along the track can the crane travel without influencing the output of the crane?
(c) Draw the crane's operating cycle diagram independent of the travelling time along the track.

### Tower crane exercise (2)

Figure 29.30 shows a plan view of the present site layout for the construction of a rectangular high-rise building using a tower crane. Criticise this arrangement.

Information
 (i) 500 litre concrete batcher sited 40 m from tower crane.
(ii) Load moment of tower crane at 40 m radius is 30 mt.

## SOLUTION

*Present method*

(a) Three entrances to the site (two not guarded) are required.
(b) A 500 litre batching plant is too large for the safe working load at the required radius.

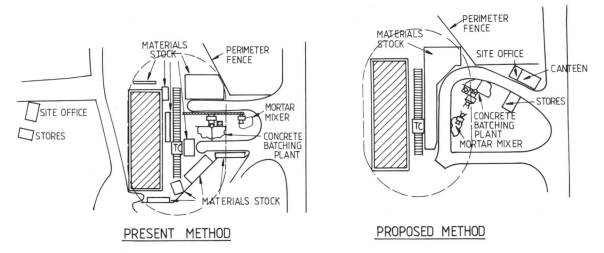

**PRESENT METHOD**          **PROPOSED METHOD**

**Fig. 29.30**   Example of a tower crane site layout

(c) The mortar mixer is outside the reach of the crane.
(d) Materials are too spread over the site.
(e) Office and stores are too far from the site.

*Proposed method*

(a) Stores and office are located near the site entrance and exit.
(b) A through road now connects the entrances and exit.
(c) The mortar mixer is located within reach of the crane.
(d) The materials compound is arranged along crane track.
(e) The mixer is reduced to 375 litre to match the tower crane capacity.
(f) The third entrance is eliminated.

# 30

# GANTRY (PORTAL) CRANE

The gantry crane is usually associated with stockyard work, for example precast concrete manufacturing where lifting facilities are required to cover both the casting and storage areas, for which a portal structure is ideally suited. The portal legs are mounted upon rail track and the bridging beam is positioned to span the full width of the manufacturing/stockyard area. The structure is of lattice frame construction, with one pair of legs rigidly mounted to the bridging beam to give stability and designed to resist both horizontal and vertical forces, while the other pair of legs is of a more slender construction and carries only the vertical component of the load force. The portal beam supports an electrically powered hoist travelling on wheels along the bottom flange, generally arranged to cantilever some distance out from the legs with height altered by changing the number of sections. Models are available with capacities ranging from 5 to 30 tonnes. Purpose-built cranes are designed to suit specific needs.

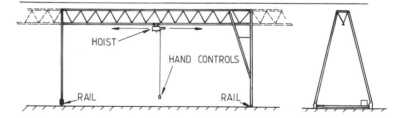

**Fig. 30.1**  Gantry or portal crane

## Typical data

| | |
|---|---|
| Clear span | 25–50 m |
| Cantilever, each side | 5–15 m |
| Clear height below hook | 5–10 m (may be increased for specific requirements) |
| Wheel base | 10–15 m |
| Travel speed | 30–40 m/min |

## Uses

The working height is a limiting factor in comparison with, for example, tower cranes on high-rise construction, but occasionally where the high capital cost of the gantry crane can be justified on a single contract or over a series of similar contracts, adoption of this method of materials handling is very efficient.

This type of crane has occasionally proved quite economic for the construction of precast concrete buildings (Fig. 30.2), large foundations of uniform shape, dry docks, etc., usually where working space is at a premium.

The portal crane has also found an application in the slightly different form of a beam launching crane in bridge deck construction.

Figure 30.3 shows the gantry being used to span two bridge piers for placing the deck beams. Figure 30.4 illustrates the alternative use of the gantry as a bridging beam to provide temporary support for precast concrete bridge units raised from barges below.

**Fig. 30.2** Portal crane application in high use connection

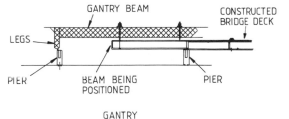

**Fig. 30.3** Gantry crane as a beam launcher

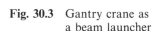

**Fig. 30.4** Gantry crane as a temporary support in bridge construction

# 31

# HOIST

Before the almost universal introduction of cranes, the hoist was normally the device selected to carry out the task of raising materials above ground level. Today the method is mostly restricted to uses in building construction, such as low-rise housing and multistorey structures, for handling small and light loads of less than 1 tonne, e.g. bricks, mortar, roof tiles and similar building materials. The hoist has maintained a presence in civil engineering work, in the form of a passenger/materials elevator to transport personnel and materials quickly on massive structures, such as dams, power station buildings, chimneys, cooling towers, etc. It can also be used to supply concrete to supplement a tower crane in high-rise construction. In this latter situation, the rate of concreting may be increased by using the hoist skip to deposit concrete in a wet hopper at the working level. The hopper in turn feeds the tower crane skip, thus allowing more time for the actual concrete placing, by reducing the lifting period. In this manner at heights of about 60 m, production can be stepped up by over 50%. Even so, both in high- and low-rise buildings, a combination of the tower crane and concrete pump is often preferred with the trend for low-rise construction more towards the use of the fork-lift truck, which is undergoing rapid development as a construction industry machine. However, for the construction of one to perhaps a dozen houses, the hoist is likely to remain economic.

The range of hoists available to suit different tasks is roughly divided into mobile and fixed hoists.

## 31.1 MOBILE HOISTS

These are light, mobile and robust and can be transported from site to site on a truck or towed. Once on site they are easily erected and may be operated by unskilled labour. The smaller models are used, for example, on housing where the operating height is less than 7 m.

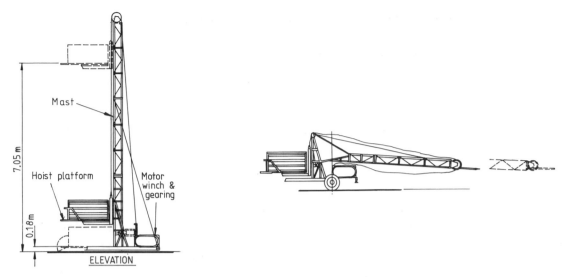

**Fig. 31.1** Arrangement for transporting 1 tonne portable materials hoist

The unit consists of a base frame, a winch unit powered by a small self-contained diesel engine, and two mast sections, which are folded over for transporting. The base incorporates four screw jacks to level up the unit when in the operational mode.

Simple arrangement keeps maintenance costs low which combined with the low purchase price has ensured popularity with building firms.

A range of free-standing hoists is available for lifting materials weighing 250–1000 kg to landing heights of about 15 m.

The equipment has proved economical in the past, but suffers the disadvantage that materials must be transported to and from the landings by other means. Quite often this involves manhandling, and other forms of moving materials are seriously eroding the hoist's advantages, most notably the fork-lift truck and concrete pumps.

## 31.2    FIXED HOIST

Fixed hoists are designed for operation in one location for a long period. They are most commonly used in high-rise construction, such as office blocks, chimneys, cooling towers, etc. In the case of multistorey buildings, it is often more economic to use a materials/passenger hoist after completion of the structural frame by tower crane, particularly where external scaffold is not involved, as the crane cannot place the materials at the required floor levels very easily. Also the construction of chimneys, lift shafts, cooling towers, sometimes involves concrete pumps and so tower cranes would not be required, the inclusion of a fixed hoist to transport labour, materials and equipment quickly, then becomes essential.

## 31.3   MATERIALS/PASSENGER HOISTS

These hoists are designed for loads of 500–1500 kg or about twenty passengers. In general the cage size is about 2.5 m high × 3 m wide × 1.25 m broad, which may be either a single front-mounted cage or two side-mounted cages (Fig. 31.2). The most popular system for the majority of construction projects is a unit having a maximum operating height of 100 m with a cage of capacity 1000 kg or twelve passengers, and hoisting speed of 40 m/min. However, with the mast suitably braced at roughly 18 m centres such hoists can be operated at heights up to 600 m.

The cathead (i.e. sheaves at the top of the mast over which the winch rope passes from cage to winch), is mounted independently on the vertical mast sections allowing the tower to be extended whilst the hoist remains operational. As no counterweight is employed the cathead may be raised without disturbing the hoist reeving. The extension sections are raised into place by a small jib powered from an electric motor housed on the cage as shown in Fig. 31.3.

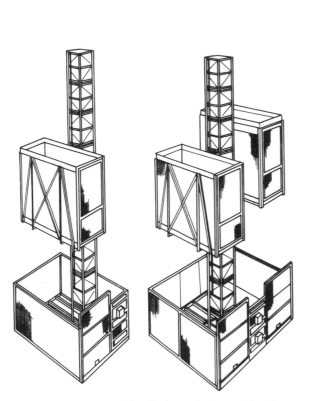

**Fig. 31.2**   Fixed-position hoist – single and double cages

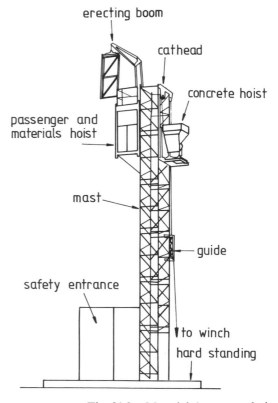

**Fig. 31.3**   Materials/passenger hoist

Offering similar advantages a rack-and-pinion mechanism provides a popular alternative system particularly for access platform applications in fixing cladding, etc, on high-rise construction (Fig. 31.4).

### Hoist data

| | |
|---|---|
| Capacity | 500–1500 kg |
| Cage speed | 40–90 m/min |
| Electric winch | 25–50 hp (18–38 kW)) |
| Voltage/frequency | 380/50 V/Hz |

## 31.4   CONCRETE HOIST

The concrete hoist is proving to be a very suitable supplement to the tower crane for concreting on the construction of high-rise structures. The hoist consists of a concrete skip, fixed either inside the mast or externally like the materials/passenger cage (Fig. 31.3). When the required height is reached the concrete is discharged, sometimes automatically, into a storage hopper (wet hopper), and then distributed by wheelbarrows or transferred into the crane's concrete skip. Skips are available in the range $\frac{1}{2}$–1 yd$^3$ (0.38–0.76 m$^3$) capacity, the larger version is designed to carry 2000 kg at speeds up to 100 m/min, or heavier loads more slowly, with an electrically powered winch of about 100 hp (76 kW). Ideal output data are given in Fig. 31.5 for a 1 yd$^3$ (0.76 m$^3$) capacity skip. The production is likely to be reduced in practice due to general factors limiting the efficiency.

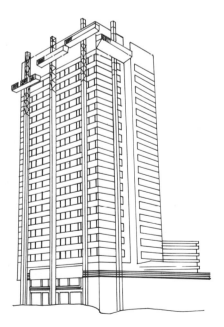

**Fig. 31.4**   Access platforms

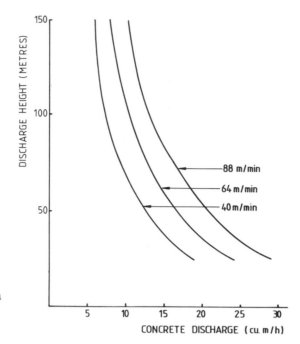

**Fig. 31.5** Production data for a concrete hoist

## 31.5 HOIST EXERCISE

The construction site shown in Fig. 31.6 illustrates the layout of building materials and access roads for the purposes of servicing two hoists for the erection of a low-rise building. Comment on and criticise the present layout in relation to the positioning of both the materials and hoists.

### SUGGESTED SOLUTION

(A) Criticism of existing site layout (Fig. 31.6):

(a) Both hoists have separate scaffold staging, causing increased costs.

(b) Materials are not stockpiled near hoists.

(c) Entrance to the site is too narrow for trucks to pass.

(d) Stores are located behind the batching plant so obscuring storeman's view and the check point is separated from the stores.

(e) Concrete and mortar mixers are located too far from the hoists, and some distance from the temporary road.

(f) Stockpiles are dispersed and hinder unloading and policing.

(g) Temporary roads are long and narrow.

(h) Some stores areas are difficult to reach, and cannot be seen from stores.

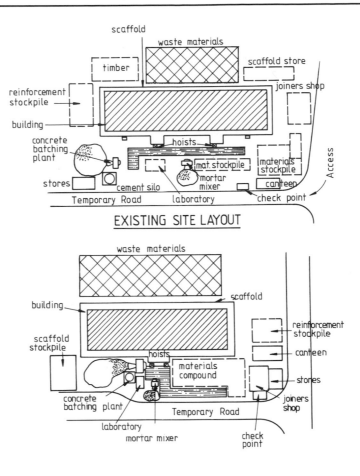

**Fig. 31.6** Example of hoist site layout

(B) Suggested improved layout (Fig. 31.6):

    (a) Both hoists are housed in a common scaffold.

    (b) Batching plants have direct discharge into dumpers onto the hard-standing.

    (c) The temporary road is shorter and wider.

    (d) The access has been widened near the site entrance.

    (e) The stores is located to give a good view of all materials stockpiles, and is sited near the temporary road. The check point is also near the stores.

    (f) Aggregate bins can be directly accessed from the hard standing.

    (g) A compound is provided to police non-bulk materials.

# 32

# FORK-LIFT TRUCK

Since the early 1960s the fork-lift truck has made a great impact on materials handling in the construction industry and in particular on general building. On many sites, such as housing estates, much time and effort is wasted in moving materials from the initial delivery point to the subsequent places where they are required. Deployment of the fork-lift truck provides both a transport and lifting facility and so obviates the need for a hoist or even cranage of any kind. Even when a hoist (and therefore dumper trucks also) is required, the fork-lift is often a better choice than the dumper truck, especially as many materials are increasingly being packaged to suit fork-lift handling. Today the fork-lift truck has the ability to transport horizontally, and vertically at up to 10 m and to raise loads of several tonnes, although for construction site work trucks within the range 1000–5000 kg are more typical. Some machines now have forward reach capabilities and other sophistications as shown in Fig. 32-1.

## 32.1 FORK-LIFT TRUCK ASSEMBLY AND OPERATION

The machine comprises a chassis supporting a diesel engine over the rear axle and a lifting mast at the front. The machine can be supplied with two- or four-wheel drive. In the two-wheel drive arrangement the drive is usually applied through the front wheels via a torque converter and power shift gearing. Both the steering and mast operation are hydraulically controlled. For construction work the machine is robustly constructed and supplied with large wheels and correspondingly large, wide tyres to cope with the rough and uneven terrain.

The advantage of the vehicle is that it can unload from any side of a delivery wagon. The forward extending truck in particular can often reach across the full width of the vehicle, for example where limits are imposed on working space on one side.

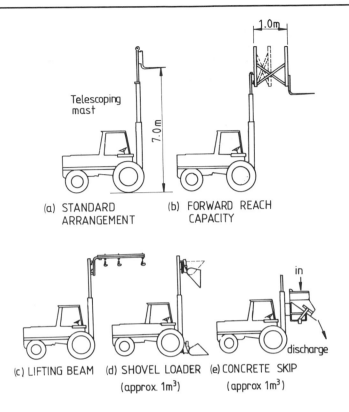

(a) STANDARD
    ARRANGEMENT

(b) FORWARD REACH
    CAPACITY

**Fig. 32.1** Fork-lift truck
options

(c) LIFTING BEAM

(d) SHOVEL LOADER
    (approx. 1m³)

(e) CONCRETE SKIP
    (approx 1m³)

Once the materials have been unloaded they may be stacked in convenient stockpiles or transferred directly to the place of work, e.g. bricks to scaffold levels, thus avoiding double handling.

The load is carried as near to the ground as possible to lower the centre of gravity and reduce the risk of overturning, the load only being raised when the discharge point is reached. Because of the faily uneven terrain on the construction site the mast is equipped with hydraulic rams to move the load radially and thus avoid the need for too much manoeuvring of the vehicle itself.

Increasingly the telescopic boom version of the machine (Fig. 32.2) is proving popular and facilitates loads to be placed well inside a building at heights up to 10 m.

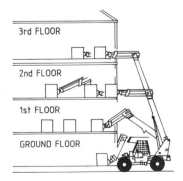

**Fig. 32.2** Telescopic boom
fork-lift truck

Rapid developments are also taking place with extended booms for access platforms. Typical applications include cladding, painting, street lighting maintenance etc. The demarcation of forklifts and mobile cranes by different manufacturers for these aspects is becoming increasingly blurred.

## 32.2  STANDARDISED LOADS

The fork-lift truck requires standardised loads to operate economically. This is being increasingly assisted by the specification of standard components for building work, e.g. bricks, blocks, roof tiles, paving slabs, kerbstones, timber, joinery, bagged cement, roof trusses, drainage pipes, lintels, scaffolding tubes, packaged in economic quantities to reduce breakage and wastage. Many of these items, furthermore are delivered on wooden pallets at the factory, which are later returned when freed at the construction site.

# 33

# GENERAL POINTS ON THE SELECTION AND USE OF LIFTING EQUIPMENT

## 33.1 COMPARISON OF ALTERNATIVE METHODS OF MATERIALS HANDLING

### Hoist

The hoist is a cheap and rapid method of lifting materials, but it does not load, stack, carry or place materials and usually the materials need additional transport by some other means, such as the dumper plus manhandling.

### Tower crane

The tower crane will off-load from delivery vehicles, lift, carry and stack, place and pour concrete, etc., but requires a banksman/dogman and is slow to manoeuvre. It tends to lose its advantages over the fork-lift truck for low-rise building construction such as housing except for placing materials inside the structure. It is, however, ideally suited to foundation construction and engineering work, and high-rise construction because of its clear span of outreach.

### Mobile and truck crane

These have similar characteristics to the tower crane, but greater lifting capacity. Firm ground, good access roads and clear unobstructed outreach are required.

### Fork-lift truck

The fork-lift truck will off-load, lift, carry and stack and does not require well-prepared roads, but can only lift around the sides of the structure. It is limited to heights of about 10 m.

## 33.2  BOOM SELECTION FOR CRANES

### Cantilever-booms

Cantilever-booms give greater clearance where headroom is limited, particularly when handling bulky loads. Lifting capacity is generally lower than for a comparable size strut-boom crane, because of the need to carry self-weight and for permanent site based cranage, the crawler strut-boom crane tends to be more economical for construction work duties, the cantilever being used mainly on wheeled vehicles in stockyards where manoeuvrability and headroom are at a premium.

### Strut-booms

Strut-booms are used with derricks, crawler and truck-mounted cranes. They have high lifting capacity with long reach but are cumbersome and require dismantling to increase the boom length or to add fly-jibs. Dragline and grabbing crane, duties are suitable options.

### Fly-jib

A fly-jib may be attached to either a strut- or cantilever-boom to provide extra reach in the offset position. The arrangement is particularly useful for placing concrete reinforcement, and other light loads in restricted situations.

### Telescopic-boom

The telescopic-boom is a variation of the cantilever, offering a quickly operational and variable length boom, for mobile and truck-mounted cranes but cannot normally handle dragline and grabbing duties. The self-weight of the telescoping rams reduces the weight/radius capacity compared to the strut-boom, but offers increased manoeuvrability. It is an expensive type of crane.

## 33.3  BOOMS AND SAFE LOAD INDICATORS

To prevent the driver of a crane exceeding the permitted working load under normal working conditions, a safe load indicator is installed and calibrated to warn of overload on the hook.

The signalling device takes the form of a coloured flashing light and a bell mounted inside the driver's cab.

The indicator mechanism itself is usually mounted on the boom and must be fitted to all cranes of 1 tonne lifting capacity or over. There are several types of system and equipment approved according to the particular country and safety standard in operation.

### 33.4 LIFTING ATTACHMENTS FOR CRANES

Figure 33.1 shows the commonly used devices attached to the lifting hook of cranes of all types.

*Slings* (a) are used widely for moving bundles of reinforcement and timber, steel and concrete beams, precast concrete units, etc. The safe working load for two-legged slings is generally quoted at 90° as the lifting capacity varies with the angle of the legs. For example, a two leg wire sling of 19 mm diameter will carry 5280 kg at 60°, 4300 kg at 90° and 3050 kg at 120°. The capacity of the equivalent single leg sling is 3050 kg.

*Lifting beam* (b) – sometimes it is preferable to design and fabricate a steel beam for lifting slender and fragile items, e.g. precast concrete beams.

*Lifting brackets* (c) and *pincers* (d) – increasingly materials are delivered as standard packaged items, e.g. bricks banded in units of 1000. Specially designed brackets can usually recoup the initial fabrication costs from the reduced unloading time.

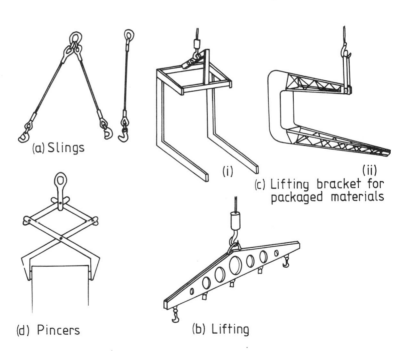

(a) Slings

(i)

(ii)

(c) Lifting bracket for packaged materials

**Fig. 33.1** Crane lifting attachments

(d) Pincers

(b) Lifting

### Concrete skips

Figure 33.2 shows concrete skips. The bottom dumping skip (b) is selected for mass concrete work where positioning is not critical. For placing concrete in thin walls, beams, columns, etc., the side-dumping skip (a) is preferred. Sometimes the tipping skip (c) is preferred. It is filled from dumper trucks standing at ground level, thus the skip can be loaded whilst a second full skip is hooked on to the crane.

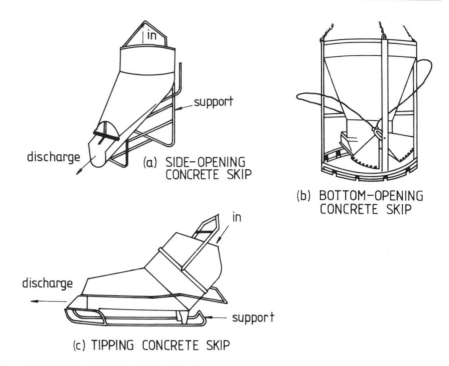

discharge

(a) SIDE–OPENING CONCRETE SKIP

support

(b) BOTTOM–OPENING CONCRETE SKIP

in

discharge

support

(c) TIPPING CONCRETE SKIP

**Fig. 33.2**  Concrete skips for application with a crane

## 33.5  SAFETY REQUIREMENTS

It is essential to appoint a banksman/dogman to guide the crane operator and unhook loads, etc., especially when operating with restricted vision, such as tower cranes engaged on high-rise construction. The accepted signals are shown in Fig. 33.3.

1.  ALL MOVEMENTS
2.  LUFFING OR DERRICKING BOOMS
3.  TELESCOPING OR TROLLEY BOOMS
4.  ALL CRANES
5.  MOVING CRANES
6.  ALL CRANES

① LIFT    LOWER    ④ SLEW LEFT    SLEW RIGHT

② BOOM UP    BOOM DOWN    ⑤ TRAVEL TO ME    TRAVEL FROM ME

③ EXTEND BOOM OR TROLLEY IN    RETRACT BOOM OR TROLLEY OUT    ⑥ STOP    EMERGENCY STOP

**Fig. 23.3**  Banksman signals for crane operations

# 34

# AERIAL CABLEWAY

The cableway is a form of overhead crane, with a cable substituted for rigid girders, so permitting a greatly increased span. For this reason, it has found successful application for the construction of concrete dams, in particular. The cableway has also proved ideal for the construction of river barrages, bridges and viaducts, and is occasionally economic for handling materials in quarries and stockyards, but the gantry crane tends to be favoured for these latter duties.

## 34.1  CABLEWAY HOISTING AND ROPING SYSTEMS

The simplest roping arrangement shown in Fig. 34.1 consists of a track cable firmly anchored at each end to heavy steel towers. This cable provides the support for a wheeled carriage, made of a light frame containing two wheels for running on the track cable and two further sheaves to guide the hoist line. The carriage is winched between the towers by a looped travel rope which is played out and wound in simultaneously, causing the carriage to ride along the track line. The hoist rope on the carriage is independently operated from a hoist winch, the line is fed from the winch over a sheave in the head tower, to the first sheave on the carriage, down and under the hook block

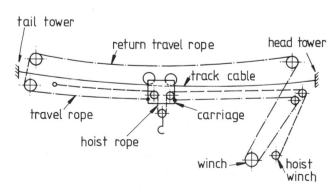

**Fig. 34.1**  Simple cableway roping method

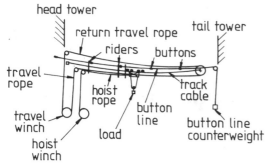

**Fig. 34.2**    The Lidgerwood roping method

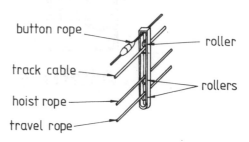

**Fig. 34.3**    Button rope and rider used by the Lidgerwood roping method

sheave, up and over the second carriage sheave and is finally anchored to the tail tower. The load may thus be raised and lowered at any position between the towers.

This method is generally only practical for short spans as the hoist rope tends to sag and get entangled with the other lines but the arrangement allows high travel and hoisting speeds and is consequently quite economical.

To overcome the roping problems on long spans Lidgerwood in the USA developed a system of rope carriers (Fig. 34.2). In this method, an extra cable is provided fitted with buttons spread out along its span between the towers. The buttons are of increasing size working away from the head tower. As the carriage moves from the head tower a rider is trapped at each of the buttons. The amount of sag in the hoist and travel ropes is thereby significantly reduced, as the self-weight of each rope is transferred on to the track cable through the riders, which also relieves the hoist and travel winches from the need to cope with the dead weight of sagging cables. In this way the possibility of the cables becoming entangled is much reduced.

This Lidgerwood design, however, requires the hoist line to be 'dead-ended' on the carriage rather than at the tail tower, since the riders are gathered in on the return trip to the head tower. The hoist drum is therefore arranged to pay out and wind in, in co-ordination with the travel of the carriage.

## 34.2    TYPES OF CABLEWAY

The aerial cableway is an expensive item of equipment whose capital cost can only be justified if the depreciation extends over the full duration of the contract. As this is unlikely, construction firms usually specialise in the type of construction work where specific types of cableway are required so that continuity of work is available. Cableways are made in a variety of types, custom engineered to suit the required conditions of the construction site. The more common types are the fixed cableway, luffing cableway, radial travelling cableway and parallel cableway.

### Fixed cableway

The head and tail towers are held in a fixed position as shown in Fig. 34.4 and may be sturdy framed structures or simply light guyed masts depending upon the span and loads to be carried. Only a straight line area may be serviced and fixed cableways are consequently used mainly for transporting materials over rivers and ravines rather than for construction work.

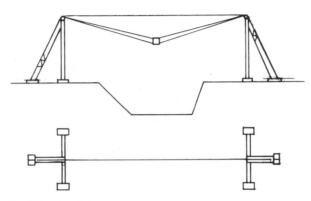

**Fig. 34.4**   Fixed cableway

### Luffing cableway

For the construction of relatively long and narrow structures, such as bridges, the luffing cableway is sometimes suitable. The head tower is fixed, while the tail mast is mounted on a ball pivot base so that the mast may sway up to 20° from the vertical. The mast is guyed loosely at the side and tied firmly at the rear, thereby allowing the cableway to cover the radial area circumscribed by the luffing movement.

Typically a system with 30 m high masts should, at the widest points, serve an area 9 m times the length of the span.

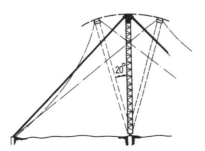

**Fig. 34.5**   Luffing cableway

### Radial cableway

A typical arrangement is illustrated in Fig. 34.6. The head tower is secured and counterweighted at a fixed position, whilst the tail tower or mast is mounted on a travelling carriage on rail tracks, which are set radially about the head mast. The angle of travelling is usually less than 40°. For angles greater than 20°, the head mast must have two stays to cope with the high horizontal force in the cable acting unsymmetrically through the tower. A 40° traverse provides a significant improvement upon the luffing-boom, and makes the system most suitable for the construction of long and relatively wide dams and bridges.

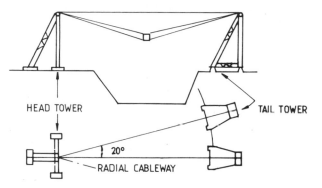

**Fig. 34.6** Radial moving cableway

### Parallel moving cableway

Where a cableway system is chosen to construct a very wide dam, then the only alternative is the fully moving cableway, whereby the head and tail towers are both mounted on rails and travel parallel to each other. The system is cumbersome and requires two operators to co-ordinate both units. A more economic arrangement can sometimes be achieved by setting up a pendulating mast (Fig. 34.7).

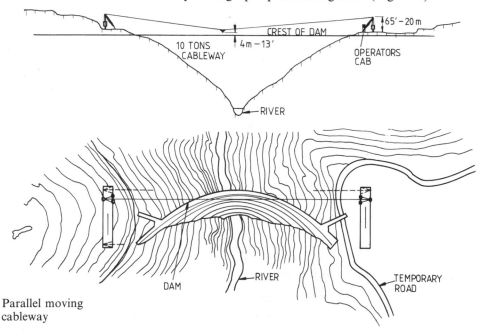

**Fig. 34.7** Parallel moving cableway

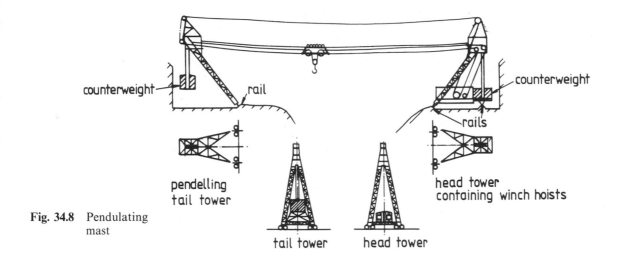

**Fig. 34.8** Pendulating mast

## 34.3 CABLEWAY FEATURES

### Winches

The winch is mounted on the head tower, but the controls are positioned so that the operator can obtain a clear view of the work. Sometimes on dam construction, as the dam height increases the driver's view gets obscured and radio communication is then necessary.

The winches are electrically powered with speed-regulated motors.

### Output data

| | |
|---|---|
| Tower travelling speeds (on rails) | $V_T$ up to 12 m/min |
| Carriage speed | $V_c$ up to 360 m/min |
| Raising load | $V_R$ up to 100 m/min |
| Lowering load | $V_L$ up to 170 m/min |

The maximum span of a cableway is dependent upon the tensile strength of the wires of the cable, the weight of the cable and the permissible amount of sag. With 75 mm diameter cable of breaking strength of 500 tonnes, a 5 tonnes load could be handled on a span of 1500 m. Generally, however, spans are within the range of 60–1000 m.

### Working loads

For construction work and particularly when placing concrete in dam building, typical loads are 3–10 tonnes. Cableways have been designed for 10–20 tonnes and even 50 tonnes on a 1000 m span, but usually 25 tonnes is the practical limit with spans up to 900 m.

### Cableway cycle time

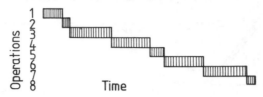

**Fig. 34.9** Cycle time for cableway

The typical activities involved in cableway operation are shown in Fig. 34.9 as follows:

(1) Fill and attach concrete skip $\quad t_1 = 40\text{–}60$ s

(2) Raise load $\qquad\qquad\qquad\quad t_2 = \dfrac{L_T}{V_R} + 10$ s

(3) Carriage travel $\qquad\qquad\quad\; t_3 = \dfrac{L_C}{V_C} + 10$ s

(4) Lower load $\qquad\qquad\qquad\; t_4 = \dfrac{L_W}{V_L} + 10$ s

(5) Empty skip $\qquad\qquad\qquad\; t_5 = 20\text{–}30$ s

(6) Raise skip $\qquad\qquad\qquad\; t_6 = \dfrac{L_W}{V_R} + 10$ s

(7) Carriage travel $\qquad\qquad\quad\; t_7 = t_3$

(8) Lower skip $\qquad\qquad\qquad\; t_8 = \dfrac{L_T}{V_L} + 10$ s

When the carriage travel and load hoisting can be performed at the same time, then the cycle time shown in Fig. 34.9 is reduced to

$$\text{cycle time} = t_1 + t_5 + \text{max. of} \quad \begin{array}{c} t_2 + t_4 + t_6 + t_8 \\ \text{or} \\ t_3 + t_7 \end{array}$$

$L_T$ = hook height above loading point;
$L_W$ = hook height above discharge point;
$L_C$ = distance between load point and discharge point.

In general, for concreting work the daily output is likely to reach only 60% of the theoretical and over a month perhaps only 50%. For other types of work, such as handling precast concrete units for bridge decks output is likely to be much lower.

# 35

# BELT CONVEYOR

The belt conveyor is proving economical for the transport of large quantities of material over long distances. In the past the belt conveyor was mainly associated with tunnelling and ore extraction in the mining industry, but in recent years this method of materials transport has been offered to the construction industry as an alternative to trucks and rails, for example in the placing of concrete for the construction of the foundations of massive industrial plants.

## 35.1  BELT CONVEYOR ASSEMBLY

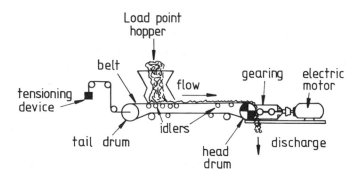

**Fig. 35.1**  Belt conveyor arrangement

The belt conveyor consists of an endless flexible, moving belt supported intermittently by idler rollers (Fig. 35.1). The belt is driven from a head drum by an electric motor, and returns around a tail drum. Slack in the belt is taken up by a tensioning device, either from the tail drum or with a vertical gravity tensioner as shown in Fig. 35.2.

The whole arrangement is supported on a light steel frame.

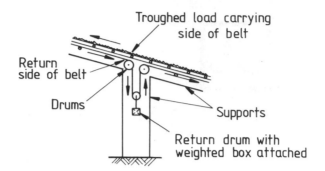

**Fig. 35.2**   Belt tensioning device

### The belt

The belt must be strong and hard wearing. The most common form consists of plies of cotton/nylon duck (sacking) vulcanised with rubber to provide strength in tension. The top and bottom surfaces of the belt are covered with a layer of rubber, PVC or similar material to protect the belt from abrasion and impact loading.

The thickness of the outer layer depends upon the type of material to be carried, as shown in Table 35.1 for a rubberised surface.

**Table 35.1**

| Material | Top surface thickness (mm) | Underside surface thickness (mm) |
|---|---|---|
| Sand | 2 | 1 |
| Aggregate | 4–5 | 2 |
| Ores, sharp stones | 5–8 | 2–3 |
| Maximum | 16 | 8 |

The belt is formed into a trough shape to increase the carrying capacity, and reduce spillage (Fig. 35.3). Before the introduction of cotton/nylon and cotton/terylene fabric belts the idlers were set at 20° to 30°, but these new belt fabrics now provide sufficient flexibility and robustness to allow angles up to 55°, thereby significantly reducing spillage.

The spacing of the idlers along the belt depends upon the weight of material to be carried together with the belt construction and troughing angle. The return idlers can be spaced more widely and act mainly to reduce belt sag.

### General data

(a) *Idling rollers* – generally the carrying idlers are set at 1.0–1.5 m centres which close to 300–400 mm near the loading point. For the return idlers, the centres are approximately 2–3.5 m. The roller diameter is typically 50–250 mm, depending upon the belt width.

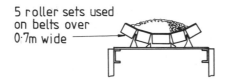

5 roller sets used
on belts over
0.7m wide

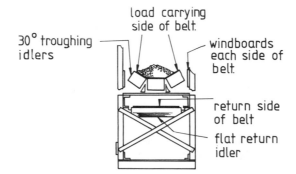

load carrying
side of belt

30° troughing
idlers

windboards
each side of
belt

return side
of belt

flat return
idler

**Fig. 35.3** Arrangements of belt troughing idlers

**Fig. 35.4** Two-motor driven conveyor belt

(b) The size of the drive motor depends upon the length and width of the belt and the density of material carried. For very long lengths and heavy material two motors or more may be required, as shown in the arrangement in Fig. 35.4.

(c) *Gradient* – most modern belts are capable of holding the material in a stable condition at an inclination of between 15° and 25°.

(d) *Drive and tail drums* – the distance between these two drums defines the operating length on the conveyor. For most uses the design is tailor made to the specific application and the length of conveyor and can range from a few metres up to 10 km or more. One of the longest systems in use is 100 km, assembled in 11 km sections.

(e) Belt widths vary from 400 to 3000 mm, with 1 m as the typical requirement.

## Output

The output from a conveyor belt is given by the equations:

$$Q_v = 3600 \times V \times A$$

or

$$Q_t = 3600 \times V \times A \times D$$

where
$Q_v$ = theoretical production in m³/h;
$Q_t$ = theoretical production in tonnes/h;
$V$ = belt speed in m/s;
$A$ = cross sectional area through material on the belt in m²;
$D$ = material density in tonnes/m³ (i.e. loose density).

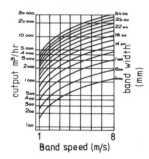

**Fig. 35.5**   Production data for a conveyor belt

The loose volume carried per metre on the belt is governed by the width of the belt, which is itself influenced by the particle size of the material being carried.

The theoretical outputs for specific band widths are shown plotted against band speed in Fig. 35.5. A bandspeed of about 1.5 m/s is typical.

For coarse material of particle size 50 mm or more the band width required is approximately as follows:

band width = 4 × particle size + 15 cm (uniform lumps)

band width = 2 × particle size + 20 cm (non-uniform lumps)

For non-uniform lumps particle size is taken to be the greatest lump dimension.

## 35.2   PORTABLE CONVEYOR

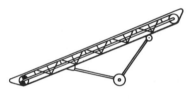

**Fig. 35.6**   Portable conveyor

For moving materials on the construction site, flexibility is required. The expensive and almost permanent conveyor system previously described is unsuitable except for tunnelling or similar work. The portable conveyor however, is a relatively inexpensive, and versatile device which can be of use on many construction sites, where the transport or disposal of bulky materials is required.

The arrangement shown in Fig. 35.6 consists of a portable frame supporting a conveyor belt, rollers and electric motor. The length of the conveyor is usually of fixed length, about 3–6 m, but models are available up to 20 m.

Applications are constantly widening, although as yet not a common item on the construction site, where the flow path requires regular repositioning.

(a) Materials movement is independent of surface shape, form and condition.
(b) The conveyor can be used on load-bearing soils, over road junctions, streams and even small rivers, where other methods would be inconvenient and more expensive.
(c) A conveyor requires little labout and maintenance.
(d) Transporting materials by conveyor is relatively independent of the weather.

However, large and sharp, fragmented material such as broken rock tends to damage the belt. Also where a long conveyor is installed a breakdown brings the whole system to a halt. Although this can be avoided where portable units are used.

### 35.3   HANDLING OF CONCRETE

Concrete conveyors are provided with narrow belts of 300–400 mm width thereby reducing the load per unit length is reduced compared with the standard conveyor. The unit is as a consequence light, easy to install and manoeuvre. High belt speeds 2–3 m/s are typical, which assists in keeping the belt clean as the concrete is thrown off on reaching the head drum, often augmented with a scraper (Fig. 35.7) or brush (Fig. 35.8).

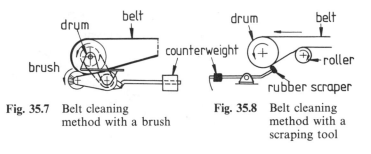

**Fig. 35.7**   Belt cleaning method with a brush

**Fig. 35.8**   Belt cleaning method with a scraping tool

#### Concrete delivery on a 400 mm wide belt

| Belt speed (m/s) | Delivery rate (m³/h) |
|---|---|
| 2 | 50 |
| 3 | 115 |

These delivery rates compare favourably with concrete pumps and only suffer the disadvantage that the gradient is limited to 15° or so. The problems of quick positioning are also being overcome, for example by means of the truck-mounted telescopic boom, whereby two or more conveyors are telescoped out.

### 35.4   CONVEYORS FOR EARTHWORKS

Usually the most economic method of moving excavated earth on the construction site is by scraper, loader shovels, trucks or lorries. However, research work has been carried out to assess the feasibility of using a conveyor system in conjunction with large draglines and bucket excavators on, for example, motorway construction (Fig. 35.9) and alternatively bringing fill over a long distance to be spread by conventional earthmoving plant (Fig. 35.10).

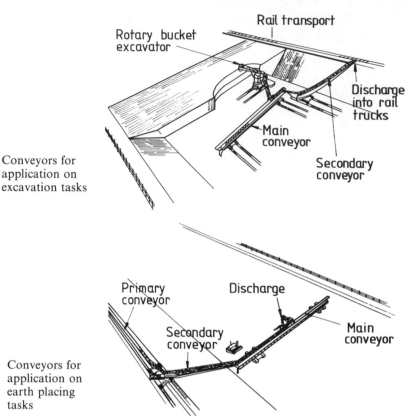

**Fig. 35.9** Conveyors for application on excavation tasks

**Fig. 35.10** Conveyors for application on earth placing tasks

# 36

# MONO-RAIL

The mono-rail, before the advent of reliable concrete pumping methods, was used extensively on sites where concrete or other materials needed transport over short distances through congested areas where trucks, etc., could not pass. Today the method is perhaps more of an alternative to a conveyor system and is probably still preferred because of the flexibility allowed in stopping the wagons wherever required.

The transporting skip consists of a self-travelling tippable car, which drives on a single track by making contact with the top of the rail. The unit is steadied by means of four idler rollers spring loaded against the flange of the rail.

The car can be started by unskilled labour and halted at a selected position by a tripping mechanism.

The track is assembled from short sections of rail mounted on tressles, which either rest on a firm base or are bolted to structural steelwork, for example in concrete placing of a bridge deck, the system is used up to about 300 m, but this can be increased where the track is looped. The cars thereby are started at different times to maintain a steady concrete delivery rate.

The car is about $\frac{1}{2}$ m$^3$ capacity and will travel up an incline of 1 in 20 at 2 m/s.

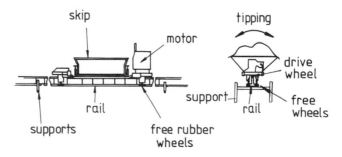

**Fig. 36.1** Mono-rail materials handling system

**Exercise**

The exercise involves the generation of alternative methods of producing, transporting and placing concrete from the selection of equipment described in previous chapters. The aim is to encourage rational plant selection by assembling a method plan and costing out the various alternatives.

PROBLEM

Your company has recently obtained a sub-contract for the construction of the concrete floor for a process factory as shown in Fig. 36.2. As the planning engineer you are required to provide a method plan and statement together with a cost estimate for the production, transporting and placing of the concrete.

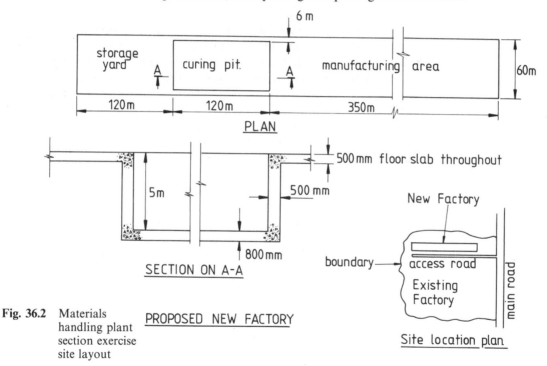

**Fig. 36.2** Materials handling plant section exercise site layout

# 37

# CONCRETE PUMP

Concrete pumps were first introduced into the construction industry in the 1930s, but until the development of hydraulic pumps in the 1960s the mechanical and maintenance problems were responsible for the general lack of interest. Today the concrete pump is accepted as a much easier, quicker and cheaper method of placing concrete compared with many of the other methods of handling.

## 37.1 MECHANICAL CONCRETE PUMPS

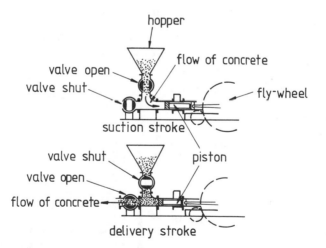

**Fig. 37.1** Mechanical concrete pump

The early pumps were fixed position, which had flexibility only at the discharge end of the pipe. Typically these were 150–200 mm diameter and consequently were too heavy and cumbersome for construction site use.

The pump (Fig. 37.1) consists of a feeding hopper, inlet and outlet valves and piston, all driven by a fly-wheel. The method causes high wear and tear on the valves and a jerky flow of concrete. In addition the concrete mix must be rich and fatty, if blockages are to be avoided.

When the pump is operating satisfactorily the following outputs are possible.

| Output | m³/h | 12 | 20 | 40 |
|---|---|---|---|---|
| Number of pistons | | 1 | 1 | 2 |
| Pipe diameter | mm | 125–150 | 150–180 | 150–180 |
| Cylinder diameter | mm | 160 | 195 | 195 |
| Power | kW | 20 | 35 | 45 |
| Horizontal distance | m | 300 | 300 | 300 |
| Vertical height | m | 40 | 40 | 40 |

## 37.2   HYDRAULIC CONCRETE PUMPS

The introduction of the hydraulic concrete pump has had the greatest bearing on the acceptance of concrete pumping methods.

The pumps may be either truck-mounted or towed as a separate unit. Oil is the working fluid, used to operate either single (Fig. 37.2) or more commonly dual pumping cylinders.

Concrete is drawn from the hopper into the cylinder on the suction stroke, and discharged along the pipe on the pressure stroke. A diesel engine powers the hydraulic pump but occasionally an electric motor may be used.

This method of pumping provides several advantages over the earlier mechanical pumps as follows:

(a) The pump is designed with a long piston stroke, requiring fewer strokes with a consequent reduction in static friction encountered by the concrete.

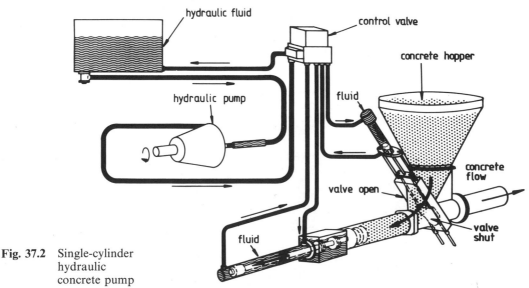

**Fig. 37.2**   Single-cylinder hydraulic concrete pump

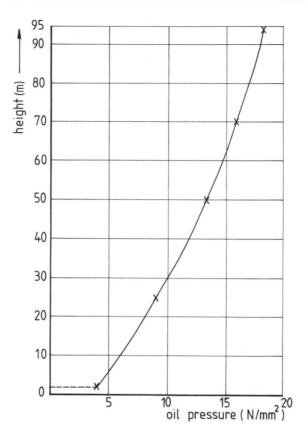

**Fig. 37.3** Relationship between oil pressure and concrete pumping height

(b) The pressure exerted on the concrete may be varied to control vertical pumping (Fig. 37.3).

(c) The piston is accelerated at the start of the delivery stroke and then maintained at a steady speed, so producing a smooth and even discharge.

The new dual cylinder models have the ability to deliver concrete at up to 120 m³/h with a pipeline diameter of 150 mm. Such volumes, however, would cause handling difficulties on most sizes of concrete pour and the usual pipe bore is 100 mm diameter, giving a corresponding output of up to 35–45 m³/h. Some models include speed regulation by varying the oil flow and pressure, allowing output to be adjusted down to 20–25 m³/h or less. Hydraulic pumps have been used to deliver concrete, through over 600 m of pipeline to heights exceeding 300 m or so. Bends in the pipeline reduce the effective pumping distance by about 10 m for each 90° bend, 6 m for 45° and 3 m for $22\frac{1}{2}°$.

**Pump operation**

(a) Before pumping is started a grout mixture should be flushed through the pump and pipeline, to provide lubrication. The best

**Fig. 37.4**  Rubber plug for cleaning pump delivery pipe

results are achieved by initially inserting a plug into the pipeline, which is then forced along with the grout.

(b) The concrete in the hopper should be agitated, to keep the concrete moving and the concrete level must be constantly topped up to prevent air pockets forming in the pipeline.

(c) A compressed air supply nearby is useful for clearing obstructions in the pipeline.

(d) Pumping tends to be more efficient with concrete having a slump of 50–75 mm, with a high percentage of fines. The design of pumpable mixes is fairly specialised, however, and the reader is advised to seek further information, e.g. from the Cement and Concrete Association.

(e) To clean out the pipeline a section of pipe is removed and the hopper is pumped clean. A plug is inserted into the remaining pipe and forced along the line by compressed air or water pressure (Fig. 37.4).

## 37.3    PUMP MOUNTING

### Truck-mounted pump

The demand for the occasional concrete pour such as in high-rise construction, bridge decks, etc., coupled with the advantages of ready-mixed concrete, has led to the development of truck-mounted concrete pumps for short-term hire, e.g. the duration of the concrete pour. Many sizes of mobile, truck-mounted hydraulic concrete pumps are now available, supporting booms with a reach of up to 50 m (Fig. 37.5), depending upon the restrictions placed on axle loads for travelling on public highways.

### Trailer pump

As the size of buildings has increased, many structures require access beyond the reach of truck-mounted boom pumps. The trend in such situations is to use a trailer concrete pump in conjunction with a portable hydraulic boom, fixed either to a mast (Fig. 37.7a) or to the structure (Fig. 37.7b).

## 37.4    SQUEEZE PUMP

A flexible rubber pipe secured half circle in an air-tight drum is squeezed by two diametrically opposed rollers successively flattening and squeezing the concrete towards the output. The operation takes place in a vacuum at 0.8–0.9 bar and thus the pipe regains its shape allowing more concrete to be sucked from the mixing chamber.

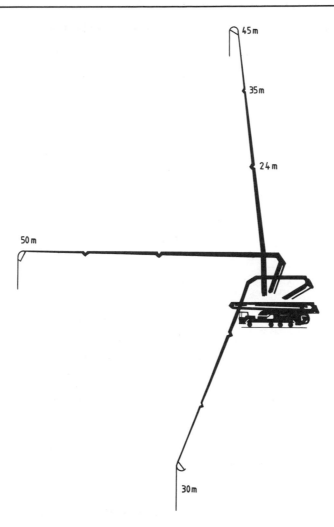

**Fig. 37.5**  Truck-mounted
concrete pump

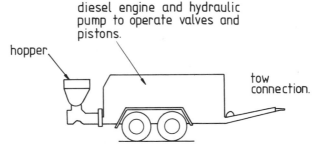

**Fig. 37.6**  Trailer-mounted
concrete pump

The unit may be lorry-mounted, with the hydraulic pump to drive the rotor coupled to the truck's diesel engine or alternatively the pump may be towed and powered with its own diesel engine.

The system requires little preparation time and provides a continuous pumping action. However, only very workable concrete can be pumped. The pipe is limited to 75–100 mm diameter.

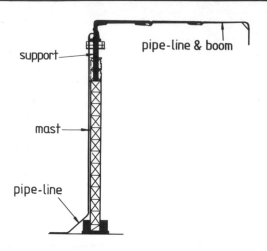

**Fig. 37.7(a)** Portable mast and boom

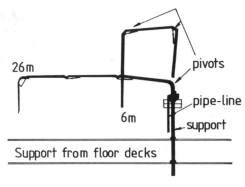

**Fig. 37.7(b)** Portable hydraulic boom

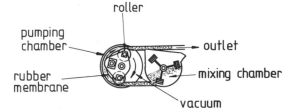

**Fig. 37.8** Squeeze pump

### Data

| | |
|---|---|
| Pipe diameter | 75 mm |
| Max. output | 28 m³/h |
| Pumping length, or | 130 m |
| Pumping height | 40 m |
| Power | 80(60) hp(kW) |
| Self-weight | 6000 kg |
| Travelling speed | 90 km/h |

## 37.5 PNEUMATIC PUMP

The apparatus consists of a compressor, compressed-air cylinder and pneumatic concrete pump. Concrete is charged into the pressure vessel of capacity 250–1000 litre. The vessel is then sealed and compressed at 2.5–3 bar from the storage cylinder. The concrete is then in effect blown along the delivery pipe into a wet hopper.

Delivery pipes of 150–180 mm are typical and bends must be securely anchored to cope with the fast moving concrete (0.5–1.0 m/s). Approximately twenty to forty charges per hour are possible at the maximum delivery distance (about 300 m along the horizontal).

The compressor must be capable of supplying approximately fifteen times the volume of concrete to be pumped, e.g. 1 m³/h of concrete = 15 m³/h of compressed-air.

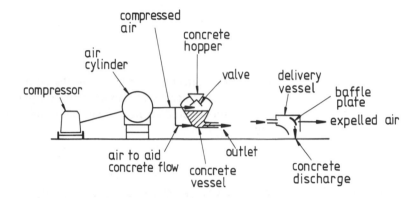

**Fig. 37.9**  Pneumatic concrete pump

## 37.6  CEMENT PUMP

Many modern concrete batching plants transfer dry cement from a storage silo into the mixing hopper.

The pump to execute the transfer is shown in Fig. 37.10. The cement is drawn from the hopper on to a screw feed which is rotated at approximately 1000 cycles per minute and fed into a chamber. The cement particles are blown along the delivery pipe by compressed air and may be transported horizontally up to 1500 m and vertically to heights of 30 m. The delivery pipe diameter range is 30–200 mm and correspondingly requires 0.5–3 kW for each tonne/h of delivered cement.

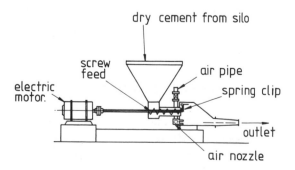

**Fig. 37.10**  Powder cement pump

# CONCRETE PRODUCTION AND EQUIPMENT

This section concentrates mainly on the equipment needed for mixing and batching concrete, transportation and placing methods having largely been covered earlier under materials handling. The chapters include:

## SAND AND AGGREGATE PRODUCTION

Many types of construction involve the production of concrete in regions distant from readily purchasable sources of materials. Construction engineers and managers are then faced with the task of setting up local quarrying and crushing facilities. Crushing equipment, screening and washing processes are described together with a worked example illustrating the calculations and decisions required in producing particular sized materials from crushed rock.

## CONCRETE EQUIPMENT

An introduction to concrete materials and technology is provided focusing on handling and batching concrete, concrete mixers, central mixing plants and production output information.

## PRESTRESSED CONCRETE

Both the longline and post-tensioned methods of manufacture and assembly are described together with guidelines provided on quality control and selection of materials.

# AGGREGATE PRODUCTION

Aggregate is used for hardcore, asphalts and macadams, concrete, mortar, general fill, etc. The sizes of the constituent parts of the aggregate depend upon the grading required, for example 150 mm down to 3 mm can be obtained by crushing rock.

The rock itself is usually obtained by blasting operations and transported to the crushing equipment. Modern crushers are capable of handling large blocks of over 1 m maximum dimension, with the choice of suitable machinery depending upon the costs of reducing the size of the fragments at the quarry and the method of transport to the plant. Following crushing, washing and sorting into various sizes is commonly carried out.

Sand, unlike crushed aggregate, can be produced artificially, but is normally obtained from natural deposits which may be subsequently mixed to give the required gradings.

Specifications for aggregate types and gradings are beyond the scope of this work, as these depend upon the use. For help in selecting materials, codes of practice are available for concrete, road works, fill material, etc.

## 38.1 EQUIPMENT FOR AGGREGATE PRODUCTION

The capacity for the crushing plant is selected to produce a given rate of tonnage output. The equipment required is as follows:

1. Feeders and hoppers.
2. Primary crusher.
3. Secondary or reduction crusher.
4. Conveyors to transport the material between processes.
5. Screens and washing plant for separating, grading and directing the sizes to the bins.
6. Storage bins and discharge hoppers.

## 38.2   CRUSHING PLANT

The crusher takes the raw material and breaks it into small sizes. To do this the machine is set to the maximum size required. However, the rocks will fragment into a range of sizes down to dust, which can be fed away for further reduction or screening and sorting.

The machine is usually described in terms of its crushing ratio, i.e. the size of the feed opening to the setting for crushing (Fig. 38.1). The range of crushing ratios for various types of machine are shown in Table 38.1.

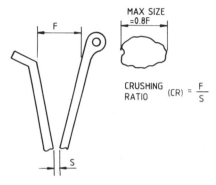

CRUSHING
RATIO   (CR) = $\dfrac{F}{S}$

**Fig. 38.1**   Jaw crusher

**Table 38.1**   Crushing ratios

| Crusher | Smaller units | Larger units |
|---|---|---|
| Jaw | 4–8 | 3–10 |
| Gyratory | 5–8 | 4–6 |
| Impact | 5–25 | 5–50 |
| Cone | 3–14 | 4–14 |
| Twin-roller | 1–4 | 2–8 |

It can be seen from Fig. 38.2 that there are several different types of crushing machine, for handling different size feed and material. The output from each varies between types for a given crushing ratio and

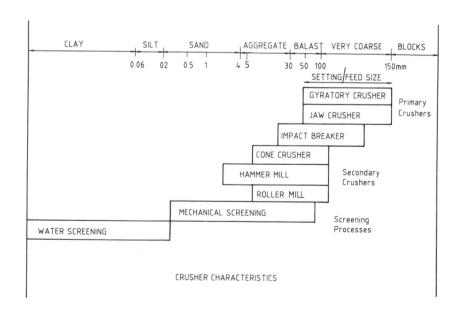

**Fig. 38.2**   Crushing machine characteristics

therefore careful costing is necessary when several stages of reduction are required. For primary crushing, i.e. breaking material down from large pieces, either a jaw or gyratory type is preferred while for secondary or further reduction, the cone, hammermill/or roll crusher is usually selected.

## 38.3  PRIMARY CRUSHERS

### Jaw crusher

The jaw crusher is a simple but effective method, having a rectangular shape on plan and triangular from the side elevation. Three walls are fixed, while the fourth is a moving plate, so forming a jaw-type structure. Material is fed in at the top and passes out through the base. There are three types of jaw crusher, namely:

(i)  Double toggle.
(ii)  Single toggle.
(iii)  Double jaw.

The crusher size is designated by the dimensions of the top opening.

### DOUBLE TOGGLE JAW CRUSHER

The machine was first developed by Blake in England and introduced in the 1850s. It is robust, durable, and works on the following simple principle. One side of the jaw (a) is held stationary, while the other (b) is attached to pivots at the top and bottom (c and d). The top pivot (c) simply acts as a suspension point, crushing motion being obtained by a rotating eccentric shaft raising and lowering the rod (e) connected by two toggles to the jaw (b) at (d) and a reaction point (h). The design principle is very effective with the crushing force increasing towards the top of the jaw. Also the opening and closing movement of the jaws allows small fragments to fall quickly through the bottom and not restrict the flow.

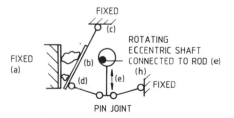

**Fig. 38.3**  Double toggle jaw crusher

### SINGLE TOGGLE JAW CRUSHER

This type of jaw crusher has only a single toggle at the bottom of the moving jaw connected to a reaction point. Motion is produced with an

**Fig. 38.4** Single toggle jaw crusher

eccentrically rotating shaft, located at the top joint, so giving both vertical and horizontal movement, consequent greater attrition results in more fines than the Blake-type machine.

## DOUBLE JAW CRUSHER

A few machines have been made with both jaws able to move but have not proved popular.

## DATA FOR JAW CRUSHERS

**Table 38.2** Dimensions, settings, capacities and power for portable jaw crushers

| Item | Feed opening (mm) | | | | | |
|---|---|---|---|---|---|---|
| | 1.1 × 1.2 | 0.9 × 1.15 | 0.75 × 1.05 | 0.65 × 1.00 | 0.5 × 0.9 | 0.45 × 0.8 |
| Closed settings (mm) | 125–300 | 100–250 | 120–200 | 75–200 | 50–180 | 30–180 |
| Capacity (t/h) | 275–750 | 180–600 | 125–575 | 100–400 | 40–250 | 40–130 |
| Max. power (kW) | 230 | 160 | 150 | 120 | 75 | 45 |
| Weight (tonnes) | 75 | 70 | 55 | 55 | 30 | 25 |
| Max. output/max. power ratio | 3.2 | 3.75 | 3.8 | 3.3 | 3.3 | 2.9 |

## GRADING CURVE FOR CRUSHED MATERIAL

The crushed material will not be of uniform size, but graded. The proportions of each size depend upon the setting, type of raw material and crusher. Typical values for single and double toggle machines are shown in Fig. 38.5 for hard non-brittle rock. The maximum size of material fed in should not exceed about 0.8 × the feed opening (Fig. 38.1).

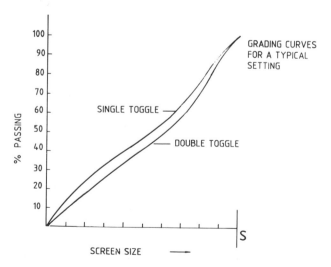

**Fig. 38.5** Grading curve obtained with a jaw crusher

## REDUCTION RATIO

It can be seen from Table 38.2 that for the jaw crusher the reduction ratio can be varied in the range 3–10 on a typical machine.

### Gyratory crusher

The crusher comprises a hardened steel head mounted vertically on a shaft, all contained within a cast iron or steel housing, lined with hardened steel plates (Fig. 38.6).

The head has a long conical shape, with a through shaft suspended in a bearing at the top, and an eccentric base connection to gearing. Thus as the cone is rotated the gap between itself and the walling changes from a maximum to minimum each cycle. The rock is fed into the chamber at the top and as it moves downward, undergoes crushing, finally emerging through the bottom gap.

The crusher size is designated in terms of the top diameter of the housing $D$ and the distance between the housing core at the same level, $d$. The setting, $S$, can be stated either for the maximum (open) or minimum (closed) gap at the base, but it is more usual to quote the larger value. The setting may be altered by raising or lowering the core or by introducing thicker wall plates.

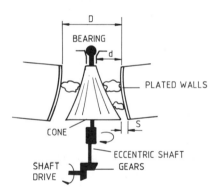

**Fig. 38.6** Gyratory crusher

## DATA FOR GYRATORY CRUSHERS

**Table 38.3** Dimensions, settings, capacities and power of gyratory crushers

| Item | Feed opening (m) (d × D) | | | | | |
|---|---|---|---|---|---|---|
| | **1.4 × 4.0** | **1.0 × 3.6** | **0.75 × 2.5** | **0.5 × 2.00** | **0.4 × 1.5** | **0.33 × 1.1** |
| Open settings (mm) | 150–180 | 125–165 | 100–125 | 85–115 | 75–100 | 50–75 |
| Capacity (t/h) | 650–800 | 500–650 | 300–400 | 200–250 | 150–200 | 80–150 |
| Max. power (kW) | 225 | 210 | 130 | 95 | 75 | 50 |
| Weight (tonnes) | 45 | 45 | 40 | 35 | 35 | 30 |
| Max. output/max. power ratio | 3.55 | 3.1 | 3.1 | 2.6 | 2.6 | 3 |

### GRADING CURVE OF CRUSHED MATERIAL

Like the jaw crusher, the maximum size of material fed in should not exceed about 0.8 × the feed opening. The final grading curve is similar to that obtained from the jaw crusher. Figure 38.7 illustrates typical gradings for brittle and tough rocks, respectively.

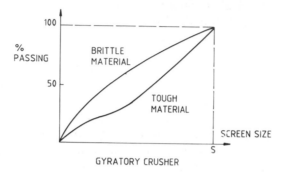

**Fig. 38.7**  Grading curve obtained with a gyratory crusher

### REDUCTION RATIO

It can be seen from Table 38.3 that for the gyratory crusher the reduction ratio ($d/S$) can be varied on a given machine, typically 4–6 for large crushers and 5–8 for smaller units.

## 38.4   CRUSHERS SUITABLE FOR PRIMARY AND REDUCTION CRUSHING

### Hammermill

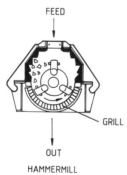

**Fig. 38.8**  Hammermill

The hammermill comprises hammers linked to a rotor or several rotors for a large unit, all contained in a strong housing (Fig. 38.8). The hammers are rotated at high velocity, striking the lumps of material as they enter the chamber and breaking them up into smaller pieces. The base of the chamber is lined with a grating plate allowing grinding to take place and as a consequence more fines are produced compared with most other methods. The hammer size is designated similar to the jaw crusher in terms of the dimensions of the feed opening. The setting may be varied by adjusting the grate bars. Some variation in the production capacity can be obtained by altering the hammer velocity, e.g. 35–50 m/s circumferential velocity.

An alternative type of hammermill is the impact breaker (Fig. 38.9). The material struck by the hammer is thrown against breaker plates, bars or grates situated above the rotor and so broken into smaller pieces. There is no bottom grating and therefore the amount of splintering is largely dependent upon the impact velocity, typically 15–30 m/s circumferential velocity. The size of the feed material should be much smaller than the feed opening and introduced into the

FEED

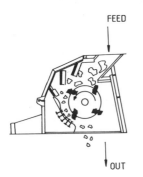

OUT

**Fig. 38.9**  Impact breaker

chamber at a rate which will not cause clogging, and therefore initial screening is necessary to remove oversize pieces. The method is suited to soft rocks that are also not too abrasive, otherwise wear and tear will be excessive. Generally the impact breaker is used for primary crushing while the hammermill is preferred for reducing crushing duties. Typical data for a hammermill are shown in Table 38.4.

### Reduction crushers

Material which has passed through primary crushing will generally have fragments distributed down from at least 50 mm in size even when small machines are used. Thus, for concrete, aggregate, etc., further reduction in size is required. Commonly material larger than a specified size is retained by screens after passing through the primary crusher, and subsequently reduced by further crushing in another machine. The crushers available for this purpose are cone crusher, roll crusher, rod mill and ball mill.

### CONE/GYROSPHERE

The cone/gyrosphere crusher operates on a similar principle to the gyratory crusher described earlier, and is designed to produce large quantities of fine aggregate. The crushing head is much broader but shallower than the gyratory crusher and is connected at the base through eccentric bushing and gearing to the drive and rotated at a

**Fig. 38.10**  Cone crusher

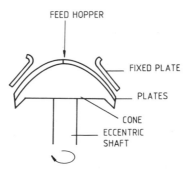

FEED HOPPER

FIXED PLATE

PLATES

CONE

ECCENTRIC SHAFT

high rate of revolutions. The machine is designated in terms of the maximum diameter of the crushing head (cone), but it is also customary to quote the feed opening dimension. The cone crusher has straight sides to the head while the gyrosphere tends to be more curved. The setting may be adjusted by altering the relative position of the housing rings and head plates.

**Table 38.4** Hammerill/impact breaker data

| Item | Feed opening (mm) | | | | | | | | | | |
| | 225 × 325 | 275 × 535 | 275 × 830 | 300 × 850 | 300 × 1050 | 445 × 1100 | 445 × 1360 | 445 × 1625 | 445 × 1890 |
|---|---|---|---|---|---|---|---|---|---|
| Maxstone (mm) | 250 | 75 | 100 | 100 | 150 | 150 | 250 | 250 | 250 |
| *Setting (mm) | 2–65 | 2–65 | 2–65 | 2–65 | 2–65 | 2–65 | 2–65 | 2–65 | 2–65 |
| Capacity (t/h) | 5–35 | 12–75 | 19–120 | 25–145 | 32–190 | 41–250 | 51–320 | 60–300 | 20–440 |
| Max. power (kW) | 45 | 95 | 152 | 190 | 228 | 266 | 380 | 456 | 532 |

* Hammermill only.

## DATA FOR THE CONE CRUSHER

**Table 38.5**  Dimensions, settings, capacities and power of cone crushers

| Item | Max. cone dia. (mm) | | | |
|---|---|---|---|---|
| | **600** | **900** | **1200** | **1650** |
| Feed opening (openside) (mm) | 90 | 130 | 190 | 280 |
| Minimum discharge setting(s) (mm) | 6–25 | 10–40 | 13–50 | 20–65 |
| Capacity (t/h) | 13–50 | 28–105 | 63–210 | 160–470 |
| Max. power (kW) | 20 | 50 | 75 | 185 |
| Max. output/max. power ratio | 2.5 | 2.1 | 2.8 | 2.5 |

*Grading curve*
Figure 38.11 illustrates a typical grading curve for the cone/gyrosphere crusher, and has a similar shape to that obtained for the gyratory machine used for primary crushing.

*Reduction ratio*
Table 38.5 indicates that for cone/gyrosphere crushers the reduction ratio can be varied on a given machine between 4–14 approximately.

## ROLL MILL

Roll mills are used for similar duties to the cone crusher and produce fine aggregate or even coarse sand (particle size less than 3 mm). The machine comprises two counter rotating rollers, and material is fed through the gap between the two for crushing. Usually one roller has a fixed axis while the other can be adjusted to give the required setting,

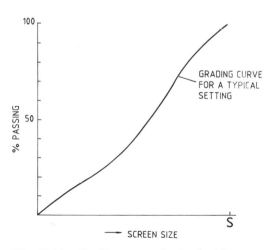

**Fig. 38.11**  Grading curve obtained with a cone crusher

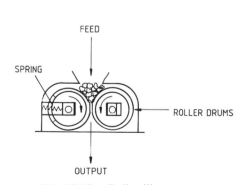

**Fig. 38.12**  Roll mill

it is also spring loaded to prevent damage when oversize non-crushable material is encountered. Usually the roller surface is smooth, but corrugations on one or both are available depending upon the material for crushing. The maximum size of material that can be put through the system is proportional to the roller diameter. If pieces are too large they cannot be nipped by the rollers and would simply fall out. The maximum size stone, $B$, for input can be approximately determined from the formula: $B = 0.085 \times$ roller diameter + setting (all in inches).

The crusher is usually designated in terms of the setting, the diameter and length of the rollers.

*Grading curve*
The distribution of particle sizes is similar to that obtained from jaw or cone crushers. Coarse sand can be obtained from the lowest setting.

*Reduction ratio*
Table 38.6 indicates that the reduction ratio for a typical roll mill can be varied from about 1–4 for small crushers and 2–8 for larger versions. A comparison of roll and cone crushers (Table 38.6) shows that for a given output/power ratio the cone crusher accepts larger size

**Table 38.6**  Dimensions, settings, capacities and power for roll mills

| Item | Roller diameter × roller length (mm) | | | | | | |
|---|---|---|---|---|---|---|---|
| | 400 × 400 | 600 × 400 | 750 × 450 | 750 × 500 | 1000 × 500 | 1000 × 600 | 1400 × 600 |
| Setting (mm) | 5–60 | 5–60 | 5–60 | 5–60 | 5–60 | 5–60 | 5–60 |
| Capacity (t/h) | 15–140 | 15–140 | 15–160 | 20–200 | 20–200 | 20–200 | 20–240 |
| Max. power (kW) | 24 | 25 | 55 | 75 | 75 | 75 | 115 |
| Roller speed (rev/min) | 120 | 80 | 60 | 60 | 50 | 50 | 40 |
| Max. output/max. power ratio | 5.5 | 5.5 | 2.9 | 2.7 | 2.7 | 2.7 | 2.1 |
| Max. feed size (mm) | 20–75 | 30–85 | 35–90 | 35–90 | 45–100 | 45–100 | 65–120 |

feed material and has a greater crushing ratio. Therefore fewer crushing stages would be necessary for the same size of input material.

## BALL MILL AND ROD MILL

These units are mainly used to produce uniform sand from 15 mm size aggregate (approximately). The rod mill consists of a rotating drum containing 25–50 mm diameter rods occupying about 30% of the total volume. They are slightly shorter than the length of the cylinder. The aggregate is fed in at one end of the drum, broken down into small grains as it is rotated and crushed by the rods, finally emerging at the other end through a sieve. The feed material can be introduced wet or dry.

An alternative arrangement is to use steel balls instead of the rods (Fig. 38.13).

Output:  1–200 m³/h
Power:   2–3 kW/m³ per h

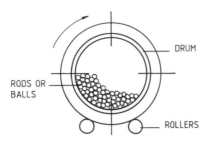

**Fig. 38.13**  Ball/rod mill

## 38.5  SCREENING OF AGGREGATES

The process of sorting aggregate into different sizes is illustrated in the example shown in Fig. 38.14. It can be seen that before each stage of crushing, material smaller than the crusher setting is allowed to pass

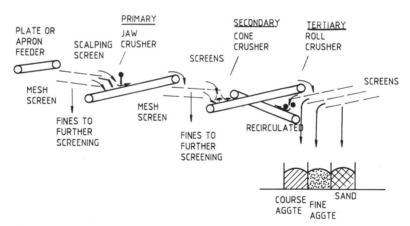

**Fig. 38.14**  System for crushing and screening

through one or more screens. In this way the capacity of the crushing units can be kept smaller than the actual quantity of material passing through the whole system. Commonly double or triple deck screens are used except at the primary stage where grizzle bars or scalping screens are needed to cope with large lumps of rock. Conveyors transport the crushed material between stages or to stockpiles.

### Apron feeders and grizzles

Material for the primary stage of crushing is usually delivered in very large lumps depending upon the quarry and mode of transport to site.

Thus robust equipment is needed to receive the dumped rocks and feed them at a steady rate into the crusher (Fig. 38.15). Two types of feed are commonly used:

(i) Vibrating plate feeder.
(ii) Apron feeder.

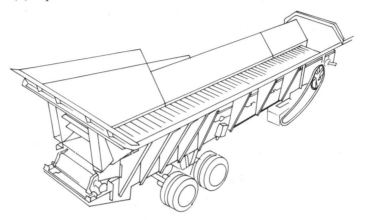

**Fig. 38.15**   Mobile feeder unit

## VIBRATING PLATE

The feeder pan comprises a heavy steel plate, side liners and hopper all mounted on heavy duty coil springs. The vibrating unit consists of two parallel counterweighted offset shafts geared together (Fig. 38.16b). Vibrating motion is produced by the interaction of the forces induced by the rotating shafts. In positions (2) and (4) all forces are acting against each other while for (1) and (3) the forces are in the same direction. Thus is these conditions the motion is up or down, and sideways, with one revolution producing a straight line reciprocating motion at 45° angle to the plane of the plate. Material is therefore alternatively lifted and conveyed forward.

This type of system is commonly selected for material obtained from a gravel pit.

FEED END

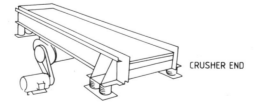

CRUSHER END

**Fig. 38.16(a)**   Vibrating plate feeder

## APRON FEED

The apron feed is necessary when quarried material is involved. The equipment is constructed to withstand heavily dumped loads and

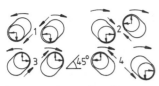

**Fig. 38.16(b)** Vibrating plate mechanism

consists of a series of steel pans forming a continuous chain supported by steel idler rollers. The 'belt' is drawn by a head roller linked through worm reduction gearing to an electric or diesel motor. A variable rate of feed is available on most models through gear changes, but in any case at least 25% more capacity than that of the crushing unit is desirable.

Apron feeders are manufactured for a range of duties commonly described as standard duty, heavy duty and super heavy duty

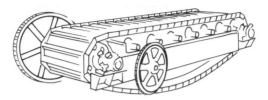

**Fig. 38.17** Apron feeder

## GRIZZLES (SCALPING UNIT)

A grizzle is usually placed at the end of the apron feed and used to screen out material smaller than the crusher setting or if constructed as a double tier unit the top deck may be arranged to direct oversize material away from the crusher. The equipment is generally referred to as a 'scalping unit' and comprises a spring-mounted frame with heavy duty parallel bars forming the top deck and a tough perforated plate for the lower deck. The unit usually has a vibrating mechanism and is placed at a slight angle downwards towards the crusher. The material simply passes along the bars and is discharged at the end of the run into the crusher while the lower deck feeds directly on to other crushing stages (Fig. 38.14).

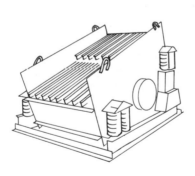

SINGLE STEP
VIBRATING GRIZZLY
FEEDER

**Fig. 38.18** Grizzle/scalping units

**Table 38.7** Vibrating plate feeders/grizzles

| Item | Size (length × breadth) (m) | | | | |
| --- | --- | --- | --- | --- | --- |
| | 3.5 to 4.8 × 0.9 | 3.5 to 4.8 × 1.0 | 3.5 to 6.0 × 1.2 | 4.8 to 6.7 × 1.5 | 4.8 to 6.1 × 1.8 |
| Capacity (t/h) | 90–270 | 155–335 | 225–400 | 270–450 | 350–600 |
| Power (kW) | 10 | 15 | 20 | 25 | 35 |
| Vibration (rev/min) | 1400 | 1400 | 1400 | 1400 | 1400 |
| Weight (kg) | 3000 | 3500 | 4500 | 6000 | 8000 |

**Table 38.8**   Belt feeders

| Item | Belt width (m) | | | | | |
|---|---|---|---|---|---|---|
| | 0.6 | 0.75 | 0.9 | 1.05 | 1.2 | 1.35 |
| Belt speed (mm/s) | 50–150 | 50–150 | 50–150 | 50–150 | 50–150 | 50–150 |
| Capacity (t/h) | 50–150 | 75–225 | 110–325 | 150–450 | 200–580 | 275–725 |
| Power (kW)* | 2–5 | 2–5 | 2–5 | 4–7.5 | 4–10 | 4–10 |
| Weight (kg) | 2500 | 2800 | 3500 | 4100 | 4900 | 5600 |

* (i) Power depends upon length of belt (2.5 m min. to 6 m). (ii) A separate grizzle is needed, single or double deck. Dimensions 1.5 × 0.9 to 3.0 × 1.5 m are available. Power 4–15 kW.

## 38.6   SCREENING AND SCREENS

Once the material has passed through the primary crusher it, together with the bypassed product, is either temporarily stockpiled or directly transported to the next crushing phase. Before this takes place, however, particles smaller than the setting are screened out. Further separation into single sizes can also take place at this stage by using double or triple deck screens. These are commonly of the vibrating type or occasionally a revolving drum.

### Inclined vibrating screens

The screen is generally made from interwoven wires to form a mesh with regular size openings through which smaller particles fall leaving the large fragments behind to be subsequently transferred to storage bins. Screens may be arranged in tiers with each layer or deck having progressively smaller openings.

The whole unit is set on an incline with the material rolling off each deck being directed to separate storage bins. The screening motion is provided by a vibratory mechanism (Fig. 38.20) in the form of a rotating heavy duty eccentric shaft.

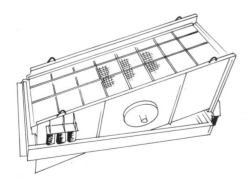

**Fig. 38.19**   Inclined vibrating screen

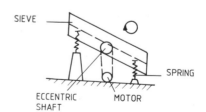

**Fig. 38.20**   Vibrating screen mechanism

### Horizontal vibrating screen

These are similar to the above, but the decks are mounted horizontally and therefore afford considerable savings in head room. The vibratory mechanism is obtained by means of two counter rotating eccentric shafts, as described for the griddle. Thus the material is moved up and down and along the decks.

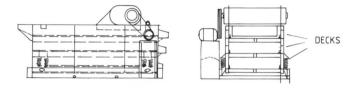

**Fig. 38.21** Horizontal vibrating screen

### Revolving drum screen

The drum rotates on a sloping axis, with material entering at the upper end and discharging progressively along the drum through perforations in the sides. The size of the perforations increase in stages along the drum length with the smaller material screened first into an outer drum as shown. The inclusion of a solid section extends the sieving time for the coarse material. Oversize material is simply allowed to escape at the lower end.

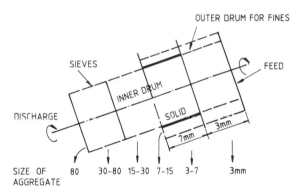

**Fig. 38.22** Revolving drum screen

### Capacity of screens

Manufacturers usually state the quantity of material that can be passed through a single screen deck based on a standard aggregate, moisture content, etc. In practice, however, such conditions are unlikely to occur and therefore adjustments to these volumes are needed if a correct balance between crushing units, transporting systems and screening is to be maintained. This is best done by trial and error gained from experience with similar aggregates and screening units but approximate adjustment factors can be applied for shape and size of the aggregate, number of decks in the unit, moisture content, spraying, etc.

## DECK CORRECTION FACTOR

Not all the material which is of a size capable of passing through the screen openings actually does so and some will be retained, thus reducing the capacity (i.e. efficiency) of the screens. While this is normally assumed and included in manufacturer's output data for a single deck, when more decks are added the reduced efficiency of the system should be taken into account. Typical adjustment factors are:

| No of decks | Factor ($x$) |
|---|---|
| 1 | 1.00 |
| 2 | 0.90 |
| 3 | 0.75 |
| 4 | 0.60 |

These values, however, are only a guide as the type and shape of aggregate will have an additional influence.

## SIZE OF AGGREGATE FACTORS

(i) Undersize aggregate: manufacturers usually base the screen capacity on the assumption that 45% of the aggregate falling on to the deck will be half the size of the openings in the deck. If this proportion increases then the screen will be able to handle more material and vice versa as shown below:

| % Aggregate coming on to screen less than half screen opening size | Factor ($y$) |
|---|---|
| 10 | 0.5 |
| 20 | 0.7 |
| 30 | 0.8 |
| 45 | 1.0 |
| 50 | 1.2 |
| 60 | 1.5 |
| 70 | 1.8 |
| 80 | 2.3 |
| 90 | 3.0 |

Again these factors should be tempered by other variables such as shape and type of aggregate.

(ii) Oversize aggregate factor: manufacturers also base their capacity figures on the assumption of 25–30% of the material coming onto a screen remaining there otherwise a small adjustment is required.

| % Aggregate coming on to screen greater than screen size | Factor ($z$) |
|---|---|
| Less than 30 | 1.0 |
| 40 | 0.95 |
| 50 | 0.90 |

| | |
|---|---|
| 60 | 0.85 |
| 70 | 0.8 |
| 80 | 0.75 |
| 90 | 0.7 |

### SHAPE OF AGGREGATE, SPRAYING, ETC.

Moist aggregate with fines reduces the efficiency compared to the dry condition, on the other hand where water is continuously sprayed on to the aggregate as it is screened the rate of throughput can be increased. This technique is particularly suitable for medium fine mesh size, but is less effective for very fine and coarse material. When water is at a premium then the aggregate is best dried. This can be achieved with heated screens.

| Screen opening size (mm) | Spraying factor (D) compared to dry conditions of screening |
|---|---|
| 1 | 1.25 |
| 2 | 1.50 |
| 3 | 1.75 |
| 4 | 1.8 |
| 5 | 1.9 |
| 6 | 2.0 |
| 8 | 1.9 |
| 10 | 1.75 |
| 12 | 1.5 |
| 18 | 1.0 |

The capacity of the screen can be described as

$$W^1 = Q \times A \times X \times Y \times Z \times D$$

where

| | |
|---|---|
| $W^1$ | = volume per hour, e.g. m$^3$ per hour; |
| $Q$ | = manufacturer's capacity rating, e.g. m$^3$ per hour per m$^2$; |
| $A$ | = surface area of the screen, e.g. m$^2$; |
| $X, Y, Z, D$ | = adjustment factors, e.g. units. |

Typical manufacturer's capacity values are shown in Table 38.9.

### WASHING

Generally crushed aggregate can be adequately washed during the screening phase by applying water from sprays placed over the screen deck (Fig. 38.23). Usually 50–150 litres per m$^3$ of aggregate at 2–3 bar pressure is sufficient to wash away any clay or similar material adhering to the crushed fragments. Aggregate or sands (especially when obtained in the natural state from a pit) having more than about 3% by weight of clay or other contaminating material should be passed through separate washing equipment before screening. The choice of

**Table 38.9** Capacity of Vibrating Screens
Capacity in tons per hour passing through 1 ft$^2$ (0.093 m$^2$) of screen

| | Size of clear square opening (in) (1 in = 25.4 mm) | | | | | | | | | | | | | | | | | | | | | | | |
|---|---|---|---|---|---|---|---|---|---|---|---|---|---|---|---|---|---|---|---|---|---|---|---|---|
| | 0.0117 | 0.0165 | 0.0234 | 0.0331 | 0.0469 | 0.0787 | 0.093 | 0.125 | 0.131 | 0.185 | $\frac{1}{4}$ | $\frac{3}{8}$ | $\frac{1}{2}$ | $\frac{5}{8}$ | $\frac{3}{4}$ | $\frac{7}{8}$ | 1 | $1\frac{1}{4}$ | $1\frac{1}{2}$ | 2 | $2\frac{1}{2}$ | 3 | 4 | 5 |
| USS mesh size | 50 | 40 | 30 | 20 | 16 | 10 | 8 | 7 | 6 | 4 | | | | | | | | | | | | | | |
| Sand | 0.144 | 0.183 | 0.226 | 0.282 | 0.36 | 0.45 | 0.57 | 0.69 | 0.73 | 0.90 | — | — | — | — | — | — | — | — | — | — | — | — | — | — |
| Stone dust | 0.120 | 0.152 | 0.188 | 0.235 | 0.30 | 0.375 | 0.475 | 0.56 | 0.595 | 0.75 | — | — | — | — | — | — | — | — | — | — | — | — | — | — |
| Coal dust | 0.091 | 0.115 | 0.142 | 0.178 | 0.226 | 0.284 | 0.36 | 0.43 | 0.45 | 0.57 | — | — | — | — | — | — | — | — | — | — | — | — | — | — |
| Gravel | — | — | — | — | — | — | — | — | — | — | 1.08 | 1.40 | 1.68 | 1.94 | 2.16 | 2.36 | 2.56 | 2.90 | 3.20 | 3.70 | 4.05 | 4.30 | 4.65 | 4.90 |
| Crushed stone | — | — | — | — | — | — | — | — | — | — | 0.88 | 1.19 | 1.40 | 1.60 | 1.80 | 1.96 | 2.12 | 2.40 | 2.68 | 3.10 | 3.38 | 3.60 | 3.86 | 4.07 |
| Coal | — | — | — | — | — | — | — | — | — | — | 0.68 | 0.88 | 1.04 | 1.21 | 1.36 | 1.48 | 1.60 | 1.83 | 2.00 | 2.31 | 2.53 | 2.69 | 2.91 | 3.06 |

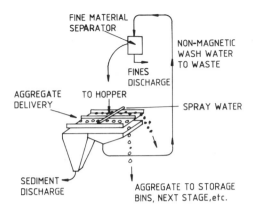

**Fig. 38.23**  Screen washing system

plant is dependent upon the size of the aggregate and amount of undesirable material present. The common methods are as follows:

(i)  Simple wash drum.
(ii)  Screw classifies.
(iii)  Drum scrubbing.

*Simple wash drum*
This is a rotating horizontal drum similar to the revolving drum used for screening (Fig. 38.22), but with only a single coarse screen at the discharge end, the dirty water is simply led away or recycled after cleaning. The capacity depends upon the diameter and drum length but units capable of washing 200 m³/h of aggregate are available. The method is commonly used for producing clean washed chippings and grits. Typical data are given in Table 38.10.

**Table 38.10**  Rotary screen and wash drum

| Item | Size of drum (dia. × length) (m) | | | |
|------|-----------|-----------|-----------|-----------|
| | **0.8 × 3.8** | **1.0 × 4.2** | **1.2 × 5.0** | **1.5 × 5.8** |
| Capacity (m³/h) | 20–25 | 40–50 | 70–85 | 160–190 |
| Water required (litres/s) | 25 | 50 | 75 | 125 |
| Power (kW) | 5 | 8 | 12 | 20 |
| Rotation (rev/min) | 16 | 14 | 12 | 10 |

*Screw classifier*
The screw classifier is generally used for washing pit sands and comprises a feed box, settling hopper and rotating screw set at an incline. Water and sand are constantly added at the feed box with the dirty water overflowing to waste or recycled. The sand particles are washed as they travel along the screw, emerging wet but suitable for transport away in trucks.

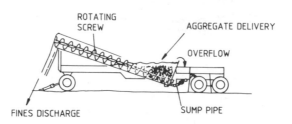

**Fig. 38.24**   Screw classifier

The rotation speed of the screw must be slow if the type of material to be washed is fine, otherwise particles passing the No. 50 sieve will be washed out.

Where screening is involved, the screw classifier is generally positioned to receive the screened fines and the screen wash water. Typical data are shown in Table 38.11.

*Scrubbing drum*
For ores, stones and gravels and very dirty sands, washing generally requires a paddle drum. The equipment comprises a rotating drum fitted with paddles, spiral blades, etc. The aggregate enters at one end of the drum and leaves at the other, while water is introduced under pressure in the opposite direction. The whole unit can be positioned horizontally, but the retention time can be altered by providing an incline. Typical data for scrubbing washer drums are shown in Table 38.12.

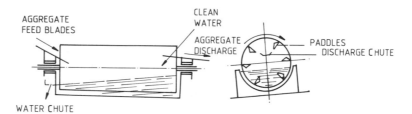

**Fig. 38.25**   Scrubbing drum

## RECOVERY OF FINES FROM WASH WATER

Fines capable of passing the No. 100 sieve will generally be lost in the wash water. However, when it is desirable to retain this material for subsequent blending, fine particles in 30–200 mesh size can be recovered by a cyclone technique (Fig. 38.26). The sand-bearing water is fed under pressure, tangentially into the top of a conically shaped drum. Near the walls centrifugal forces and boundary flow cause the sand particles to travel towards the outlet, while the secondary flows produce a circulating motion along the length of the cone together with a central vortex which is lead to an overflow pipe.

**Table 38.11**  Capacities in tph of screw classifiers

| Material | Screw speed % of Max. | Size of classifier (screw dia. × length) (mm) | | | | | | | |
|---|---|---|---|---|---|---|---|---|---|
| | | 500 × 5750 | 600 × 6750 | 750 × 7500 | 900 × 7500 | 1000 × 8500 | 1200 × 9750 | 1300 × 10 500 | 1800 × 11 000 |
| Very coarse sand | 100% | 30 | 45 | 70 | 100 | 140 | 185 | 290 | 420 |
| Coarse sand | 80% | 25 | 35 | 55 | 80 | 110 | 150 | 230 | 335 |
| Medium sand | 60% | 20 | 28 | 40 | 60 | 90 | 110 | 175 | 250 |
| Fine sand | 40% | 12 | 18 | 27 | 40 | 55 | 75 | 115 | 170 |
| Very fine sand | 20% | 6 | 10 | 13 | 20 | 30 | 40 | 60 | 85 |
| Water (litre/s) | | | | | | | | | |
| 100 mesh | — | 35 | 42 | 48 | 60 | 95 | 120 | 130 | 150 |
| 200 mesh | — | 10 | 10 | 12 | 14 | 24 | 30 | 35 | 38 |

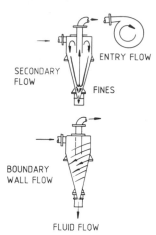

**Fig. 38.26** Cyclone method of fines recovery

**Table 38.12** Scrubbing drum data

| Item | Size of drum (dia. × length) (m) | | |
| --- | --- | --- | --- |
| | **1.8 × 3.2** | **2.4 × 4.3** | **3.0 × 5.3** |
| Washing time (mins) | 1–35 | 1–35 | 1–35 |
| Capacity (t/h) | 50–180 | 120–340 | 225–750 |
| Water requirements (litres/s) | 14–65 | 35–150 | 63–250 |
| Power (kW) | 60 | 150 | 380 |

## TYPICAL CYCLONE DATA

| | |
| --- | --- |
| Capacity | 40–100 litre/s |
| Delivery | 1–1.5 bar |
| Pump power | 15–30 kW |
| Pump speed | 650–850 rev/min |

The alternative to the cyclone method is to use settling tanks or basins (Fig. 38.27) but regular emptying to remove the sediment is needed.

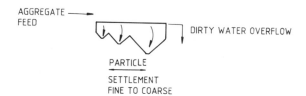

**Fig. 38.27** Settling tanks for fines recovery

## Conveyor system

A belt conveyor is used to transport aggregate between crushing units, stock piles, screens and storage bins. The selections of capacities should be determined from a careful analysis of the whole system with input to output related to usage rates of aggregate and sand. It is not enough to simply select the conveyor on the basis of the capacity of the crusher since individual units often carry smaller quantities to various screening and stock piles locations. Examples and data on conveyors are described in Chapter 35.

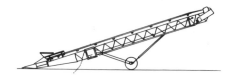

**Fig. 38.28** Portable conveyor

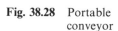

**Example of an aggregate production solution**

PROBLEM

Quarry rocks of 350 mm size are delivered to a crushing plant at a rate of 50 t/h. The sizes required are

$1\frac{1}{2}$ in (38 mm) to $\frac{3}{4}$ (19 mm)
$\frac{3}{4}$ in (19 mm) to fines.

Select appropriate crushers and screens.

SOLUTION AS FOLLOWS

**Primary crushing**

Maximum allowable size of material is $0.8 \times$ feed opening. Thus minimum feed opening required $= 350/0.8 = 438$ mm.

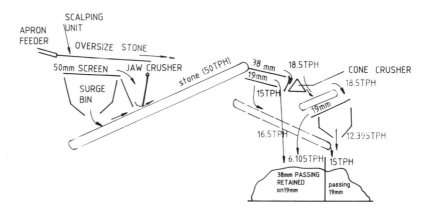

**Fig. 38.29** Crushing and screening example

**Alternative options**

Interpolating from Table 38.2 for the $0.45 \times 0.8$ m jaw crusher:

Closed setting = 50 mm     output = 52 t/h
Closed setting = 180 mm     output = 130 t/h

From Table 38.3 for $0.33 \times 1.1$ gyratory crusher:

Open setting = 50 mm     output = 80 t/h
Open setting = 75 mm     output = 150 t/h

Thus the jaw crusher is selected with a 50 mm setting to produce the required output of 50 t/h.

From Fig. 38.30 illustrating an analysis of product from jaw crushers it can be seen that for a 50 mm setting 63% of material passes the $1\frac{1}{2}$ in (38 mm) screen and can be further screened to produce the $1\frac{1}{2}$ in (38 mm) to $\frac{3}{4}$ in (19 mm) and $\frac{3}{4}$ in (19 mm) to $\frac{3}{16}$ in (5 mm) size products. The 37% of oversize material or 18.5 t/h is directed to a secondary crusher.

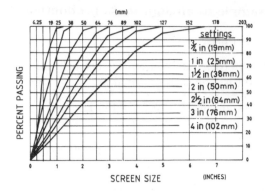

**Fig. 38.30** Jaw crusher product sizing graph

## Secondary crushing

Typically the cone crusher could be selected as the secondary crusher. It can be seen in Fig. 38.30 that the oversize 18.5 t/h consists of pieces up to 76 mm size. Thus the feed opening of the crusher should be approximately 75/0.8 = 94 mm.

Table 38.5 which gives typical cone crusher data indicates that a unit with a 600 mm diameter cone is adequate as follows:

| | |
|---|---|
| Feed opening | 90 mm |
| Discharge setting | 6–25 mm |
| Capacity | 13–50 t/h |

Figure 38.30 shows that with a $\frac{3}{4}$ in (19 mm) setting all the stone would pass the $1\frac{1}{2}$ in (38 mm) screen and so can be screened into the two required grades of material.

## Screen sizing

### SCREENING AFTER PRIMARY CRUSHER

$$\text{Screen capacity} = Q \times X \times Y \times Z \times D$$

*Top deck* (38 mm screen)
From Table 38.9 for crushed stone passing the 38 mm ($1\frac{1}{2}$ in) screen

$$Q = 2.68 \text{ t/h per ft}^2 \text{ (29 t/h per m}^2\text{)}.$$

Factor $X$: for top deck = 1.00.
Factor $Z$: 37% of material to be screened is larger than 38 mm, i.e. 37% oversize. Factor = 0.965.
Factor $Y$: from Fig. 38.30:
    30% of the total material coming onto the screen is less than half screen size (19 mm).
    Therefore factor = 0.8.
Factor $D$: = 1.0.

If 50 t/h are delivered then the required top screen area is

$$\frac{50}{29 \times 1 \times 0.8 \times 0.965} = 2.23 \text{ m}^2$$

*Lower deck* (19 mm screen)

Factor $X$: for lower deck 0.9.
Factor $Z$: from Fig. 38.30 30% of 50 tonnes passes the 19 mm sieve, i.e. 15 tonnes; 63% passed the 38 mm sieve, i.e. 31.5 tonnes, therefore 16.5 tonnes remain on the 19 mm sieve i.e. 16.5/33.5 = 49%.
Factor = 0.91.
Factor $Y$: from Fig. 38.30 15% of the total material coming onto the screen is less than half screen size (9.5 mm).
Factor = 0.6.
Factor $D$: = 1.0.
From Table 38.9 for crushed stone passing the 19 mm ($\frac{3}{4}$ in) screen:
Factor $Q$: = 1.80 t/h per ft$^2$ (19.35 t/h per m$^2$).

If 31.5 t/h pass the top deck then the required bottom deck screen size is

$$\frac{31.5}{19.35 \times 0.9 \times 0.6 \times 0.91 \times 1.0} = 3.3 \text{ m}^2$$

Commonly a twin-deck vibrating screen would be chosen and in this case the dimensions would be dictated by the bottom screen requirement. Thus 1.2 × 3.0 m screens would be adequate.

## SCREENING AFTER SECONDARY CRUSHING (19 mm screen)

18.5 t/h is passed through the secondary crusher which has a 19 mm ($\frac{3}{4}$ in) setting.

Factor $Q$: = 1.8 t/h per ft$^2$ (19.35 t/h per m$^2$).
Factor $X$: = Single deck = 1.0.
Factor $Y$: From Fig. 38.31 35% of the total material coming on to the 19 mm screen is less than half screen size.
Factor = 0.86.
Factor $Z$: 33% (6.105 t/h) of material to be screened is larger than 19 mm. Factor = 0.98.
Factor $D$: 1.0.

If 18.5 t/h is passed then the required screen area =

$$\frac{18.5}{19.35 \times 1.0 \times 0.86 \times 0.98 \times 1.0} = 1.13 \text{ m}^2$$

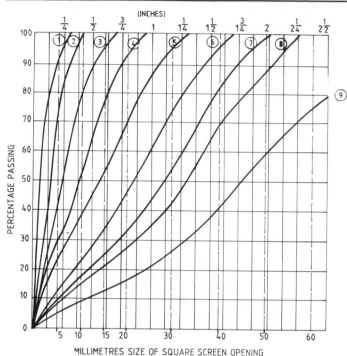

| CURVE No | SETTING | |
|---|---|---|
| 1 | $\frac{1}{8}$ in | 3.2 mm |
| 2 | $\frac{1}{4}$ in | 6.5 |
| 3 | $\frac{3}{8}$ in | 10 |
| 4 | $\frac{1}{2}$ in | 12.5 |
| 5 | $\frac{3}{4}$ in | 19 |
| 6 | 1 in | 25 |
| 7 | $1\frac{1}{4}$ in | 32 |
| 8 | $1\frac{1}{2}$ in | 38 |
| 9 | 2 in | 50 |

**Fig. 38.31**   Cone crusher product sizing graph

## Portable screening equipment

Fixed plant is usually set up where crushed stone is required over a long period and the transporting costs to the customers are economical. However, the many instances where rock is available for crushing in the nearby vicinity has led to the development of portable equipment. The early developments were fully self-contained portable units comprising crushing, conveying and screening. More modern plant, however, has concentrated on providing individual processing stations mounted on their own chassis. Thus a single station might comprise the screens or feeder and crushing machine. Portable conveyors are used to connect individual units and storage bins.

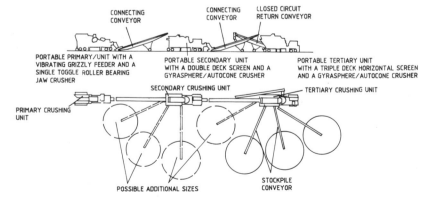

**Fig. 38.32**   Portable crushing and screening equipment

# CONCRETE PRODUCTION

## 39.1  CONCRETE DESIGN

Concrete consists of a mixture of graded sand, graded coarse aggregate, Portland cement and water. By itself cement mixed with water causes a chemical reaction producing a hardened paste after about 45 min, called the 'initial set'. At not more than 10 hours this plastic phase is over and the final set obtained. Provided sufficient water is present for the reaction to continue, the compound will gain further strength until about 28 days when almost the final strength is reached, the rate of gain depending upon the surrounding temperature.

The addition of aggregate particles simply serves to act as filler material bound together by the cement paste. Thus the strength and grading of these materials are influential in controlling the strength of the hardened concrete.

## 39.2  CEMENTS

Portland cement klinker is commonly manufactured in a rotating kiln by gradually heating and finally burning a finely ground mixture of calcium carbonate (chalk, limestone) and siliceous material (clays). Different types of Portland cement can be produced with various properties by modifying the chemical composition and final grinding of the klinker to conform to particular National Standards including specification for fineness, strength, setting time and limits of impurities. The most common Portland cements are:

(i)  Ordinary Portland Cement (OPC) – a basic cement used for general concrete. Fineness: 30 mm$^2$/g.
(ii) Rapid hardening Portland – a finer cement used to give high early strength.

(iii) Low heat Portland cement – used for massive concrete pours such as dams, to reduce the heat of hydration generated during the chemical reactions.

(iv) Sulphate-resisting Portland cement – made by adding iron compounds during the kiln process to produce a material less affectable to acid waters and other injurious salts.

(v) Other cement varieties include, coloured, blast furnace, pozzolanic, masonry, waterproof, hydrophobic, high alumina and oil well cements.

(vi) Pulverised fuel ash may be combined with OPC to produce structural concrete. The resulting mix gives a slightly increased workability.

## 39.3   AGGREGATES

Aggregates occupy about 75% of the volume of concrete and are obtained either from naturally occurring deposits of sands and gravels, including sea-dredged material, or are produced by crushing quarried rock. Granite, basalt and the harder types of limestone and sandstone are commonly used. They should be free of impurities, clay, silt and degradable materials. A simple quality test for sand can be carried out by filling a 200 ml measuring cylinder up to the 100 ml mark and then adding water up to 150 ml mark. After shaking the mixture and allowing about 3 hours for settling, the thin layer of silt formed should not exceed 10% of the sand volume.

### Grading of aggregates

To obtain good strength, concrete should contain particles ranging in size to facilitate an interlocking of the stones. In achieving this the aggregates can be obtained in single sizes and then combined to produce suitable gradings but are more commonly supplied already graded. Fine aggregate or sand is that material passing through a 5 mm ($\frac{3}{16}$ in) sieve, while coarse aggregate is retained on this size of sieve (Table 39.1). The maximum size of coarse depends upon the job requirements, for example reinforcement congestion would dictate small size aggregate, whereas mass concreting might be suitable with 75 mm or more stone size.

## 39.4   CONCRETE MIXES

### Water/cement ratio

While the quality and grading of the aggregate are important, the strength of concrete depends mainly on the water/cement ratio and the type of cement, the water cement ratio being defined as the weight of

**Table 39.1**  Coarse and fine aggregates

| Sieve size (mm) | Percentage passing | | | | | | | |
| --- | --- | --- | --- | --- | --- | --- | --- | --- |
| | Graded aggregate | | | Single-sized aggregate | | | | Fine aggregate |
| | 40 to 5 mm | 20 to 5 mm | 14 to 5 mm | 40 mm | 20 mm | 14 mm | 10 mm | Overall limits |
| 50 | 100 | — | — | 100 | — | — | — | |
| 37.5 | 90–100 | 100 | — | 85–100 | 100 | — | — | |
| 20.0 | 35–70 | 90–100 | 100 | 0–25 | 85–100 | 100 | — | |
| 10.0 | 10–40 | 30–60 | 50–85 | 0–5 | 0–25 | 0–50 | 85–100 | |
| 5.00 | 0–5 | 0–10 | 0–10 | — | 0–5 | 0–10 | 0–25 | 100 |
| 2.36 | — | — | — | — | — | — | 0–5 | 90–100 |
| 1.18 | | | | | | | | 30–100 |
| 600 $\mu$m | | | | | | | | 15–100 |
| 300 $\mu$m | | | | | | | | 5–70 |
| 150 $\mu$m | | | | | | | | 0–15 |

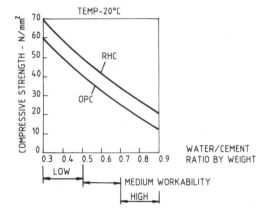

**Fig. 39.1**  Compressive strength and water/cement ratio

water in a mix divided by the weight of cement. The higher the w/c ratio, the weaker the concrete (Fig. 39.1) becoming significantly pronounced at about 0.6 w/c and above. Indeed, a water content of only 25% of the weight of cement is needed to combine chemically, the higher amounts being required to provide workability. High water/cement ratios also increase permeability.

### Degree of compaction

The strengths shown in Fig. 39.1 assume a fully compacted concrete. Unfortunately during mixing air is entrapped and unless sufficient vibration is provided the ultimate strength will be seriously affected as illustrated in Fig. 39.2.

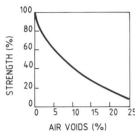

**Fig. 39.2**   Compressive strength and voids ratio

## Workability

Workability describes the degree of stiffness of the mix and can be directly altered by changing the water/cement ratio. However, the shape, size and texture of the aggregate also have some effect. For example sea-dredged rounded stones generally produce more workable mixes than crushed angular material: also less water is required for a given workability by increasing the maximum size of stone.

### TESTS FOR WORKABILITY

(i) *Slump test* (Fig. 39.3). A conical mould is filled in three equal layers and rodded twenty-five times per layer. The top is struck level, the mold removed and the slump measured. Low workability is usually indicated by a 25 mm slump and less, medium by 25–75 mm, and high greater than 75 mm. A collapsed or sheared slump should be repeated.

**Fig. 39.3**   Slump test

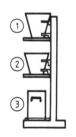

**Fig. 39.4**   Compaction factor test

(ii) *Compaction factor test* (Fig. 39.4). Concrete is dropped from vessel (1), into (2) and finally into cylinder (3). The final surface is smoothed level and the cylinder and contents weighed. The cylinder is then refilled and rammed to full compaction.

$$CF = \frac{\text{weight of partially compacted concrete}}{\text{weight of fully compacted concrete}}$$

(iii) *Vebe test* (Fig. 39.5) is used for very stiff concrete mixes and consists of vibrating a standard slump test specimen placed in a specified vessel. The workability is expressed in seconds of time required to level the cone, e.g. 10 mm slump is approximately equivalent to 12 s, 30 mm ≡ 6 s, 60 mm ≡ 3 s, 180 mm less than 1 s.

(iv) *DIN test* (Fig. 39.6) is designed for very high workability mixes, i.e. those almost of flowing consistency. A shortened conical mould is filled in two layers and rodded ten times. The mould is then removed and the foundation board lifted, tilted and dropped fifty times in 15 s. The average spread of concrete denotes the workability measure.

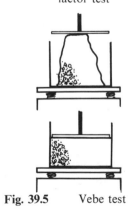

**Fig. 39.5**          Vebe test

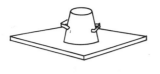

**Fig. 39.6**   DIN test

## Proportioning of constituent materials

The cement, water and aggregates are usually defined in terms of kilograms of each material per cum of concrete, determined by one of various well established standard methodologies, e.g. BRE method in the UK supported with trial mixes. Typically: 250 kg/m$^3$ cement, 125 kg/m$^3$ water, 1500 kg/m$^3$ aggregate.

Determining the ideal combination of coarse and fine aggregate in particular is best achieved by trial and error, making reference to appearance, workability and cement content in achieving the required 28-day strength. Alternatively proportions may be roughly gauged from specified grading curves, charts, tables, etc.

## Moisture content of the aggregate

Generally aggregates when arriving at site are wet from the washing operation. Further changes in the moisture content are likely to take place during storage out in the open atmosphere. The weight of this water should be deducted when measuring out the quantities during batching and may be ascertained from weighing wet and dried samples or by proprietary chemical methods. Unfortunately, some of the water may be absorbed as pore water in the aggregate and while contributing to the water/cement ratio would tend to produce less workability than expected.

### EXAMPLE OF THEORETICAL ADJUSTMENTS

Consider a 1 cum batch of concrete

| Sand | Aggte. | Cement | Water |
|------|--------|--------|-------|
| 600 kg | 1200 kg | 300 kg | 150 kg |

Moisture content for sand = 5%

$$\text{weight of water} = \frac{600 \times 5}{100} = 30 \text{ kg}$$

Moisture content for coarse aggregate = 3%

$$\text{weight of water} = \frac{1200 \times 3}{100} = 36 \text{ kg}$$

Batching weights (approximately)

| Sand | Aggte. | Cement | Water |
|------|--------|--------|-------|
| 630 kg | 1236 kg | 300 kg | 84 kg |

*Mixing water* should be free from impurities such as organic materials and injurious chemicals. Tests for water for making concrete are laid down in various National Standards.

### 39.5  CONCRETE STRENGTH

#### Setting time

Concrete, unlike cement paste, is not described in terms of initial and final setting times, the set being more practically determined according to the time trowelling or re-vibrating cease to be possible. This condition is reached after about $\frac{1}{2}$ hour in warm air temperatures and 4 hours or so when cold. The inclusion of a retarding admixture can further extend this period.

#### Strength gain

Strength continues to improve provided sufficient water is supplied to the set concrete to ensure that drying out during the heat of hydration stage is prevented, otherwise the chemical reactions which allow strength to be developed, will cease. The process is known as 'curing'.

*Shrinkage* occurs as concrete hardens, being greater in wet mixes. Curing helps to limit its effects and therefore concrete in the actual structure should be thoroughly and continuously moistened for some considerable period after placing.

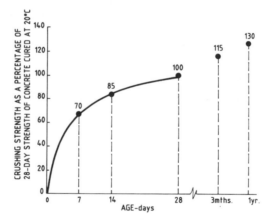

Fig. 39.7  Compressive strength gain with age

#### Test specimens

Concrete strength is usually measured according to various National Standards by carrying out a compression test on samples taken from the mix. A cube or cylinder specimen is crushed to destruction (Fig. 39.8) and the sustained maximum load per unit area measured. The 28-day value is normally selected to represent the characteristic compressive strength. Depending upon the required concrete properties grades from 2.5 to 60 N/mm² can be achieved for normal mixes, but with suitable admixtures and strong aggregates up to 200 N/mm² is possible. Unfortunately, 28 days is often too long to wait and methods of accelerated curing of test specimens have been developed. They principally involve placing the test specimens and mould into a tank of

**Fig. 39.8**   Typical crushed test cube

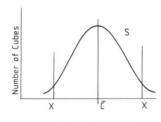

**Fig. 39.9**   Normal distribution of crushed test cube results

heated water shortly after making and leaving it for a period before crushing, normally 24 hours or so depending upon the temperature. Typically accelerated tests have been developed for 35°C, 55°C and 82°C whereby the early and 28-day results are correlated and thereafter used as a gauge for all further testing of specimens.

Test specimens may also be obtained as core samples taken for the actual concrete structure or component or alternatively by non-destructive means such as ultrasonic, gamma-ray, etc. Other surface investigation methods include the Schmidt hammer, internal fracture and pull out tests.

## INTERPRETING TEST RESULTS

When a large number of test samples are crushed to destruction, the usual pattern of strength is a normal distribution (Fig. 39.9). To ensure that no results fall below, say, a permitted minimum crushing strength the average strength of the mix would need to be considerably raised. In practice however a small number of results are allowed to fall below the minimum, calculated from the following formula:

Minimum crushing strength = average crushing strength
$-2 \times$ standard deviation that is $x = \bar{c} \pm Zs$

where

$$s = \text{standard deviation} = \left[ \frac{\sum (c - \bar{c})^2}{n - 1} \right]^{1/2}$$

and
- $\bar{c}$ = average strength of all test results;
- $c$ = each individual test result of strength;
- $n$ = number of results;
- $Z$ = standard normal variable, obtained from statistical tables, e.g. 1.96 for 95% confidence level.

### EXAMPLE

A large number of test results produce a standard deviation ($s$) a 1.5 N/mm², and average strength 10 N/mm². If 5% of results are permitted to fail, what should be the minimum design strength of the mix?

$$x = \bar{c} - Zs$$
$$= 10 - 1.96 \times 1.5 = 7.06 \text{ N/mm}^2$$

That is, if the specified minimum strength were 7 N/mm² then a mix with an average strength of nearly 10 N/mm² would be necessary if not more than 5% of test results are not to fall below the specified strength.

### Admixtures

Chemicals are available for addition to the mix to act as set retarders, set accelerators, strength accelerators, water reducers, air-entraining

**Table 39.2** Typical uses of some admixtures

| Use | Set retarder | Set accelerator | Strength accelerator | Water reducer | Air-entraining agent | Water repellent | Thickening agent | Superplasticiser | Pigment |
|---|---|---|---|---|---|---|---|---|---|
| Chemical resistance | | | | * | * | | | | |
| Pumping | * | | | * | * | | * | * | |
| Early formwork removal | | | * | | | | | | |
| Workability aid | | | | * | * | | | * | |
| High early strength | | * | * | * | | | | * | |
| Permeability reduction | | | | * | * | * | | | |
| Mortars | | | | | * | | | * | * |
| Precast cladding | | | * | | * | * | | | * |
| Frost resistance | | | | * | * | | | | |
| Mass construction | * | | | | | | | * | |
| Waterproofing | | | | * | * | * | | | |
| Improved surface finishes | | | | | * | * | | * | |
| High strength | | | | * | | | | * | |
| Hot weather | * | | | * | | | | | |
| Cold weather | | * | * | | | | | | |

agents, water repellents, thickening agents, superplastisicers, pigmenters. The amounts and use of such additives vary depending upon the application and manufacturer.

## 39.6 OTHER TYPES OF CONCRETE

Concrete can be made from various materials to suit specific needs. The following are typical examples: clinker, no-fines, light-weight aggregate concrete, high strength, dense aggregate concrete.

## 39.7 CONCRETE REINFORCEMENT

Cutting and bending of reinforcement can be carried out on site or ready supplied. The steel bars are fixed together by tie wire or welding.

Mats of steelwork are supported by 'stools' or 'chairs'. Individual bars may be given sufficient cover of concrete by providing plastic or simple concrete spacer blocks attached to the reinforcement at points facing the formwork. In normal practice mild steel is usually specified but occasionally high tensile steel may be stipulated by the designer.

## 39.8 DEFECTS IN CONCRETE

### Bleeding

When the solids settle under the action of gravity bleeding results to leave a layer of water, and consequently a porous and weak final surface. Increased cement and/or the use of an air-entraining agent in the mix will help to counter the effect.

### Segregation

When the cement paste separates from the aggregate segregation arises and is generally caused or aggravated by rough handling during the placing operation. The problem is particularly acute with excessively sanded mixes, when the coarse aggregate tends to separate out to the bottom during vibration.

### Sulphates

Concrete exposed to sulphates, e.g. sea water, contaminated ground, etc., will undergo chemical change, expand and may crack, the process being mainly caused by the calcium hydroxide and tricalcium aluminate hydrate in the cement reacting with sulphate ions to produce solid products greater in volume than the original.

### Acids

Any acidic liquid in contact with concrete will produce some dissolution, the degree of harm being dependent upon the type and concentration of acid present. In general, however, acid attack in most grounds is not a major problem.

### Salts

Dissolved salts can penetrate concrete, any subsequent crystalline growth may then exert considerable forces and cause damage.

### Carbonation

The hydroxides in concrete may become neutralised near the surface by the action of carbon dioxide in the atmosphere, and destroy the alkaline protection to the embedded reinforcement. A potential difference could then be established between two such regions and metal ions transferred. Any water and oxygen present then causes a reaction to produce ferrous hydroxide, i.e. corrosion in the carbonised layer. The depth of cover should thus be sufficiently deep to reduce such an effect.

### Chlorides

The presence of chlorides in the concrete can produce an electrolytic solution and thereby induce corrosion in reinforcement by causing local potential drop in adjacent parts of any embedded steel where chlorides are present. The maximum total chloride content of aggregates, and cement are laid down in National Standards but should not exceed 0.3% by weight of cement, the only reliable means of reducing the problem is to provide cathodic protection to the steel. Sea-washed aggregate should also be carefully scrutinised for chloride content.

### Alkali-silica reaction

Most cement contains small amounts of sodium and potassium alkalis, e.g., $Na_2O$, $K_2O$, which react in the presence of water and some forms of silica aggregate, for example, opal reacts to produce a gel and subsequent disruptive expansion and associated cracking. A low alkali cement containing less than 0.6% by weight of $Na_2O$, etc., is recommended to reduce or at least limit the effect.

### 39.9 PRACTICAL CONSIDERATIONS

### Water control

1. Cover all exposed concrete surfaces with damp sand, hessian, etc., and keep moist.
2. Maintain formwork in place for as long as practically possible within the work programme (the minimum period is often specified in the contract and will generally vary according to the nominal grade of concrete, and whether the structure is columns, walls, slabs, foundations, etc.). Thereafter drape damp hessian over the exposed surfaces and keep wet.
3. Alternatively, spray any exposed concrete surface with a waterproof membrane to help hinder the drying out process. Thereafter keep the concrete continuously wet.
4. Curing should always continue for from 3 to 7 days after formwork removal according to the prevailing temperature conditions. In any case a strength of at least 2 $N/mm^2$ is needed to resist freezing forces and 10 $N/mm^2$ to support temporary working loads.

**Temperature control**

## COLD CONDITIONS

First and foremost fresh concrete should be prevented from freezing by applying one or more of the following:

1. Heat the water and aggregates prior to mixing. The latter can be achieved with steam pipes placed in the stockpiles.
2. Insulate and heat the formwork, usually obtained with electric cabling.
3. Heat the surrounding air in enclosed situations.
4. Transport concrete in insulated or heated vehicles and containers.
5. Introduce suitable admixtures (see Table 39.2).

## HOT CONDITIONS

In hot climates evaporation needs to be controlled to avoid rapid stiffening, shrinkage and plastic cracking. The following preconditions should be taken:

1. Insulate formwork.
2. Provide shading to reflect sun rays and cover all exposed surfaces with wet hessian, etc.
3. Fog spray.
4. Cool the concrete with embedded pipework. The heat extracted can be theoretically determined by the formula:

$$N = Q\gamma_w S_w t$$

where  $N$  = heat quantity extraction rate, e.g. kcal/hr;
$Q$  = cooling fluid passed, e.g. $m^3/h$;
$\gamma_w$  = specific weight of cooling fluid, e.g. water = $1000 \ kg/m^3$;
$S_w$  = specific heat of cooling fluid, e.g. water = 1 k cal/kg °C;
$t$  = exit and entry temperature difference of cooling liquid (°C). Initially the exit temperature will be that of the fresh concrete and should not exceed 30–35°C. (15°C for mass concrete).

*Note*: total heat to be extracted = $WS_c t$

where  $W$ = weight of concrete
$S_c$ = specific heat of concrete.

5. Introduce blocks of ice, solid $CO_2$ or liquid nitrogen into the mix and cool the aggregate; the temperature of the mix should be kept below 35°C or so.
6. Introduce suitable admixtures (see Table 39.2).

## 39.10 CONCRETE BATCHING AND MIXING

### Batching

A hopper is filled with the specified weight of each material obtained from the storage bins and the contents transferred to the mixing drum. Cement is usually pumped from a silo, weighed and discharged separately, similarly with admixtures. When all the ingredients are in the mixing drum the required quantity of water is added from a water-measuring tank, water meter, etc. Depending upon the sophistication of the equipment the process may be manual with the operator having to watch the scale dials carefully, or alternatively on some modern central mixing units automatic gate controls close as the specified weight of material is loaded. Regular servicing and checking of the weighing mechanisms is crucially important to maintain batching accuracy.

### Mixing

The sequence of discharging materials and water to the mixing drum varies depending on the type of equipment and is best determined by trial and error judged by the uniformity of the concrete produced. In modern plants roughly 1 min is required for charging the drum and a further 1 to $1\frac{1}{2}$ min needed for mixing. The choice of mixer being either drum or pan types.

### DRUM MIXERS

These comprise a bladed vessel, usually of conical shape having one opening both to receive and discharge the ingredients, rotating on an inclinable axis (Fig. 39.10) to facilitate discharge away from the loading point. Small to large units are available capable of producing mixed batches in the range from 5 to 100 ft$^3$ (140–2800 litres), with outputs of from 5 to 120 yd$^3$/h (4–90 m$^3$/h). The mixing action is produced by helical baffles moving the material from the charging end to the base of the drum. Reversing the rotation causes a counter reaction to discharge the contents. Rotation speeds of up to 10 rev/min are normal. An alternative form of this type of mixer is *non-tilting* with the drum rotating on a horizontal axia and having two openings (Fig. 39.11) one to receive materials, the other to discharge the mix. Reversing action is not required and as a consequence are normally of simpler design, having a relatively lightweight support frame. However, the discharge time is longer than for the tilting type. The size of vessel should be capable of holding at least 10% more than its nominal capacity. The drum volume itself is often up to four times the nominal capacity to allow sufficient space for the mixing action.

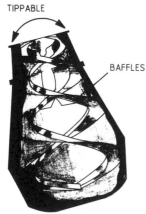

TIPPABLE

BAFFLES

**Fig. 39.10** Tilting drum mixer

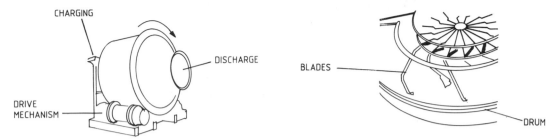

**Fig. 39.11**   Non-tilting drum mixer          **Fig. 39.12**   Pan mixer

## PAN MIXERS

The mixing vessel is mounted on a vertical axis (Fig. 39.12) with the contents entering at the top and discharging through a sliding trap in the base and down a chute into transportation vehicles. Mixing is achieved with blades rotated within the vessel at 14–30 rev/min which on some models is also rotated counter to the blades. Pan sizes vary from about 5–90 ft³ nominal capacity (140–2500 litres) capable of producing 5–130 yd³/h (4–100 m³/h).

### Central mixing plants

When large quantities of concrete are required, a central mixing facility with well prepared storage bins, and semipermanent equipment is usually installed. Good access points for distributing materials and transportation of concrete are essential.

## 39.11   HANDLING AND TRANSPORTING CONCRETE

### Wheelbarrows

These are one or two wheeled pneumatic tyred barrows of capacity from 0.1 m³ to 0.2 m³.

1. Output per man on a 50 m return run is 1 m³/h. Gang strengths of up to six men are common.
2. When using the two wheel type, output may increase to $1\frac{1}{2}$ m³/h but due to the increased weight, difficulty in turning and transporting up slopes, efficiency is generally reduced.

### USES

Chimneys, cooling towers, floor construction at ground level or in multistorey construction, confined areas.

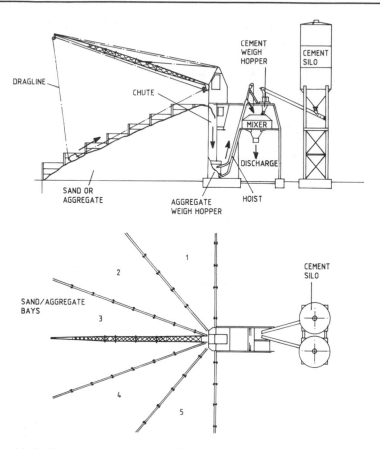

**Fig. 39.13** Central mixing plant

## Hoisting with a crane (Figs. 23.5, 24.6, 26.1, 29.3)

Three types of crane are used:

1. Crawler-mounted – used for general all-purpose work such as pouring concrete in foundations, retaining walls, floor slabs, culverts.
2. Derricks – used for pouring concrete in pump houses, caissons, and basement construction where erection costs can be justified, such as work continuing for long periods.
3. Tower cranes – used for concreting floors, columns, multistorey construction, ground level work where the area to be covered is large.

### BUCKETS AND SKIPS

These range from $\frac{1}{2}$ to 8 m³ capacity, the common types being $\frac{1}{2}$ and 1 m³ skips. The bottom of the skip is fitted with a gate to allow the discharge to be manually controlled. The skip is hung from a crane.

### OUTPUTS USING CRANES

Ground level work        $\frac{1}{2}$ m³ skip 8 to 12 m³/h
                         1 m³ skip 14 to 20 m³/h

Above ground level   $\frac{1}{2}$ m³ skip 5 to 8 m³/h
                      1 m³ skip 10 to 14 m³/h

## Concrete pumps

### HYDRAULIC PUMP

Concrete is placed in the pump hopper and forced along a
100–150 mm diameter steel pipe to the point of discharge.

*Uses* – large foundations, multistorey construction, floors slabs.
*Outputs* – (a) single cylinder, from 12 to 15 m³/h; (b) double cylinder,
   from 15 to 50 m³/h.
*Typical range* – 400 m horizontally or 35 m vertically.
*Losses* – A 45° bend loses 6 m horizontally.

Note: Small diameter (50 mm) pipe is now on the market. Such
pumps are commonly mounted on trucks and are hydraulically
operated. Generally for pumping purposes concrete should have a
slump of about 75 mm. Pumping can be improved with lubricating
additives (retarders) and PFA.

### PNEUMATIC PUMP

This consists of a cylinder containing the concrete. Compressed air is
then introduced forcing the concrete to the place of discharge.

## Belt conveyors

These can now offer comparable costs with other forms of placing
method. The concrete is simply transported from the mixer to the
discharge point by the conveyor belt.
   *Output* – depends upon the speed and size of the belt, transportable
unit now on the market gives from 7 to 16 m³/h.

## Dumper

Used for transporting concrete from the mixer to the works and have
capacities ranging upwards from 1 m³.

## Transit mixers (trucks)

Mobile trucks (Fig. 39.14) are very popular for delivering ready-mixed
concrete to site and are usually of the inclined-axis revolving-drum
type, discharging by a counter rotating action. Mixed concrete may be
loaded at a central batching plant or mixing undertaken while
travelling. Nominal capacities are available from approximately 2–6 m³
depending upon the permissible axle load and weight of truck allowed
on the highway.

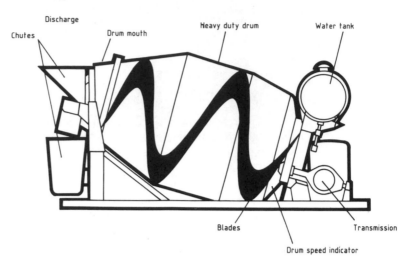

**Fig. 39.14** Ready-mixed concrete truck

### Mono-rails

(Figure 36.1) – consist of concrete hoppers mounted upon a single rail. A series of these can be used to deliver concrete very efficiently on many types of works such as the concreting of bridge decks or similar where conventional access might be difficult.

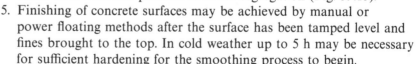

## 39.12 PLACING CONCRETE

1. Concrete should be placed within 30 min of mixing.
2. Forms should be blown clean of debris and oiled to aid subsequent removal before concreting operations begin.
3. Concrete should be spread evenly in 300 mm layers to avoid segregation.
4. Concrete should not be dropped more than 2 m vertically and a chute used wherever possible to avoid segregation (Fig. 39.15).
5. Finishing of concrete surfaces may be achieved by manual or power floating methods after the surface has been tamped level and fines brought to the top. In cold weather up to 5 h may be necessary for sufficient hardening for the smoothing process to begin.

**Fig. 39.15** Placing concrete from a skip

*Tremmie tubes* – where the height of placing concrete exceeds 2 m, a tremmie pipe should be used, (Fig. 39.16).

### Concreting under water

Either use a tremmie or a bottom-opening skip. The concrete should be carefully placed otherwise the fines will be washed out to form a layer of latence on top of the concrete. Even so, in large pours this layer may be up to 400 m deep. A high slump concrete (150 mm), and about 25% more cement giving a water/cement ratio of about 0.5 may be necessary.

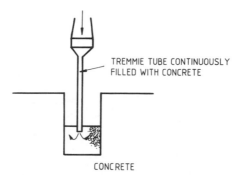

**Fig. 39.16** Placing concrete with a tremmie tube

TREMMIE TUBE CONTINUOUSLY FILLED WITH CONCRETE

CONCRETE

### Compaction

Concrete must be consolidated to remove voids which seriously reduce the strength of the hardened concrete, e.g. by up to 30% with 5% voids and 25% voids the concrete might only have 10% of the expected strength of the fully compacted value.

Compaction may be achieved with internal vibrating pokers inserted at $\frac{1}{2}$ m points as the concrete is placed, and slowly withdrawn when air bubbles cease to come to the surface. External vibrators clamped to the outside of the shuttering are an alternative system.

### Construction joints

Such joints are required to combat shrinkage, contraction and expansion and are normally required as follows:

*Vertical joints* – concrete floors and ground slabs are normally broken up by construction joints into bays or strips about 6 or 7 m wide and up to 20 m long. Walls are normally divided into panels with vertical construction joints at about 10 m centres.

*Horizontal joins* – are normally placed in walls and columns at 2 to 3 m centres vertically.

All joints should be formed level and thoroughly cleaned before the next concrete is placed.

## 39.13 WET CONCRETE PRESSURE

Fresh concrete placed in formwork for slabs (Fig. 39.17), beams (Fig. 39.18), walls (Fig. 39.19) and columns (Fig. 39.20) will exert a pressure on the sides and base unit set. The formwork thereafter is left in place until sufficient strength has been obtained in accordance with the specification laid down by the designer of the structure.

During the setting period the concrete will behave similar to that of a liquid contained in a vessel and thus the formwork needs to be sufficiently strong to both resist the 'fluid' forces and limit deflections to reduce the possibility of leaks. Facing formwork is usually made of timber boards or plywood sheeting suitably framed with bracing, all assembled into practical sized units to suit the method of site handling,

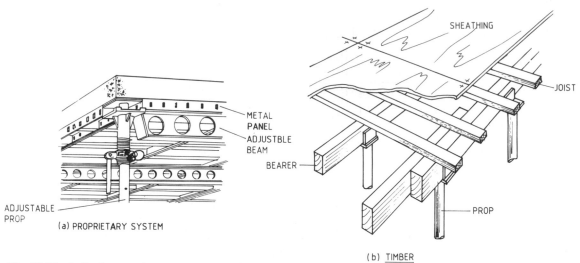

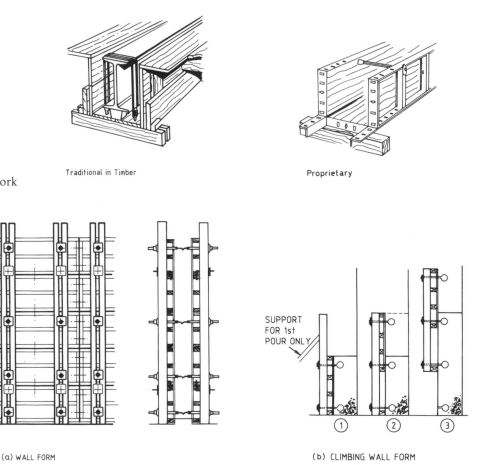

**Fig. 39.17** Soffit formwork to a concrete slab

(a) PROPRIETARY SYSTEM

(b) TIMBER

**Fig. 39.18** Beam formwork

Traditional in Timber

Proprietary

(a) WALL FORM

(b) CLIMBING WALL FORM

**Fig. 39.19** Wall formwork

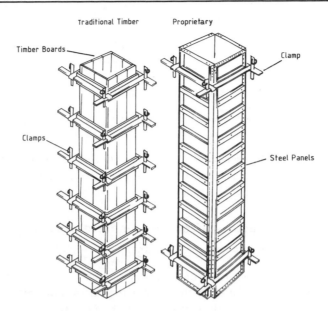

**Fig. 39.20** Column formwork

erecting and removal. Commonly a proprietary system made up of steel panels may be chosen when many uses are required as plywood normally deteriorates rapidly after six or so uses. The whole system is assembled manually and opposing faces cross tied (Fig. 39.21). In designing the formwork the weight of the ancillary items all may need

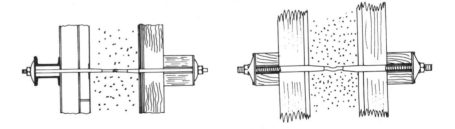

**Fig. 39.21** Formwork cross-ties

to be taken into account as additions to the forces exerted by the wet concrete itself. Typical values are:

| | |
|---|---|
| Self-weight of concrete | 25 kN/m³ |
| Steel reinforcement | 80 kN/m³ |
| Solid softwood timber | 5.5 kN/m³ |
| Solid hardwood timber | up to 10 kN/m³ |
| Proprietary panel formwork | from 0.5 to 0.85 kN/m² |
| Timber formwork | 0.5 kN/m² |
| 19 mm thick plywood | 0.11 kN/m² |
| Superimposed load of workmen | 1.5 kN/m² |
| Superimposed load of equipment | 7 kN/m² |
| Horizontal impact loads | 10% of total static load |
| Winds loads | varies |

### Pressure formulae

The maximum pressure, ($P_{\text{max}}$), would not be greater than the total head, $H$, of concrete (excepting above factors); i.e.

$$P_{\text{max}} \text{ not greater than } \gamma H$$

where $\gamma$ = concrete density at 25 kN/m² depending upon mix properties.

In practice $P_{\text{max}}$ may be reduced depending upon the rate of pour, temperature, vibration, weight of concrete, slump and arching, subject to the formulae used.

### Factors affecting concrete pressure

### RATE OF PLACEMENT

The setting of the concrete varies with time, therefore the faster the rate of pour the greater the horizontal pressure exerted on the face of formwork. Conversely, if the rate were slow enough the pressure at the bottom of the formwork would not be greater than at the top.

### EFFECT OF TEMPERATURE

Concrete at low temperatures sets more slowly than at high temperatures. Indeed for a given rate of pour the pressure exerted can be from 50 to 75% higher in cold weather compared to that under warm conditions.

### EFFECT OF WORKABILITY

The stiffer the concrete the better the ability to stand unsupported, thereby reducing pressure – a very workable mix behaves like a liquid.

### Effect of Arching

Concrete placed between narrow gaps demonstrates similar characteristics to other granular materials and some arching will develop. The effect being to reduce pressure.

The combination of all these influences is to alter the normal triangular hydrostatic pressure design, with the maximum value shifted away from the base (Fig. 39.22).

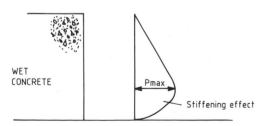

**Fig. 39.22** Concrete pressure diagram

### USA formulae

$$P = 150 + \frac{9000R}{T}, \text{ where } R \leq 7 \text{ ft/h}$$

$$P = 150 + \frac{43\,400}{T} + \frac{2800R}{T}, \text{ where } R > 7 \text{ ft/h}$$

where  $P$ = maximum lateral pressure (in lb/ft$^2$);
  $R$ = rate of placement (in ft/h);
  $T$ = temperature of concrete in form (in °F);
  $H$ = maximum height of concrete in the form (in ft).

### Other formulae

In the UK the effect of arching (Fig. 39.23) is also included in some formulae, the pressure selected being the minimum value obtained from the following:

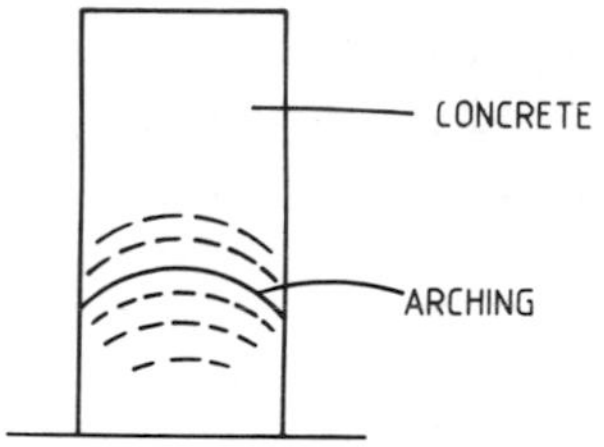

**Fig. 39.23**  Arching of wet concrete

(i)  Pressure head obtained by depth.

$$P = \gamma H \text{ kN/m}^2$$

where  $H$ = height of wet concrete (in m)
  $\gamma$ = density of wet concrete (in kN/m$^3$)

(ii)  Arching effect.

$$P = 3R + \frac{d}{10} + 15 \text{ kN/m}^2$$

where $d$ is the width of the section (in mm). The formula only being valid for $d > 500$ mm. $R$ is the rate of placement in m/hr.

(iii)  Stiffness effect. $S$ varies with both slump and temperature, as determined from tables. Typically for 15°C concrete:

$$P = \gamma RS + 5 \text{ kN/m}^2$$

50 mm slump,  $S = 1$;
75 mm slump,  $S = 1.5$;
100 to 150 mm slump  $S = 1.75$.

The latest research (Clear and Harrison) outlines revised formulae to include the effects of cement type, admixtures, size and shape of formwork. In general, however, the results are not dissimilar to other formulae.

The formula is given as:

$$P_{\text{max}} = D(C_1\sqrt{R} + C_2 K\sqrt{H - G\sqrt{R}}) \text{ or } DH \text{ kN/m}^2$$

whichever is the smaller

where:

$C_1$  = 1.0 for walls
$C_1$  = 1.5 for columns
$C_2$  = 0.3–0.6 depending upon the cement type and admixture,
$OPC$ = 0.3
$D$   = Self weight of concrete (kN/m³)
$H$   = Height of wet concrete (m) or vertical pour height if same
$k$   = temperature coefficient = $\left(\dfrac{36}{T + 16}\right)^2$
$R$   = Rate of concrete placement (m/hr)
$T$   = Temperature of concrete in form (°C)

The report (108) also converts the formula to easy to use Tables.

## Example

Calculate the maximum pressure produced when placing 75 mm slump concrete at a rate of 2m/h, shuttering 2.5 m high and 250 mm wide. Assume the temperature of the concrete in the formwork is initially 15°C and vibration is continuous.

## SOLUTION

*UK formulae*

(i)   By height     $P = 25 \times 2.5 = 62.5 \ \text{kN/m}^2$

(ii)  By arching    $P = 3 \times 2 + \dfrac{250}{10} + 15 = 56 \ \text{kN/m}^2$

(iii) By stiffness  $P = 25 \times 2 \times 1.5 + 5 = 80 \ \text{kN/m}^2$

The figure of 56 kN/m² is selected as arching effect produces a favourable result in such a thin section.

*USA formula*

$$P = \frac{150 + 9000 \times 2 \times 3.24}{59} = 1138 \ \text{lb/ft}^2$$

$$= 1138 \times \frac{47.88}{1000} = 55 \ \text{kN/m}^2$$

# 40

# PRESTRESSED CONCRETE

## 40.1 INTRODUCTION

Normal structural concrete members rely on the reinforcing steel to take up tensile stress as load is applied, allowing the surrounding concrete to crack when its tensile limit is exceeded. Such an effect is particularly undesirable in water-retaining structures and those subject to alternate freezing and thawing. To overcome this problem the concrete can be compressed such that tension is just obtained when the design load is applied.

Today precompressing (more commonly referred to as 'prestressing') has been extended to general applications such as bridge beams (Fig. 40.1) and decks (Fig. 40.2) foundations and piles (Fig. 40.3) subject to impact loading, suspension bridge anchorages (Fig. 40.4), rock bolting (Fig. 40.5), portal frames (Fig. 40.6) and for specialised work such as dam strengthening (Figs. 40.7 and 40.8).

**Fig. 40.1**  Simply supported beam

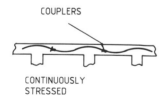

**Fig. 40.2(a)**  Continuous deck with tendon coupling

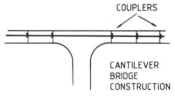

**Fig. 40.2(b)**  Bridge deck with tendon couplings

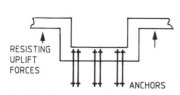

**Fig. 40.3(a)**  Basement anchors

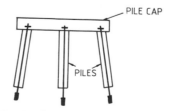

**Fig. 40.3(b)**  Foundation anchors

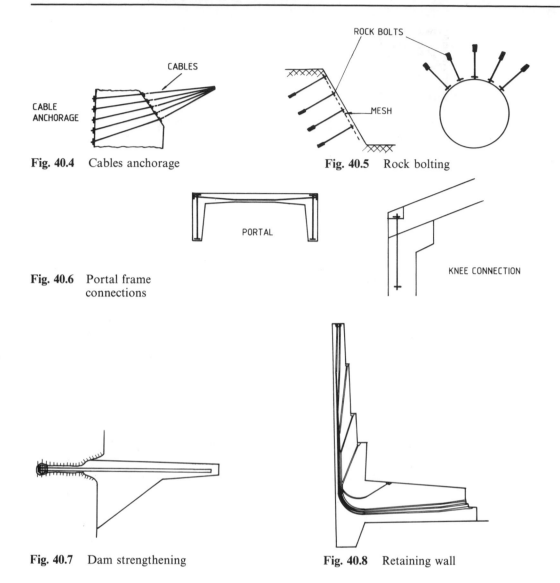

**Fig. 40.4**  Cables anchorage

**Fig. 40.5**  Rock bolting

**Fig. 40.6**  Portal frame connections

**Fig. 40.7**  Dam strengthening

**Fig. 40.8**  Retaining wall

## 40.2  DESIGN PRINCIPLE

The principle of prestressing requires the concrete to be initially forced into a state of compression which is then counteracted by the working load placed on the member. For example a beam subjected to bending would be stressed as shown in Fig. 40.9, on applying a uniform compressive stress of equivalent magnitude over the full face of the beam, the bottom tensile stress would be nullified while the top compressive stress doubled.

Furthermore, it can be seen in Fig. 40.10 that by applying the prestressing force eccentrically a 'hogging' effect is produced, thus

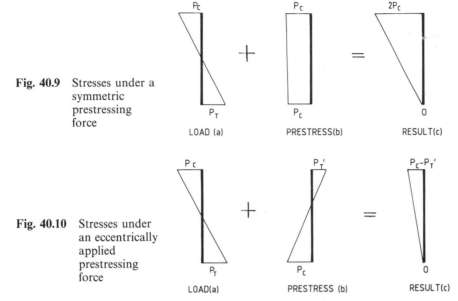

**Fig. 40.9** Stresses under a symmetric prestressing force

**Fig. 40.10** Stresses under an eccentrically applied prestressing force

relieving some of the compression in the top fibres. In practice, the member will sometimes only be loaded with its own weight and therefore stability is necessary in both this condition and under live load.

The four cases of stress which must be kept within the critical strength limits of the materials used are as follows:

1. The bottom compressive stress in the concrete under self-weight.
2. The top tensile stress in the concrete under self-weight.
3. The bottom tensile stress in the prestressing tendon under self-weight and live load.
4. The top compressive stress in the concrete under self-weight and live load.

## 40.3 TENDON PROFILE

In the case of beams produced by the 'long line' process the tendon is cast horizontally in the lower section of the member. The bending moment, however, will be parabolic as shown in Fig. 40.11. Consequently the eccentricity of the applied force may cause the critical stresses to occur as described in (2) and (4) above, near the end supports where there is little bending. This effect can be reduced by choosing a post-tensioning method whereby the eccentricity of the cable is altered to match the reduction in bending moment as shown in Fig. 40.12 for a simply supported member. Thus at positions of zero bending moment the prestressing force can be arranged to produce uniformly distributed compressive stress across the face. To achieve this the cable is positioned in the member by means of ducting firmly

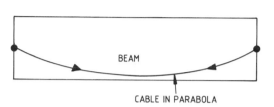

**Fig. 40.11** Bending moment diagram with a uniformly distributed load

**Fig. 40.12** Cable in parabola to match eccentric prestressing force

attached to the shear reinforcement stirrups and cage. Joints in the ducting must be well taped to avoid penetration of the grout during subsequent filling of the void.

## 40.4 METHODS OF PRECOMPRESSING (PRESTRESSING)

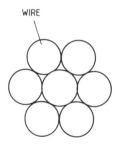

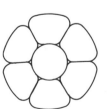

**Fig. 40.13** Strand configuration

High tensile steel (HTS) bars, wires, or strands are stretched, and anchored at the ends of the concrete member. Alternatively, the concrete may be cast around the tendon to provide bond between the steel and concrete after hardening. The latter is usually the basis of pretensioning and the former post-tensioning.

## 40.5 MATERIALS

### Steel

Many materials have been tried for prestressing, but cold drawn high tensile alloy steel wire or cold worked bar have proved the most reliable, both being made from hot rolled rod. The tendon itself may be in the form of a single wire or bar, a strand made up of a bundle of wires spun together, or cable, which is a bunch of strands.

### Strand

A strand usually comprises seven wires (Fig. 40.13), occasionally nineteen, six being helically wound on to a straight single wire core. Typical performance characteristics of both strand and cable are given in Table 40.1.

**Table 40.1**  Typical Design Forces

**Strand, comprising seven wires**

| Strand diameter (mm) | Characteristic strength (kN) |
|---|---|
| 12.5 | 165 |
| 15.0 | 225 |
| 18.0 | 370 |

**Cable, comprising seven-wire strand**

| Strands/strand diameter (No.) (mm) | Permissible tension force (i.e. 70% characteristic strength (kN) for standard strand)* |
|---|---|
| 4(12.5) | 450 |
| 4(15) | 650 |
| 4(18) | 1050 |
| 7(12.5) | 800 |
| 7(15) | 1150 |
| 7(18) | 1800 |
| 12(12.5) | 1400 |
| 12(15) | 1950 |
| 12(18) | 3200 |

* Higher tensile strength can be obtained with super and drawn strand.

### Bar

High tensile steel bar is obtainable smooth or ribbed and supplied in lengths up to 18 m. Available diameters are usually in the range 20–50 mm. The corresponding permissible prestress loads at 70% of the characteristic failing load are 228–875 kN. Threads are *rolled* on at the ends of the bar for anchorage or coupling purposes. The ribbing, however, will satisfactorily serve the purpose in the case of the ribbed bar.

## 40.6  PERMISSIBLE PRESTRESS LOAD IN THE TENDON

Unlike ordinary mild steel, cold drawn or worked high tensile steel does not have a clearly discernible yield point (Fig. 40.14) and thus some other method of determining the permissible prestress load is needed. The British Standard uses the concept of a proof load (Fig. 40.15), this being defined as the load which produces a particular permanent increase in the length of the wire – 0.1% is the commonly adopted elongation. The load is determined by drawing a line starting at the 0.1% elongation point and proceeding parallel with the elastic section of the graph until the deformed point of the curve is reached – this point is the 0.1% proof load, and in practical terms is

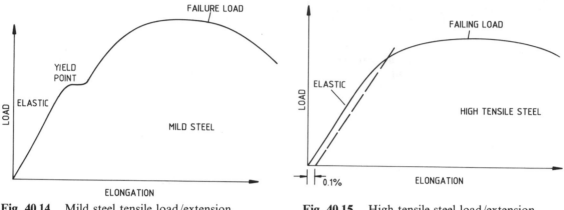

**Fig. 40.14**  Mild steel tensile load/extension characteristics

**Fig. 40.15**  High tensile steel load/extension characteristics

approximately 80% of the characteristic failing load of the specimen. The applied prestress tension should normally fall within this value. It should be noted that wire or strand is nowadays generally supplied as 'relaxed', i.e. the curvature resulting from the drawing process is straightened by stress-relieving heat treatment.

## 40.7   CONCRETE

The concrete for prestressed concrete applications should demonstrate low shrinkage, creep and elastic deformation, combined with high strength in compression. These characteristics can be obtained by using a low water/cement ratio together with selection of good quality and well graded aggregates. Precast concrete units commonly do not require congested cages of reinforcement and low workable mixes can be satisfactorily handled.

### Pretensioned concrete method

High tensile steel wires are tensioned between two anchorages and then surrounded by concrete in a formwork mould. When the concrete has gained enough strength the ends of the wires are released from the anchorages. Immediately the wires seek to return to their untensioned length, but being restrained by the bond with the hardened concrete, compressive force is transferred in the member. Some elastic deformation takes place, and slowly over time both shrinkage and creep will act to reduce the tension in the wires and thus the level of compression in the concrete.

While pretensioning is suitable for making individual or specialised units the most common application is in mass production of standard items in a factory environment, where conditions can be controlled and high quality achieved.

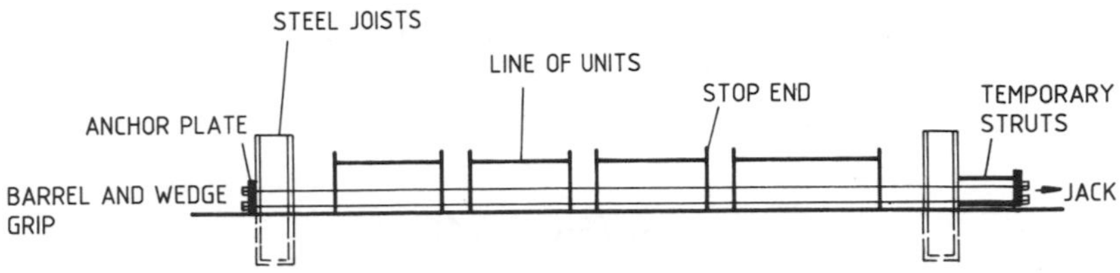

STEEL JOISTS

LINE OF UNITS

STOP END

ANCHOR PLATE

TEMPORARY
STRUTS

BARREL AND WEDGE
GRIP

JACK

**Fig. 40.16** Long line
method of
casting

The usual method is the 'long line' system (Fig. 40.16). The moulds
are placed end to end with a space between each stop end and the wire
stretched between two steel joists firmly fixed to the floor bed. The ends
of the wires are passed through templates and grip fittings attached.
One of the anchor plates is then jacked away from its steel strut to take
up the slack and temporary props positioned. Each wire is
subsequently tensioned as required with a hydraulic jack and anchored
off.

When the concrete has reached sufficient strength the exposed wires
between the units are cut. The stress at these points will suddenly be
reduced to zero, rapidly increasing to the prestress tension as the bond
takes effect. As a result lateral swelling occurs to form an ideal cone
shaped anchorage. Generally shearing reinforcement in the form of
stirrups is required along this short transfer length of tendon.

### Post-tensioned concrete methods

The concrete may be cast either as the complete unit or alternatively in
segments before the tendon is stressed. Usually the tendon is introduced
after the concrete has hardened by casting in duct tubes arranged to
give the appropriate profile (Fig. 40.17). Occasionally, for example
when using bar, the tendon is cast with the concrete either in a sheath
wrapped in tarred paper or with a bitumen coating to prevent bonding.
When the concrete has reached the 28 day strength post-tensioning can
generally proceed, using one of several proprietary systems.

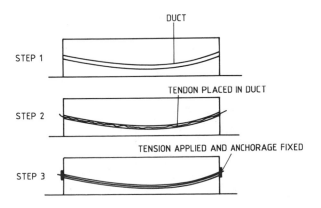

DUCT

STEP 1

TENDON PLACED IN DUCT

STEP 2

TENSION APPLIED AND ANCHORAGE FIXED

STEP 3

**Fig. 40.17** Duct profile

With the exception of bar they are all very similar and differ mainly in the form of anchorage and arrangement of the tendon into wire, strand or cable.

### Bar (Fig. 40.18)

The two common systems are:

(i) Macalloy bar – smooth bar.
(ii) Dividag bar – ribbed bar.

The usual form of anchorage transfers the tension in the bar through a nut and washer to a solid plate, bearing against the concrete.

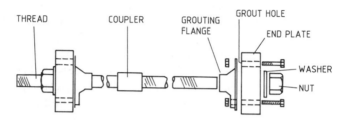

**Fig. 40.18** Threaded bar and nut system

### Tensioning procedure

The jack (Fig. 40.19) comprises a hydraulic pump powered by a diesel motor. The anchor nut-tightening device is integral with the jack housing and consists of a hexagon pinion manually turned using a ratchet spanner.

Before commencing tensioning, the end plate washers and anchor nuts are positioned. The hexagon insert is placed over one of the nuts

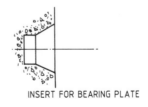

INSERT FOR BEARING PLATE

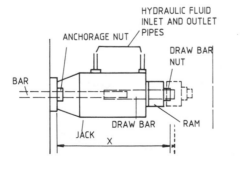

**Fig. 40.19** Tensioning jack for the bar system

and the draw bar screwed on to the projecting bar thread. The jack is then slid on to this bar and the draw bar nut screwed against the ram. A small amount of load is applied to take up slack and looseness and the initial dimension $x$ measured. Further load is subsequently applied and the anchor nut tightened until full load is reached. The increased value is measured and compared with that expected for the actual load. Any difference outside the manufacturer's quoted tolerance should be referred back for advice.

On completion the jack is removed and the duct grouted as a precaution against corrosion of the tendon.

## 40.8 BARREL AND WEDGE GRIP SYSTEMS

The majority of manufacturers have adopted strand systems rather than bar. The anchoring method generally consists of barrels and conical grooved wedge grips (Fig. 40.20a) butting against a strong cast iron

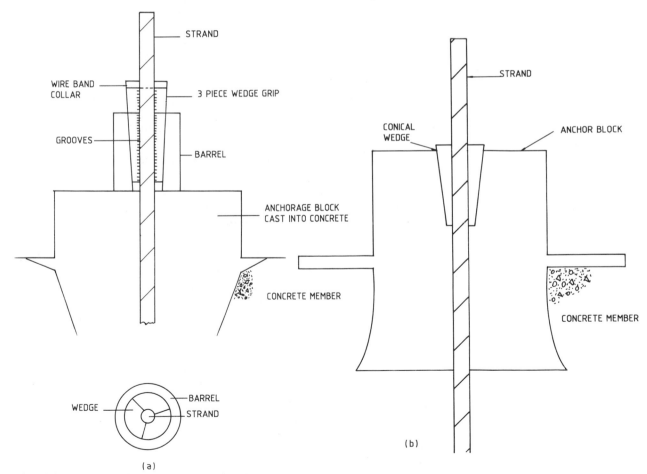

**Fig. 40.20** Barrel and wedge systems of anchorage

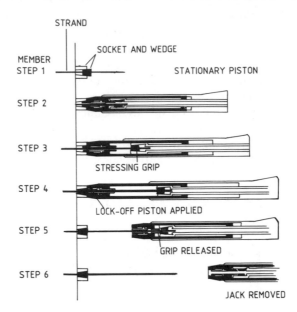

**Fig. 40.21** Tensioning jack for single-strand barrel and wedge system

distribution plate cast in the concrete. Alternatively, the conical wedges may directly engage the anchor block as shown in (Fig. 40.20b).

The wedge, tapered on the outside, is usually in two or three segments held together with a wire band collar and is grooved on the inside to aid grip with the tendon. Where several strands are involved, each is passed through the anchor block to produce a regular symmetrical pattern, the strands being gathered together to form a cable all housed in a single duct.

To achieve post-tensioning, the strands are either stressed individually in the order specified by the manufacturer, or more commonly all together as a cable. In the former case the jack is relatively light and man-handleable, for the latter because of its weight. Expensive lifting assistance is required, but the stressing time ought of course to be shorter. The typical stressing procedure shown in Fig. 40.21 requires a single strand to be passed through the jack, and either a temporary barrel and wedge fixed behind the ram, or alternatively the tendon is gripped inside the jack. The slack in the grips and tendon is then taken up (extension $e_1$ and $e_2$ in Fig. 40.22) and the required load applied. The extension of the strand $e_3$ is measured and compared with expected value. Any significant difference should be referred back to the manufacturers for advice. If everything is satisfactory, a ram inside the jack is activated to push the wedge into its barrel. The stressing force is then released causing the wedges to tighten, and increase the grip on the strand. A small loss of tension will occur during this phase ($e_4$). The temporary wedge and barrel can then be removed and the jack transferred to the next strand. A similar procedure can be adopted for stressing the strands simultaneously (Fig. 40.23). Single-strand jacks are available in the range 50–300 kN and 50–500 mm strokes, cable-stressing jacks up to 10 000 kN and bar jacks 700–1600 kN.

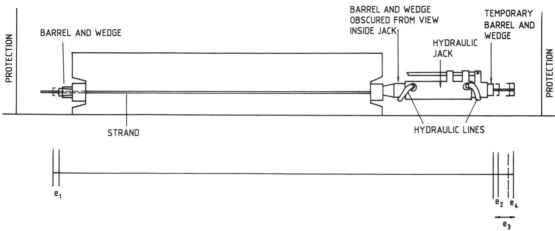

**Fig. 40.22** Tensioning procedure

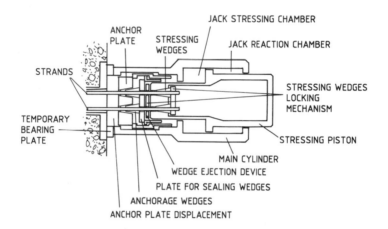

**Fig. 40.23** Tensioning jack for multi-strand barrel and wedge system

Many systems are available from different manufacturers all very similar. A few of the more popular ones are listed below.

1. CCL Systems Ltd, (Fig. 40.24). Monostrand and multiforce systems. Strands stressed singly or simultaneously.
2. Losingers Systems Ltd. VSL. Strands stressed simultaneously.

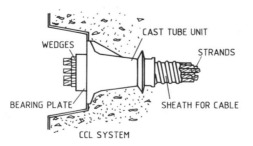

**Fig. 40.24** CCL system

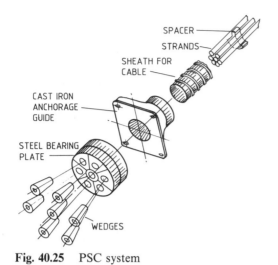

**Fig. 40.25**  PSC system

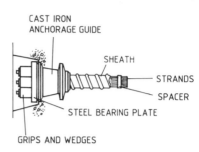

**Fig. 40.26**  SCD system

3. PSC Fregssinet Ltd (Fig. 40.25). Monostrand and monogroup systems. Strands stressed singly or simultaneously.
4. Pilcon Engin, Ltd (Fig. 40.26). SCD monogrip and multigrip systems. Strands stressed singly or simultaneously.
5. Stronghold International Ltd. Stronghold system. Wire or strands stressed simultaneously.

## 40.9  THREAD AND NUT WITH WIRE SYSTEM

In the BBRV system (Fig. 40.27), individual wires are passed through the anchor head and button ended. A choice of two types of anchorage is available – the anchor head is either threaded internally to receive a temporary pulling bar (a), or threaded externally to receive a pulling sleeve (b), the latter being more appropriate for larger anchorages. The tensioning method is similar to that of the bar system and thus the tendon length needs to be accurately determined to allow for the extension. In the final condition the tension is transferred from the anchor head on to a distribution plate.

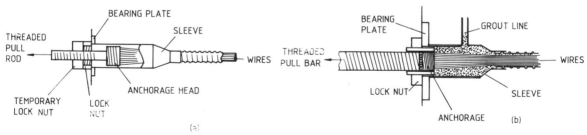

**Fig. 40.27**  BBRV system

### Couplings

Occasionally tendons require coupling as demonstrated in Fig. 40.28. Where a bar is used the method is quite simple requiring only a threaded sleeve (Fig. 40.28), but for strand, usually the tendons are arranged around integral anchorage and counter anchorage heads (Figs. 40.29 and 40.30).

The various manufacturers generally have purpose made systems for such joints.

**Fig. 40.28**    Bar coupling

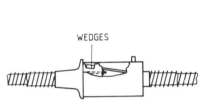

**Fig. 40.29**    Wedges and strands coupling

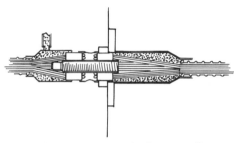

**Fig. 40.30**    BBRV wires coupling

## 40.10    LOSSES AFTER STRESSING

A loss in tension in the wires takes place as a result of:

|  |  | *Loss* % |
|---|---|---|
| (i) Shrinkage of the concrete | up to | 5 |
| (ii) Creep of the concrete | up to | 10 |
| (iii) Elastic strain in the concrete | up to | 5 |
| (iv) Creep or relaxation of the steel | up to | 5 |
| (v) Losses during the tensioning and anchoring operations | Total | 20–30% |

### Shrinkage

As concrete hardens, shrinkage takes place, partly caused by the chemical reactions and partly by drying out. Practice has demonstrated that low water/cement ratios produce less shrinkage, for example

| *Water/cement ratio* | *Shrinkage (mm/mm of length)* | |
|---|---|---|
| 0.4 | $200 \times 10^{-6}$ | ca. 5% of initial prestress |
| 0.7 | $800 \times 10^{-6}$ | |

About half of the shrinkage takes place in the first 3 weeks and is virtually complete after 6 months.

### Creep

When concrete is held under a sustained compressive stress, permanent deformation occurs causing a reduction in length. The amount of reduction is proportional to the stress in the section and increases with time. Up to 10% loss of stress may be caused by creep.

### Elastic strain

As load is applied to the concrete a reduction in length of the member occurs, the amount being related to the modulus of elasticity of the steel and concrete and is directly proportional to the stress applied. Elastic deformation takes place before creep (Fig. 40.31) with the latter eventually reaching a maximum value. In pretensioned concrete, stress is immediately transferred to the concrete and may eventually result in up to 4–5% loss of prestress. In post-tensioned work, where the stress is applied in stages, strand by strand, progressive shortening occurs, causing some loss of stress in the previously tensioned tendons, about 2% loss of prestress is normally assumed. Where post-tensioning is applied to the strands simultaneously at one of the anchorages, then the effect of elastic deformation is nullified, since the force in the wire is under direct control at the jacking position. Thus no allowance is needed for prestress loss.

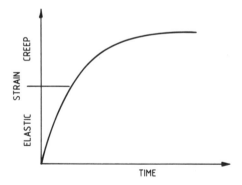

**Fig. 40.31** Creep and elastic deformation

### Relaxation of the steel

Tests have shown that steel under a tensile load loses some of its stress over a period of time. The rate of loss decreasing with time such that over a period of about 1000 hours the limit is virtually reached. The amount of creep is greater for wires in the 'as drawn' condition (i.e. delivered in coils with a natural curvature relating to the capstan size of the wire-drawing equipment) compared with heat treated straightened wire. Up to 5% should be allowed for steel relaxation, when the actual stress in the tendon is about 70% of the breaking load (normal recommendation). Slightly less relaxation is exhibited by bar.

### Loss before and during tensioning

In post-tensioned work, friction between the tendon and the duct prevents the full tension at the jacking point reaching positions further along the duct especially where curved profiles are involved, the degree of seriousness depending upon the duct material. Some loss of prestress also occurs in transferring the load to the anchorage.

### Loss of stress caused by jointing

In post-tensioned work, the final member may comprise several smaller precast units butted together (Fig. 40.32). To avoid movement problems after stressing, the units should be cast back-to-back, numbered and erected in an identical sequence.

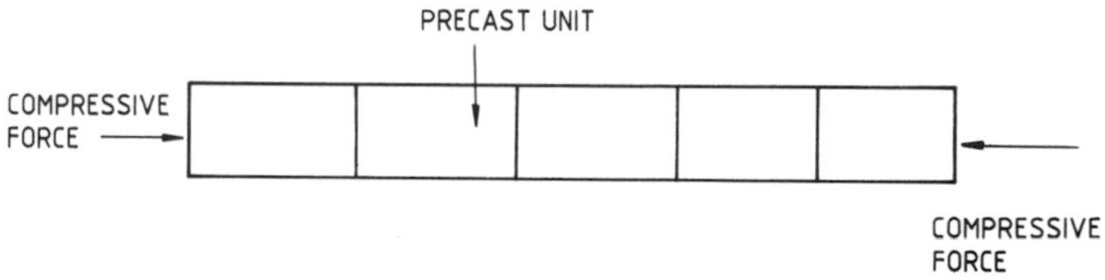

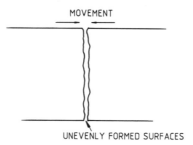

**Fig. 40.32** Assembled precast post-tensioned units

Alternatively the member can be temporarily supported in the final working position and in the case of a wide joint, filled with *in situ* concrete (Fig. 40.33). For a narrow gap high strength quick setting resin may be suitable.

MOVEMENT

**Fig. 40.33** Joint movement with post-tensioned units

UNEVENLY FORMED SURFACES

# ROAD PAVEMENT CONSTRUCTION AND EQUIPMENT

Both bituminous and concrete laying equipment are described, including modern paving trains. Batching, mixing and recycling of macadam and asphalt are also included. Some of the processes and equipment used in soil stabilisation are considered.

# 41

# FLEXIBLE PAVEMENT CONSTRUCTION

## 41.1 INTRODUCTION TO ROAD PAVEMENTS

A highway or road is composed of layers of selected materials which act to transfer and distribute vehicle loads to the subgrade. The objective being to limit the stresses within the capacity of the subgrade. This can be achieved with either a flexible or rigid pavement (Fig. 41.1).

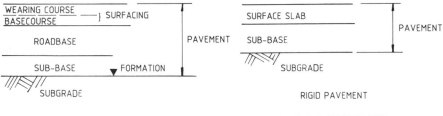

**Fig. 41.1** Flexible and rigid pavement layering

The flexible method relies on distributing the stresses by means of the pavement thickness, whereas a rigid pavement comprises a concrete slab utilising bridging action to span weak areas of the subgrade. As a consequence flexible pavements are commonly much deeper.

## 41.2 FLEXIBLE PAVEMENT DESIGN

The soil foundation formed after either completion of excavation works for a cutting or the formation of an embankmant is called the 'subgrade'. Usually the surface of the subgrade is referred to as the 'formation level'. The flexible pavement is then built up in layers as follows:

### Surface course

The surface course is usually constructed to produce a comfortable ride for traffic, and should ideally be hard wearing, resistant to weather attack, and display good properties with respect to skidding. In addition adequate water-proofing to the lower layers is essential and can be obtained by careful selection of surface materials.

On highly used roads it is common practice to separate the surface course into two layers – wearing and base layers. This method of construction facilitates maintenance, repairs and, in particular, replacement of the wearing layer. However, for minor roads the two are usually combined. The surface courses are generally made from tar or bitumen-bound aggregate.

### Roadbase

The roadbase acts to distribute the wheel loads through to the subgrade and therefore performs a most important function in the design of the pavement. High quality materials are required, such as hardcore, graded granular materials laid either wet or dry, cement or bituminous stabilised soil or gravel, or lean mix concrete. Depending upon the material, the method of laying can either be carried out using a paving machine or by simple spreading with bulldozers or graders combined with adequate compacting.

### Sub-base

This is an extension of the roadbase and will be required depending upon the quality of the subgrade and loading on the pavement. Compacted granular materials to aid drainage are common and may be of lesser quality than those used in the roadbase. Nevertheless the sub-base must be stronger than the subgrade.

## 41.3   RIGID PAVEMENT DESIGN

### Surface slab

This forms a structural plate and is formed from high quality strong concrete, sometimes reinforced with steel bars or mesh. Common practice often allows for expansion and contraction by incorporating joints at intervals passing through the slab. Laying is generally carried out with a paving machine.

### Sub-base

When the subgrade cannot provide uniform support to the slab because of the effects of frost action, poor drainage, swell and shrinkage, bad practice during earthmoving operations, construction traffic, etc., a sub-base is usually interposed between these two surfaces.

The sub-base may be formed of granular materials or a weak-mix concrete with the top surface sealed with a thin bituminised layer to prevent water absorption from the concrete layer. Laying can be carried out from a paver or simply by spreading, grading and rolling with earthmoving equipment.

### Subgrade

The subgrade forms the natural foundation to the pavement and is the result of either excavation works or the forming of an embankment. Thus it is important that adequate compaction takes place before constructing the other layers.

Maximum compaction is achieved at some optimum water content for the particular soil (Fig. 41.2). This value may be determined in the laboratory, with the subsequent standard of compaction to be achieved on site specified at some proportion of this value (Proctor test).

The design thickness of the pavement may be determined, by several methods, for example California Bearing Ratio (CBR) of the soil. The laboratory test uses a disturbed sample which is re-formed and compacted in a standard mould to a state anticipated at the construction site. A penetration load test is then performed and the CBR value obtained. A similar procedure could be carried out *in situ* on the actual subgrade.

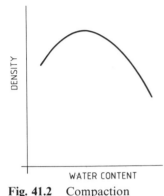

**Fig. 41.2** Compaction curve

### 41.4 BITUMINOUS LAYING METHODS

Macadams and sometimes bituminous materials can be spread by hand and raked level before trial rolling. This technique, however, is only suitable for small scale works such as patching, footpaths, domestic work, etc. For road surfacing a paving machine is commonly selected.

### The paving machine

Paving machines were introduced during the 1930s and are capable of laying a wide range of material consistencies and thicknesses. The width of the machine may be altered to suit the road design and its production potential is considerably greater than manual methods.

The paving machine comprises two parts, the tractor unit and the screed. The tractor, mounted on either tracks or wheels, supports the engine, transmission, controls, receiving hopper, feeders and spreader augers. The screed is pivoted at the rear of the tractor system and is simply towed on long levelling arms. The screed unit carries the compacting tamper, thickness adjuster, heaters and screed plate.

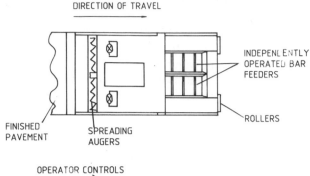

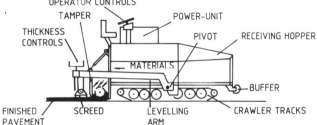

**Fig. 41.3**   Bituminous paving machine

## LAYING ACTION

Paving material is dumped into the receiving hopper from trucks positioned in front of the moving machine. It is subsequently conveyed to the spreading augers, deposited, tamped (variable 0–25 cycles/sec) and then passed under the screed plate.

The screed is hung from pivoted levelling arms. As the tractor pulls the whole unit forward, the plate lines up parallel with the direction of pull (Fig. 41.4). The depth of the material may be changed by adjusting the angle of the plate.

However, because the levelling arms are quite long, the reaction will not be immediate, and approximately 3–4 m of travel is required before equilibrium condition is achieved. This has the advantage of reducing the effect of irregularities which the tracks or wheels may meet, and therefore the finished surface will be much smoother than the travelling surface.

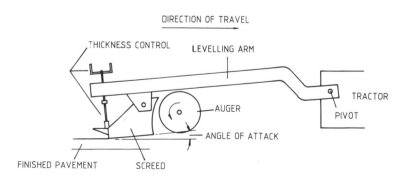

**Fig. 41.4**   Laying action

## AUTOMATIC SCREED CONTROL

The unpaved surface may have undulations and irregularities which could be undesirable if reproduced on the riding surface. Modern machines, however, are equipped to adjust the pivot height of the screed with respect to a guide line, thereby allowing the shape of the surface to be controlled.

There are three main types of automatic control system:

*Stringline or laser*

A stringline accurately set out between support pegs acts as a guide for a sensor mounted on a pivot arm of the paving machine. Usually the stringline is only required at one side of the machine as a built-in crossfall control mechanism operated automatically or manually sets the height of the opposite pivot arm. Stringlines are not particularly suitable for producing a flat surface because of sag and the latest machines would alternatively be equipped with laser sensing devices to do a similar job.

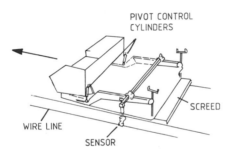

**Fig. 41.5**  Stringline control

*Shoe*

A kerb line or previously laid surface acts as the reference line for a 'shoe', which in turn activates the sensors on the pivot arm.

*Averaging beam*

The averaging beam comprises two pairs of feet, each pivoted to a stiff beam. The beam is independently pulled along, to match the travel of the paver, and a sensor on the paver's levelling arm is activated on

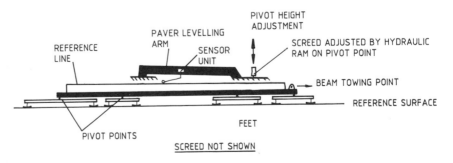

**Fig. 41.6**  Averaging beam control

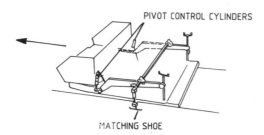

PIVOT CONTROL CYLINDERS

MATCHING SHOE

**Fig. 41.7** Matching shoe control

touching the reference wire attached to the top of the beam. Thus the irregularities are bridged resulting in improved smoothness of the paved surface.

## WHEELS OR TRACKS

On undulating surfaces, tracks ride on the crests of the irregularities, and provide similar advantages to the averaging beam, whereas wheels tend to reproduce the shape of the riding surface. However, with a bump such as a manhole cover the wheeled version produces only a small isolated deflection of the screed, while with tracks, the effect is reproduced along the full length of the tracks as the machine passes the irregularity. Also in the case of manholes it is sensible practice to lay material over the cover plate which can then be lifted flush later.

### Paving practice

1. The base surface should be clean, firm and free from excessive irregularities.
2. A tack coat of bituminous emulsion (0.5 kg/m$^2$) is required to give adequate adhesion to a non-bituminous based surface, or to prevent water seepage if the base course is of porous material.
3. Work uphill to give better feed control for the lorries.
4. When using a hot mix some heating of the screed prevents material sticking to the plates of the paving machine.
5. Run the paver at a travelling speed to match the rate of delivery of material, frequent stopping and starting may produce surface irregularities.
6. A thin wearing layer should not be laid on a wet surface if good adhesion is to be achieved.
7. Never allow the roller to remain stationary on the compacted surface, otherwise dents, wheel ruts, etc. may occur.

### Laying width

Pavers are manufactured with an approximate 3 m standard width screed, but extension augers and screeds are available to produce a carpet up to 12 m wide. Common practice is to lay a third or half the pavement width, then turn the machine around and lay the next strip.

This keeps the cost of the operation down (fitting extensions is expensive) and allows traffic to keep flowing without diversions. The main disadvantages are the risk of damage to the edge of the newly laid carpet by passing traffic and faulty jointing of the strips.

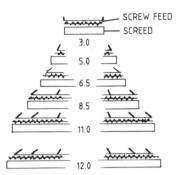

**Fig. 41.8** Laying width adjustment

## Paving data

| | |
|---|---|
| Paving widths | 3–12 m |
| Paving thickness | up to 300 mm |
| Paving speed | 1–30 m/min |
| Travelling speed | 25 km/h |
| Output up to | 500 t/h |
| Bearing pressure under tracks | 0.05–0.15 N/mm$^2$ |
| Bearing pressure under tyres | 0.05–1 N/mm$^2$ |
| Transport of hot mixes | not longer than $\frac{1}{2}$ h |

## Compaction

Traditionally, adequate compaction has been assumed by specifying the method of compaction. This is described as an 8–10 mg smooth wheeled static roller, with a roll width of not less than 450 mm or a multi-wheeled pneumatic tyred roller of the same weight, the surface layer always being rolled by a smooth wheeled type.

Rolling generally proceeds longitudinally, starting from the side working toward the centre of the strip, overlapping the preceding run by approximately half the roller wheel width. Rolling continues until all the roller marks have disappeared. The roller wheels must be kept moist to prevent adhesion. In more modern practice, however, the resultant density is specified to a percentage value of that obtained during laboratory design tests. This has caused investigations into more efficient compacting methods, especially using vibration and has allowed the specification of thicker construction layers. The principal rolling methods are summarised in Table 41.1.

**Table 41.1**   Compaction methods

| Roller type | Action | Advantages | Disadvantages |
| --- | --- | --- | --- |
| Static weight | Static weight | Produces a smoother, flatter surface with increasing stiffness of the material. Gives a smooth even surface | Effective only for shallow thickness |
| Vibration rollers | Static weight and dynamic action | Effective compaction for thick layers Good interlocking of aggregate | Difficult to determine the most suitable frequency and speed of travel |
| Rubber tyred rollers | Static weight and kneading action | Reasonably good interlocking locking of aggregate Effective compaction for thick layers Good surface seal Ability to vary contact pressure Uniform compaction | Expensive to purchase Experienced driver needed |

### Effect of temperature

Ideally the temperature of the material at the time of compacting should be approximately 50–60 °C above the softening point of bitumen (ring-and-ball test). For most of the common types of bitumen and tar, softening occurs at about 25 °C, thus compaction should have been completed before the materials temperature falls to 80 °C or so. Since the mixing temperature for asphalt and macadam is between 100–150 °C, during cold weather little time is available to achieve good compaction results. Figure 41.9 illustrates the approximate time needed to compact various thicknesses when the outside temperature is near

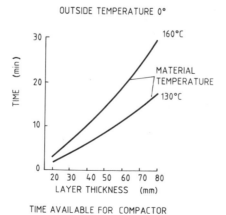

**Fig. 41.9**   Effect of temperature on compaction time

freezing. It can be seen that for thin layers cooling is very rapid. Hot boxes can be used to maintain the temperature at a high level for several hours, but evaporation of the oils can result in segregation of the aggregates.

### Number of compactors

The required number of passes to achieve compaction is usually determined in the laboratory and the required density subsequently specified for the construction phase. However, the conditions for achieving acceptable compaction results depend greatly upon the temperature of the material. For example, rolling of hot, soft material, especially asphalt, produces tracks, while cool harder material is difficult to compact. Thus the size and number of compactors must be carefully balanced with the capacity of the paving machine, and the time available before cooling is complete.

## 41.5 SURFACE DRESSING

As a road gets used the surface may suffer degradation due to a combination of traffic and weather conditions. Ultimately the wearing

CHIPPINGS SPREAD ON SURFACE          INITIAL EMBEDMENT INTO BINDER FILM

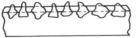

**Fig. 41.10** Surface dressing cross section

FINAL EMBEDMENT INTO EXISTING PAVEMENT

layer may have to be completely removed for example on major highways, or alternatively be given a new surface dressing as is commonly adopted on secondary roads and housing estate streets.

The surface dressing process consists of the application of a film of bitumunous or tar binder on to the existing road surface followed by a layer of stone chippings to be subsequently rolled in. The process gives:

(i) Improved non-skid wearing surface.
(ii) Waterproofing of existing surface.
(iii) Arrests surface disintegration.

### Chippings

In order that adequate adhesion to the binder can be obtained, the chippings should be strong, durable, dry and free from dust, otherwise complete failure of the surface may occur on heavily used roads. An adhesion agent can be added to reduce this risk.

### Application of surface dressing

The surface must be clean and free from: loose fragments, stone, dust, mud, etc. Also soft spots and depressions should be made good before the surface dressing is applied.

The bitumen or tar binder is sprayed uniformly across the old surface at a rate dependent upon the size and shape of the chippings, the hardness of the surface and the expected traffic density. The spraying equipment comprises a mobile tank fitted with pressure and rate of flow controls, together with a temperature monitor.

Common practice is to lay surface dressing equal to one traffic lane width, while additional lanes may be overlapped or butt jointed. The chippings are usually spread mechanically (Fig. 41.11) at a rate dependent upon the size, shape and specific gravity of the stones. A generous quantity is spread completely to cover the binder film. The final operation involves rolling the chippings into the old surface, preferably with a rubber-tyred or rubber-rimmed roller to avoid crushing.

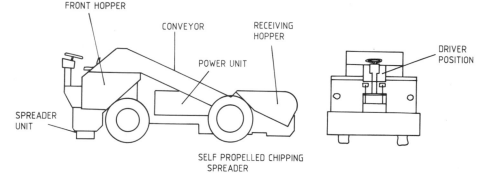

**Fig. 41.11** Surface dressing spreader

### 41.6 REPAVING

The removal of the wearing surface can be achieved with either hot or cold planning techniques, and the material reprocessed in a heater (drum mixer) and used again if necessary.

### Hot planing

The equipment comprises a self-propelled or pulled wheel-mounted burner and separate cutting drum. The layer to be removed is heated by means of propane-fired infra-red heaters, the softened material subsequently being scraped away with the planning drum to leave a relatively smooth, even surface for repairing. The loosened material is usually limited to about 40 mm working depth. The machine is rather cumbersome weighing roughly 30 tonnes.

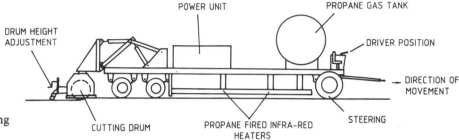

**Fig. 41.12**  Hot planing machine

## Cold planing

Cold planing is becoming increasingly popular, because of its ability to cut to greater depths (typically up to 300 mm) than the hot technique and produces no fumes.

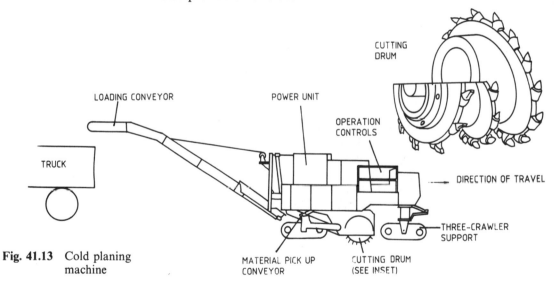

**Fig. 41.13**  Cold planing machine

The equipment simply comprises a wheeled or tracked cutting drum and elevator to transport the material to trucks for disposal. The planing action is achieved with tungsten carbide tipped teeth. Drums capable of spanning widths up to approximately 4 m are available. Like paving machines, automatic slope and grade controls can be included to remove the pavement to $\pm 3$ mm or so. The power unit is commonly in the range 250–600 kW with a production capability of 300–800 t/day, working at up to 50 m/min for shallow cuts of 25 mm. However, speeds as low as 5 m/min are not unusual for deep cuts. Such machines lie in the weight range 20–50 tonnes.

## 41.7 RECYCLING OF ASPHALT SURFACE MATERIAL

Recycling of existing surface material is becoming increasing common as the cost of bitumen products rise. The technique, however, dates back to the 1940s and takes place as either a hot or cold process, using material obtained from the planing process. Recycled pavement has been laid as base courses, and top or wearing courses for secondary roads.

### Hot recycling

The more commonly used equipment is a drum mixer, similar to Fig. 42.8. A 'mid-entry' method is favoured whereby the reclaimed asphalt pavement is introduced downstream from the burner and mixed with the heated new aggregate which has passed through the flame. After adding sufficient binder and mixing, the material is transferred to a storage bin.

A reclaimed asphalt/fresh aggregate ratio of 70/30 is achievable, but 50/50 is the usual practical limit to avoid air pollution problems – mixing temperatures exceeding 140 °C (280 °F) produces unpleasant quantities of blue smoke. The appropriate mixing temperature depends largely on the moisture content of the reclaimed material. The finished material is usually laid with conventional paving plant.

### Cold recycling

Cold recycling is generally an in-place method using material obtained from cold planing to which an admixture or bitumen rejuvenator (e.g. foamed bitumen) is added. To achieve successful results at least 95% of the planed material should pass through a 50 mm screen.

A common method of working involves spraying the pulverised surface with the additive followed by mixing during a second pass of the machine. However, some modern equipment allows the additive to be introduced during either the grinding or mixing phases (Fig. 44.6). Subsequently reshaping and compacting can be performed using conventional rollers. The technique is more appropriate to secondary roads with the recycled surface subsequently given a sealing of chippings or new wearing surface. A more conventional method of recycling requires distribution of the blended material as a windrow behind the planer to be picked up by a paving maching and re-laid in the normal manner or spread by bulldozers or graders. However, the best results are usually achieved by a travelling pugmill mixer/paver machine, fed from strategically positioned stockpiles of material crushed to the desired size with pulverising equipment (hammermill). Any additive is blended in the pugmill, the contents thoroughly mixed and relaid to grade and slope. The method is suitable for base courses up to 350 mm deep.

# BITUMINOUS-BASED MATERIALS FOR FLEXIBLE PAVEMENTS

The surface layers and often the road base of a flexible pavement are made from a combination of a binder material and aggregates. There are several mixtures in common use as outlined below.

## 42.1   HOT ROLLED ASPHALT

This consists of fine aggregate and a bitumen binder to which various proportions of coarse aggregate may subsequently be added to suit the particular circumstances. For example 30% coarse aggregate (i.e. that returned on a 2.36 mm sieve) is typically used for the wearing course, but higher proportions are commonly selected for base courses and road bases.

Coated chippings are generally rolled into asphalt wearing surfaces to improve skid resistance.

## 42.2   DENSE TAR MACADAM

Tar is used as the binder and laid as a hot material. Like hot rolled asphalt the proportion of coarse aggregate may be varied. Rolling produces a dense, impervious surface.

## 42.3   COATED MACADAMS

These contain little fine aggregate and either tar or bitumen may be used as the binder or coating agent. As a result the material has an open texture and depends on an interlocking of the aggregate particles for strength and stability. This method is suitable for all-weather roads lightly trafficked, garage forecourts, footpaths, etc.

Coated macadams may be laid hot, using a high-viscosity binder, or cold with a low-viscosity binder, such as 'cutback bitumen'. The latter method allows the material to be stockpiled and used at intervals as required without detrimental effect.

## 42.4   PRODUCTION OF ASPHALTS AND MACADAMS

The plant required to manufacture asphalts (Fig. 42.1) and macadams (Fig. 42.2) basically provide the following operations:

  (i) Drying the aggregate and removing dust.
  (ii) Screening the aggregate into the required sizes and sorting into storage bins.
  (iii) Heating the aggregate to the appropriate temperature.
  (iv) Heating the binder to the appropriate temperature.
  (v) Providing fine filler material.
  (vi) Weighing, batching and mixing.

Asphalt contains a high proportion of fines, i.e. sand, which tend to segregate from the coarse aggregate during movement after drying, while macadam being largely a coarse material does not suffer this effect. Thus for the asphalt-making process the screening, sorting and weighing is usually the last operation before mixing.

### Drying process

In order to obtain an efficient coating of the aggregate with the binder, the surface moisture content must be reduced by drying to less than 1% by weight. The aggregate is delivered by belt conveyor or feed hopper

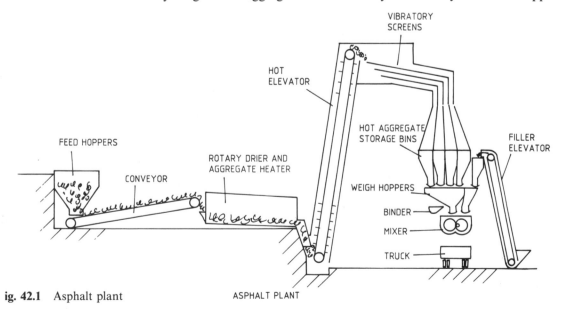

**ig. 42.1**   Asphalt plant

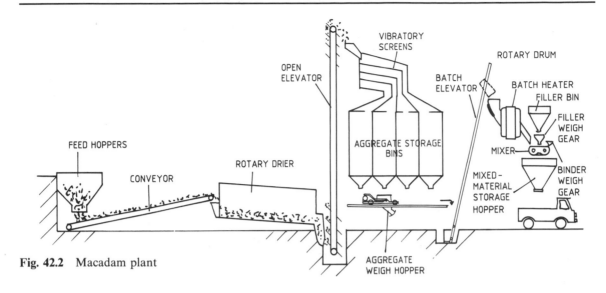

**Fig. 42.2** Macadam plant

from the storage bins to the drier. The most common type consists of a rotating drum, fitted internally with lifter blades to cause a cascading motion as the aggregate is propelled forward. It is usual to have the drum on a slight downward slope. Heating is commonly provided by an oil burner located at the discharge end of the drum so directing the hot gases towards the oncoming aggregate. The high percentage of fines in an asphalt mix necessitates the drying unit to be much larger than that for macadam. Furthermore in the case of macadam, heating of the aggregate to the mixing temperature is carried out as a separate process and therefore the temperature in the drying process may be relatively low (a little above 100 °C).

### Heating the aggregate

For asphalts, heating and drying are performed in a single operation, after which the aggregate is screened into separate sizes and temporarily stored in small hot hoppers before being combined together and mixed with the binder. This is necessary to avoid the fines separating out during movement. Macadam, however, is stored in bins immediately after drying and then proportioned into the batch size and transported to a separate rotary heater drum for mixing.

### Heating the binder

The bitumen or tar tar binder is heated in an insulated tank either with steam or hot oil circulating through pipes located inside the tank.

## DRYING AND HEATING CONTROL DATA

(i) Hot laid asphalt – drying/heating temperature:

| *Binder* | *Aggregate* |
|---|---|
| 135 °C (275 °F) | 150 °C (300 °F) $\pm$ 10 °C |

(ii) Cold laid asphalt – drying/heating temperature:

| | *Binder* | *Aggregate* |
|---|---|---|
| Straight bitumen | 125–160 °C | 80–120 °C |
| Cutback bitumen | 90–125 °C | 50– 75 °C |

(iii) Macadam (drying temperature 100 °C +); heating temperature:

| | *Binder* | *Aggregate* |
|---|---|---|
| Straight bitumen | 95–150 °C | 65–150 °C |
| Cutback bitumen | 65–130 °C | max. 70 °C |

## CAPACITIES OF PRIMARY DRIERS

| Production | 50–500 t/h. |
|---|---|
| Oil consumption | 5–15 kg/t (depending upon moisture content). |

### Dust extraction

Excessive dust may reduce the efficiency of the plant and in any case would cause difficult working conditions when the aggregate is moved from process to process. The dust is extracted during the drying phase using fans to direct it to a collector, and then removed by spraying with a stream of water. The resulting sludge is deposited at the base of the tank and subsequently removed and dumped. Approximately 95% of the dust can be removed in this way.

The cyclone collector (Fig. 42.3) is a more modern process and gives a dry product which can be fed back into the system as filler material. The dust-laden gas enters tangentially at the top of a conically shaped

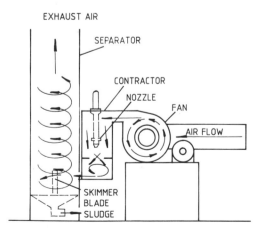

**Fig. 42.3**  Cyclone collector for fines

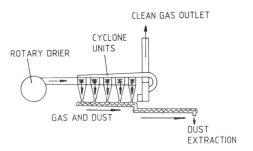

**Fig. 42.4** Cyclone collectors in series

drum, causing the particles to be thrown centrifugally on to the walls. A clean air vortex is consequently formed up the centre of the cone while the dust falls to the base for collection. Approximately 60–90% of the dust may be removed, depending upon the particle size. Where a single unit is insufficient the cyclone collectors may be arranged in series (Fig. 42.4). Furthermore a wet collector can be added as a secondary unit if necessary.

## Proportioning and mixing

The apparatus for screening and storing sand and aggregate is described in the section on aggregate production (see page 438). For macadam screening, storing and proportioning are carried out immediately after drying but before heating, while for asphalt this operation takes place after the combined drying/heating operation. Both weight and volumetric methods of proportioning are used, although for accuracy weigh batching is favoured. First the aggregate is drawn from the storage bins and transferred into a weigh hopper in the required proportions until enough material to produce a particular batch size has been weighed out. After weighing it is discharged into the mixer and the binder added. Proportioning of the binder may be weighed out similar to the aggregate but in a separate trough. Pumping the hot binder and monitoring through a flow meter is an alternative method, however this technique is unsuitable for binders containing abrasive matter such as lake asphalt.

Generally the last process in the sequence involves introduction of the filler, but unlike the aggregate and binder, is stored in silos and pumped or conveyed to a storage bin and subsequently weighed in a separate hopper before transfer to the mixer.

## Mixing

A batch mixer consists of a steel drum, fitted with twin counter rotation (approximately 40 rev/min) parallel shafts supporting a series of paddles (Fig. 42.5). The material is discharged through the base.

Continuous mixers are longer and the material enters at one end and discharges at the other, thus the paddles are designed to keep the mix rolling forward. It is customary to provide heating equipment to the

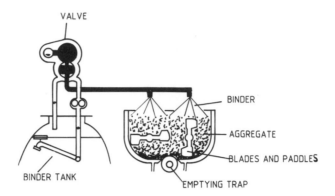

**Fig. 42.5**   Batch mixer

casing. In batch mixing approximately 20–40 s mixing time per batch is usual.

## 42.5   MODERN DEVELOPMENTS IN BITUMEN-COATED MATERIAL

### Production

The objective of the several new methods is to increase the rate of production and to reduce the dust problems associated with the older techniques. Figure 42.6 illustrates the use of the drying drum after the mixing process. The aggregates are screened and batched in the normal way and fed into the mixer, while the filler and hot bitumen are added. The mix is then transferred into the drying drum now called an 'activator', and the water in the aggregate is evaporated. An additive is generally included during the mixing phase to enhance the separation of the water and bitumen.

Figure 42.7 shows an alternative arrangement where the mixer unit is completely eliminated, mixing takes place in the drying drum together with the filler and bitumen. This method does not remove all the water, leaving a final moisture content of 1–3%. A binder additive is usually included.

These new methods are well suited to handle recycled asphalts and macadams (Fig. 42.8).

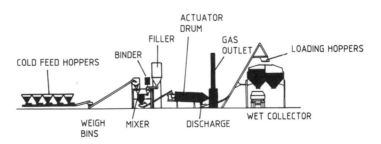

**Fig. 42.6**   Drying drum after the mixing process

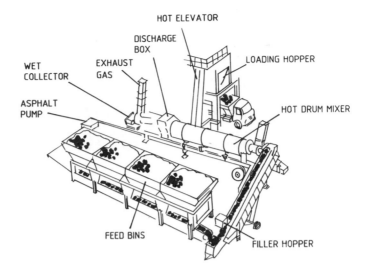

**Fig. 42.7** Production plant with combined mixer/drying drum

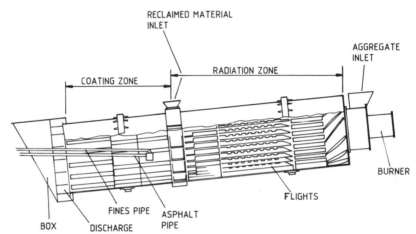

**Fig. 42.8** Mixer/drying drum

## PRODUCTION DATA: LARGE UNITS (ASPHALT)

| Output (t/h) | 25 | 50–100 | 100–120 | 120–150 | 150–200 | 200–250 | 250–300 |
|---|---|---|---|---|---|---|---|
| Drier size (m) | $1 \times 6$ | 1 @ $8 \times 2$ | 2 @ $8 \times 2$ | 2 @ $8 \times 2$ | 4 @ $10 \times 2$ | 8 @ $10 \times 2$ | 10 @ $10 \times 3$ |
| Mixer capacity (kg) | 250 | 600–1000 | 1300 | 1500 | 2000 | 2500 | 3000 |
| Filler silos (t) | | $1 \times 25$ | | $2 \times 25$ | | | $5 \times 60$ |
| Floor area (m²) | 1000 | 1500 | 2000 | 3000 | 2500 | 2500 | 3000 |

*Note*: smaller plant down to 5 tonnes per hour is available. For macadam production a secondary but smaller heater is necessary.

# 43

# CONCRETE PAVEMENT CONSTRUCTION

Mechanised methods for laying concrete pavements were introduced during the 1930s but the popularity of bitumen products has largely overshadowed concrete. The techniques for forming concrete roads with mechanised equipment may be classified as:

1. Paving trains.
2. Slip-form pavers.

The need to give adequate compaction, careful surface finishing, and to cope with jointing and grooving has demanded more complicated pavers than those used for asphalt and macadam.

## 43.1  TRANSVERSE JOINTS

Joints in the pavement are required to facilitate expansion and contraction caused by temperature and moisture changes.

### Expansion joints

These are usually incorporated at about 100 m centres, while those for contraction are required at about 5–10 m centres. The type of expansion joint (Fig. 43.1) may be the same for both reinforced and unreinforced slabs and comprises dowel bars cast into the concrete,

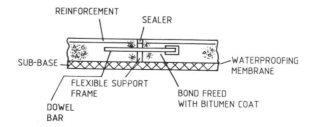

**Fig. 43.1**  Expansion joint

with one half coated in bitumen and encased in a loose sheath to allow movement. Any reinforcement in the deck must be discontinued at the joint. During construction the bars are supported on fibrous boarding secured to the sub-base – the material being flexible to accommodate the expansion movement. The top of the groove is filled with joint sealant.

Recent developments in laying pavements during the summer months, however, have indicated that expansion joints can be eliminated as there is usually sufficient shrinkage at the contraction joints to allow for any subsequent expansion.

### Contraction joints

These are usually induced by forming a groove in the top of the slabs with temporary filler bar (Fig. 43.2). The resulting plane of weakness subsequently cracks thereby providing freedom for contraction. When the slab is unreinforced, dowel bars must be inserted to prevent vertical movement and like expansion joints, one half of the dowel must be free to move and is therefore coated with bitumen (Fig. 43.3). When the concrete has reached sufficient strength the filler bar is removed and replaced with a sealing compound.

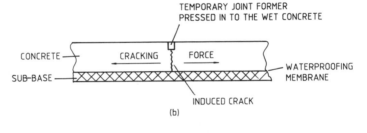

**Fig. 43.2** Contraction joint for unreinforced slabs

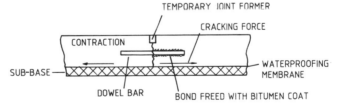

**Fig. 43.3** Contraction joint reinforced slabs

## 43.2 LONGITUDINAL JOINTS

Dowelled longitudinal joints are provided when the slab is concreted in more than one strip (Fig. 43.5) or more than about 5 m wide. They serve to prevent the slabs drifting apart and/or differential vertical movement. When the road is laid as a single carpet and is wide enough to require joints, these may be formed in a similar manner to the previously described dowelled contraction joints but with the bars assembled in a continuous cage and placed in position ready for the concreting operations. Otherwise they generally need to be cranked through 90° tied to the side forms, and later straightened before concreting of the abutting slab.

### 43.3 REINFORCED PAVEMENTS

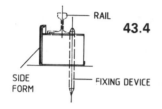

Concrete slabs reinforced with mesh are laid in two-course construction. The leading paver first places the bottom layer of concrete, followed by a work gang manually placing the sheets of mesh (or alternatively by machine). Finally the top course of concrete is placed by a second paving machine.

### 43.4 FIXED FORM PAVING TRAIN

**Fig. 43.4** Edge formwork for use with paving train system

Paving train equipment rides on rail track mounted on steel side formwork positioned along the pavement edges (Fig. 43.4). The forms are usually fixed in position a day or so in advance and not removed for several hours after concrete placing. Thus sufficient length must be in position to meet the planned rate of advance of the paving training.

The main units and sequence of operations in forming two course construction are shown in Fig. 43.5.

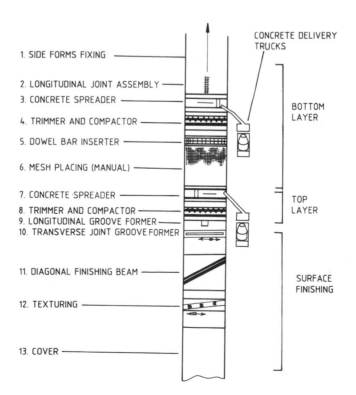

**Fig. 43.5** Paving train equipment

### Sequence of operations

1. Forms and guide rails are set, waterproof membrane laid and side joint longitudinal tie bars fixed in position.
2. Longitudinal central joint reinforcement assembled and positioned.
3. Bottom course of concrete spread from travelling hopper.
4. Bottom course trimmed and vibrated mechanically (if required).
5. Transverse dowels inserted mechanically.
6. Mesh reinforcement manually placed onto bottom layer of concrete.
7. Top course of concrete spread by second travelling hopper.
8. Top course trimmed vibrated and screeded by second compacting machine.
9. Longitudinal surface joint groove cut in wet concrete, temporary filler strip inserted.
10. Transverse joint formed with similar machine to (5).
11. Top surface given a final screeding with oscillating or vibrating finishing equipment.
12. Surface wire brushed with texturing equipment to provide skid resistance.
13. Groove filler strips replaced by flexible sealing material after concrete hardening.

## 43.5   PAVING EQUIPMENT

### Travelling spreader

Spreading the concrete may be performed with a blade spreader, screw spreader or box-hopper spreader (Fig. 43.6). The latter is usually preferred on paving trains because of its ability to handle fairly stiff concrete mixes. It can also be loaded from the side and thereby avoid the need to bring trucks onto the paving area. The box-hopper contains approximately 3 cm of concrete separated segmentally to aid distribution. The spreading action takes place as the hopper is travelled transversely from one side of the pavement to the other during forward movement of the whole unit. The depth of the layer is set by trial and error to allow for compaction during the subsequent operations.

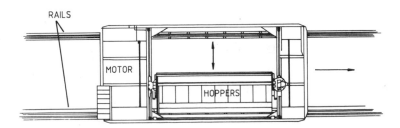

**Fig. 43.6**   Travelling box spreader

DATA

| | |
|---|---|
| Typical working speed | 0–20 m/min |
| Travelling speed range | 0–60 m/min |
| Power | 20–30 kW |
| Width (various) | 3–13 m |
| Operation | 1 operator |

### Screeding and compacting machine

The machine comprises a row of strike-off paddles or screw blades mounted in front of the compacting beam adjusted to level the surface to the required height by pushing excess concrete forwarded to the sides. (Usually the depth set is 15–20% greater than the final slab thickness to allow for compaction.) Compaction is obtained with a transverse compacting plate, whose vibrations and amptitude may be varied by trial and error to suit the mix consistency (Fig. 43.8). For most mixes the plate is commonly vibrated at from 50 to 75 Hz.

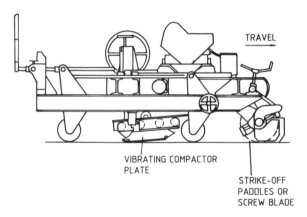

**Fig. 43.7**  Screeding and compacting machine

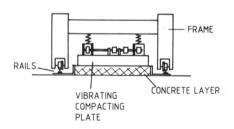

**Fig. 43.8**  Compacting mechanism

DATA

| | |
|---|---|
| Working speed | 0–12 m/min |
| Power | 20–30 kW |
| Paving depth | up to 350 mm |
| Vibrating compacting beam | 50–75 Hz |
| Widths | 2.50–12 m |

### Dowel bar insertion

Dowels for longitudinal and expansion joints are normally manually positioned ahead of the paving train as described earlier. However, the fixing of transverse joints dowels can usually be mechanised as shown in Fig. 43.9. The unit follows immediately behind the first compacting machine releasing and vibrating dowels into the bottom course concrete.

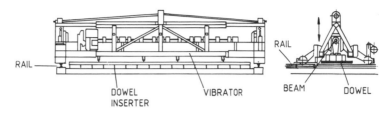

**Fig. 43.9** Dowel bar insertion machine (transverse joints)

DATA

| | |
|---|---|
| Working speed | 0–30 m/min |
| Frequency | 60–70 Hz |
| Widths | 2.5–12 m |
| Power | 8–10 kW |

### Joint forming

Longitudinal joint forming takes place after the top course compacting operation. A groove is mechanically cut into the wet concrete with a ploughing tool and temporarily filled with plastic strip. Transverse joints are formed separately using a machine almost identical to the dowel inserter unit (Fig. 43.10), but with the dowel insertions replaced by a blade spanning the full width of the strip. The machine can also be arranged to install expansion type joints, whereby a gap corresponding to the filler board width is first vibrated into the concrete and then removed. The filler board and dowels are subsequently positioned manually and vibrated down to the required depth by the machine.

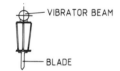

**Fig. 43.10** Transverse joint grooving plate

### Laying mesh reinforcement

Where a reinforced deck is required the concrete must be laid in two-course construction. The mesh sections are stored on a wagon travelling directly behind and to the side of the first paving train and manually placed on top of the first course layer. Gradually, however, this operation is being updated by a mesh-laying and depressing machine following the first paving train.

### Final shaping and finishing

On major roads and airfields a high quality surface without irregularities is often specified. This can be achieved by a final finishing unit travelling behind the joint-grooving machine. Two commonly used methods are: (i) parallel screed and oscillating screed units and (ii) vibrating and oscillating screed unit.

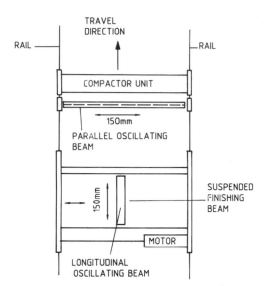

**Fig. 43.11** Oscillating finishing beams

A parallel finishing screed forms part of the compactor unit and is simply towed along so avoiding any subsequent joint-grooving operations. The beam or blade is oscillated parallel across the line of travel at up to 70 strokes/min with approximately 150 mm strokes. The planing angle may be varied to accommodate different concreting consistencies and travelling speeds.

When better quality finishing is necessary an independently propelled longitudinal finishing beam is also included in the paving train (Fig. 43.11) behind the grooving operations. The beam is suspended from a bridge and moved to and fro between the side forms at up to 15 m/min, while also being oscillated parallel to the centre line of the movement and has the advantage of minimising disturbance to joint-groove strips.

The vibratory/oscillating finisher consists of a vibrating beam mounted just in front of an oscillating screed. The vibrating effect removes high spots in the concrete and recompacts, while the oscillating beam produces the smooth surface finish (Fig. 43.12). The usual configuration is the diagonal mode set at up to 60° to the line of travel to minimise disturbing filler strips (Fig. 43.13). The whole unit is normally self-propelled, but when filler strip type of jointing is absent, a simple tow behind the compacting machine can be arranged (Fig. 43.14).

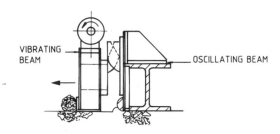

**Fig. 43.12** Vibratory/ oscillating finishing equipment

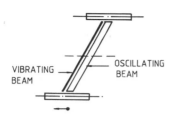

**Fig. 43.13** Diagonal mode for ideal finishing

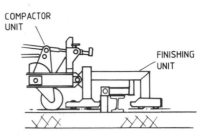

**Fig. 43.14** Compactor-towed finishing beam configuration

DATA

| | |
|---|---|
| Travelling speed | 0–15 m/min |
| Stroke | 150 mm |
| Oscillating frequency | 40–70 strokes/min |
| Vibration frequency | 60–70 Hz |
| Power (self-propelled) | 20 kW |

### Texturing

The type of textured finish varies depending upon designer's preference. For example, a simple brushed surface can be applied by walking a brush across the pavement or alternatively by raking or tooth combing. The equipment comprises a bridging truss to support the mechanism and/or operator.

The final task involves covering the concrete to give protection against the elements. The covers are mounted on the rails and sufficient area is necessary to allow concrete to harden sufficiently to prevent rain damage.

### Sawed joints

While many concrete roads are designed with wet grooved construction joints, considerable inconvenience can be avoided, for example problems of disturbing the filler strip in wet concrete by sawing the joint in hardened concrete with a diamond impregnated rotary bladed machine (Fig. 43.15). Depths up to 200 mm can be cut with blades varying in thickness from 2 to 25 mm, and 250 to 500 mm in diameter. A rotation speed of 3000–4000 cycles/min is usual.

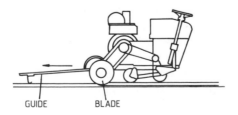

**Fig. 43.15** Joint sawing machine

PAVING TRAIN DATA

Component parts (two-layer construction) and labour.

(i) First concrete spreading unit (one operator).
(ii) Screeding and compaction unit (including dowel insertion, one operator).
(iii) Mesh wagon (crane driver, truck driver, two labourers, one unloading, one operator on mesh depressing unit).
(iv) Second concrete spreading unit (one operator).
(v) Screeding and compacting unit (one operator).
(vi) Jointing unit (one operator).
(vii) Finishing units (one operator).
(viii) Texturing units (from one to two operators).
(ix) Fixing side forms (two labourers).
(x) General labourers for tidying up.

## Output rate

Actual output $\qquad Q_a = v_p \times b \times d \times y$
Theoretical output $\qquad Q_t = v_t \times b \times d$

where $b$ = breadth of pavement;
$d$ = depth of concrete in the pavement;
$Q$ = concrete quantity by volume;
$y$ = efficiency factor depending upon the control of the units, supply of materials, gradient, crossfall and camber of the deck, etc. – approximately 0.3 to 0.8;
$v$ = speed of the train; up to 20 m/min ($v_t$) can be obtained but usually 1–3 m/min ($v_p$) is more likely because of the problems of concrete supply.

Units are manufactured to produce pavement widths from 3 m to 12 m, assembled in modules to allow flexibility.

## 43.6 SLIP-FORM PAVING METHODS

The need for, and accuracy required in setting side forms for the conventional concrete paving train is time consuming. Also the constituent parts of the train demand careful control requiring a relatively high number of operators. These disadvantages have largely been removed with the development of the slip-form paving machine during the past 30 years, whereby a 'carpet' of concrete is put down without the need for semipermanent side forms. However, much stiffer mixes, with slumps of 30–40 mm are necessary for the concrete to maintain its shape without the side supports.

Several manufacturers make a variety of slip-form pavers broadly classified into: (a) track-mounted paving train with integrated side forms, and (b) track-mounted conforming plate paving machine.

### Integrated paving train

In this method of paving, the side forms are part of the machine and simply act to give shape to the concrete. The various stages in the conventional paving train are incorporated into the unit arranged in succession from the front to the rear of the paver as follows:

(i) Spreading the concrete from a hopper or screw auger with compaction provided by built in or internal vibrating units.
(ii) Striking off the concrete to the required depth with a primary screeding beam or spreading plate.
(iii) Finishing the concrete surface with transverse oscillating float beams. Final smoothing can be given by a suspended static float plate at the rear of the paver.

Two-course paving is usual. A mesh-depressing machine is located in front of the rear paving unit, finishing beams are not required for the first-course paver. Bar insertion equipment for both longitudinal and transverse units and grooving machines can be included in the train as required.

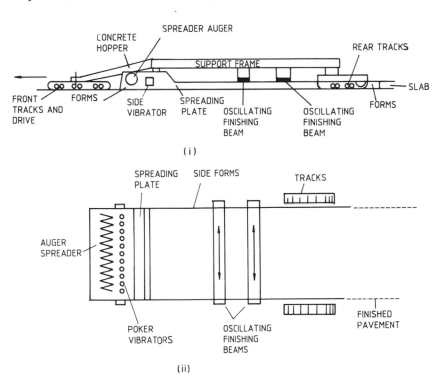

**Fig. 43.16** Integrated slip form paving machine

## PLACER-SPREADER MACHINES

In single-course paving the concrete is often spread in windrows up to 30 m ahead of the paver to increase production. The unit is similar in construction to the paver and distributes the concrete with a screw auger, the uncompated depth being set by means of a screeding/strike-off plate. The concrete is usually fed to the auger with a conveyor belt, which on modern machines is designed to move along the length of the auger, so ensuring uniform distribution.

### Conforming plate paver

The conforming plate paver is designed to produce in a single course the width and the thickness of the pavement by extruding the concrete beneath a plate mould as the unit is moved forward. Concrete in the hopper is kept topped up to a surcharge level fed by conveyor from the side and controlled with a hydraulically adjustable metering plate. In the larger versions a screw spreader is sometimes mounted just in front of this plate and acts to distribute the material more effectively. Compaction is obtained by means of a row of poker vibrators or a vibrating beam mounted across the hopper.

Transverse joint dowels can be injected with a plunger mechanism through tubes from magazines placed in front of the concrete receiving hopper. However, because of the complexity involved with a machine equipped with this sort of facility, mechanical problems have been a

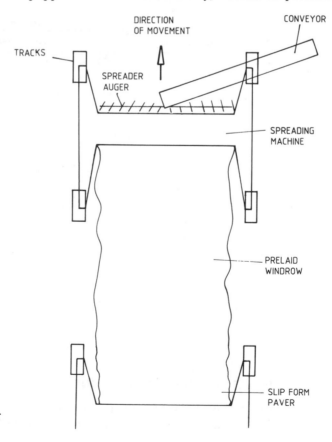

**Fig. 43.17**  Placer-spreader unit

common feature, and as a consequence manually pre-set dowels are still quite usual.

Longitudinal joint dowels can also be mechanically inserted (including the temporary joint forming strip fed into a groove cut behind the conforming plate) as the paving unit moves along. Surface finishing prior to joint grooving is carried out with a transverse oscillating beam running side to side across the top of the projecting side forms. A final stationary hanging float beam is sometimes attached immediately behind this beam to give a fine finish. To avoid complications transverse joint groove forming operations are usually sawn after the concrete has hardened.

## SLIP FORM PAVER DATA

| | |
|---|---|
| Power up to | 300 kW |
| Road widths | up to 10 m (machine width adjusted by adding inserts manually, or by built-in hydraulic equipment) |
| Working speed | up to 0–5 m/min depending upon the ability to supply concrete |
| Pavement thickness | up to 300 mm |

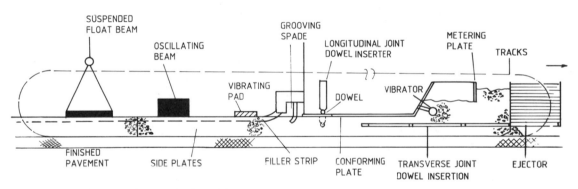

**Fig. 43.18**  Conforming plate paving machine

### Line and level

Guide wires or laser beams accurately set out each side of the paver just beyond the tracks are the common means of level control. Sensors attached to the four corners of the paving unit, either raise or lower the side forms or conforming plate.

### Texturing and curing the concrete

The concrete surface must be textured to improve skid resistance and like conventional paving trains this process is commonly obtained by means of wire brushing raked either manually or mechanically propelled from a bridging unit. Good practice usually requires the finished surface to be sprayed with a curing compound and the wet concrete protected from rainfall or intense sunlight with travelling tentage. Approximately 40–60 length is necessary for 1–2 m/min travelling speed.

## PRODUCTION OUTPUT EXAMPLE

Although paving machines are capable of travelling at 20 m/min or more, the logistical problems of supplying large quantities of concrete commonly result in no more than about 1000 m of pavement being laid during a typical day of production. (That is, 8 hours with the average travelling speed of the paver being about 2 m/min.) Thus for a 9 m wide road 200 mm deep and a 50% efficiency factor.

$$Q = \frac{1000}{8} \times 9 \times 0.2 \doteqdot 225 \text{ m}^3/\text{h}$$

The time between mixing and placing should not exceed 45 min. Good practice would normally require the transport distance from batching plant to paving unit to lie within 3 km.

# SOIL-STABILISED PAVEMENT CONSTRUCTION

During the past 30 years or so, soil-stabilisation methods have been developed to increase the strength and durability of soils, to improve impermeability, to resist frost attack and reduce volume variations. The relative success in achieving these properties has led to applications in road construction, embracing the wearing surface, foundation layers and subgrade. Soil stabilisation may be achieved by:

1. Mechanically by straightforward compaction.
2. Physical reactions caused by the addition of water to cement or lime, or the solidification of hot bitumen or drying out of cut-back or emulsified bitumen, mixed with the soil.
3. Chemically by the action of two or more chemicals mixed with the soil to produce a new compound.

Figure 44.1 illustrates the application of various methods and additives.

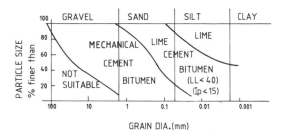

**Fig. 44.1** Soil stabilisation materials

## 44.1 STABILISATION WITH CEMENT (RIGID PAVEMENT)

Cement-stabilised soil achieves its strength in a manner similar to concrete, in which the hydrated cement matrix bonds the soil particles together. However, unlike concrete the cement paste is not designed completely to fill the voids, but only to coat the particles, with the bonds only forming at the points of contact (Fig. 44.2).

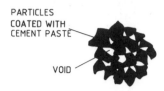

PARTICLES
COATED WITH
CEMENT PASTE

VOID

**Fig. 44.2**    Stabilisation
with cement

It can be seen from Fig. 44.1 that cement can be used to stabilise a wide range of soils from gravel down to clay. However, coarse gravels are difficult to mix. Also fine soils are only suitable if the liquid and plastic index of the natural soil is less than 40 and 15, respectively, otherwise intensive mixing becomes difficult to achieve.

### Quantity of cement required

Different types of soil require varying quantities of cement depending upon the grain size and surface area, usually the finer the material the greater the cement quantity as typified in Table 44.1. The range of values for a particular soil type is governed by the need to achieve strength, frost resistance, etc. The technique is suitable for producing rigid type pavement.

**Table 44.1**    Cement quantities and soil type

| Soil type | Cement as % of the combined dry weight | 7 days compressive strength (N/mm$^2$) |
|-----------|------------------------------------------|------------------------------------------|
| Sand | 4–6 | |
| Silty sand | 6–10 | |
| Silt | 6–12 | 1–4 |
| Clay | 10–10 | |

Water is added during the mixing phase to bring the moisture content up to the optimum level for maximum density during compaction (Proctor test). From 1 to 2% additional water should be added to allow for evaporation.

## 44.2    STABILISATION WITH LIME (FLEXIBLE PAVEMENT)

Lime commonly takes two forms, quicklime (CaO) and slaked lime (CaOH$_2$). Unlike cement, lime does not produce a bonded cellular structure with soil, but simply undergoes chemical changes, such that the voids are filled to form larger granules. The technique is suitable in clayey sands, silts and clays. The precise chemical reactions which take place are as yet not fully understood. In general terms the calcium oxide forms calcium hydroxide when mixed with water causing the production of heat, as a result much of the free water in the soil is evaporated leaving only that absorbed in the lattice structure of the clay component. This water is gradually drawn out by the calcium ions, removing sodium and potassium ions at the same time. As a consequence the clay begins to flocculate and subsequently coagulate to form a gel between the remaining grains of the soil and a crystalline structure is thereby obtained.

The overall effect causes an increase in the plastic limit of the soil, while the liquid limit is barely changed. Thus the 'plastic index' is reduced, with the soil becoming more granular and therefore capable of compaction with conventional equipment. Water absorption and permeability are reduced, leading to improved frost resistance. Also the resulting material can be reworked without losing its properties, and furthermore remains stable in wet conditions. The technique is well suited for use in stabilising road subgrades or even for the road base itself. As with cement, lime-stabilised soil has water added during the admixing phase to achieve maximum possible compaction (determined by the Proctor test). If the soil to be stabilised has a greater moisture content during construction than the optimum determined from the Proctor test, then powered quicklime is preferred, otherwise slaked lime is usually adequate.

Quantity required – 3 to 9% of the combined dry weight.

## 44.3 STABILISATION WITH BITUMEN OR TAR (FLEXIBLE PAVEMENT)

Bitumens and tars can be used with almost any type of soil, but for practical mixing reasons very coarse material and soils with a liquid limit and plastic index greater than 40 and 15, respectively, are usually not treatable. Unlike cement, bitumen or tar stabilisation does not rely on a chemical reaction, the strength being obtained by adhesion of the soils grains resulting from the initial tacky coating which subsequently hardens to form the bonds. After compaction a strong waterproof material is obtained. A good quality surface finish is difficult to obtain and therefore the technique is usually restricted to road base courses or minor roads. The binder is generally applied as:

(i) Emulsions of bitumen or tar.
(ii) Hot bitumen or tar.

### Emulsions

A bitumen emulsion is designed to evaporate quickly, leaving a thin coating around the grains as shown in Fig. 44.3. However, for this effect to be efficient the soil should have only 4–12% water content, very dry or very wet soil can be mixed but only with considerable difficulty.

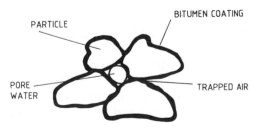

**Fig. 44.3** Stabilisation with tar or bitumen

### Hot bitumen and tar

Hot tar is applied at 90–100 °C, while 70–80 °C above softening point is sufficient for hot bitumen, either binder can be used with most types of soil. The best strength and compaction is obtained with soils having a continuous and proportionate reduction in grain size through the grading curve. The viscosity of the binder, should be varied to suit the grain size, for example, the larger the grain size, the more viscous the binder, and vice versa.

The main disadvantage with a hot binder is that its temperature will sharply fall on contact with the colder and wet soil particles, viscosity will increase and thus the coating process may not be satisfactory. Better results can be obtained by using an emulsifier containing dust particles. Pozzolana added in 1:1 proportion to the binder is usually sufficient. The precise quantities should be determined from test samples depending upon the specification for strength, compaction and water absorption.

### Bitumen quantity

| | | |
|---|---|---|
| Coarse sand | 4–6% | |
| Medium-to-fine sand | 4–7% | } by combined dry weight |
| Silt | 5–8% | |

## 44.4 STABILISATION WITH CHEMICALS (VARIOUS)

Although both cement and lime stabilisation methods involve a chemical change, chemical stabilisation usually refers to more 'esoteric' materials. These are beyond the scope of the discussion but the more widely known varieties are:

(i) Calcium and sodium chloride.
(ii) Phosphoric acid.
(iii) Polymers and resins.
(iv) Calcium acrylate.
(v) Aniline furfural.
(vi) Sulphate liquor.
(vii) (Reynolds Road Packer 233).

## 44.5 CONSTRUCTION OF A SOIL-STABILISED LAYER

Soil stabilisation may be carried out as follows:

1. Central mixing and subsequent laying with a paving machine.
2. Mixed-in-place with a train of machines.

### Central mixing

The soil is excavated with conventional earthmoving equipment, loaded into trucks and transported to a central mixer. Mixing plant suitable for concrete, asphalt or macadam production (according to the admixture) is commonly used depending upon the binder, filler, etc. The mix is then transported back to site transferred to a paving machine and spread in up to 300 mm deep layers. Output from 200 to 400 t/h are possible with large batching equipment.

### Mixed-in-place

Mixed-in-place involves the following series of operations as illustrated in Fig. 44.4.

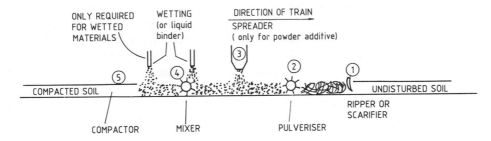

**Fig. 44.4**   Mixed-in-place soil stabilisation sequences

   (i) Loosening the soil (top soil having been removed).
  (ii) Pulverising followed by levelling and grading.
 (iii) Spreading the stabiliser (and filler).
 (iv) Mixing and wetting (or adding liquid binder).
  (v) Compacting.

In hard compact soils, the layer to be stabilised may need to be initially loosened with a ripping or scarifying method before using a blade-type pulverisor.

### SPREADING

In the case of a powdered stabiliser the material is spread in about a 2 m wide carpet from a tractor or truck pulled spreader (about 5 m$^3$ capacity), which is continuously topped up by pumping from a tanker. A liquid binder, however, is usually added during the mixing phase.

### PULVERISING AND MIXING

Fine compact material should be thoroughly pulverised before mixing takes place. Generally the same machine can be used for both operation. Bladed versions (Fig. 44.5) are suitable for soils, but more

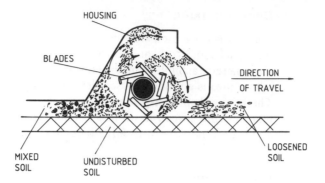

**Fig. 44.5**  Pulverising and
mixing unit

modern equipment can cold plane, pulverise and mix both blacktop
and concrete pavements.

The unit comprises a segmental tungsten carbide tipped toothed
rotor accommodating different cutting widths (Fig. 41.13) and has an
appearance similar to the road-milling machine. In contrast to
blade-type machines the rotor turns counter to the travel direction and
the whole unit is self-propelled. A liquid binder can be applied directly
by the unit during mixing but a bitumen stabiliser is best heated to
about 180 °C to avoid hardening during metering.

### MIXER UNIT DATA

| | |
|---|---|
| Cutting width | 1.5–2 m |
| Mixing depth | up to 300 mm |
| Working speed | up to 0.5 km/h |
| Rotor speed | 0–150 rev/min |
| Power | 250–600 hp (190–450 kW) |

Up to 2000 m$^2$ per day can be treated.

### COMPACTION

Final compaction is performed with convention rollers, either of the
smooth wheel static type, sheep's foot or pneumatic tyred versions.

### 44.6  COMBINED UNITS

Machines have been developed combining all the functions of
pulverising, adding the binder, mixing, spreading, levelling and
compacting in a single unit, however, little information is available on
their applications.

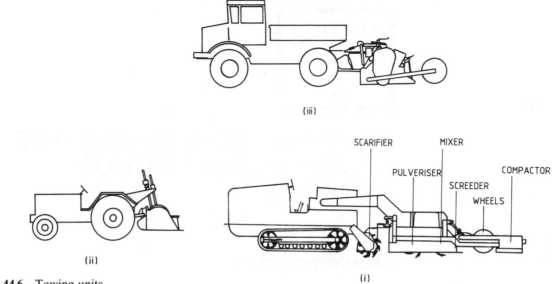

(iii)

(ii)

SCARIFIER     MIXER

PULVERISER     COMPACTOR

SCREEDER

WHEELS

(i)

**Fig. 44.6**   Towing units

# WELDING TECHNOLOGY AND BRIDGE CONSTRUCTION METHODS

## WELDING TECHNOLOGY

Many aspects of construction involve the use of welded steel plate and girders. Much of this work is often carried out exposed to the vagaries of weather and the choice of welding method must therefore be carefully selected according to the conditions. Basic gas welding and manual methods are described, together with details given on the $CO_2$ shielded and submerged arc processes. Some of the specialised techniques such as plasma arc, laser, ultrasonic, electron beam, thermit, explosion and induction welding are briefly mentioned together with brazing. Guidelines on good practice, inspection and defects testing are also covered.

## BRIDGE DECK CONSTRUCTION

In recent years new forms of bridge design and use of materials have been introduced leading to lighter, more economical solutions and increased spans. In this extensive chapter, (Chapter 46), systems of bridge classification according to type of structure and method of construction are presented, the emphasis being directed towards establishing the criteria for selecting the appropriate temporary works. Specifically covered are simple beam decks, progressive placing, push-out method, stepping formwork systems, cantilever methods, stay cable and suspension bridges.

# 45

# WELDING TECHNOLOGY

## 45.1 INTRODUCTION

Welding is the joining of at least two pieces of metal by: (i) raising the temperature at the joint to melting point, or (ii) cold welding using pressure. Category (i) is referred to as 'fusion', while (ii) can be achieved in a number of ways including forge welding and resistance welding.

Fusion welding is particularly suitable for construction applications in joining metal pipes, beams, reinforcing bars, plates, fittings and fastenings, etc., in most metals including steel, cast iron, alloys and coated metals both in the ordinary atmosphere and underwater.

Methods for inspecting and testing welds without the need for destruction has also progressed to ensure that good quality is obtained and the design strength achieved. This aspect is especially important in construction work, where much of the welding uses manual methods exposed to the elements – the opportunities for automatic welding set-ups are rarely obtainable on site.

## 45.2 WELDING PROCESSES

### Forge welding

Forge welding was developed in earlier times and is the technique practised by blacksmiths whereby two pieces of metal are heated in a fire and joined by hammering. The method can be used for simple butt, tee-butt and scarf welds. Considerable craft skill is required in judging the correct temperature to achieve a satisfactory joint, and is little used in the construction industry. Modern versions of this method include hot pressure, cold pressure and friction welding which have applications in manufacturing.

### Resistance welding

Resistance welding was probably invented by E. Thompson in the 1870s. The technique relies on the heat which occurs when a current flows through a resistance, for example, the two pieces of metal to be joined (Fig. 45.1). When the metal in the vicinity of the joint reaches a certain temperature the pieces can be forged together by applying pressure.

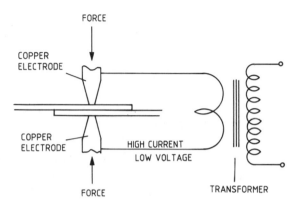

**Fig. 45.1** Resistance welding

**Table 45.1** Summary of welding processes

| | Aluminium | Copper/alloys | Iron | Magnesium | Nichel/alloys | Precious metals | Steels | Titanium | Tungsten | Zinc | Lead |
|---|---|---|---|---|---|---|---|---|---|---|---|
| Forge | Most metals | | | | | | | | | | |
| Resistance | √ | √ | √ | √ | √ | √ | √ | √ | √ | √ | √ |
| Gas (fusion) | √ | √ | √ | × | √ | √ | √ | × | × | √ | × |
| Shielded manual arc | √ | poor | √ | × | √ | × | √ | × | × | × | × |
| CO₂ shielded | √ | √ | √ | √ | √ | × | √ | √ | × | × | × |
| Submerged arc | × | × | poor | × | poor | × | √ | × | × | × | × |
| Plasma arc | √ | √ | √ | √ | √ | √ | √ | √ | √ | √ | √ |
| Laser beam | Mostly used in welding special steels such as titanium | | | | | | | | | | |
| Ultrasonic welding | | | | | | | | | | | |
| Electron beam | | | | | | | | | | | |
| Thermit | √ | | √ | √ | | | √ | | | | |
| Explosion | Mostly used for welding dissimilar metals | | | | | | | | | | |
| Induction | Mostly used for joining steel pipes | | | | | | | | | | |
| Brazing | √ | √ | √ | × | √ | √ | √ | × | × | × | × |

Note: √ = yes, × = no.

According to Joules law the heat produced $J$ is governed by the following expression:

$$J = I^2 R T$$

where   $I$ = the current (in amps);
        $R$ = the resistance (in ohms);
        $T$ = the time the current is applied (in seconds);
        $J$ = the heat produced (in Joules).

While $I$ and $T$ can be carefully controlled, $R$ depends upon the pressure between the electrodes, the types of material being welded, its thickness and surface condition. The method is mostly used for welding thin plates, and so relatively low voltage is required (20 V to 50 V) to produce high current from 2000 to 100 000 A. The pressure applied at the point of contact between electrodes and plates is approximately 100 N/mm$^2$ and the current is maintained for about half a second.

Modern equipment can be set to produce a welding time of as little as $\frac{1}{1000}$ second thus allowing the use of currents at the upper end of the range giving a consequent increase in the welding rate.

The current should only be applied when the full pressure has been obtained (Fig. 45.2) and left on only sufficient time to achieve forging heat otherwise burning of the electrodes will take place.

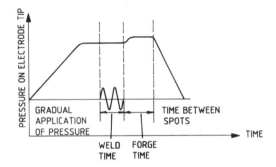

**Fig. 45.2**  Resistance welding pressure–time requirements

The technique is suitable for most types of metal including, steel, copper, aluminium, zinc, lead, silver, gold and metal alloys.

Steels pose few problems for plates up to 15 mm thick, and welds have been made in 50 mm plate. Aluminium, brass and copper have lower resistances than steel and therefore require higher currents. The maximum plate thickness for these metals varies up to approximately 5 mm.

The method is widely used in the manufacturing industry and in steel fabrication workshops. However, the difficult working conditions and variety of joint shapes found on construction sites are rarely suitable for resistance welding.

## TYPES OF RESISTANCE WELDING

This method is used in many forms including spot welding, seam welding, projection welding and flash butt welding.

*Spot Welding*

Spot welding is commonly adopted for joining thin sheets. The equipment comprises two stem-type copper electrodes, moved from spot to spot applying current and force to each.

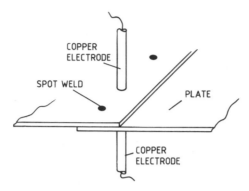

**Fig. 45.3**   Spot welding

*Seam Welding*

Seam welding is similar to spot welding except that the stem electrodes are replaced by wheels. They are run quickly along in a straight line to make a 'closely-spaced' row of spot welds. This technique in effect produces a continuous weld and can be used for pressure vessels.

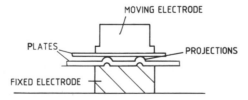

**Fig. 45.4**   Seam welding

*Projection Welding*

Protrusions are pressed on to one of the sheets at predetermined locations such that when the electrodes are positioned the applied force causes the weld to take place exactly at the projection and be flattened in the process.

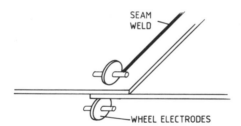

**Fig. 45.5**   Projection welding

*Flash Butt Welding*

Butt welding relies on the high localised heat caused by the resistance between two touching surfaces when electrical current is passed and is especially suited to welding rails, hoops, rings, etc.

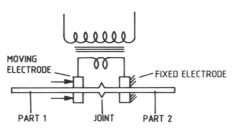

**Fig. 45.6** Flash butt welding

The equipment consists of two electrode clamps each gripping a piece of material – one clamp is movable while the other is fixed. To obtain a simple butt weld the two pieces are brought together, the current applied and the force increased when the junction reaches sufficient temperature.

Flash butt welding is a slightly different technique for higher quality butt welds, the essential difference being that the current is switched on before the pieces are brought together. As the two parts approach, intense heat is produced and arcing takes place, such that when full pressure is applied molten metal is extruded to create a high quality joint.

## METAL WITH PROTECTIVE COATINGS

If the coating will conduct electrical current, resistance welding can be achieved. Thus galvanised, electroplated, tin-plated or sheradised surfaces are all weldable, however some of the coating may be lost near the vicinity of the weld.

Non-conducting surfaces such as paint, would have to be removed before a weld could be effected.

**Table 45.2**  Resistance welding data

| Item | Spot welding | Seam welding |
|---|---|---|
| Sheet thickness (mm) | 0.5–15 | 0.5–10 |
| Power (kW) | 5–75 | 5–75 |
| Welding time | 0.5–10 sec/weld | 60–10 mm/sec |

## 45.3  FUSION WELDING METHODS

### Gas welding

Gas welding is often chosen because the equipment is inexpensive and easily obtainable. Also the method is versatile and can be used for welding many different shapes and positions of joint, for a wide variety of metals up to about 25 mm thickness. Indeed the technique is an alternative to electric arc welding, the main disadvantage being that greater skill is required and the variability in the quality of the weld.

Oxy-acetylene is the gas mixture commonly used as this produces 3000–3500 °C flame, however for metals with a low melting point, a lower temperature is needed to avoid the metal melting too quickly. For example a hydrogen-oxygen flame or oxygen-propane flame at about 2650 °C is more suitable for alloys.

## OXY-ACETYLENE

Oxygen and acetylene gases react together during combustion to produce a high-temperature flame. The immediate reaction produces carbon monoxide and hydrogen:

$$C_2H_2 + O_2 = 2CO + H_2$$

The highest temperature is reached just in front of this part of the flame. Oxygen continues to react with the carbon monoxide and hydrogen as they move out into the surrounding air to produce carbon dioxide and water vapour:

$$2CO + O_2 = 2CO_2$$

$$2H_2 + O_2 = 2H_2O$$

When the acetylene and oxygen are burnt in equal proportions by volume, the flame is described as neutral. Oxidising takes place when there is an excess of oxygen. The effect can be lessened by using a deoxidising agent such as silicon with the welding rod material to form a protective layer of silica ($SiO_2$) on the metal plate surface. Conversely, when there is an excess of acetylene the flame is carburising or reducing, causing the metal to absorb carbon from the flame. Non-ferrous alloys and carbon steel sometimes have these requirements. For welding of most steels a neutral flame is preferred.

## SHAPE OF THE OXY-ACETYLENE FLAME

A neutral flame is produced by first lighting the acetylene to give a yellow smoking flame. The gas pressure is then increased until the smokiness disappears, so providing the signal to introduce the oxygen. Its pressure is subsequently raised until a sharp white cone appears at the tip of the pipe. A neutral flame (Fig. 45.7a) has a clear white inner cone surrounded by a uniform and even outer light blue flame. When oxidised acetylene is present a ragged feathered white plume will be noticed around the cone within the outer flame (Fig. 45.7b). Excess oxygen, however, causes a reduction in the size of the cone combined with a very streaky appearance to the outside plume (Fig. 45.7c).

## GAS CYLINDER STORAGE

Acetylene is unstable when compressed below about 1 bar ($0.1 \text{ N/mm}^2$) and may detonate and explode. However, if the storage cylinder is filled

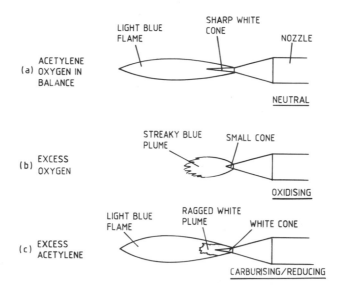

**Fig. 45.7** Oxy-acetylene flame settings

with a porous substance, such as asbestos, and the acetylene gas dissolved in acetone (which will take up many times its own volume of acetylene) the mixture is stable and made safe for transport.

Typically the gas is compressed to about 1.4 $N/mm^2$ and stored in a strong steel cylinder.

Oxygen is also stored in a steel cylinder but compressed to about 17 $N/mm^2$. The gas pressure at the torch must therefore be reduced, because both gases are mixed together and delivered at a common pressure, generally at from 0.01 to 0.1 $N/mm^2$ (1–14.7 $lb/in^2$) depending upon the type of torch and the size of the tip. Pressure reduction takes place through regulating valves, delivered in equal proportions by volume from the mixing part of the blow pile or torch. A typical set up is shown in Fig. 45.8.

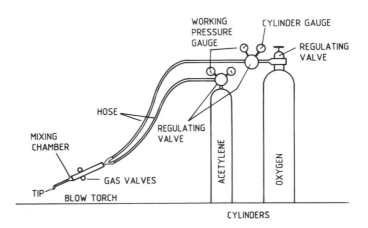

**Fig. 45.8** Oxy-acetylene welding equipment

### BLOW PIPE DESIGN

When using gas cylinders a high pressure blow pipe must be used (Fig. 45.9a). However, if acetylene is produced on the premises at low pressure then it is necessary to incorporate an injector into the torch to deliver oxygen at a fairly high pressure (from 0.1 to 0.2 N/mm²) to help speed up the acetylene flow (Fig. 45.9b). This type of blow pipe is known as a 'low pressure' torch and can also be used with the high pressure cylinder system if a high pressure torch is not available. However a high pressure torch *cannot* be used with the low pressure system (i.e. low pressure acetylene production) as it is unsafe to do so.

**Fig. 45.9**  Oxy-acetylene welding blow pipe (torch)

## Welding methods

Assuming the welder is right-handed, the welding technique is known as either 'leftward' or 'rightward' movement.

### LEFTWARD MOVEMENT

The method is mainly used for plate under 5 mm thick and non-ferrous metals. The torch is held in the right hand and given a side-to-side motion across the joint, moving slowly forwards but trailing the filler rod held in the left hand. Hence the direction is away from the right hand (i.e. away from the body) so giving leftward movement.

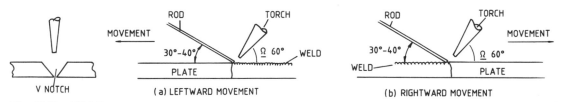

**Fig. 45.10**  Welding movements

### RIGHTWARD MOVEMENT

The filler rod is trailed behind the torch thus there is less tendency for the filler material to be pushed forward and spilled on to heated parts of the plate. Also a better annealing action takes place. The method is more appropriate for plates thicker than 5 mm, up to 15 mm plate can be welded in one pass thereafter the weld should be built up in runs and layers.

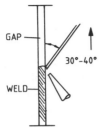

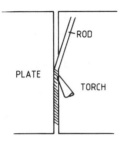

**Fig. 45.11** Vertical welding method

# VERTICAL WELDING

Common practice adopts the leftward movement working from the base upwards (Fig. 45.11). Up to 5 mm plate can be welded without any special preparation of the joint (i.e. V-shaping). When two operators are working each side of the plate, then this limit may be increased to about 15 mm plate. Thereafter either a rightward technique, with a prepared joint working from the top downward or the bottom upwards is needed.

# OVERHEAD WELDING

Either a leftward or rightward technique may be applied but considerable skill is necessary. The method relies on surface tension acting on the molten pool. The blow pipe is held almost vertically.

# PREPARATION OF THE JOINT

The configurations shown in Fig. 45.13 are preferred for butt joints. The adjoining surfaces should be uniform, level and cleansed of oxides and grease.

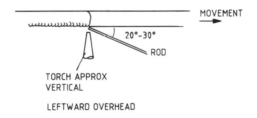

**Fig. 45.12** Overhead welding method

**Fig. 45.13** Weld joint design

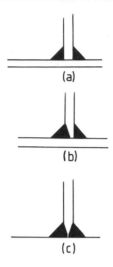

(a)

(b)

(c)

**Fig. 45.14**   Fillet welds

## LINDE WELDING

This method is often used for manual welding of pipes and tubes and uses a carburising flame. In this way any iron oxide present is reduced and the carbon absorbed, thereby providing improved fusion to give a stronger weld. Also the increase in carbon causes a lowering of the melting point and so leads to more rapid welding. The rod should contain silicon and manganese to aid deoxidisation and fluxing.

## FILLET WELDS

These require a slightly larger size of blow torch than for a butt weld with a similar plate thickness. Good practice requires the legs of the weld to be of equal length (Fig. 45.14a).

Where the weld is to be subjected to high loading then a single V should be considered (Fig. 45.14b).

For very high loads the double V is more appropriate (Fig. 45.14c).

## WELDING RATES FOR STEEL

|  |  | *(One operator)* |
| --- | --- | --- |
| *Plate thickness* | *Rate* | *Acetylene* |
| *(mm)* | *(m/h)* | *(litres/h)* |
| 2 | 7 | 100 |
| 6 | 3 | 350 |
| 10 | 2 | 500 |
| 20 | 0.5 | 1500 |

## 45.4   OTHER FORMS OF WELDING

It is not always necessary to produce a fusion weld, for example a fairly strong unfused joint can be obtained in cast-iron, steel, copper, brass, tin-bronze and aluminium-bronze by using a low melting point filler rod such as nickel-bronze, manganese-bronze or silicon-bronze. The technique requires a concentrated heat to be applied causing melting of the rod (about 900 °C) which then adheres to the plate surface. However, all surface impurities must be thoroughly removed, and a flux is generally required to assist in this process.

### Aluminium welding

Fusion welding of aluminium requires special care, as the surface coating of aluminium oxide, always present, has a higher melting point than pure aluminium. The heat should thus be applied gradually to avoid a sudden 'run' or 'holing' of the metal plate. A clean surface and use of a flux to remove the oxide is needed to reduce this effect.

### Brazing

Brazing is similar to soldering but uses capillary forces to cause the filler material to flow into the joint. The process involves heating the metals to a higher temperature than the brazing material by playing the blow torch over plate surface, and is thus best placed in the joint or along the joint.

Dissimilar metals can be joined in this way, including alloys. The most commonly applied brazing materials are copper, zinc, copper-alloy or silver-alloys; aluminium, however, requires alum-silicon.

The surfaces should be free from oxides, grease, etc., and a flux is normally required. The resulting joint is often as strong as the brazing metal.

### Fluxes

A flux is mainly used to remove oxide from the surface of the metals to be jointed and to shield the molten metal from further oxidisation. The use of a flux, however, is not universal and may not be necessary for steel, iron, lead, and copper. For most other metals and alloys, borax is the commonly applied flux except for aluminium where a special aluminium flux is necessary.

The flux can be applied as a powder or paste either directly in the joint or more commonly by dipping the welding rod.

## 45.5 GAS CUTTING

Metal can be severed by raising its temperature to ignition level and then applying oxygen to cause combustion and burning.

Generally an oxy-acetylene flame in the neutral setting is first applied to raise the temperature level. Pure oxygen is then introduced as a separate stream. The metal is immediately oxidised and blown away by the further heat produced from the chemical reaction.

For the cutting to be successful, the melting point of the metal must be above its combustion temperature, and any oxides present should not have a melting temperature above that of the base metal. Most steels and steel alloys possess such properties and so can be cut in this way.

Non-ferrous metals, however, commonly have resistant oxide coatings which prevent successful application of the technique.

For applications underwater, the torch is designed to provide an air bubble surrounding the flame. This bubble is maintained by a flow of compressed air issuing around the tip of the torch.

## 45.6 ARC WELDING

Several different forms of electric welding have been developed in recent years including manual-arc, inert-gas shielded arc, $CO_2$ shielded arc

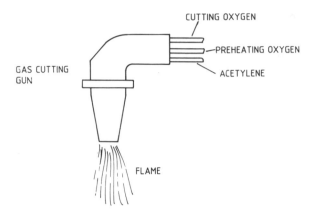

**Fig. 45.15** Gas cutting torch

and submerged arc welding. The choice of method for construction work, depends upon many factors including the opportunities to use automatic processes, weather conditions, quality of weld required, type of metal, speed, position of welding, etc.

### Shielded manual metal arc welding

The method in principle is quite simple, an electrode consisting of a metal rod acts as the filler material and is usually charged positively ($+$) while the plate to be welded is charged negatively ($-$). By momentarily touching the two units together the circuit is connected and current flows. On slight withdrawal of the electrode an arc is formed in the gap between plate and electrode as electrons stream across from $-$ve to $+$ve. Heat and intense light are generated in the process, the electrode and plate melted and globules of the filler rod transferred to the weld pool (Fig. 45.17). The method is easier to use than oxy-acetylene welding and produces a relatively good weld. Also less skill is required.

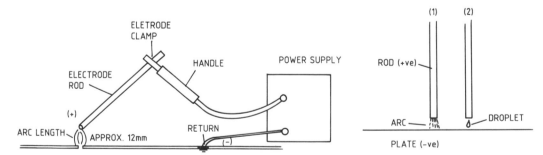

**Fig. 45.16** Shielded manual arc welding

**Fig. 45.17** Arc welding principle

### ELECTRODES

The electrode rod is made of metal appropriate for the plate to be welded, the length varying between 200 and 500 mm depending upon the diameter and the need for stiffness.

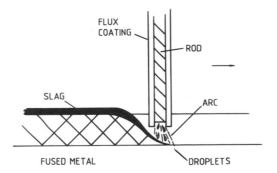

**Fig. 45.18**  Electrode rod

The rod is normally coated with flux which serves to produce a shield of gas as it burns (Fig. 45.18), thereby partially protecting the weld from oxygen and nitrogen in the atmosphere and also to act as an ionising agent. Several compounds possess these qualities including: (i) cellulose to provide a gaseous shield; (ii) manganese to assist as a deoxidising agent; (iii) oxides of iron to control the fluidity of the slag; (iv) silicates to give strength to the covering; and (v) calcium fluoride to reduce the uptake into the weld of hydrogen from any water vapour present – hydrogen bubbles trapped in the weld lead to porosity and cracking.

If excessive porosity is to be avoided the moisture content of the flux covering should thus be kept to the minimum, especially if the coating is intended to produce a low hydrogen weld. Fluorspar coating composition is best suited for this purpose.

The rods should therefore be stored in an oven and when taken out to site maintained in a protected environment until ready for use.

## WELDING METHOD

*Current/voltage*

Either a d.c. or a.c. power source is suitable and if mains voltage is used a transformer is required to bring the voltage down to at least 70 V for striking up the arc, with a further control down to 15 V or so to maintain the arc. The heat developed across the arc is proportional to the square of the current which should be capable of increase up to several hundred amperes depending upon the plate thickness and rod diameter.

In manual welding it is almost impossible for the welder to hold the arc length constant, as a consequence the voltage drop will change. The power supply should therefore be of the 'constant' current type (Fig. 45.19) whereby variations in arc voltage produce relatively small changes in current.

Welding can be performed in almost any position but generally welding on the horizontal is preferred. As soon as the arc is struck the welder must begin to move the rod imparting a weaving side-to-side motion to ensure good fusion with the plate sides (Fig. 45.20). Bad technique can cause defects as illustrated in Fig. 45.21.

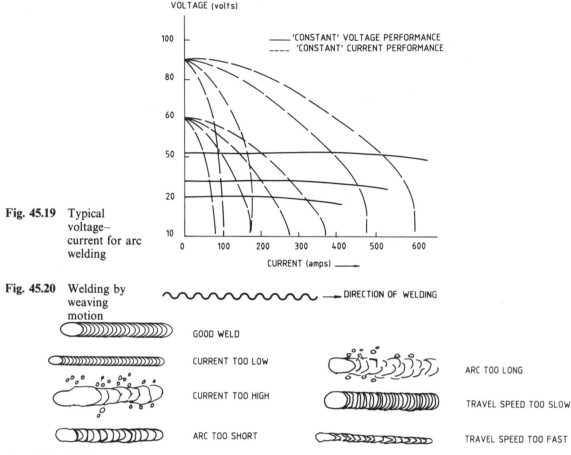

**Fig. 45.19**  Typical voltage–current for arc welding

**Fig. 45.20**  Welding by weaving motion

**Fig. 45.21**  Weld defects

For thick plate which can only be welded from one side, it is good practice to tack on a backing plate to ensure that a good root weld is obtained and to give local strength when the metal is in a soft condition.

Immediately after the run of weld has been completed the slag must be thoroughly chipped away before further runs are built up. Otherwise intrusions, holes and a generally poor weld will result.

*Preparation of the Joint*

The preparation of the joint should ensure that grease and oxides are cleaned away and that the edges to be joined are uniform and level with each other. For butt joints configurations shown in Table 45.3 are preferred.

For fillet welts the plate thickness and corresponding rod diameter are similar to those given in Fig. 45.13 for butt welds.

Welds should be built up in runs (Fig. 45.22) as indicated in Table 45.3.

**Table 45.3** Data for butt welds

| Plate Size | Shape | Diameter of Electrodes | No. of Runs | Current (amps) |
|---|---|---|---|---|
| 3 mm | NO GAP | 2 mm | 1 | 100–120 |
| 6 mm | 1–6mm | 4 mm | 1 | 150–180 |
| 12 mm | 60° 3mm | 5 mm | 2 to 3 | 200–250 |
| 18 mm | 60° 3mm | 6 mm | 3 or more | 300–400 |

(i) ROOT WELD     SINGLE V FILLET WELD     ROOT WELD / BUTT WELD

**Fig. 45.22** Layering of weld runs

(ii)

## APPLICATIONS

Manual arc welding can be applied to steels and steel alloys, stainless steel, aluminium, copper, bronze, nickel and cast iron. Some may require preheating. The method is simpler to use than most other techniques.

## BRONZE WELDING

As with oxy-acetylene welding, the manual arc method can be used to bronze-weld steel, etc. using a bronze electrode, an unfused joint will be produced.

### $CO_2$ shielded arc welding

The effect of the burning flux in the manual arc welding method shields the welding zone from the surrounding atmosphere and thus reduces the oxidising reaction with the molten metal. In $CO_2$ welding this shield is provided by a separate stream of gas. The earliest attempts to do this simply used a permanent tungsten electrode and a surrounding jacket of inert gas, the filler wire being held manually as in the oxy-acetylene process. In modern equipment, however, the filler wire acts as the electrode and is automatically fed through the gum as welding progresses (Fig. 45.23). Unfortunately, carbon dioxide ($CO_2$), has a tendency to break down into carbon monoxide and oxygen at high temperatures, resulting in oxidation of the iron, and therefore must be countered with flux-covered or cored electrode. The use of inert gas such as argon, called metal inert gas method (MIG), provides an alternative to $CO_2$.

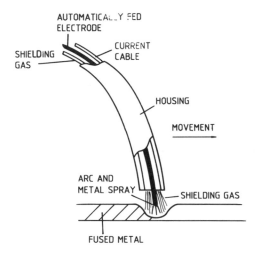

**Fig. 45.23**  $CO_2$ shielded arc welding

## WELDING METHOD

The process is generally semi-automatic using a hand gun although fully automatic equipment is available. The gas is led from a storage cylinder through flexible tubing and allowed to emerge as a stream through the shroud to shield the arc. The plate should be prepared with a V-joint as for the manual arc method and welded on the horizontal wherever possible, although vertical welds can be made.

The amount of current applied depends upon the type of metal and plate thickness as shown in Table 45.4. For plate up to about 12 mm thick the globular transfer method is mostly selected (Fig. 45.17). The

**Table 45.4**  Spray transfer method

| Plate thickness (mm) | Electrode diameter (mm) | Current* (A) | Travel speed (mm/min) | $CO_2$ gas (mph) | No. of passes |
|---|---|---|---|---|---|
| 2 | 1.0 | 300 | 500 | 1 | 1 |
| 5 | 1.5 | 400 | 900 | 1 | 1 |
| 9 | 2.5 | 400 | 650 | 1 | 2 |
| 13 | 2.5 | 500 | 400 | 1 | 3 |
| 20 | 2.5 | 500 | 350 | 1 | 4 |
| 25 | 2.5 | 500 | 300 | 1 | 6 |

\* Either a.c. or d.c. current.

gap between the plates should be parallel and about 5 mm if a good root weld is to be obtained. A ceramic backing tile can help to do this.

The current used in $CO_2$ welding is significantly higher than that required in the manual-arc process, and reduces the size of the droplets or globules virtually to a spray. For thick plate the wire may be permitted to touch the molten pool (dip method) causing the arc to short-circuit and thereby increase the current. In this way the tip of the wire is melted and drawn into the weld pool and the arc momentarily re-established while more wire is fed through to cause a further short circuit and so the cycle is repeated. In contrast to shielded manual metal arc welding, a 'constant' voltage power supply is generally preferred, the reason being seen from Fig. 45.19 where a small change in arc length (voltage) causes a large change in current, as a consequence the burn off rate quickly alters but the arc length is restored by means of the rate of wire feed preset to deliver at a constant speed depending upon the plate thickness, etc. If a 'constant' current source is used, the voltage variation causes only very small current changes of insufficient magnitude to affect the rate of burn and the arc length must then be kept constant automatically by adjusting the rate of wire feed.

## APPLICATIONS

The technique can be applied to most steels, stainless steel, aluminium, magnesium, nickel, nickel–copper alloy and bronze.

The main advantage over manual arc is the continuity of welding, avoiding the need to change electrodes. The process is therefore quicker and produces less weld defects. However, when used in the open air with exposure to the elements, wind speeds exceeding about 25–30 knots tend to destroy the shielding effect of the gas and produce unsatisfactory welds. The equipment is also much more expensive.

### Submerged arc welding

The method is generally automatic and mostly used to weld mild and low-alloy high strength steel including stainless steel and some non-ferrous alloys. Work is usually performed on the horizontal and is limited to fillet welds and long, parallel butt or seam welds. Variations

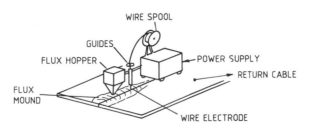

**Fig. 45.24** Submerged arc welding

in the gap dimension or irregularities lead to defects. The joint must be perfectly clean with a machined or flame cut finish.

## THE WELDING METHOD

The principles are similar to other arc-welding methods whereby an arc is struck between the tip of the electrode and plate. The filler wire forms the electrode and is fed automatically from a trolley-mounted reel as welding proceeds. Unlike $CO_2$ arc welding, a gas shield is not provided instead the arc is kept continuously submerged below a granular flux deposited from a hopper as the trolley advances. In this way, fumes, spatter and intense light are virtually eliminated. Any flux not used can be collected for re-use.

The flux shields the molten pool from atmospheric oxygen and nitrogen with a covering of molten slag and produces a smooth uniformly shaped bead. Either a.c. or d.c. current may be used, but d.c. provides better control of the bead shape and welding speed, especially when a constant current type power source is used.

When low hydrogen uptake into the weld is desired (i.e. to avoid porosity) then the flux granules or powder must be dry. The flux is manufactured for specific applications and contains deoxidising compounds and sometimes substances for introducing alloys into the weld.

The joint to be welded should be prepared as for the other arc welding methods and typical welding data are given in Table 45.5.

**Table 45.5**  Automatic submerged arc welding

| Plate thickness (mm) | Electrode diameter (mm) | Current (A) | Voltage (V) | Travel speed (mm/min) | No. of passes |
|---|---|---|---|---|---|
| 2 | 2 | 300 | 25 | 1800 | 1 |
| 5 | 3 | 450 | 30 | 1200 | 1 |
| 9 | 3 | 500 | 35 | 900 | 2 |
| 13 | 5 | 700 | 35 | 600 | 2 |
| 20 | 6 | 900 | 35 | 500 | 2 |
| 25 | 6 | 1000 | 35 | 400 | 2 |
| 35 | 6 | 1100 | 35 | 250 | 2 |

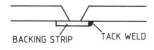

BACKING STRIP          TACK WELD

**Fig. 45.25**  Backing strip and tack weld

A metal backing strip is normally required to give a sound root weld as the metal remains molten for a long period (Fig. 45.25). Also the flux hides the root of the weld and good guidance control equipment therefore is necessary.

## APPLICATIONS

The process permits much higher current with consequently more intense heating than other arc welding methods and is thus considerably faster. However, this can be disadvantageous causing blow holes when the variable parameters, such as fit-up, current, voltage, atmospheric conditions, travel speed, etc., cannot be accurately controlled.

The technique is mostly suited to long straight runs such as girder plates. Also large diameter pipes can be effectively welded by keeping the welding equipment fixed while rotating the pipe (Fig. 45.26).

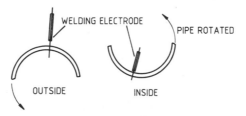

**Fig. 45.26**  Welding large-diameter pipes

## 45.7  ARC CUTTING

Arc cutting can generally be carried out with identical equipment to that suitable for arc welding. The finished edge, however, is usually rough, lumpy and unsuitable for preparing parts for welding. The oxy-fuel blow torch method should be used for this circumstance.

The simplest and most readily applied cutting method is manual metal arc with a conventional metal electrode. Extra high current produces excessive heat and the weld pool becomes large and unmanageable compared to the welding mode, causing the molten metal to fall away. As in welding, the electrode is consumed. Typical cutting data are:

| | |
|---|---|
| Plate thickness | 5–10 mm |
| Electrode diameter | 3–5 mm |
| Current | 200–350 A |
| Travel speed | 175–600 mm/min |

## 45.8  SPECIALISED WELDING METHODS

In recent years the following several welding methods have been developed for specialised needs, e.g. welding special steel such as titanium, fine welds, welding dissimilar plate thicknesses, etc. Few have general application in the construction industry at present.

### Electron beam welding

A stream of electrons at high velocity is focused into a fine beam (approximately 0.2 mm diameter) and concentrated on the materials to be joined. If welding is allowed to take place in a vacuum, plate thicknesses up to 150 mm are possible, but this is reduced to about 50 mm in open air.

The mode of action relies on the conversion of the kinetic energy of the electrons to produce heat on striking the base metal. Because the beam is so fine, butt joints must be well finished and very accurately aligned. Most metals can be welded with this method.

### Laser beam welding

A typical simple method of light amplification by stimulated emission of radiation shown in Fig. 45.27 works on the principle of imparting a flash of light into a synthetic rod-shaped ruby crystal silvered at both ends. As the flash is pulsed, chromium atoms in the ruby reflect quantities of light between the mirrors cumulatively to build up a beam of light of great intensity along the axis of the rod. A small hole in the silvering at one end allows the beam to emerge. It is subsequently focused on to the target to be welded. The light beam, monochromatic and issued at the centre frequency of the spectrum, is red in colour. The resulting electromagnetic radiation directed on to the surface to be joined is designed to have sufficient concentrated energy to melt the metal. Most metals can be welded with the technique. Laser technology is undergoing rapid development and is beyond the scope of this discussion.

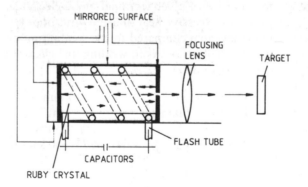

**Fig. 45.27**  Laser beam welding

### Explosion welding

Explosion welding involves placing the plates to be joined above one another separated by a gap. The bottom plate is detonated, so causing the plates to collide and weld.

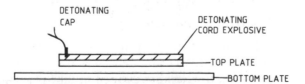

**Fig. 45.28**  Explosion welding

### Ultrasonic welding

Pressure is applied to the plates while very high frequency vibrations are imparted. This produces very small deformations in the plates causing frictional heat build up and subsequent welding. Melting does not take place, but welded joints as strong as the parent metal can be formed.

### Thermit welding

Iron oxide and aluminium powders when ignited burn at approximately 2500 °C in the following reaction:

$$\underset{\text{slag}}{} \quad \underset{\text{steel}}{}$$
$$8\,\text{Al} + 3\,\text{Fe}_3\text{O}_4 = 4\,\text{Al}_2\text{O}_3 + 9\,\text{Fe} + \text{heat}$$

To effect a weld using these products, the groove between the plates is surrounded with a sand mould and the plates preheated to about 200 °C. The heated filler metal is then fed into the joint from a crucible causing melting of the parent material and fusion.

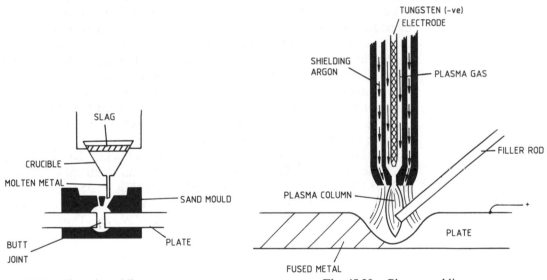

**Fig. 45.29**  Thermit welding                **Fig. 45.30**  Plasma welding

### Plasma welding

An electric arc, when constricted by argon or helium, generates temperatures in the range 10 000–25 000 °C thereby giving faster welding speeds and better quality welds compared to tungsten inert gas (TIG) welding methods. However, the filler rod is not integral with the electrode and therefore some advantage is lost compared to the semi-automatic and automatic methods such as $CO_2$ shielding.

### Induction welding

The method has proved feasible in welding large diameter pipes. The principle requires the butt joint to be surrounded by an electromagnetic coil to induce high local current densities and subsequent raising of temperature in the metal. Each tube is clamped and the joint pulled together. The applied pressure quickly completes the weld. High frequency current and cooling of the coil is necessary.

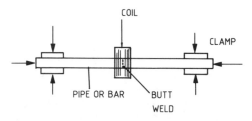

**Fig. 45.31**   Induction welding

## 45.9   WELDING PLASTICS

Thermoplastic materials such as PVC, polystyrene, polypropylene, melt at quite low temperatures and become fluid at 100–200 °C and therefore relatively light-weight welding equipment can be used compared to that used for the welding of metals. Typical construction applications include joining sealing strip, water bar, pipes etc., using hot plates (Fig. 45.32), rollers (Fig. 45.33), or a welding torch (Fig. 45.34). In the case of plates and rollers, the pieces of strip are simply placed above one another or lapped, or chamfered and butted together for seaming. Electrical elements heat the plates to welding temperature and pressure is applied to complete the joint.

For butt and fillet joints or for welding plates thicker than 2 mm the torch method is usually more applicable. Either an electrical filament (Fig. 45.35b) or hot gas passing through a coil (Fig. 45.35a) is used to heat a stream of air or $CO_2$ to about 300 °C. Welding proceeds similar to the oxy-acetylene method using a filler rod of plastic. Typical data are given in Table 45.6.

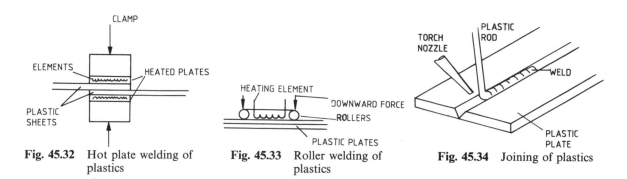

**Fig. 45.32**   Hot plate welding of plastics

**Fig. 45.33**   Roller welding of plastics

**Fig. 45.34**   Joining of plastics

**Fig. 45.35**   Welding torch for plastics

**Table 45.6**   Hot gas butt welding of plastics

| Sheet thickness (mm) | Filler rod diameter (mm) | Travelling rates (m/h) | Joint strength as % of parent plastic |
|---|---|---|---|
| 3–5 | 2 | 12–18 | 50–60 |
| 15–20 | 4 | 12–18 | 50–60 |

## 45.10   STRESS RELIEVING OF WELDS

### General comments

During welding, heat causes expansion and if on contraction movement is restricted then distortions and residual stresses are likely to remain. Thus wherever possible welding should take place working from a point where the member is fairly fixed through to a free end. If this cannot be obtained then some stress relieving afterwards should be attempted, especially for pressure vessels and those parts subject to regular impact loading.

Several methods of stress relieving can be applied including furnaces and heating coils, e.g. induction heating and resistance heating. Typically for carbon steels, a re-heat temperature in the range 550–650 °C is usual, with subsequent cooling at a rate of about 150 °C per hour. Unfortunately, this elaborate procedure is virtually impossible to perform on the often large and awkward members found on construction sites and therefore good practice to ensure minimum distortion is of paramount importance.

### Preheating of welds

The common welding methods generate large temperature gradients between the local welding area and the rest of the metal. As a consequence rapid expansion and contraction with associated stressing and hardening may cause subsequent cracking and fracturing on those parts subjected to shock loading. These effects can be reduced by preheating the member in a furnace or with electrical coils but for construction work the only practical solution is generally to play a flame over the weldment area some distance on both sides of the run and bring the parts up to the preheat temperature. After welding, slow cooling or even stress relieving is desirable on critical work.

### 45.11 INSPECTION AND TESTING OF WELDS

The quality and therefore the strength of welds can be influenced by many factors, including the weld technique, compatibility of the metals, storage of consumables, skill of the welder, atmospheric conditions, cleanliness of the metal surface, alignment of welds, travel speed, slag intrusions (especially on multipass welds), preheating, stress relieving, etc., and not least the importance given by management to inspection and testing.

The methods used to minimise or detect faults arising from the above are broadly as follows:

1. Good practice.
2. Visual inspection.
3. Non-destructive testing (NDT).
4. Destructive testing.

#### Good practice

(i) Clean weld surfaces.
(ii) Correct alignment and root opening.
(iii) Parts firmly secured – usually by 10–20 mm long tack welds.
(iv) Preheating.
(v) Weld craters to be filled.
(vi) Cracks to be repaired before subsequent passes laid.
(vii) Surface slag to be removed before subsequent passes.
(viii) Broken tack welds to be removed and replaced.
(ix) Filler material to be compatible with weld plates.
(x) Root fusion should be complete – use a backing plate where possible.
(xi) Position weld on the flat if possible.
(xii) Keep consumables in the dry.
(xiii) Shield the welding area from the elements.

#### Visual inspection

CHECK PRIOR TO WELDING

(i) Welding procedures, specifications, documents, and appropriate techniques are clearly set out and comprehendable.
(ii) Weld details, drawings, and sketches are clear and orderly.
(iii) Welders are capable and experienced.
(iv) Access platforms and working space are adequate and safe.
(v) Weld preparation of joints, edges, etc., has been performed and surfaces are clean.
(vi) Appropriate electrodes, fluxes, shielding gases, etc., are available.
(vii) Dry storage facilities are available for consumables.
(viii) Tolerances, backing plates and tack welds have been properly carried out including adequate preheating.

## CHECK DURING WELDING

(i) Welding equipment is performing properly and safely including current, voltage and travel speed.

(ii) Preheating temperatures are being achieved.

(iii) Slag surfaces are thoroughly removed after each weld run and any additional defects corrected.

(iv) Test samples are taken.

## CHECK AFTER WELDING

(i) Dimensions, alignments, etc., meet with requirements on the drawings.

(ii) Unacceptable defects are identified and repaired.

(iii) Cleaning and dressing of weld surfaces are properly performed and the results inspected.

(iv) Non-destructive tests are carefully executed and problems put right.

(v) Post-heat treatment provided where necessary.

### Non-destructive testing (NDT)

Non-destructive testing is applied in detecting weld defects such as flaws, fine cracks and other discontinuities which cannot be adequately detected by visual inspection. The five common methods used are:

(i) Dye-penetrant testing.

(ii) Fluorescent-penetrant inspection.

(iii) Magnetic particle testing.

(iv) Radiographic testing.

(v) Ultrasonic testing.

## DYE-PENETRANT TESTING

The surface of the weld is thoroughly cleaned with a solvent to remove dirt, grease, paint, etc. A visible dye is sprayed or poured on and sufficient time is allowed for the liquid to be drawn into cracks, holes, etc., by capillary action. The surface is then wiped clean of the dye and allowed to dry. Application of a white absorbant powder acts like 'blotting paper' to reveal the position of any surface cracks or flaws.

## FLUORESCENT-PENETRANT INSPECTION

A liquid fluorescent under ultra-violet light is used instead of a dye penetrant, to produce a yellowish discoloration at the positions where the liquid has penetrated into cracks, etc. The method is slightly more sensitive than the dye technique.

## MAGNETIC PARTICLE TESTING

A current is passed between probes mounted on the length of the steel or iron part to be tested for defects. Magnetic lines of force form around a circular field in a plane at right angles to the direction of the current and any duct or crack at right angles to the field become the poles of a magnet. If the surface of the part is covered with magnetic particles, either dry or fluid-based, these will be attracted to the poles and so reveal the defect. However, only surface defects can be found in this way, and usually a machined finish is necessary for good results. Defects below the surface can be detected by passing moving search coils connected to a galvanometer over the magnetised area to detect any change in the magnetic field in the specimen.

## RADIOGRAPHIC TESTING (X-RAY METHOD)

A silver halide film becomes exposed when X-rays (or gamma-radiation) are directed on to its surface (Fig. 45.36). However, if an object is placed between the X-ray source and film, some of the energy is absorbed by the object and a shadow produced. Thus when the film is developed, differences in the thickness of the object will be indicated by dark and light areas, i.e. thinner parts will absorb less energy than thicker parts and show up as dark areas. By careful examination of developed film, a trained observer can detect slag intrusions, gas holes and most cracks with the exception of those facing right angles to the X-rays. To determine the level of detail on the exposed film a small piece of material (the penetrameter) the same as the specimen and about 2% of its thickness is placed on the surface on the specimen during X-raying, if after developing the penetrameter is visible then defects of this size should be detectable.

Because gamma-radiation can be a health hazard, the X-ray machine must be insulated within enclosing walls of concrete or steel and therefore the specimen in general must be taken to the equipment. Where this is impracticable, special screening is necessary, with filming taking place when the workmen are away from the site.

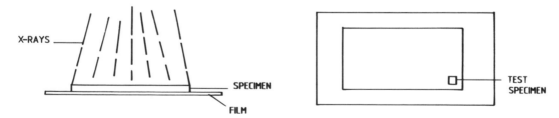

**Fig. 45.36**   Radiographic testing (x-ray method)

## ULTRASONIC TESTING

An ultrasonic probe transmitter sends very high frequency waves into the specimen which are reflected at discontinuities, such as the

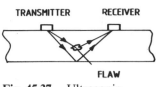

**Fig. 45.37** Ultrasonic testing

specimen surfaces or flaws, and converted by a receiver probe to electrical signals suitable for visual display on an oscilloscope.

In a perfect specimen only the initial pause and return surface echo would be visible, but where a flaw is present its echo will also be returned as shown in Fig. 45.37. To detect all flaws in the weld it is necessary to scan along the whole length, but care must be exercised if cracks parallel to the direction of the beam are not to be missed.

To ascertain the size of the flaw, the probe is calibrated with a steel block pierced by a small drill hole. Test echoes indicate the amplitude of the echo from a hole of known size.

Ultrasonic testing is presently the most sensitive of all the techniques, is simple to operate, safe, and can be used on all materials. However, high levels of interpretative skill are needed.

### Destructive testing

Destructive testing commonly involves the cutting out of a section of the welded plate or at least preparing samples typical of the actual weld. The common tests are:

| | |
|---|---|
| Tensile (Fig. 45.38) | Test to breaking or yield point. |
| Bending (Fig. 45.39) | Two tests: (i) weld face in tension and tested to yield point; (ii) weld root in tension tested to fracture point. |
| Impact (Fig. 45.40) | Brittleness tested either by the Charpy or Izod impact methods. |
| Hardness (Fig. 45.41) | Typically involves pressing a small hardened steel ball into the surface of the specimen. The common methods use Rockwell, Vickers and Brinell equipment. |
| Fatigue (Fig. 45.42) | Specimen subjected to alternate tension and compression. |

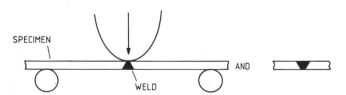

**Fig. 45.38** Tensile testing

**Fig. 45.39** Bending test

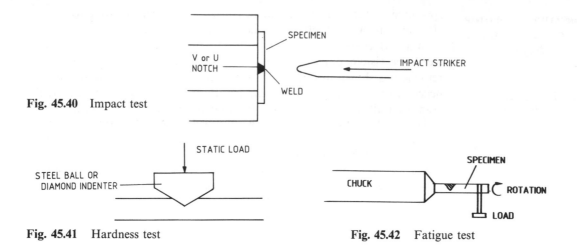

**Fig. 45.40**   Impact test

**Fig. 45.41**   Hardness test                    **Fig. 45.42**   Fatigue test

## 45.12   UNDERWATER WELDING

In recent years there has been demand for underwater welding, particularly for offshore structures. Generally the quality of weld will not be as good as welds conducted in the normal atmosphere, mainly due to the take up of hydrogen into the molten weld metal and of course poor visibility and working conditions. In addition, the quenching effect of the water tends to produce brittleness. Conventional manual metal arc methods have proved reliable techniques to date, the electrode, however, must be waterproofed either by wrapping in tape or sealing with a coating of sodium silicate. The torch and all leads must be well insulated and connections should not come into contact with the water. Also the current must only be switched on when the electrode is in contact with the weld surface, thus good communications are necessary with the operator of the power source at the surface.

For better quality work the welder needs to work in the dry, for example in a special compressed air chamber provided with fume removal, air supply, power and communications equipment.

# 46

# BRIDGE CONSTRUCTION METHODS

## 46.1 INTRODUCTION

Bridges are a vital part of the infrastructure in most countries, linking trading and manufacturing regions, towns, etc., enabling people, goods and materials to be moved quickly and economically.

In earlier times the most successful and prolific bridge builders were the Romans. Their empire was based on trading and was well defended by the military who demanded a comprehensive network of roads and bridges. Many of the bridges constructed were made of wood and have long since disappeared, however a few examples of stonework viaducts remain in good condition throughout Europe. The technology at this time allowed only very short spans between piers, and in the case of stonework, arches and abutments (Fig. 46.1) were the only feasible means of achieving strength and stability. Nevertheless, with careful siting of piers, quite wide stretches of water were successfully bridged, e.g.. the River Thames in London.

**Fig. 46.1** Stonework arch bridge

After the collapse of the Roman Empire, European societies were overtaken by the Dark Ages, and roads decayed. Not until the setting up of turnpike trusts in Britain between 1750 and 1850 did road transport re-emerge as an important means of communication. Once again, the builders, like the Romans, turned to wood and stone for construction material and it was not until the advent of railways in the second half of the nineteenth century that serious attempts were made to use steel and iron. Rapid industrialisation during this period led to the development of stronger materials which, coupled with advances in structural analysis and design, allowed bridges to carry heavier loads over much wider spans than hitherto. The leading pioneers of this era included John Rennie, Thomas Telford and George Stephenson – of special note are Rennie for Waterloo Bridge, Telford for the six-hundred foot span iron-chained suspension bridge over the Menai Straits linking Anglesey to the North Wales Coast, and Robert Stephenson (George's son) for the Britannia Bridge also across the Menai Straits.

During this time advances in bridge engineering were also taking place in continental Europe and especially in the United States of America as rail-roads were constructed. In particular, timber and steel trusses were a feature of American developments. During the twentieth century road transport has regained its dominance and bridge technology has progressed to meet the needs of large-scale transport networks culminating in suspension bridges capable of spanning 1000 m or more.

While spans have been continuously increased other developments have also been taken place, especially the use of prestressed concrete enabling much of the complex falsework and centring essential with *in situ* concrete to be avoided. Also the change to welded cellular steel

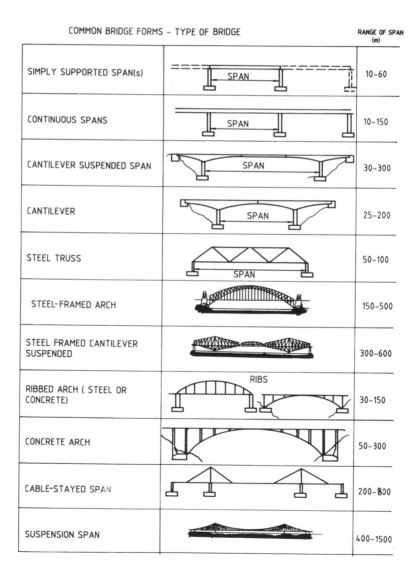

**Fig. 46.2** Bridges classified by structural configuration

boxes assembled to form the bridge deck, has allowed more off-site fabrication compared with trusses and framed arches. As a consequence, construction times have been reduced and indeed more aesthetically pleasing designs have emerged. Thus today bridges of many shapes and sizes can be seen throughout the world. Choosing the appropriate type of bridge for a particular site and function is a difficult task as many alternative solutions to a bridge problem are often possible.

Various classification systems have been devised to aid the selection process, for example structural form categorising bridges into trusses, arches, beams, slabs, multipiered bridges, etc. (Fig. 46.2). The objective is to try and relate type of bridge to a corresponding range of spans and heights. However, the conditions prevailing at the site so influence the choice of construction method, for example, availability of materials, ability to use floating craft, cost and skill of labour, etc., that a classification taking these factors into account would be more informative. Investigations by Wittfoht and Lampart involving many bridges constructed throughout the world indicate that the majority of modern bridges would fall into the broad categories given in Table 46.1.

**Table 46.1**  Common bridge construction methods

| Type of bridge according to method of construction | Span (m) | Length of bridge (m) |
|---|---|---|
| 1. Prefabricated steel or precast concrete beams lifted into place over the full span | | |
|    (i) standard beams | 10–40 | |
|    (ii) non-standard beams | 40–250 | No limit |
|    (iii) steel trusses | 50–100 | |
| 2. Prefabricated steel or precast concrete beams progressively placed by gantry girder method | 30–60 | 300–600 |
| 3. Precast concrete or prefabricated steel deck pushed out (constant radius deck only) | 25–60 | 300–1200 |
| 4. Prefabricated or precast elements or *in situ* concrete structures supported by falsework founded directly on the ground | 5–125 | 5–125 |
| 5. *In situ* concrete deck constructed on stepping formwork | 20–60 | 300–600 |
| 6. *In situ* or precast concrete segments constructed in cantilever and progressively post-tensioned | 50–150 | No limit |
| 7. Prefabricated steel or precast concrete segments stayed by cables | 75–800 | No limit |
| 8. Prefabricated steel or precast concrete segments suspended from cables | Span between towers 400–1500 | No limit |

Note

Values are typical ranges. Where long bridges are involved, many spans may be necessary, and use of simple short span systems may be more economical than a few long spans.

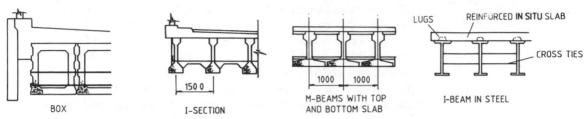

**Fig. 46.3** Precast beam sections

## 46.2 PREFABRICATED STEEL OR PRECAST CONCRETE BEAMS ERECTED IN PLACE OVER THE FULL SPAN

### Standard elements

In many countries there is sufficient demand for concrete products to enable firms to specialise in the sole manufacture of precast concrete elements. As a consequence, high production volumes are possible with a corresponding saving in costs. Bridge engineers have taken advantage of this market by specifying designs which utilise standard beams of relatively short span, 10–40 m being the typical manufactured range.

### Precast beam sections

The common sections now available are:

  (i) I-beam.
 (ii) Inverted T-beam.
(iii) Hollow box beam.
(iv) M-beam.

In general the units are prestressed and pretensioned on the long line process, although ordinary reinforced members are not precluded.

### I-BEAMS

Normally these are simply supported between the bridge piers and individually separated by evenly spaced transverse diaphragms. The deck is formed *in situ* and thus generally requires extensive formwork.

### INVERTED T-BEAMS

These are butted together and act compositely with concrete between and over the units, formwork only being required to form parapets, kerbs, etc.

### M-BEAMS

Inverted T-beams are used with permanent formwork to cast the slab and produce a voided composite structure.

## HOLLOW BOX BEAMS

The beams are butted against each other and transversely post-tensioned. The deck is added *in situ* and usually is not designed to act compositely with the beams.

## STEEL BEAMS

Steel continues to feature in simply supported spans using I-sections. The choice of material largely depending upon the relative costs of steel and concrete. Like concrete elements the slab is often designed to act compositely with the beams and therefore extensive temporary formwork is a requirement.

### Handling prestressed concrete elements

**Fig. 46.4** Lifting beam for precast units

Pretensioned single-span members are designed as simply supported spans and usually cannot act as cantilevers. Thus units should always be lifted at the hooks cast in place by the manufacturers. Where these have not been provided then the lifting point of slings should be near the ends. A lifting beam would be preferable (Fig. 46.4). Beams may be stacked for temporary storage, with timber spacing blocks located near the ends and positioned vertically above one another (Fig. 46.5).

**Fig. 46.5** Stacking arrangement of pretensioned units

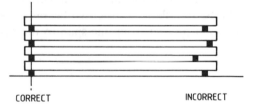

CORRECT          INCORRECT

### Beam erecting

Placing standard beams between abutments is usually quite straight-forward using conventional cranes, such as crawler or truck-mounted boom types. The units are generally brought in by trailer and temporarily stored near the crane. The crane itself can be positioned between the abutments to suit the achievable lifting radius. Alternatively, for a rail or river crossing or when difficult ground is encountered, the beams may be placed from the bridge approach road (Fig. 46.6). However, a large crane would be required to provide lifting capacity at about 20–30 m radius and methods would thus for economical reasons be generally limited to light steel beam bridges.

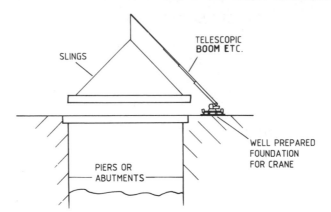

**Fig. 46.6** Beams erected by crane

## Composite construction

In composite design the deck acts as the compression member while the beams largely serve to take tension. During construction the deck concrete is placed *in situ* and therefore must be both supported and allowance made for deflection of the beams, if an unsightly line along the bridge is to be avoided.

The problem is particularly acute with steel beams which deflect markedly under load and therefore a camber is normally induced before concreting takes place. The simplest method requires a frame slung under the deck with a post located below the centre point of each beam (Fig. 46.7). Screw or hydraulic jacks are used to put in the desired camber and released when the concrete has reached the required strength. The frame, constructed as a truss with a working platform to aid both the jacking and shuttering operations, is hung from the bottom flange of the two beams suspended from a roller device to aid transfer from one span to the next.

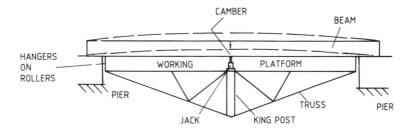

**Fig. 46.7** Composite construction method

## Non-standard beams

Many bridges are designed to achieve particular aesthetic lines, for which relatively short, standard concrete beams are generally inappropriate, whereas purpose-made units can be formed to shape as required. Although the beams for such bridges would be much heavier than standard units, modern developments in cranage have made lifting by this method quite feasible. For example, strut boom truck-mounted cranes are available with capacities over 1200 tonnes and the more versatile hydraulic-boom truck-mounted cranes are currently capable of lifting over 800 tonnes, but these are expensive to hire and set up.

For work over water, crane-mounted pontoons and shear legs with capacities over 2000 tonnes can sometimes prove more economical in long multi-span bridges where several sections are being constructed simultaneously (Fig. 46.8). Unfortunately, however, strong currents and

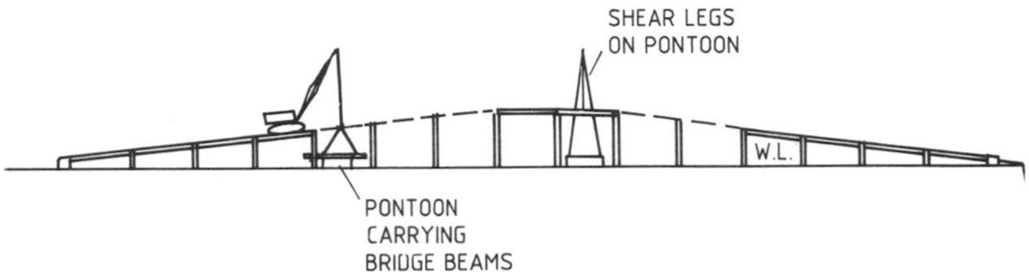

**Fig. 46.8** Beam erection by crane and/or pontoon

tidal waves often present severe difficulties in both the launching and transporting operations. Also a suitable prefabrication area and loading wharf are generally required and therefore bridges through step valleys are normally unsuited to this mode of construction, although with several minor bridges it has proved viable to prefabricate units at a distant yard and tow the members several miles to the bridge location. For large spans involving heavy pre-assembled trusses or prefabricated box sections beyond the capacity of normal cranage, specialised lifting methods are generally required, for example, box units weighing over 2500 tonnes and 300 m long have been successfully raised from temporary bunds (Fig. 46.9). Where this would not be possible, some form of temporary support must be provided at the piers (Fig. 46.10).

The method facilitates fabrication under factory conditions, away from the site, giving the potential for improved quality control and productivity. Towing to position can be planned to coincide with favourable weather windows with the minimum erection period exposed to the vagaries of the elements.

## 46.3  BEAMS PROGRESSIVELY PLACED BY A GANTRY GIRDER

Where many regular simply supported spans predominate, for example long bridges of 600 m or more, then construction of the deck working out from the abutments using a gantry girder is sometimes more economical than using conventional cranage or pontoons.

The gantry technique is suitable for spans of 30–60 m involving beam units up to 250 tonnes, with the gantry girder itself often weighing up to 150 tonnes or more for beams of this size.

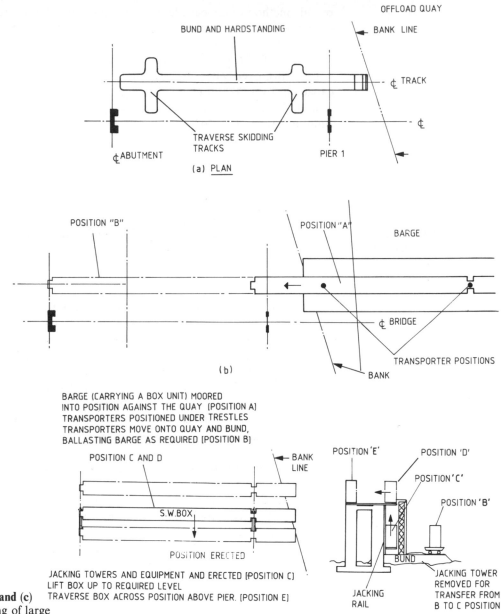

**Fig. 46.9(a), (b) and (c)**
Specialised lifting of large
and heavy units

Wherever practicable, the casting or fabricating yard(s) should be set up on the approaches to the bridge to minimise transport problems. Typically the previously placed beams are used to support the transport bogies. Figure 46.11 shows three views of a commonly adopted procedure. The beams are loaded on to two wagons and then transferred eo either bogies or hangers running on rails along the previously erected deck. The gantry then distributes the units into place

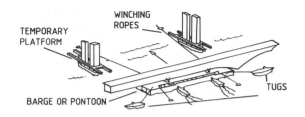

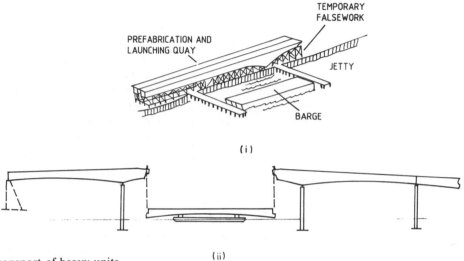

(i)

(ii)

**Fig. 46.10**    Barge transport of heavy units

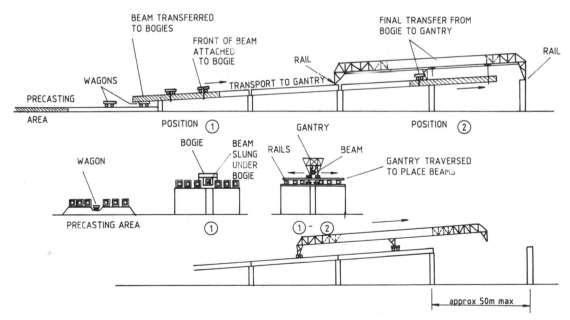

**Fig. 46.11**    Progressive placing of precast concrete/steel beam by gantry method

between the piers. Rails temporarily fixed across the deck and on the far pier are provided for traversing the gantry. When all units are in position, the girder is winched forward on rail-mounted bogies over the next span. Thus the gantry needs to be slightly longer than twice the span.

Finishing of the roadway deck follows up behind the units.

## 46.4  INCREMENTAL LAUNCHING OR PUSH-OUT METHOD

In this form of construction the deck is pushed across the span with hydraulic rams or winched (Fig. 46.12). Decks of prestressed post-tensioned precast segments, steel or box girders have been erected. Usually spans are limited to 50–60 m to avoid excessive deflection and cantilever stresses, although greater distances have been bridged by installing temporary support towers (Fig. 46.13). Typically the method is most appropriate for long, multispan bridges in the range 300–600 m, but, much shorter and longer bridges have been constructed. Unfortunately, this very economical mode of construction can only be applied when both the horizontal and vertical alignments of the deck are perfectly straight, or alternatively of constant radius. Where pushing involves a small downward grade (4–5%) then a braking system should be installed to prevent the deck slipping away uncontrolled and heavy bracing is thus needed at the restraining piers.

Bridge launching demands very careful surveying and setting out with continuous and precise checks made of deck deflections. A light aluminium or steel launching nose forms the head of the deck to provide guidance over the pier. Special teflon or chrome-nickel steel plate bearings are used to reduce sliding friction to about 5% of the weight, thus slender piers would normally be supplemented with braced columns to avoid cracking and other damage. These columns would generally also support the temporary friction bearings and help steer the nose.

In the case of precast construction, ideally segments should be cast on beds near the abutments and transferred by rail to the post-tensioning bed, the actual transport distance obviously being kept

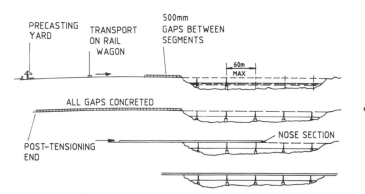

**Fig. 46.12**  Incremental launching/ push-out method

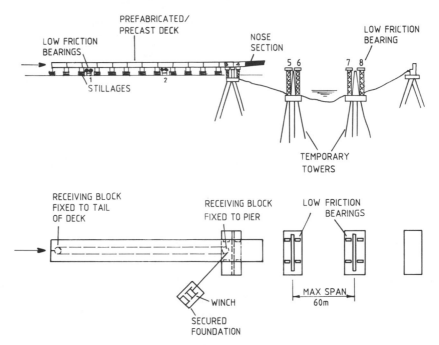

**Fig. 46.13** Temporary support for push-out method

to the minimum. Usually a segment is cast against the face of the previously concreted unit to ensure a good fit when finally glued in place with an epoxy resin. If this procedure is not adopted, gaps of approximately 500 mm should be left between segments with the reinforcement running through and subsequently filled with concrete before post-tensioning begins. Generally all the segments are stressed together to form a complete unit, but when access or space on the embankment is at a premium it may be necessary to launch the deck intermittently to allow sections to be added progressively. The corresponding prestressing arrangements, both for the temporary and permanent conditions would be more complicated and careful calculations needed at all positions.

The principal advantage of the bridge-launching technique is the saving in falsework, especially for high decks. Segments can also be fabricated or precast in a protected environment using highly productive equipment.

For concrete segments, typically two segments are layed each week (usually 10–30 m in length and perhaps 300 to 400 tonnes in weight) and after post-tensioning incrementally launched at about 20 m per day depending upon the winching/jacking equipment.

## 46.5  CONSTRUCTION WITH FALSEWORK

### Beam decks

While bridges designed for standardisation or regularity are often more economically built with precast concrete or prefabricated steel sections other considerations such as aesthetics may be overriding requirements. In cases where beam depth varies and geometry is complicated, or pier design is awkward, the most feasible construction solution generally involves falsework to support either an *in situ* concrete or a precast concrete deck, or combination of both (Fig. 46.14). Bridges built at

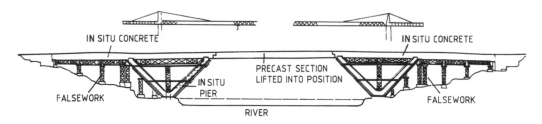

**Fig. 46.14** Complex falsework for *in situ* concrete bridges construction

heights of 30–40 m above ground are practicable and spans over 100 m are not uncommon.

Not withstanding these factors, much simpler bridges, such as those normally considered appropriate for construction with standard precast elements (beams) are frequently constructed *in situ* because the falsework and formwork would involve little more than scaffold tubes and timber or proprietary shuttering systems, so enabling a local contractor to carry out the work (Fig. 46.15). Furthermore, falsework

**Fig. 46.15** Standard falsework for *in situ* concrete deck

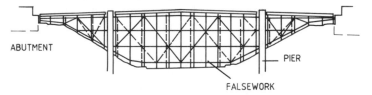

supports avoid the need to design the permanent structure to carry imposed loads generated only during the construction phase.

Unfortunately the technique is labour-intensive and wasteful of shuttering unless many re-uses can be obtained. Also great care has to be taken in preparing the foundations for supports, checking that members are plumb, timbers well seated and joints tight before concreting begins. There is also more likelihood that an *in situ* concrete finish will vary because of the difficulties of quality control on site. For these reasons precast elements supported on falsework and subsequently post-tensioned have been commonly chosen as an alternative to *in situ* work (Fig. 46.16). Prefabricated box segments have proved suitable for multispan bridges, up to about 40 m span. Over 100 m of completed deck per week can be achieved with this method. A further advantage lies in the ability to drape the prestressing cables to give the most efficient use, and the post-tensioning is performed for

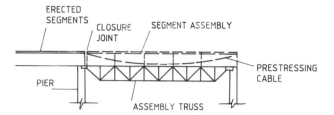

**Fig. 46.16** Beam falsework for precast post-tensioned concrete deck

each span compared with the stage stressing required for the alternative of cantilever construction.

## Concrete arches

Concrete arches have been popular because of their inherent pleasing aesthetics and the shape is also ideal to take up compressive stress. Unfortunately, however, complicated falsework is usually required, whether the construction is in precast or *in situ* concrete. A common construction procedure involves driving temporary piles into the river bed to act as foundations for the falsework. Depending upon the weight to be carried the support system is made up of single scaffold tubes and timber beams, or specially designed struts and Warren girders for heavier loads.

A complete soffit of formwork is prepared between the abutments (Fig. 46.17) and concreting of the deck carried out in the sequence shown in Fig. 46.18 thereby keeping movement of the falsework to the minimum. After completion of the arch the piers and general superstructure are constructed either conventionally with the aid of falsework (Fig. 46.19) or with precast units placed in position by a crane mounted barge.

In deep ravines it is often not feasible to support the arch formwork from piled foundations, because of the depth and awkwardness of the site. A common solution has been to hang the falsework from the completed abutments as shown in Fig. 46.20 enabling the arch to be progressively concreted section by section, with perhaps the middle third built on centring assembled at the bottom of the valley and hoisted into place (Fig. 46.21).

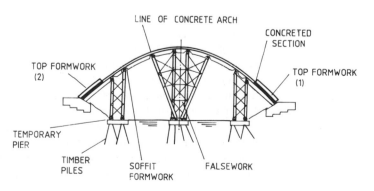

**Fig. 46.17** Falsework for *in situ* concrete arch

**Fig. 46.18** Pour sequence for *in situ* concrete arch

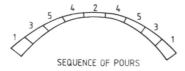

**Fig. 46.19** Falsework for deck supported off completed arch

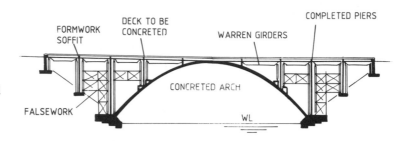

**Fig. 46.20** 'Hanging' falsework/formwork for arch construction

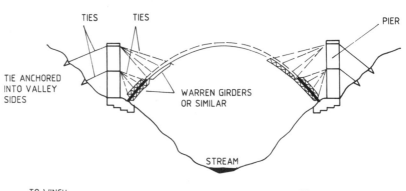

**Fig. 46.21** Construction of centre arch span

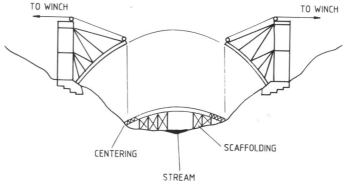

During the point of popularity of slender arch bridges built of *in situ* reinforced concrete, several major bridges were also designed as precast construction. By utilising ring compression (Fig. 46.22) the complete arch can be assembled from independent precast hollow boxes, which in the final condition behave similarly to the traditional stone arch, thus requiring no monolithic connection between sections.

A typical construction arrangement is shown in Fig. 46.23, where it can be seen that a substantial falsework structure is needed to support the precast units. Each unit is brought to the winching mast at the

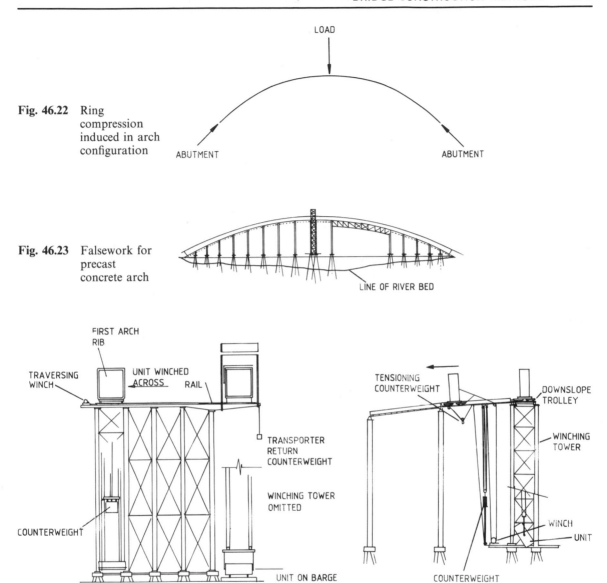

**Fig. 46.22**  Ring compression induced in arch configuration

**Fig. 46.23**  Falsework for precast concrete arch

**Fig. 46.24**  Typical erection system for precast concrete arch segments

centre of the span and raised to the crown of the arch (Fig. 46.24). Units are subsequently lowered on bogies to the abutments, gradually building up equally to the crown. Flat jacks are incorporated in the joints at the quarterpoints and after insertion of the final unit, the structure is raised into a freestanding condition above the falsework and the joints concreted.

The superstructure is subsequently added *in situ* or as precast concrete elements.

## 46.6   STEPPING FORMWORK CONSTRUCTION

While falsework standing on the ground is often viable for a single span or at most a few spans, the use of stepping formwork usually proves more economic for long bridges of 300 m or more. Individual spans between piers generally lie between 20 to 60 m, and the deck dimensions should be reasonably constant to avoid expensive formwork adjustments. This method is appropriate where a bridge is constructed over poor soil, water, or is high above ground level. Like cantilever construction, standardised formwork and equipment is used many times over, thereby avoiding the need for almost continuous re-learning as is often the case with traditional bridge-building techniques, such as steel frames, concrete arches, formwork and falsework systems, etc.

The *in situ* deck is supported on movable girders temporarily carried by the bridge columns or piers. After sufficient time has elapsed for the concrete to gain strength, the individual units are moved forward to the next span. Where possible the formwork is permanently attached to the girders and arranged to fold down during the stepping phase. However, this depends upon the shape of the cross section of the bridge and with more complicated designs, stripping and re-assembling the shutter panels may be unavoidable.

Basically two methods have proved successful: namely with or without auxiliary girders.

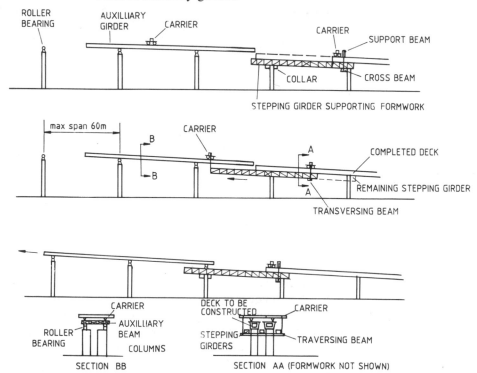

**Fig. 46.25**   Stepping formwork system for *in situ* concrete deck

### Auxiliary girders

The stepping girders are made a little longer than the casting span, with the service or auxiliary girder required only during the stepping operation. In the example shown, the forward ends of the stepping girders are supported on a collar attached to the pier or column, while the rear ends are carried on a beam hung from a cross member seated on the completed deck. After concreting, the girders are moved forward singly or in pairs, by attaching the front part to travelling bogies mounted on the auxiliary girder, while the rear is hung from rail-mounted bogies travelling on the newly completed deck. The auxiliary girder being slightly longer than two span lengths is then moved forward to bridge the next span and the cycle repeated.

### Without auxiliary girders

The stepping girders are approximately two times the span length, with the centre section made strong enough to carry the temporary loads and deck during construction (Fig. 46.26). The 'flying' sections are only needed during the stepping phase, and therefore can be of light fabrication, usually in the form of lattice beams. The complete unit is carried on collars fixed temporarily to the columns. Units are moved forward either singly or in pairs. As soon as the front section of the girder reaches the column support, the rear collar can be released and transferred to the columns on the next span. A small crane mounted on the forward section of the girders is normally used to facilitate this

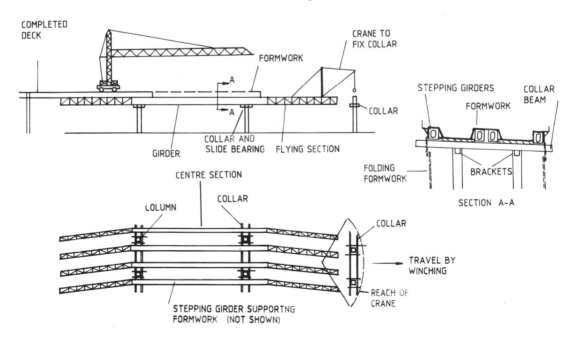

**Fig. 46.26** Stepping formwork system without auxiliary girders

operation. A separate crane travelling on the completed deck is usually selected for concrete placing and handling of materials, equipment and shuttering. A horizontal boom crane is generally the most suitable type in order that the carriage can be kept well clear of the working area of the deck. Bridges curved on plan with a radius down to about 400 m have been successfully completed. For the method to be economic it is necessary to obtain many re-uses of the formwork and falsework and therefore a divided carriageway is often favoured.

Typically 300–400 m² of completed deck surface area can be achieved in a week.

### Hanging method

The use of stepping girders mounted underneath the deck tend to lose their advantage when designs involve complicated arrangements at the column or pier heads, e.g. mushroom shapes, haunches, etc. To deal with such problems, bridge builders have occasionally used the hanging method with stepping girders. A typical example is shown in Fig. 46.27. The equipment comprises an integrated unit of formwork, yokes and girders. The sequence of operations requires the rear of the stepping girder to be supported on the previously constructed deck, the other support being provided at the next pier on a temporary base which is

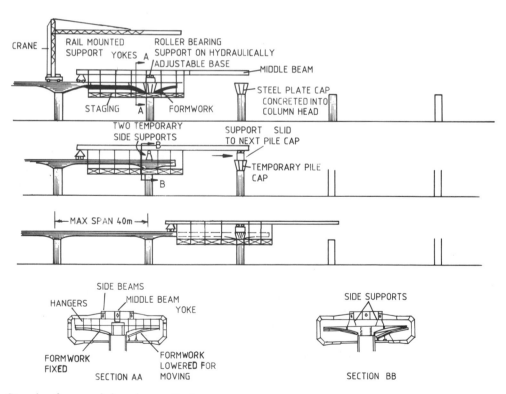

**Fig. 46.27** Stepping formwork-hanging method

subsequently removed and concreted into the deck. The centre beam of the whole unit extends over two full spans to facilitate the stepping action.

The section of deck at the column head is concreted first, followed by the mid span. Two side supports are then temporarily positioned on the completed deck, to relieve the forward support, which is then transferred to the far end and secured to the next column. The whole unit is then winched forward a full span and the cycle repeated. The formwork must be designed to hinge away from the concrete deck, and the central portion, obstructed by the column, should be capable of folding back during travelling. Output rates similar to the previously described girder methods have been achieved, i.e. 350–400 m$^2$ of deck surface area per week, stepping units weighing over 500 tonnes are typical for spans up to 40 m.

## 46.7   CANTILEVER CONSTRUCTION WITH PRESTRESSED CONCRETE

The erection of bridges span-by-span, working out from the abutments towards the centre, has proved very economical as witnessed by the popular use of stepping formwork or bridge-launching techniques with problems such as difficult terrain, and transport of materials made much easier. The advent of cantilever construction methods, during the past 25 years now allows many of these advantages to be extended to bridges with spans in the range 50–150 m. The method can be applied to *in situ* concrete, precast concrete or steel decks, and is normally carried out by balanced cantilever or occasionally by progressive placing methods.

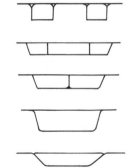

**Fig. 46.28**  Typical box section configurations

### Balanced cantilever construction

Developments in box section (Fig. 46.28) and prestressed concrete led to short segments being assembled or cast in place on falsework to form a beam of full roadway width (Fig. 46.29). Subsequently the method was refined virtually to eliminate the falsework by using a previously constructed section of the beam to provide the fixing for a subsequently cantilevered section. The principal is demonstrated step-by-step in the example shown in Fig. 46.30.

In the simple case illustrated, the bridge consists of three spans in the ratio 1:1:2. First the abutments and piers are constructed independently from the bridge superstructure. The segment immediately above each pier is then either cast *in situ* or placed as a precast unit. The deck is subsequently formed by adding sections asymetrically either side.

Ideally sections either side should be placed simultaneously but this is usually impracticable and some imbalance will result from the extra segment weight, wind forces construction plant and materials (Fig.

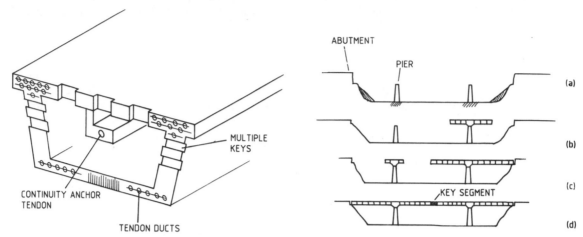

**Fig. 46.29**  Concrete box section/bridge deck

**Fig. 46.30**  Balanced cantilever construction

**Fig. 46.31**  Effect of equipment materials and wind

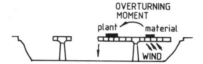

46.31). When the cantilever has reached both the abutment and centre span, work can begin from the other pier, and the remainder of the deck completed in a similar manner. Finally the two individual cantilevers are linked at the centre by a key segment to form a single span. The key is normally case *in situ*. Typical changes in bending moment during construction are shown in Fig. 46.32.

In cases where the span between the abutment and the pier is shorter than that reached out to the centre, the overturning moment must be resisted by a downward reaction, i.e. ballast or an anchor (Fig. 46.33) at the abutment end.

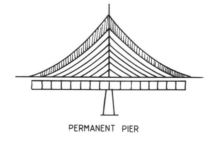

**Fig. 46.32**  Changes in bending moment by adding units

**Fig. 46.33**  Counterbalance/ support and anchor arrangements

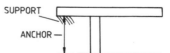

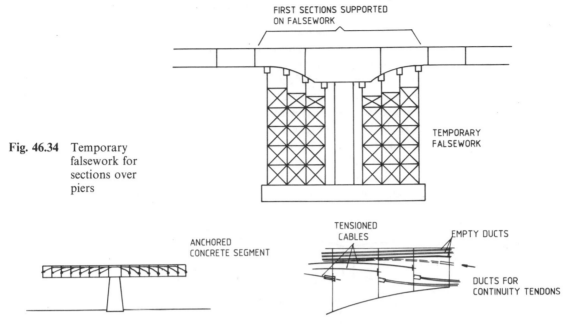

**Fig. 46.34** Temporary falsework for sections over piers

**Fig. 46.35** Precast concrete sections added in pairs and post-tensioned

**Fig. 46.36** Typical tendon and duct configuration

## APPLICATION OF PRESTRESS

The procedure initially requires the first sections above the column and perhaps one or two each side to be erected conventionally either in *in situ* concrete or precast and temporarily supported while steel tendons are threaded and post-tensioned (Fig. 46.34). Subsequent pairs of sections are added and held in place by post-tensioning (Fig. 46.35) followed by grouting of the ducts. During this phase only the cantilever tendons in the upper flange and webs are tensioned. Continuity tendons are stressed after the key section has been cast in place (Fig. 46.36).

The prestressing system would normally be one of those described in Chapter 40 thus the final gap left between the two half spans should be wide enough to enable the jacking equipment to be inserted. When the individual cantilevers are complete and the key section inserted the continuity tendons are anchored symmetrically about the centre of the span and serve to resist superimposed loads, live loads, redistribution of dead loads and cantilever prestressing forces (Fig. 46.37).

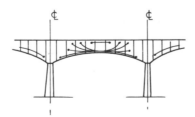

**Fig. 46.37** Continuity tendons

## THE ADVANTAGES OF CONTINUITY

The earlier bridges were designed on the free cantilever principle with an expansion joint incorporated at the centre. Unfortunately, settlements, deformations, concrete creep and prestress relaxation tended to produce deflections in each half span (Fig. 46.38), disfiguring the general appearance of the bridge and causing discomfort to drivers. These effects coupled with the difficulties in designing a suitable joint (Fig. 46.39) led designers to choose a continuous connection, resulting in a more uniform distribution of the loads (Fig. 46.40) and reduced deflection. The natural movements were provided for at the bridge abutments using sliding bearings or in the case of long multi-span bridges, joints at about 500 m centres (Fig. 46.41).

**Fig. 46.38** Deflection and distortion effect

**Fig. 46.39** Centre hinged joint

**Fig. 46.40** Centre continuous joint

**Fig. 46.41** Joints for horizontal movement

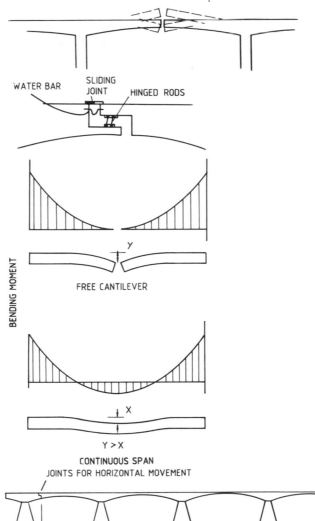

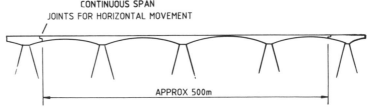

### Construction systems

Spans up to 200 m are commonly built using the balanced cantilever method, thereafter the dimensions of the box section increase quite rapidly, becoming very heavy and temporary supports impracticable. Congestion of prestressing tendons is also a problem. Four systems of construction are commonly used as follows:

1. Travelling former and cradle (Fig. 46.42).

**Fig. 46.42** Travelling former and cradle for *in situ* concrete balanced cantilever method

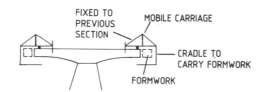

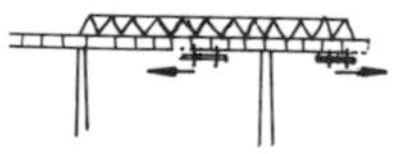

**Fig. 46.43** Travelling former with overhead gantry for *in situ* concrete balanced cantilever method

2. Travelling former with overhead gantry girder (Fig. 46.43).
3. Precast segments with mobile hoist (Fig. 46.44).
4. Precast segments with overhead gantry girder (Fig. 46.45).

Precast construction is generally much faster than the *in situ* alternative, but is only appropriate where precasting facilities can be conveniently and economically arranged. The method also tends to be a little less labour-intensive and would be favoured unless the weight of units becomes excessive and difficult to transport or launch and lift into position. Control of the bridge alignment is protracted with precasting, whereas the *in situ* method facilitates differences between actual and predicated camber to be more easily controlled during construction.

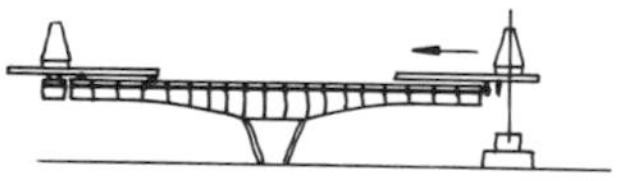

**Fig. 46.44** Precast segments erected with mobile hoist

### TRAVELLING FORMER AND CRADLE

The equipment comprises adjustable forms carried by a steel cantilevering truss and cradle temporarily anchored to the previously concreted section and mounted on rail track positioned above the webs (Fig. 46.46). The sequence of operations typified in Fig. 46.47 is as follows:

**Fig. 46.45** Precast segments supported with overhead gantry

1. Post-tension previous section, strike the formwork and move carriage.
2. Position outside form and inside flange form, fix reinforcement, ducts and cables and concrete bottom flange.
3. Fix web reinforcement, position inside formwork, fix top flange, reinforcement, ducts and cables and concrete webs and top flange.
4. Cure concrete.
5. Repeat cycle.

Crane assistance, while not essential, often proves beneficial for concrete placing and materials handling operations. An efficiently operating cycle takes approximately 7 days to complete. Thus with two

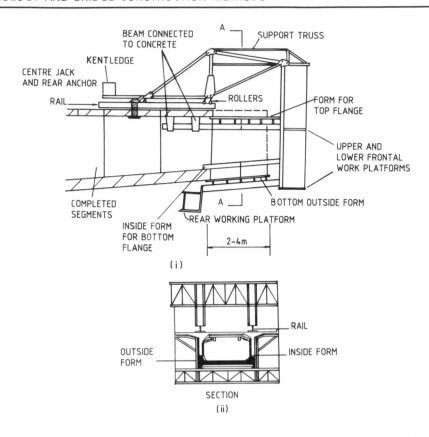

**Fig. 46.46** Travelling former and cradle

**Fig. 46.47** Concreting sequences with travelling former

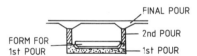

3 m long carriages operating, about 1 m per day of advance is possible. Attempts have been made to speed up this rate, for example by improving the curing period, and increasing the length of the segments. This latter method has generally proved too costly and the most economic section length seems to be 2–4 m. Other alternatives involve precasting the webs or arranging for the deck design to allow the top flange to be completed after post-tensioning of the section. A rate of advance of 2 m per day for a pair of carriages has been obtained in this manner.

## TRAVELLING FORMER WITH OVERHEAD GANTRY

Improvements in the rate of advance over cradle methods can be achieved by having larger sections carried by a gantry girder bridging a full span. However, the cost of the girder is high and only bridges 500 m or more in length can be competitively constructed in this way. Sections up to 10 m long are possible but the spans should be limited

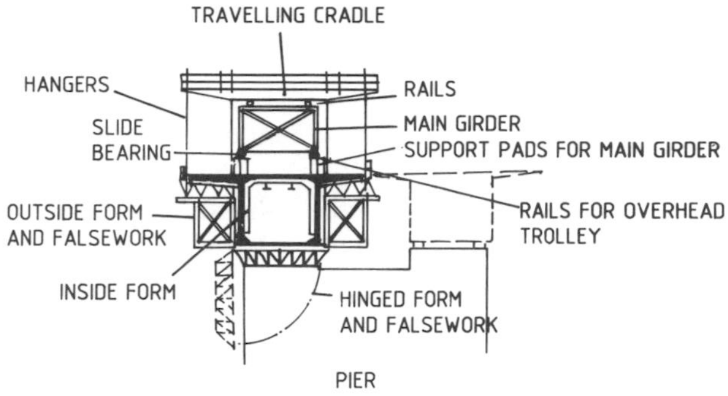

**Fig. 46.48** Travelling former and overhead gantry (cross section)

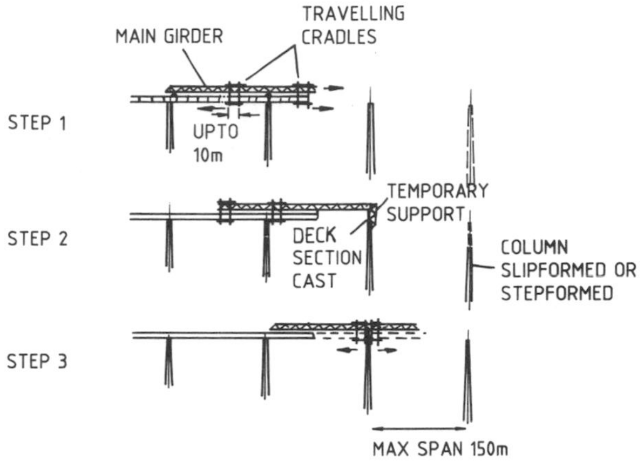

**Fig. 46.49** Travelling former and overhead gantry sequences

to less than 150 m, thereafter the girder itself rapidly increases in size, weight and capacity. The use of the gantry girder offers advantages with respect to materials handling, whereby water or ground level transport can be largely avoided for at least the deck construction phase.

Figure 46.49 illustrates a typical sequence of operations. The system comprises a main girder approximately one and a half spans in length carrying two travelling cradles. Construction of the deck section proceeds symmetrically from each column or pier head. The girder is supported in the working position on pads placed on the top flange of the deck at the web positions above each pier head (step 1). When each half span is complete and keyed together the main girder is travelled forward and temporarily propped at the next pier (step 2). The section immediately above this pier is then constructed and subsequently used to provide support for the girder during projection to the next half span point (step 3). The travelling cradles are thereafter repositioned and the cycle of operations repeated. It can be seen in the figure that the lower flange formwork is hinged to allow passage past the pier section.

The period of the construction cycle is similar to that for the independent system and approximately 20 m of advance per week is possible using 10 m long cradles.

### PRECAST SEGMENTS ERECTED WITH A MOBILE HOIST

The use of precast segments has several advantages over *in situ* concrete, for example, better quality control, manufacturing rates not dependent upon the erection programme, simpler erection methods and equipment.

The general concept, as with *in situ*, is to attach the segments in pairs to maintain symmetry either side of the pier, followed by immediate post-tensioning and grouting of the ducts. The segments can be delivered and placed in position by several methods, crane handling being most common where easy land access is available (Fig. 46.50).

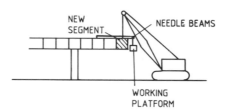

**Fig. 46.50** Precast segments supported by needle beams

Occasionally a crane-mounted barge may be used over water, but this tends to be too troublesome for most situations, as a consequence a deck-mounted lifting device is generally preferred (Fig. 46.51). Such equipment is relatively simple and typically comprises a mobile hoist mounted on two beams and anchored to the previously erected segment and projecting beyond it. The segment is raised from a barge or trailer below, moved into position, and working from a hanging platform post-tensioned as quickly as possible (in the case of erection by crane, a temporary tie beam placed across the top flange to the completed deck is usually necessary to maintain steadiness and support the working platform).

The precast units are manufactured with matching joints (see Chapter 40) and given a thin (1 mm) coating of epoxy resin glue prior to post-tensioning. However, since full shear strength does not develop immediately (and reinforcing steel is not continued through as with the *in situ* system) shear keys are normally required.

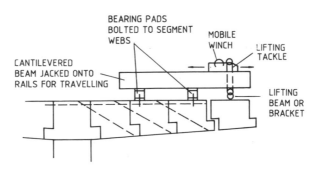

**Fig. 46.51** Needle beam equipment and post-tensioning sequences

Units 2–4 m long and weighing up to 60 tonnes are commonly erected with up to four segments put in place each day (about 40 m per week on average). A deck area of at least 5000 m² is necessary to justify the costs of the equipment.

## PRECAST SEGMENTS ERECTED WITH A LAUNCHING GIRDER

The advantage of using a launching gantry largely depends on the ease of handling materials, for example to avoid the problems of water transport by bringing the units along the previously constructed deck. Unfortunately, however, when the span exceeds 150 m or so, the weight of the deck segments requires a very heavy girder, and the method then tends to become very rapidly uneconomic. Also, unless the bridge is multispan with a deck area of at least 10 000 m², other bridge types would probably be favoured.

There are two categories of launching gantry:

1. Gantry slightly longer than a full span.
2. Gantry slightly longer than twice the span.

Figure 46.52 illustrates the first type, with the gantry supported on two trolley-mounted legs, one rests on the previously constructed cantilevering half span, A, while the other is set on the segment above the next pier, B. Segments are introduced along A, turned through 90° on the girder carriage, positioned and post-tensioned (step 1). On completion of two complete half spans the keying section is inserted to complete a full span and the gantry moved forward until the trailing leg C can be set down on the temporary bracket of the next pier (step

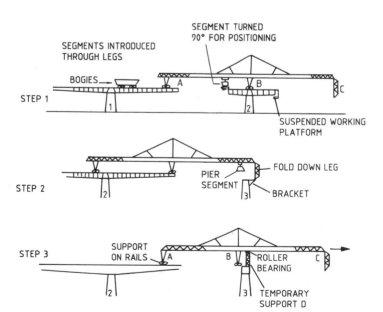

**Fig. 46.52** Precast segments erected with gantry slightly larger than full span

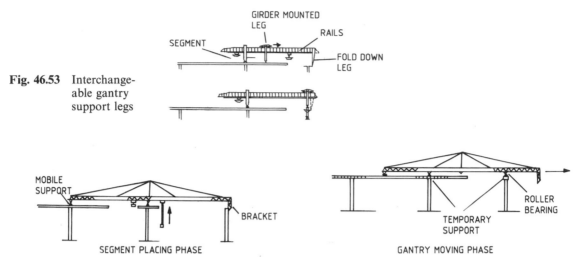

**Fig. 46.53** Interchangeable gantry support legs

**Fig. 46.54** Precast segments erected with gantry slightly longer than twice full span

2). The pier segment is then placed, a temporary support, D, erected and the final stage of the gantry manoeuvre completed (step 3). The weight of the gantry is transferred from D onto B and the next cycle of segment placing commenced. (Figure 46.53 illustrates an application of a combined design of the B and D support). Figure 46.54 shows the second type of gantry, which allows the support to be positioned above the piers during the segment erecting phase, thereby facilitating improved stability.

Gantry girders are typically used for placing segments weighing up to 100 tonnes but examples of over 250 tonnes segments exist. Rates of advance of 40 m per week (that is, four segments per day) are fairly typical. A point to consider when planning balanced cantilever construction is the construction of the segments immediately above the piers, especially when using conventional falsework as several weeks must be allowed for this phase.

### Control of stability during construction

Disequilibrium in weight of each half span and the loads carried during construction, can be stabilised by the following methods:

1. Making a rigid connection between the deck and piers.
2. Temporarily tie the deck to the piers (Fig. 46.55). Concrete wedges are used as the bearing pads and subsequently removed with the aid of the flat jacks and replaced by the permanent bearings.
3. In the case of very slender piers provide additional stiffening and support.

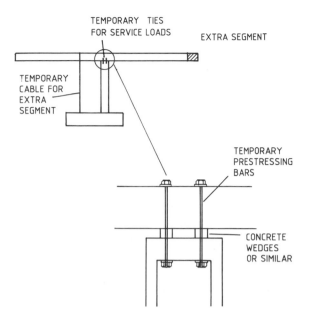

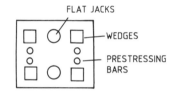

**Fig. 46.55**  Temporary ties for balanced cantilever method

## Control of deflection and camber

During construction prestress losses caused by steel relaxation, creep and shrinkage in the concrete, service loads and self-weight, non-uniform prestress, temperature, etc., cause the arm of each cantilever to deflect and distort the design profile. Compensating cambers should therefore be incorporated by making adjustments of the formwork moulds or precast segments. In particular, good quality control of the concrete is essential in order that the creep and strain can be minimised. For a detailed treatment of methods for calculating the camber adjustments see Mathivat, Bibliography.

## Precasting systems

The two principle methods are the short line and long line systems – both adopt the principle of match casting, such that the face of the previously concreted section acts as the form for its next adjoining segment. The resulting joint should be a perfect fit if the segments are then erected in the same sequence.

## SHORT LINE METHOD

A single set of shutters is used in conjunction with a permanent soffit (Fig. 46.56) to cast a single segment. The sequence of operations is as follows (Fig. 46.57).

1. Cast new segment against the old segment and steam cure overnight.
2. Take old segment to storage.
3. Release formwork from new segment.
4. Roll new segment into old segment position.
5. Reset forms for next segment and repeat cycle.

Approximately four segments per 5 day week can typically be produced. The main characterising feature of the method is that the required camber is obtained by positioning the old segment relative to the die to be cast. Thus a strong rigid unit rather than the relatively unstable formwork is adjusted. Also the soffit of the new segment can occupy a permanent geometrical position.

The setting out positions are usually determined by calculations from the theoretical casting curve, but graphical presentation offers a more convenient check, especially for comparing as-cast positions against the theoretical.

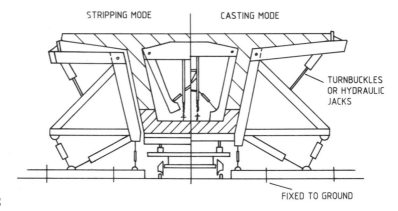

**Fig. 46.56** Formwork system for short line segment casting

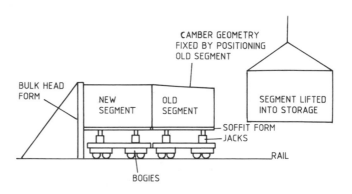

**Fig. 46.57** Short line method of segment casting

## LONG LINE METHOD

Occasionally where space permits the segments are cast as a bed equal in length to the cantilevering arm and laid out to give the geometry of the theoretical casting curve (Fig. 46.58). Rail-mounted formwork is required together with its associated curing equipment. The foundation must also be firmly secured to avoid settlement during concreting operations. Segments having obtained sufficient strength can be removed to a stockyard, freeing the bed for production of the next half span.

DIRECTION OF MOBILE
FORMWORK SYSTEM

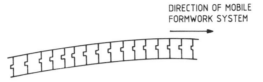

**Fig. 46.58** Long line casting method

## 46.8 PREFABRICATED STEEL OR CONCRETE SEGMENTS STAYED BY CABLES

Cable-stayed bridges are proving popular for spans up to about 800 m and when coupled with approach roads very broad crossings are bridgeable. Precast or *in situ* concrete or steel box segments may be used for the deck, which is supported by cables and stayed to a tower. There are four principle configurations, known as radiating (Fig. 46.62a), harp (Fig. 46.62d), fan and star systems. These arrangements provide compression in the deck by utilising self-weight as shown in Fig. 46.59. In this way the deck can be made up of individual segments and made to act like a prestressed beam.

Clearly cables can be spaced such that the horizontal component of force cancels out any tensile force in either the top or bottom flange, individual segments can then be assembled one-by-one and left unconnected. In the case of a concrete deck, shear forces would be resisted by shear keys and/or epoxy resin glued joints, whereas steel would normally be welded.

Where the cables are much further apart, then the tensile stresses would have to be removed by prestressing or carried in the flange of a continuous steel box girder. Vertical force components in each cable are transferred to the pylon and carried through to a foundation.

### Stability conditions

The principle aim of the structural configuration of a cable-stayed arrangement, is to prevent sideways and vertical movements of the tower/pylon and deck under asymmetrical live loading. By careful selection of the foundation types and connection of cable and girder it is possible to maintain stability of the whole structure by resisting only

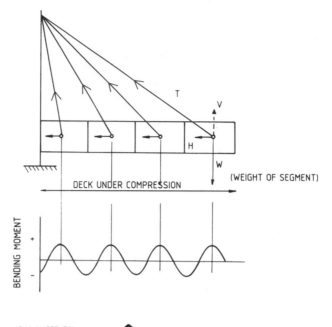

**Fig. 46.59**  Principles of the cable-stayed technique

the horizontal and vertical components of the forces generated, for example the case of instability illustrated in Fig. 46.60 with a live load W placed at point A causing rotation of the girder about the top of the pylon, is countered as illustrated in (Fig. 46.61i). Here the girder resists compressive forces, the cables are pin-pointed to the deck and tower, the tower and its foundation are hinged to the deck and the end supports are free to move horizontally but not vertically. Thus the same load at A (Fig. 46.61ii) now resolves into equilibrium of the first order and the whole system is stable.

Other stable and non-stable arrangements are shown in Fig. 46.62(a) to (f) including (c) which has a fixed earth support, horizontal movement of the girder being resisted by abutments.

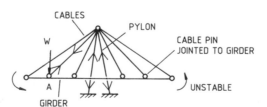

**Fig. 46.60**  An unstable arrangement

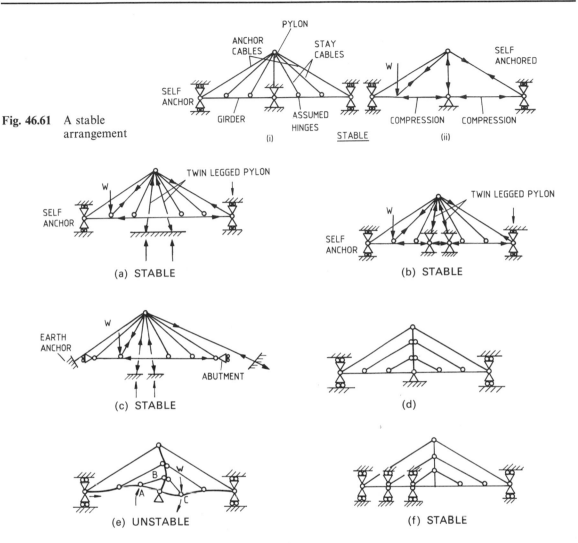

**Fig. 46.61**  A stable arrangement

**Fig. 46.62**  Examples of stable and unstable configurations

Not all systems, however, are designed as stable of the first order, for example the harp arrangement (d) and (e). Here the partial system ABC is inherently unstable unless either the deck or pylon are capable of resisting bending or alternatively supports placed under each stay for one-half deck span (f). In practice, designers often choose a mixture of stiff deck and pylon, coupled with a range of supports from the self-anchoring type to fully fixed.

**Transverse arrangement of cables**

Viewed perpendicularly to the line of the bridge, the cables are usually either arranged in a single-plane or two-plane system (Fig. 46.63). Single-plane is commonly employed with a divided road deck, and requires only a narrow pylon and pier. The deck itself generally has a hollow box cross section to provide torsional resistance across the deck width. In the two-plane system the cable can either be arranged to hang vertically or slope towards the top of the tower or pylon, the connection to the deck being through the outside edges.

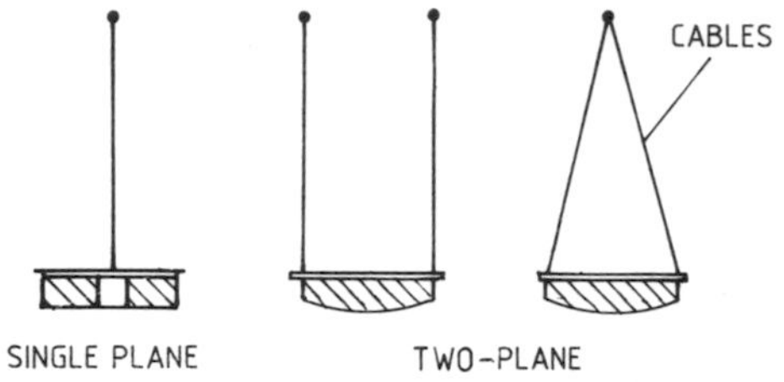

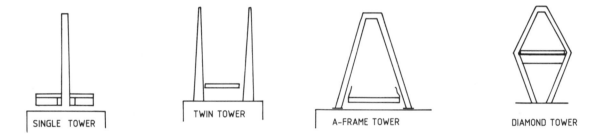

**Fig. 46.63**  Typical transverse cable arrangements

### The pylon (tower)

The pylon may be fabricated from steel plate, or precast concrete elements or occasionally in *in situ* concrete. The various configurations shown, in Fig. 46.64 illustrate the flexibility of design options available to produce good aesthetic effect.

**Fig. 46.64**  Typical tower configurations

### The deck

Like the pylon, the superstructure may be assembled in precast concrete elements, steel plate or girders, or made in *in situ* concrete. The most common form being the box section, which offers good torsional restraint (Fig. 46.66). Plate girders (Fig. 46.66) are sometimes used with a double plane system of hangers, where erection procedures require assembly in small light elements. Trusses are also an option but the high fabrication costs, expensive maintenance to counteract corrosion and poor aerodynamic characteristics now render this method relatively uneconomic.

### The cable and connections

The cable material is similar to that used for normal prestressing work and either comprises multi-strand cable (Fig. 40.24) made up of cold

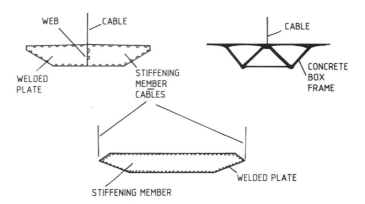

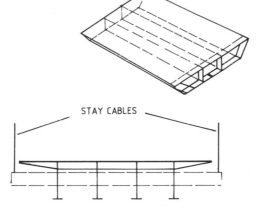

**Fig. 46.65** Typical box section deck arrangement in steel plate

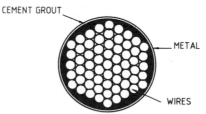

**Fig. 46.66** Girder deck

drawn wires as described in Chapter 40 or alternatively as single strand cable (mono-strand cable) consisting of parallel wires (Fig. 46.67). Diameters in the range 40–125 mm are typical. Protection against corrosion can be provided by galvanising each wire, but a more thorough practice has been to cover the cable in steel or plastic ducting and subsequently inject cement grout after positioning in place (Fig. 46.68). This latter operation is carried out after all dead loads have been applied to avoid too much cracking of the mortar.

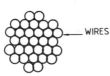

**Fig. 46.67** Parallel wire mono-strand cable

**Fig. 46.68** Protection and sheathing of mono-strand cable

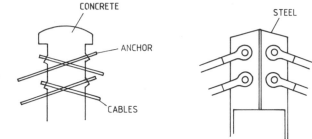

**Fig. 46.69**   Pin jointing of mono-strand cable to tower

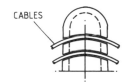

**Fig. 46.70**   Saddle for cable on tower

The cable is normally connected to the pylon with pin-type joints as illustrated in the examples shown in Fig. 46.69 or alternatively placed in the groove or guide tube of a saddle (Fig. 46.70), depending upon the design requirements.

The cable ends for the pin-type connection have either swaged (Fig. 46.71) or filled sockets. Swaging consists of squeezing a socket onto the wire in a hydraulic press and is generally used with strand having a diameter in the range 10–40 mm. Filled sockets are more suited to the larger diameter parallel wire type cable with the socket containing the whole bundle of wires. Several alternative types are manufactured differing slightly in the form of dead ending of each wire and the type of filling material. In the most simple form the wires are led through a plate at the base of the socket and finished with a button head or sockets and wedge. The inside of the socket, conical in shape, is subsequently filled with an alloy of zinc, copper, aluminium or lead, or sometimes with a cold casting compound such as epoxy resin. Thus when the cable is subject to a tension load, wedging action develops thereby increasing the grip on the wires. The deck-to-cable connection is usually of the 'free' type to accommodate adjustment. A flared arrangement is required for multi-strand cable (Fig. 46.72), while only a single socket is usually needed for mono-strand cable (Fig. 46.73). Initial tensioning of the cable to remove slack is generally carried out with a hydraulic jack similar to that used in prestressed concrete, the socket is therefore often manufactured with an internal thread for the jack connection and external thread and nut to take up the extension and other adjustments. Figure 46.74 illustrates a typical example, but other types allowing a pin joint are manufactured.

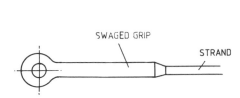

**Fig. 46.71**   Swaged socket for pin-jointed cable

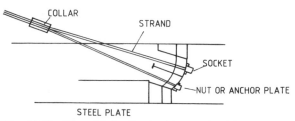

**Fig. 46.72**   Flared joint for multi-strand cable connection to deck

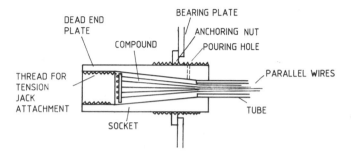

**Fig. 46.73** Connection of mono-strand cable to deck

**Fig. 46.74** Mono-strand cable/deck connection detail

## Methods of erection

The appropriate method of erection is influenced by the stiffness of the pylon, cable anchorage system, viability of installing temporary supports, maximum unsupported spans permitted by the design, ease of transporting materials, etc. However, since stability of the system largely depends upon transferring the horizontal component of the force in a cable through the stiffening girder, it is clearly necessary to have girder continuity between each pair of stays. The different procedures commonly adopted to ensure this are:

1. Erect on temporary props.
2. Free cantilever with progressive placing.
3. Balanced cantilever.
4. Push-out.

### ERECT ON TEMPORARY PROPS (STAGING METHOD)

This method is appropriate when the pylon is not designed with full end fixity to the pier or cannot be temporarily fixed, i.e. the pylon is not stable unless the anchor cable is held in position. Figure 46.75 illustrates a typical erection procedure beginning at one of the abutments. Temporary piers are first installed and the deck units progressively placed one-by-one and welded together to form short free cantilevers. A derrick-type crane mounted on rail track is commonly used for lifting and thus the weight of a unit would normally have to be significantly less than the derrick capacity (typically about 150 tonnes at minimum radius), and it may sometimes even be necessary for assembly to be carried out in sections. Prefabrication normally takes place off site, and units are erected in 5–15 m lengths. The length of free cantilever possible during the construction phase depends on the

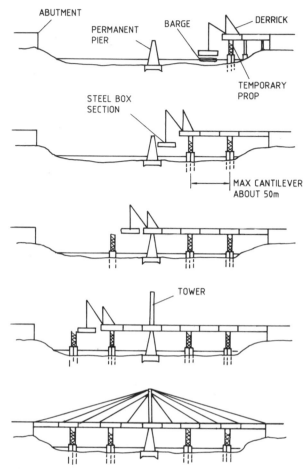

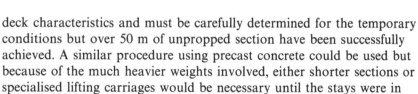

**Fig. 46.75** Staging method using temporary supports

deck characteristics and must be carefully determined for the temporary conditions but over 50 m of unpropped section have been successfully achieved. A similar procedure using precast concrete could be used but because of the much heavier weights involved, either shorter sections or specialised lifting carriages would be necessary until the stays were in position.

On completion of the deck, all the stays are connected, tensioned and the temporary piers dismantled. However, some extension of the cable is unavoidable as the self-weight of the deck is taken up. The temporary propping should therefore be erected at a height calculated to allow for this movement.

## FREE CANTILEVER WITH PROGRESSIVE PLACING

In many situations the installation of temporary supports would be difficult and expensive and cantilever construction might be considered as an alternative. Figures 46.76 and 46.77 show a typical example whereby the side spans are constructed on temporary propping

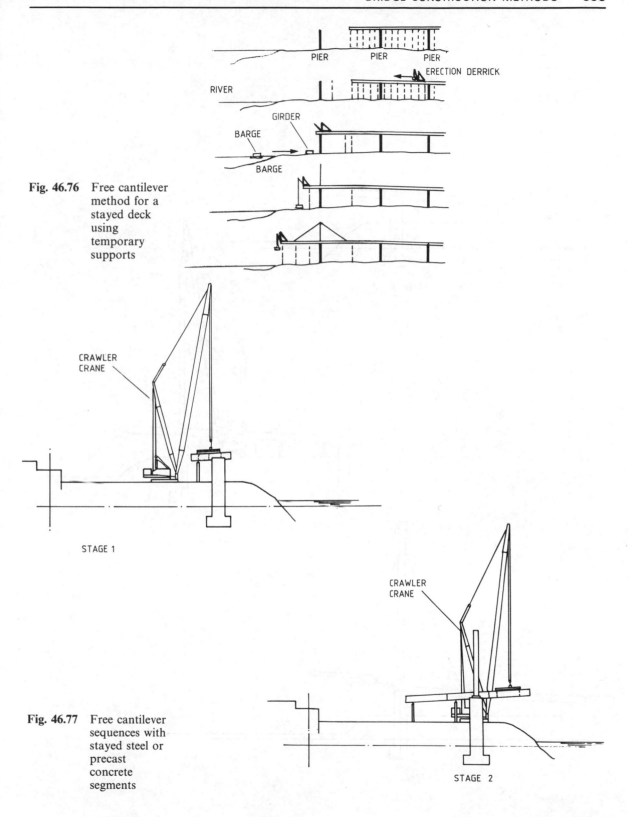

**Fig. 46.76** Free cantilever method for a stayed deck using temporary supports

**Fig. 46.77** Free cantilever sequences with stayed steel or precast concrete segments

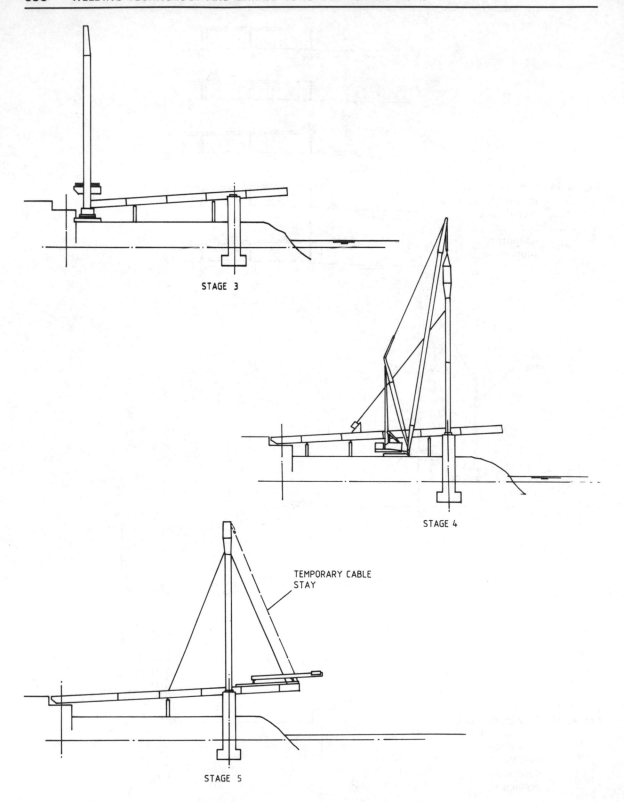

STAGE 3

STAGE 4

TEMPORARY CABLE STAY

STAGE 5

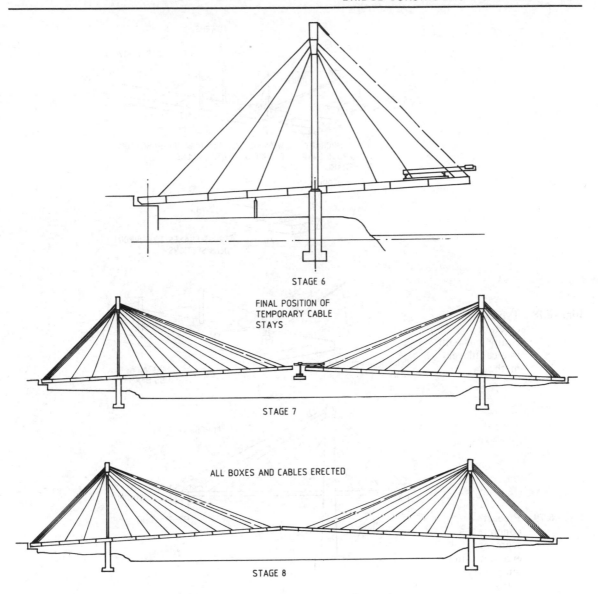

STAGE 6

FINAL POSITION OF
TEMPORARY CABLE
STAYS

STAGE 7

ALL BOXES AND CABLES ERECTED

STAGE 8

followed by the tower. This part of the bridge is often situated on the embankments where access may favour the use of cranes at ground level.

The centre span is thereafter erected unit-by-unit working out as a free cantilever from the tower or pylon. Like in the previous method, steel box sections up to 20 m long are commonly lifted either by derrick (Fig. 46.78) or with mobile lifting beams (Fig. 46.51) and welded into place. Thereafter the permanent stays are fixed each side of the tower and the bending moment caused by the cantilevering section removed. Figure 46.79 demonstrates the use of temporary stays to

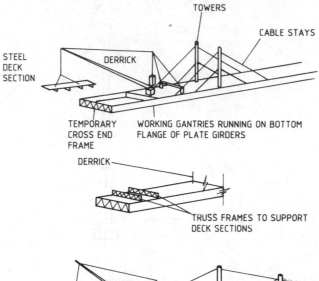

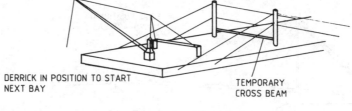

**Fig. 46.78** Typical erection equipment with free cantilever method

**Fig. 46.79** Precast concrete segment erection using temporary stays

prevent overstressing of the previously positioned permanent stay and the girder.

The provision of temporary stays is particularly important with precast concrete segments where units weighing up to 300 tonnes are occasionally erected. The normal procedure is to match cast adjacent segments and subsequently glue the joints with epoxy resin, temporary post-tensioning being applied to bring the two elements together (Fig. 46.80). The permanent cable is tensioned simultaneously as the temporary stay is released.

An *in situ* concrete cable-stayed deck constructed with a mobile carriage and formwork similar to that used in cantilever construction

(Fig. 46.81) is an alternative to steel and precast concrete, but a rate of progress of one 3–4 m section each week is very slow and thus is more commonly adopted as an alternative to stepping formwork systems on multispan bridges in the range 30–70 m between piers (Fig. 46.82).

The cable-staying technique using temporary stays only has also proved successful for multi short span bridges of the precast type (Fig. 46.83). This progressive erection method allows units to be transported along the previously constructed deck, which are then swung round and attached to lifting equipment such as swivel arm. The stays are usually tensioned with built-in hydraulic jacks, and the whole device moved forward from pier to pier as each span is erected and post-tensioned.

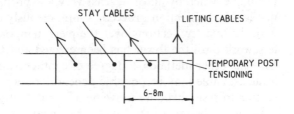

**Fig. 46.80** Precast concrete segment erection using temporary prestressing

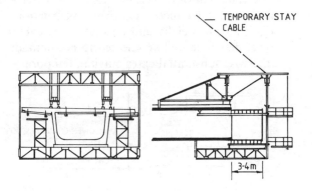

**Fig. 46.81** Mobile carriage for *in situ* concrete construction using temporary stays

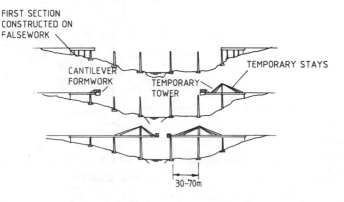

**Fig. 46.82** Mobile carriage construction with mobile stay arrangement

**Fig. 46.83** Precast segment erection with mobile stay arrangement

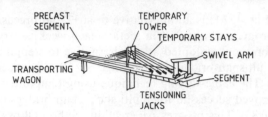

## BALANCED CANTILEVER

The occasional need to have clear uninterrupted space below the bridge, for example railway sidings, private property, etc., has forced designers and constructors to develop the balanced cantilevering technique, whereby all or at least very few props are required, as shown in Fig. 46.84. Erection proceeds simultaneously each side to the tower, with the first few sections over the piers, temporarily supported on falsework until the tower has been erected and the cables attached. Like the other methods, a degree of cantilevering beyond the last attached cable may be possible depending upon the capability of the section to resist bending movement, the potential for this possibility being much better for steel plate than heavy precast concrete segments.

An important feature of this technique is the need to have a stiff tower and fixity between the deck and tower and its foundations, because of imbalances caused by construction plant, variation in segment dead weight, and tension in the cables. Where possible, the tower design should be selected to accommodate this requirement, otherwise substantial extra staying, temporary anchor cables or a heavy

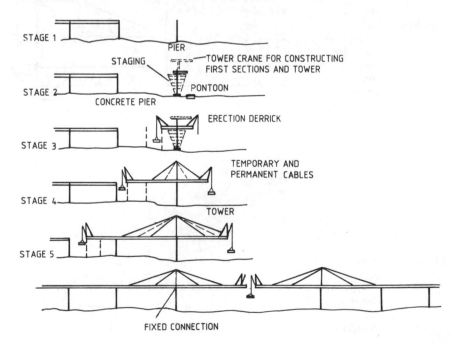

**Fig. 46.84** Balanced cantilever segment erection sequences for stayed deck

deck tower fixing clamp must be provided. Cantilever spans over 150 m each side of the tower are commonly erected, but wherever possible some propping is desirable to aid stability.

## PUSH-OUT METHOD

In some situations access beyond the abutment may not be available or deck units cannot be transported to the tower over adjoining property. To overcome these difficulties a few bridges have used the push-out method as illustrated in Fig. 46.85. The deck is assembled at one of the abutments and simply winched out over the rollers or teflon pad bearings.

A similar technique has been used with incremental launching when temporary cable stays are used rather than props (compare with Fig. 46.86).

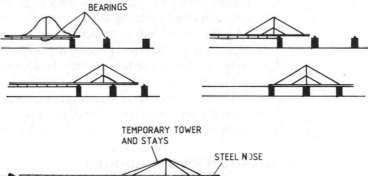

**Fig. 46.85** Push-out method using permanent stays

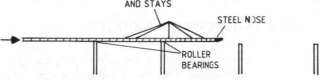

**Fig. 46.86** Incremental launching method assisted with temporary stays

## Cable erection

The majority of cable-stayed bridges are nowadays designed with monostrand cable either of the parallel wire or locked coil wire type. A complete stay is manufactured in its polyethelene tubing and delivered to site on reels. The simplest erecting procedure is to unreel the cable along the deck and hoist or lift it up to the top of the tower. Unfortunately the natural sag tends to be quite large and therefore considerable take-up has to be provided in the tensioning jack. A more satisfactory procedure is to install a guide rope and pull the cable up

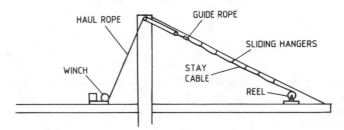

**Fig. 46.87** Stay cable erection and positioning

with a hauling rope. Intermediate supports to reduce sag are provided by intermittently spaced sliding hangers (Fig. 46.87). Tensioning is initially carried out at the deck connection end to take up the slack, final tensioning to remove bending moment in the deck and transfer dead load into the cable being supplied after all work on the newly erected section is complete (i.e. welding, post-tensioning of concrete segments, etc.). The jacking equipment is similar to that used for prestressed concrete, using the threaded bar system (Fig. 40.19). Finally the ducting is filled with pumped grout to provide protection against corrosion.

### Fabrication of steel deck units

Steel box sections are ideally suited to modern fabrication methods. In particular automatic numerically controlled cutting, drilling, milling and welding machines are positive encouragements towards manufacturing as much of the deck as possible under workshop conditions and bringing finished units to the site. Furthermore recent advances in welding technology such as submerged arc, $CO_2$, etc., (described in Chapter 44) make it possible to perform high quality welds quickly in the field thereby facilitating assembly in manageable size components without loss of performance and quality.

The time required to erect and weld deck units into place depends upon the amount, type of weld, plate thickness, etc., but a 15 m length section can be typically installed in a 2 week (10 day) period.

### 46.9  SUSPENSION BRIDGES

Suspension bridges are suited to spans exceeding 500 m, and comprise cables slung over two towers. The deck units are hung from the cables and connected together to form a stiffening girder. The cable ends are usually earth anchored while the girder generally has fixed hinges at the piers and movable ones at the towers (Fig. 46.88). Thus a live load on

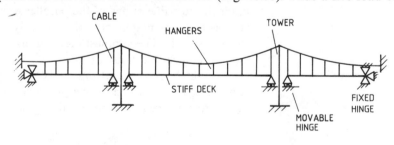

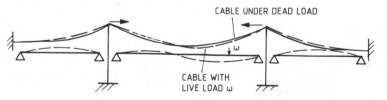

**Fig. 46.88** Principle of the suspension bridge arrangement

the centre span will cause deflection of the deck and pylons and a small change in the position of the suspension cable.

### The tower

Depending upon the height and mode of erection, towers may be shop-fabricated in steel as complete units or made up from cellular or box girder sections. Occasionally *in situ* concrete either cast lift-by-lift or slip-formed offers an alternative method.

### Cable

Except on very short spans the suspension cable is generally spun in place from individual galvanised wires, or alternatively positioned similar to the method used for cable-stayed bridges. Both systems require the wire or stands to be compacted together and then bound in galvanised wire and coated with weather-resistant paint to aid corrosion protection.

### Hangers

Hanger cables are clamped directly to the suspension cable with tightly bolted bands. The main cable wrapping only extends up to the bands as these are fixed prior to the wrapping operation.

### Deck unit

Two types of deck system are used for suspension bridges, namely:

1. Trussed girders.
2. Steel box sections.

Trussed girders were predominantly used on the earlier bridges with the members erected piece-by-piece from a crane running on the previously erected deck. Alternatively the trusses were assembled into large

**Fig. 46.89**  Typical steel box section deck

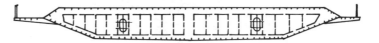

sections and lifted into place. The high costs of fabrication, inefficient use of steel inherent in the design, the large quantities of labour-intensive work and maintenance costs, however, forced designers towards cellular steel box sections (Fig. 46.89). These can be prefabricated cheaply and quickly delivered to site, complete with footpaths and railing. They are easily connected together with bolts and/or welding.

Once the units are joined, the concrete surfacing is laid and the deck finishes completed.

### Anchorages

The anchorage is fundamental to the stability of a suspension bridge, as all the load in the cable must be transferred to a fixed anchorage (a few small bridges with self-anchorages similar to cable-stayed bridges have been used). There are commonly three types:

1. Rock anchors.
2. Tunnel anchorage.
3. Gravity anchorage.

Rock anchorages simply involve drilling into the rock and grouting in large bolt-type anchors to which the strands are subsequently attached. Where suitable rock is available, a U-tunnel can be constructed and the two cables joined to form a loop (Fig. 46.90). However, the gravity anchorage has proved to be most popular with designers. The basic arrangement shown in Fig. 46.91 consists of looped over strand shoes attached to anchor bolts located in the concrete. The cable forces are thus resisted by a combination of overburden, dead weight and bearing friction.

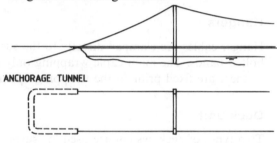

**Fig. 46.90**   Tunnel
anchorage

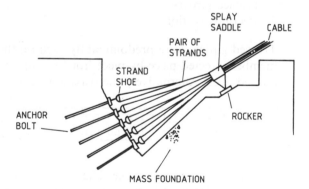

**Fig. 46.91**   Gravity
anchorage

### Saddles

The cable is usually spun in place, strand-by-strand. When the required number of wires in a strand is achieved they are grouped together with straps or bands placed at about 2 m intervals. The saddle at the top of each tower and the splay saddles are cast in steel and grooves, stepped and divided with steel spacers to aid wire location during spinning (Fig.

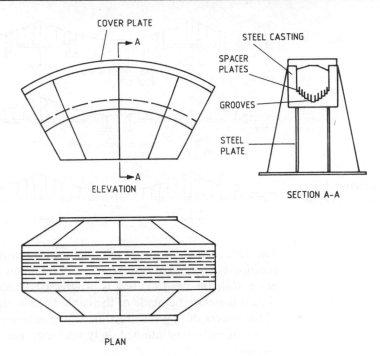

**Fig. 46.92** Saddles at the top of the towers

46.92). A cover plate is provided for protection against corrosion. The whole unit is bolted down to resist movement.

### Erection methods

The two systems most commonly selected for the deck placing sequences are:

1. Start erection from centre of main span (Fig. 46.93).
2. Start erection from the towers (Fig. 46.94).

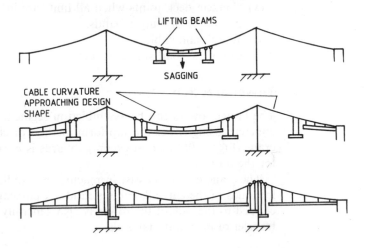

**Fig. 46.93** Erection from centre of main span method

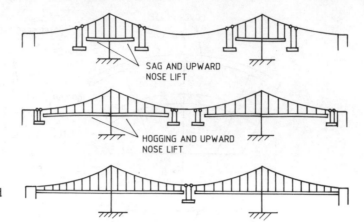

**Fig. 46.94** Erection from towers method

The deck section joints are left unconnected until the last unit is in position as significant displacement of the main cable and towers takes place while the dead loading is progressively increased. Calculations should therefore be made of the cable tension and shape, and distortion of the towers during the erection sequence to ensure tolerable limits of deformation are maintained. A typical sequence of operations is as follows:

(i) Construct piers and anchorages.
(ii) Erect towers and saddles.
(iii) Hoist catwalk cables into place and build catwalk.
(iv) Place cross bridges and fix storm ties to stabilise catwalk.
(v) Haul out tramway ropes for spinning wheels.
(vi) Erect suspension cable by air spinning method adjusting wires and strands for sag within tolerance specification.
(vii) Compact cable and temporarily clamp.
(viii) Fix permanent bands and hangers to approximate positions operating from work carriages running on the tramway cable.
(ix) Erect deck units.
(x) Connect deck joints when all units are in place and complete tightening of hanger bands.
(xi) Wrap main cable.
(xii) Complete road surfacing, etc.

## Tower erection

The bridge towers can be assembled in segments or formed from *in situ* concrete, using either the slip-form method or climbing forms. A tower crane (Fig. 46.95) or derrick on gabbards is normally placed at each leg for the purpose.

The crane can also assist in placing the saddles and hoisting up the free ends of the catwalk cable strands. A passenger/materials hoist clipped to the side of the tower is also generally provided for rapid transfer of men and materials.

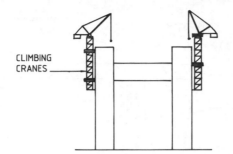

**Fig. 46.95** Tower erection with climbing cranes

### Catwalk and hauling system

Walkways (or catwalks) are provided just below each main cable to give access during the spinning and subsequent erection operation. Typically these are about 3 m wide, consisting of hardwood tread boards mounted on galvanised mesh all carried on wire strands, with two additional strands acting as handrailing. Stiffening frames are spaced regularly (approximately 50 m) along the span and cross bridges provided to link the two walks. Also two tramway support strands are stretched above the catwalk between the main towers and from towers to abutments to carry the stabilising cross beams (at approximately 90 m spacings) and spinning wheel haul rope sheaves. The whole system is stabilised by means of strong ropes anchored to the towers and abutments. The walkway strands are the first to be placed in position using barges. The free end of each is raised and connected to the top of the tower while the rest is unwound off the reel along the bed of the river, and subsequently fixed to the abutment or other tower. Adjustment is generally necessary to produce the desired sag for the spinning operation.

The rest of the catwalk equipment such as mesh, stiffening frames, main bridging beams, etc., are usually slid down the strands working from the tower tops. A small working carriage running on the handrail strands is generally set up to transport the workmen.

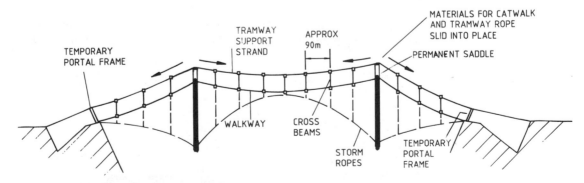

**Fig. 46.96** Catwalk and hauling system

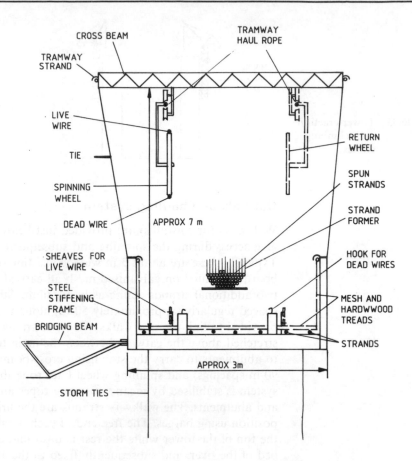

**Fig. 46.96** Catwalk and hauling system (*continued*)

The tramway support strands are next erected in a similar manner to those of the footbridge, with the cross beams also being slid into position under control from the tower tops and tied to the catwalk frames. With these operations complete the tramway haul rope can be winched along the catwalk.

## Cable erection

### REELING OPERATION

The main cable wire is generally transported to site in coils and wound on to 2 m diameter reels (Fig. 46.97) capable of holding about 10–15

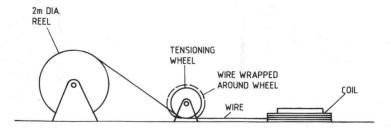

**Fig. 46.97** Reeling operation

span lengths. The end of the wire is led round a tensioning wheel (which turns against a controlled brake) and on to the reel. Ends can be spliced together with swaged couplings.

## CABLE SPINNING

The drums of wire are taken from the reeling area and set up on electrically powered unreeling machines (in the example shown a single reel is illustrated but often up to eight wires are simultaneously unreeled across the span). The end of the wire is passed over a series of pulleys housed in a high tower and led via a floating sheave to a strand shoe, where a temporary connection is made. The wire is then looped by hand over the spinning wheel and hauled with the endless tramway drive rope at 200–300 m/min to a strand shoe on the anchorage at the other band (Fig. 46.99). Thus two wires are simultaneously spun, one dead (i.e. that fixed to the shoe) and the other live. The dead wire is temporarily held in position with hooks spaced at approximately 150 m centres along the catwalk, while the live wire is run through sheaves to aid control. Meanwhile the empty wheel returns to the reeling side.

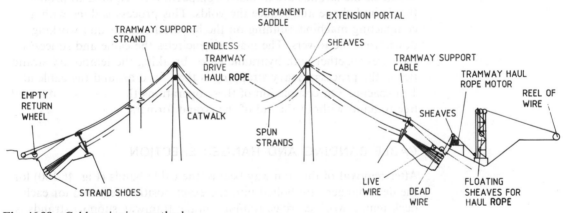

**Fig. 46.98** Cable spinning method

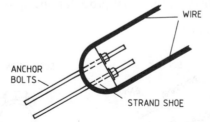

**Fig. 46.99** Cable anchorage bolt and strand shoe

Adjustment of sag for each wire generally takes place while spinning is in operation. As soon as the wheel passes over a tower the dead wire on the side span is unhooked, pulled clear of the catwalk with an electric winch, lined up with a guide wire and clamped at the top of the tower, it is then manually lifted into the strand formers located along the catwalk at about 150 m centres. The procedure is then repeated in the main span after the wheel has passed the other tower. Adjustment of the remaining side span sag takes place after the wire has been looped over the strand shoe while the wheel is stationary in ready uses for the return journey to the reeling side. During this return period, the previously live wire is released from the sheaves and adjusted in a similar manner to the dead wire.

The process is repeated until a complete strand has been made, whereafter the loose end is spliced to the temporary connection initially provided for the start end of the wire. The wires are subsequently shaken out, temporarily banded into strands and any final sag adjustments taken up by the strand shoe. The remaining strands are spun in a similar manner until the whole cable is formed. Spinning can usually only take place when wind speeds are less than about 50 km/h. The whole process typically takes up to 6 months for a 1000 m span bridge, with an average about 6 tonnes per day of spun strand.

## CABLE COMPACTION

When all the strands are in place, compaction is required to produce the desired shape and reduce the voids. This process is done with a compacting machine, running on the handrail strands and working down from the towers. The machine encircles the cable and squeezes the wires together with hydraulic jacks, breaking the temporary strand ties in the process. Heavy strapping is then fixed around the cable at 1 m spacing. On completion of these operations the cable is stable and free to hand without the aid of the strand formers.

## CABLE BANDING AND HANGER ERECTION

After removal of the tramway beams, the cable bands (Fig. 46.100) for the deck hangers are bolted into the exact position required for each deck unit, a work carriage running on the tramway support strands is generally set up to assist transport. Thereafter the deck hangers are fixed one-by-one, the usual procedure being to unreel the hanger at

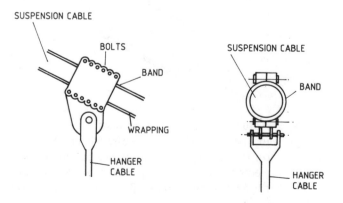

**Fig. 46.100** Cable banding and hanger connection

the top of the tower and allow it to unwind along the catwalk. It is then lowered under winched control and connected to the band. The storm ropes are finally removed and the catwalk slackened off to allow for cable movement during erection of the deck.

## DECK ERECTION

The sequence of deck erection is shown in Figs. 46.93 and 46.94. Modern practice usually prefers box sections either barged or self-floated into position below the hangers. Great care is required in prefabricating the units to ensure matching joints and prevent distortion during transportation. This can generally be managed with 15–30 m long sections weighing up to 300 tonnes. The box sections are usually lifted at four points using two support beams running on the main cable although occasionally derricks-mounted barges are used. The winches are usually placed on platforms located at the tower or on its foundation.

During the lifting operation the beams are located to the cable to avoid movement caused by the pull from the winches. Jointing of the units should normally be left until the last section is in place, since hogging or sagging will occur depending upon the erection sequence.

### Cable wrapping

The hangers are given a final tightening whereafter the cable is wrapped in galvanised mild steel wire (3–4 mm diameter) to provide corrosion protection. The wrapping machine consists of two drums encircling the cable and the wire is unwound as the unit is winched along. Wrapping can be achieved from either end of the machine in order that wire may be brought tight against a hanger band. The drum should contain just enough wire to cover between two hanger bands and allow the drums to be slit apart to pass the bands. The whole unit is hung from the tramway support strands and travels at about 0.5 m/min. The final operation requires the catwalk and other strands to be dismantled, this is a very difficult and dangerous operation requiring the workmen to raise the mesh panels by hand and winch them to the deck below.

## 46.10  FRAME-TYPE BRIDGES

Until quite recently with the introduction of welded steel box sections and prestressed concrete, bridges spanning more than about 30 m, were generally constructed in either riveted or bolted steel members to form a framed structure. The most common form being the Warren or N-girder type trusses. Indeed the method is still widely adopted in parts of the world where labour is plantiful and not too expensive to employ. Provided that the members are well made such that joints marry easily

**Fig. 46.101**   Bolts and temporary drifts in spliced connection

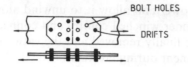

when brought together with simple drifts (pin used prior to bolting, Fig. 46.101), then relatively unskilled labour can be quickly trained to understand the comparatively simple assembly and erection procedures.

## Trusses

These are commonly pre-assembled on the river bank and transferred to pontoons or a barge-mounted crane along a well constructed jetty. The correct camber can be in-built before bolt tightening by using a template (Fig. 46.103). If the truss is assembled piece-by-piece over the river site (Fig. 46.104) then control becomes much more difficult, however, a much smaller capacity crane can be used and the transporting phase is less risky, especially in tidal or fast-flowing waters. Approximately one 30 m span can be erected and the deck concreted each month.

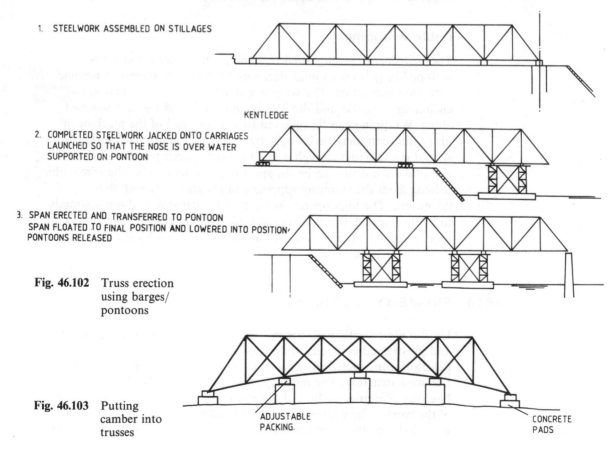

1. STEELWORK ASSEMBLED ON STILLAGES

KENTLEDGE

2. COMPLETED STEELWORK JACKED ONTO CARRIAGES LAUNCHED SO THAT THE NOSE IS OVER WATER SUPPORTED ON PONTOON

3. SPAN ERECTED AND TRANSFERRED TO PONTOON SPAN FLOATED TO FINAL POSITION AND LOWERED INTO POSITION, PONTOONS RELEASED

**Fig. 46.102**   Truss erection using barges/ pontoons

**Fig. 46.103**   Putting camber into trusses

ADJUSTABLE PACKING.

CONCRETE PADS

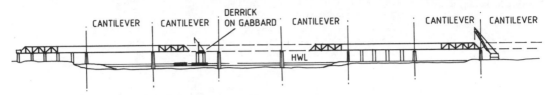

**Fig. 46.104**    Erecting trusses piece-by-piece over water

## Arch frames

A frame arch can normally be erected by building the bridge out over the water, working from either side towards the centre (Fig. 46.105). Where possible, the starting sections are tied back to the abutment or alternatively a foundation provided for temporary cables. The steel members are lifted from barges and placed into position with creeper cranes mounted on bogies running along the top string of the arch. The bridge is then erected panel-by-panel, and line and level precisely controlled by adjusting the ties at each abutment. Care is required to ensure that measurements are taken at similar temperatures, wind forces, etc. On meeting at the centre the two halves are finally joined by jacking out the temporary cables at each abutment. Today this bridge type is rarely considered, because cable-stayed or suspended cable designs are usually more economic for the large spans involved.

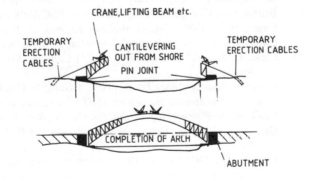

**Fig. 46.105**    Arch frame construction

# BIBLIOGRAPHY

Antill, J. M., Ryan, P. W. S. and Easton, G. (1989) *Civil Engineering Construction.* McGraw-Hill.

Atlas-Copco *Manual on Compressed Air Equipment.* Sweden.

Atlas-Copco *Manual on Rock Blasting.* Stockholm.

Barber-Greene Company (1979) *Bituminous Construction Handbook.* Illinois.

Bell, F. G. (1987) *Ground Engineers Reference Book.* Butterworth.

Bell, F. G. (1992) *Engineering Properties of Soils and Rocks.* Butterworth-Heinemann.

BOC *Handbook of Operating Instructions for Gas Welding and Cutting.* London.

Bowen, R. (1982) *Grouting in Engineering Practice.* Applied Science Publishers.

Bowen, R. (1986) *Ground Water.* Applied Science Publishers.

British Steel Piling Company. *BSP Pocket Book.* Ipswich.

British Steel Company *Piling Handbook.* BSC Publications.

Casey, H. B. (1989) *Modern Welding Technology.* Prentice-Hall.

Caterpillar Tractor Company *Caterpillar Performance Handbook.* Peoria, USA.

Clear, C. A. and Harrison, T. A. (1985) *Concrete Pressure on Formwork.* CIRIA Report 108.

Concrete Society (1986) *Formwork – A Guide to Good Practice.* London.

*Control of Substances Hazardous to Health Regulations* (1988). COSHH.

Craig, R. N. *Pipe-jacking: A State of the Art Review.* CIRIA TN112.

Davies, A. C. (1993) *The Science and Practice of Welding.* Cambridge University Press.

Day, D. A. (1991) *Construction Equipment Guide.* Wiley.

Eaglestone, F. N. (1979) *Insurance for the Construction Industry.* Witherby.

General Motors Corporation *Production and Cost Estimating of Material Movement and Earthmoving Equipment.* Terex Division, Ohio, USA.

Ginsing, N. J. (1983) *Cable Supported Bridges.* Wiley.

Harris, F. and McCaffer, R. (1991) *Management of Construction Equipment.* Macmillan.

Havers, J. A. and Stubbs, F. W. (1971) *Handbook of Heavy Construction Equipment.* McGraw-Hill.

*Health and Safety at Work.* HSE.

Heins, C. P. and Firmage, D. A. (1979) *Design of Modern Concrete Highway Bridges.* Wiley.

Heins, C. P. and Firmage, D. A. (1984) *Design of Modern Steel Highway Bridges.* Wiley.

Higgins, L. R. (1987) *Handbook of Construction Equipment Maintenance.* McGraw-Hill.

HMSO *The Work in Compressed Air Regulations*. London.
Holmes, R. (1983) *Introduction to Civil Engineering Construction*. College of Estate Management.
Huisman, L. (1972) *Groundwater Recovery*. Macmillan.
*ICE Works Construction Guides*. Thomas Telford.
Illingworth, J. R. (1992) *Construction Methods and Planning*. Spon.
International Harvester Company of Great Britain *Basic Estimating*. London.
Jackson, N. (1988) *Civil Engineering Materials*. Macmillan.
Junikis, A. R. (1987) *Foundation Engineering*. Krieger.
Karol, R. H. (1992) *Chemical Grouting*. Dekker.
Kodzi, A. (1979) *Stabilised Earth Roads*. Elsevier.
Komatsu Ltd. *Specification and Application Handbook*. Japan.
Kramer, S. R. and McDonald, W. J. (1992) *Introduction to Trenchless Technology*. Chapman and Hall.
Langefors, U. and Kihlstroem, B. (1978) *The Modern Technique of Rock Blasting*. Halstead.
McGaw, T. M. and Bartlett, J. V. (1982) *Tunnels – Planning, Design and Construction*, Vols. I and II. Ellis Horwood.
Mackay, E. B. (1986) *Proprietary Trench Support Systems*. CIRIA 95.
*Manual Handling*. HSE.
Mathivat, J. (1984) *The Cantilever Construction of Prestressed Concrete Bridges*. Wiley.
Neville, A. M. (1987) *Concrete Technology*. Longman.
Nobel Explosives Company *Blasting Practice*, 4th edition.
Nunally, S. W. (1987) *Managing Construction Equipment*. Prentice-Hall.
Oberlender, G. D. (ed.) (1986) *Earthmoving and Heavy Equipment*. Proceedings Committee of Construction and Techniques. ASCE.
Padolny, W. and Scalzi, J. B. (1982) *Modern Bridge Construction*. Wiley.
Peck, R. B. et al. (1974) *Foundation Engineering*. Wiley.
Peurifoy, R. L. and Ledbetter, W. B. (1985) *Construction Planning, Equipment and Methods*. McGraw-Hill.
Powers, J. P. (1992) *Construction Dewatering*. Wiley.
Ratay, R. T. (1984) *Temporary Structures in Construction*. McGraw-Hill.
Scherocman, J. A. (1981) *Asphalt Pavement Construction*. American Society for Materials Testing 724.
Sharp, R. D. (1970) *Concrete in Highway Engineering*. Pergamon.
Shell Petroleum (1990) *The Shell Bitumen Handbook*. London.
Sinclair, J. (1969) *Quarrying, Opencast and Alluvial Mining*. Elsevier.
Stack, B. (1982) *Handbook on Mining and Tunnelling Equipment*. Wiley.
Terzaghi, K. and Peck, R. B. (1967) *Soil Mechanics in Engineering Practice*. Wiley.
Thomson, J. (1992) *Pipejacking and Microtunnelling*. Blackie.
Tomlinson, M. J. (1986) *Foundation Design and Construction*. Longman.
Troitsky, M. S. (1988) *Cable-stayed Bridges – Theory and Design*. BSP.
Tschebotarioff, G. P. (1973) *Foundations, Retaining and Earth Structures*. McGraw-Hill.
Urbanski, T. (1984) *Chemistry and Technology of Explosives*. Pergamon.
Winterkorn, H. F. and Fang, H. Y. (1975) *Foundation Engineering Handbook*. Van Nostrand Reinhold.
Wood, S. (1977) *Heavy Construction*. Prentice-Hall.
*Work Equipment*. Health and Safety Executive.

# INDEX